T0336032

# Classical and Quantum Statistical Physics

Statistical physics examines the collective properties of large ensembles of particles, and is a powerful theoretical tool with important applications across many different scientific disciplines. This book provides a detailed introduction to classical and quantum statistical physics, including links to topics at the frontiers of current research. The first part of the book introduces classical ensembles, provides an extensive review of quantum mechanics, and explains how their combination leads directly to the theory of Bose and Fermi gases. It contains a detailed analysis of the quantum properties of matter and introduces the exotic features of vacuum fluctuations. The second part discusses more advanced topics such as the two-dimensional Ising model and quantum spin chains. This modern text is ideal for advanced undergraduate and graduate students interested in the role of statistical physics in current research. One hundred and forty homework problems reinforce key concepts and further develop readers' understanding of the subject.

**Carlo Heissenberg** is a postdoctoral scholar at the Nordic Institute for Theoretical Physics (Nordita) in Stockholm, and at Uppsala University. He received his PhD from Scuola Normale Superiore in Pisa, with a thesis on asymptotic symmetries and higher spin theories, and his research is now focused on the interface between scattering amplitudes and gravitational waves.

**Augusto Sagnotti** is Professor of Theoretical Physics at Scuola Normale Superiore and has taught the Statistical Physics course there since 2017. His research is focused on gravitational physics and conformal field theory, and his pioneering contribution led to the introduction of orientifold vacua in string theory. He is a recipient of the Humboldt Research Award.

# Classical and Quantum Statistical Physics

## Fundamentals and Advanced Topics


CARLO HEISSENBERG

Nordic Institute for Theoretical Physics and Uppsala University

AUGUSTO SAGNOTTI

Scuola Normale Superiore Pisa and Istituto Nazionale di Fisica Nucleare (INFN)

**CAMBRIDGE**
**UNIVERSITY PRESS**

University Printing House, Cambridge CB2 8BS, United Kingdom

One Liberty Plaza, 20th Floor, New York, NY 10006, USA

477 Williamstown Road, Port Melbourne, VIC 3207, Australia

314–321, 3rd Floor, Plot 3, Splendor Forum, Jasola District Centre, New Delhi – 110025, India

103 Penang Road, #05–06/07, Visioncrest Commercial, Singapore 238467

Cambridge University Press is part of the University of Cambridge.

It furthers the University's mission by disseminating knowledge in the pursuit of education, learning, and research at the highest international levels of excellence.

www.cambridge.org
Information on this title: www.cambridge.org/9781108844628
DOI: 10.1017/9781108952002



First published 2022

*A catalogue record for this publication is available from the British Library.*

*Library of Congress Cataloging-in-Publication Data*
Names: Heissenberg, Carlo, author. | Sagnotti, Augusto, other.
Title: Classical and quantum statistical physics : Fundamentals and advanced topics / Carlo Heissenberg and Augusto Sagnotti.
Description: New York : Cambridge University Press, 2021. |
Includes bibliographical references and index.
Identifiers: LCCN 2021041905 (print) | LCCN 2021041906 (ebook) |
ISBN 9781108844628 (hardback) | ISBN 9781108952002 (epub)
Subjects: LCSH: Statistical mechanics. | Quantum statistics. | Quantum theory. | BISAC: SCIENCE / Physics / Mathematical & Computational
Classification: LCC QC174.7 .H45 2021 (print) | LCC QC174.7 (ebook) |
DDC 530.13–dc23/eng/20211005
LC record available at https://lccn.loc.gov/2021041905
LC ebook record available at https://lccn.loc.gov/2021041906

ISBN 978-1-108-84462-8 Hardback

# Contents

*Preface* — *page* ix
*Acknowledgments* — xi

**Part I** — 1

**1 Elements of Thermodynamics** — **3**
1.1 The Laws of Thermodynamics — 3
1.2 Thermodynamic Potentials — 5
1.3 Comparison between $C_P$ and $C_V$ — 8
1.4 Fluctuations — 9
1.5 Stability — 10
1.6 Specific Heat and Compressibility — 11
1.7 The Ideal Monatomic Gas — 12
Bibliographical Notes — 13
Problems — 13

**2 The Classical Ensembles** — **17**
2.1 Time Averages and Ensemble Averages — 17
2.2 The Microcanonical Ensemble — 17
2.3 The Canonical Ensemble — 26
2.4 Two Examples — 31
Bibliographical Notes — 35
Problems — 35

**3 Aspects of Quantum Mechanics** — **40**
3.1 Some General Properties — 40
3.2 The Free Particle — 41
3.3 The Harmonic Oscillator — 44
3.4 Evolution Kernels and Path Integrals — 49
3.5 The Free Particle on a Circle — 53
3.6 The Hydrogen Atom — 56
3.7 The WKB Approximation — 64
3.8 Instantons — 77
3.9 The Density Matrix — 87
Bibliographical Notes — 89
Problems — 90

**4 Systems of Quantum Oscillators** — **97**
4.1 The Effect of an Energy Gap — 97

4.2 Blackbody Radiation 99
4.3 Debye Theory of Specific Heats 106
Bibliographical Notes 107
Problems 107

**5 Vacuum Fluctuations** **112**
5.1 The Casimir Effect 112
5.2 The Lamb Shift 114
5.3 The Cosmological-Constant Problem 117
Bibliographical Notes 121

**6 The van der Waals Theory** **122**
6.1 The Role of Interactions 122
6.2 The Partition Function 124
6.3 The van der Waals Equation of State 125
6.4 Phase Equilibria and the van der Waals Gas 126
6.5 The Gibbs Phase Rule 130
Bibliographical Notes 131
Problems 131

**7 The Grand Canonical Ensemble** **134**
7.1 The Grand Canonical Equations 134
7.2 Two Instructive Examples 137
Bibliographical Notes 140

**8 Quantum Statistics** **141**
8.1 Identical Particles in Quantum Mechanics 141
8.2 Identical Oscillators in the Canonical Ensemble 145
8.3 Nonrelativistic Fermi and Bose Gases 147
8.4 High-Temperature Limits 148
8.5 The Free Fermi Gas at Low Temperatures 150
8.6 Fermi Gases in Solids 154
8.7 The Free Bose Gas at Low Temperatures 159
8.8 Bosons in an External Potential 162
8.9 Atomic and Molecular Spectra 167
8.10 Some Applications 179
Bibliographical Notes 187
Problems 187

**9 Magnetism in Matter, I** **193**
9.1 Orbits in a Uniform Magnetic Field 193
9.2 Landau Levels 194
9.3 Landau Diamagnetism 199
9.4 High-$T$ Paramagnetism 202
9.5 Low-$T$ Paramagnetism 202
Bibliographical Notes 203
Problems 203

**10 Magnetism in Matter, II** 207
10.1 Effective Spin–Spin Interactions 207
10.2 The One-Dimensional Ising Model 209
10.3 The Role of Boundary Conditions 212
10.4 The Continuum Limit 215
10.5 The Infinite-Range Ising Model 217
10.6 Mean-Field and Variational Method 219
10.7 Mean-Field Analysis of the Ising Model 221
10.8 Critical Exponents and Scaling Relations 226
10.9 Landau–Ginzburg Theory 229
10.10 A Toy Model of a Phase Transition 233
Bibliographical Notes 235
Problems 235

**Part II** 243

**11 The 2D Ising Model** 245
11.1 Closed Polygons in Two Dimensions 245
11.2 Kramers–Wannier Duality 246
11.3 The Onsager Solution 250
Bibliographical Notes 254

**12 The Heisenberg Spin Chain** 255
12.1 Noninteracting Systems 255
12.2 The Spectrum of the Heisenberg Model 258
12.3 Thermodynamic Bethe Ansatz 270
Bibliographical Notes 273

**13 Conformal Invariance and the Renormalization Group** 274
13.1 Conformal Invariance 274
13.2 1D Ising Model and Renormalization Group 279
13.3 Percolation 282
13.4 The XY Model 283
13.5 $\epsilon$-Expansion and the $D = 3$ Ising Model 301
Bibliographical Notes 314

**14 The Approach of Equilibrium** 315
14.1 The Langevin Equation 315
14.2 The Fokker–Planck Equation 318
14.3 The Boltzmann Equation 320
14.4 The *H*-Theorem 321
14.5 Transport Phenomena 323
14.6 Nondissipative Hydrodynamics 325
14.7 The Emergence of Viscosity 327
14.8 The Fluctuation–Dissipation Theorem 329
Bibliographical Notes 335

**Appendix A Probability Distributions** **336**

**Appendix B Equilibrium and Combinatorics** **340**

**Appendix C WKB at the Bottom** **343**

**Appendix D Some Analytic Functions** **348**

**Appendix E Euler–Maclaurin and Abel–Plana Formulas** **352**

**Appendix F Spin and the Pauli Equation** **356**

**Appendix G The $G_{n,m}$ Operator** **358**

*References* 362
*Index* 365

# Preface

This book grew out of a 40-hour course on statistical physics for third-year undergraduate students that the senior author (A. S.) gave at Scuola Normale during the 2017–2018 and 2018–2019 academic years, and also includes a few relevant additions. Some of the lectures originated in previous courses given at the University of Rome "Tor Vergata," initially with the late Prof. Roberto Petronzio, and then at Scuola Normale. The two courses offered during the 2017–2018 and 2018–2019 academic years included 20 hours of additional lectures, given by the junior author (C. H.) and devoted to various complements and to the explicit solution of instructive problems, which also grew into important portions of the book.

A peculiarity of this book is perhaps that, while it is aimed at undergraduate students at the beginning of their third year, it also explores connections with more advanced topics without resorting, insofar as possible, to the full machinery of quantum field theory. This qualifies the end result as a hybrid, beyond the level of elementary introductions and yet below the level of many excellent textbooks devoted to statistical field theory. Optimistically, we might have blended somehow the virtues of both, but more realistically we ended up perhaps half way between this goal and the opposite end. Our choice of topics and the level of the presentation, however, were motivated by the advanced level of the undergraduate students at Scuola Normale, whose courses should provide stimulating views around and beyond the main path set out by the University of Pisa.

In order to stress its twofold nature, we have decided to divide the book into two parts. The first part (Chapters 1–10) is devoted to aspects of classical and quantum statistical physics that are more standard for a textbook of this level. These include the different ensembles and the Boltzmann, Fermi–Dirac, and Bose–Einstein statistics; the van der Waals theory; and some key magnetic properties of matter, with emphasis on phase transitions and singular behavior, but also an introduction to zero-point fluctuations. On the other hand, the second part (Chapters 11–14) explores a selection of more advanced topics, some aspects of which lie at the forefront of current research. It is meant as a reward for the students who have made the effort to get that far, and should also serve as a stimulus to go farther in more advanced texts. In Chapter 11 we describe the Onsager solution of the two-dimensional Ising model, and along the way we discuss Kramers–Wannier duality, while also stressing the utmost importance of dualities in more general contexts. In Chapter 12 we explore the one-dimensional quantum Heisenberg model, the XXX chain, thus providing a first glimpse at integrability in that context. In Chapter 13 we discuss aspects of second-order phase transitions from the vantage point of symmetry. We thus address conformal invariance, taking note of its peculiar form in two dimensions and hinting at the role of the AdS/CFT correspondence, before taking a first look at the renormalization group. We first consider the XY model, whose treatment rests only to a limited extent on the machinery of quantum field theory, and then examine the $\epsilon$ expansion for the Ising model. Finally, in Chapter 14 we discuss the approach to equilibrium, highlighting its links with fluid mechanics, before concluding with a brief discussion of the fluctuation–dissipation theorem. These last topics lie at the roots of statistical physics and are often taken as a starting point, but we inserted them in Part II since they lie, to some extent, outside our main line of development.

Almost all derivations are carried out in detail, and some mathematical results are collected in a number of appendices. Our approach to the various ensembles is direct, via the von Neumann entropy, so that the student can start getting a flavor of what the subject is about by the end of the first lecture or so. Another peculiarity of our book is the presence of a rather large chapter devoted to quantum mechanics. It contains several useful results, in particular on the WKB method, with details that are not usually available, especially in books devoted to statistical physics, but are helpful in later chapters.

We have included 140 problems, of varying levels of difficulty, all of which are in Part I, since Part II, as we have stressed, provides a preview of more advanced topics. Although several problems were at least inspired by some of the books in the bibliography, a fair portion of them were devised, over the years, for classroom tests in Rome and Pisa.

# Acknowledgments

A. S. is indebted to the late Prof. Roberto Petronzio. He shared with him, for a few years, the responsibility of a similar, albeit more elementary, course at the University of Rome "Tor Vergata," in a junior role similar to the one of C. H. at Scuola Normale, before taking it over for some time on his own. We are both grateful to the students who took the course, or other related ones, over the years for their stimulating comments. We are particularly grateful to Giuseppe Bogna, whose course notes helped us to streamline an early version of our manuscript. We are also grateful to Dr. Karapet Mkrtchyan, who used the manuscript for his lectures and helped us to correct many misprints, and to our colleagues Dr. Davide Fioravanti and Dr. Rob Klabbers for a careful reading of Chapter 12. Finally, C. H. would like to thank Prof. Loris Ferrari of the University of Bologna, who first introduced him to the subject.

The work of A. S. was partly supported by Scuola Normale, by INFN (IS GSS-Pi), and by the MIUR-PRIN contract 2017CC72MK_003. A. S. is also grateful to Prof. W. Buchmuller for the kind hospitality at DESY-Hamburg in 2019, and to the Alexander von Humboldt Foundation for kind and generous support while this work was in progress. The work of C. H. was partly supported by the Knut and Alice Wallenberg Foundation, under grant KAW 2018.0116.

# PART I

# 1 Elements of Thermodynamics

This first chapter is devoted to a quick review of some key facts of thermodynamics. We begin with the laws of thermodynamics, before turning to thermodynamic potentials and the Maxwell relations. We conclude with a discussion of stability criteria for thermodynamic systems.

## 1.1 The Laws of Thermodynamics

Let us begin with a brief recollection of the principles of thermodynamics. Our starting point are the first two laws.

**First Law:** A thermodynamic system can perform work $W$ at the expense of a variation of internal energy $\Delta U$ and a heat exchange $Q$ with the environment, and the three quantities are related according to

$$\Delta U = Q - W. \tag{1.1}$$

Note that, while $\Delta U$ depends only on the initial and final states, the two other terms depend on the specific transformation. Infinitesimally, this translates to

$$dU = \delta Q - \delta W, \tag{1.2}$$

where $dU$ is an exact differential, while $\delta Q$ and $\delta W$ are not. For most systems of interest in the following chapters,

$$\delta W = P dV, \tag{1.3}$$

where $P$ is their pressure and $V$ is their volume.

**Second Law:** For any cyclic transformation,

$$\oint \frac{\delta Q}{T} \leq 0, \tag{1.4}$$

where $T$ is the temperature of the external source. This is also known as Clausius inequality. The inequality is saturated for *reversible* transformations, that is those in which the system proceeds through equilibrium configurations.

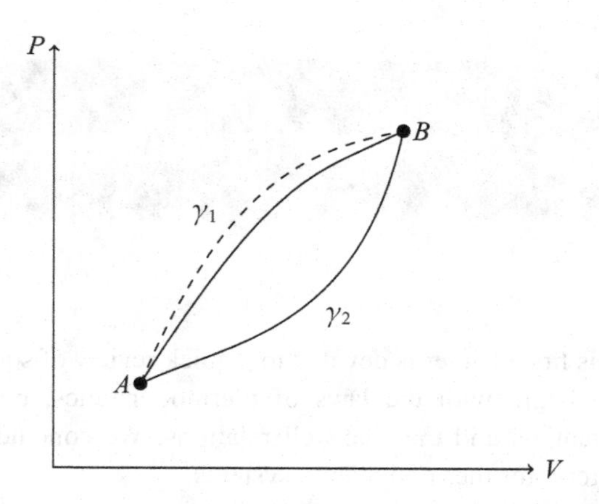


**Figure 1.1** A schematic illustration of $\gamma_1$ and $\gamma_2$ in the *PV* plane. The transformation $\gamma_1$ can be reversible, in which case the system goes through equilibrium configurations (solid line), or irreversible (dashed line).

By considering a cycle comprising a transformation $\gamma_1(A, B)$ from a state $A$ to a state $B$ and a *reversible* transformation $\gamma_2(B, A)$ from $B$ to $A$ as in Figure 1.1, one can conclude that

$$\int_{\gamma_1(A,B)} \frac{\delta Q}{T} + \int_{\gamma_2(B,A)} \frac{\delta Q_{\text{rev}}}{T} \leq 0. \tag{1.5}$$

Since $\gamma_2$ is reversible, one can also recast this relation in the form

$$\int_{\gamma_1(A,B)} \frac{\delta Q}{T} \leq \int_{\gamma_2(A,B)} \frac{\delta Q_{\text{rev}}}{T} = \Delta S(A, B), \tag{1.6}$$

where we have also defined the entropy difference $\Delta S(A, B)$ between state $A$ and state $B$.

This result has a few noteworthy consequences. To begin with, choosing $\gamma_1$ to be reversible, one can conclude that $\Delta S(A, B)$ is independent of the choice of reversible path between $A$ and $B$. Moreover, if the system is isolated during $\gamma_1$,

$$\Delta S \geq 0 \tag{1.7}$$

because $\delta Q = 0$. Finally, if $A$ and $B$ are infinitesimally separated,

$$T dS \geq \delta Q. \tag{1.8}$$

We have defined the entropy up to an additive constant, which could be associated to an arbitrary reference state. However, we can anticipate that in statistical physics the additive constant is naturally fixed via Boltzmann's formula, which we shall meet in Eq. (2.16),

$$S = k_B \log \Gamma, \tag{1.9}$$

where $k_B = 1.38 \times 10^{-23}$ J/K is the Boltzmann constant. This expression links the entropy of a given thermodynamic state to the number, $\Gamma$, of microscopic configurations (microstates) that correspond to it. In this fashion, one is setting to zero the entropy of the "ordered" thermodynamic states that correspond to a single microstate.

While we shall return to this issue in Chapter 2, one can argue that Boltzmann's formula follows from the requirement that $S$, like $U$ or $Q$, be an *extensive* quantity.[1] Indeed, a pair of independent sub-systems with $\Gamma_1$ and $\Gamma_2$ microstates build a system with $\Gamma = \Gamma_1 \Gamma_2$ microstates. Letting $S = f(\Gamma)$, and demanding that it be the sum of two entropies $S_1$ and $S_2$ of the sub-systems, the function $f$ should satisfy

$$f(\Gamma_1 \Gamma_2) = f(\Gamma_1) + f(\Gamma_2), \tag{1.10}$$

and one can show that Eq. (1.9), with a constant $k$, is the only continuous solution of this functional equation.

**Third Law:** In the $T \to 0$ limit, thermodynamic systems approach orderly states, and the entropy should have a *nonsingular* behavior, independent of pressure or volume.

This limit can be well defined only if Nernst's theorem,

$$\lim_{T \to 0} C_V(T) = 0, \tag{1.11}$$

holds, where

$$C_V = \left( \frac{\partial U}{\partial T} \right)_V \tag{1.12}$$

is the specific heat at constant volume. We shall see that quantum mechanics is instrumental to this effect. A nonzero limit for $C_V$ would indeed imply, for a reversible transformation performed at fixed volume, a logarithmically divergent

$$\Delta S = \int_{\gamma(A,B)} C_V(T) \, \frac{\mathrm{d}T}{T} \tag{1.13}$$

as the state $A$ approaches $T = 0$.

## 1.2 Thermodynamic Potentials

We can now recall a number of facts concerning thermodynamic potentials, while confining our attention to reversible transformations. The various thermodynamic potentials are characterized by different groups of natural variables. While the entropy remains constant in reversible transformations of isolated systems, the other potentials that we recall here, the enthalpy and the Helmholtz, Gibbs, and Grand potentials, remain constant when different groups of control variables are held fixed.

Let us begin from the *internal energy* $U$, for which the first and second laws imply the differential form

$$dU = T dS - P dV + \mu \, dN, \tag{1.14}$$

where we have also introduced the *chemical potential* $\mu$, a possibly less familiar thermodynamic function that is not *extensive* like $V$, $U$, $A$, or $G$, but is *intensive* like

[1] Extensive quantities grow proportionally to the size of the system, while intensive ones are independent of it.

$T$ and $P$. As we shall see, it characterizes diffusive equilibria. This expression also manifests the natural variables for $U$, which are $S$, $V$, and $N$, and the corresponding derivatives,

$$T = \left(\frac{\partial U}{\partial S}\right)_{V,N}, \quad P = -\left(\frac{\partial U}{\partial V}\right)_{S,N}, \quad \mu = \left(\frac{\partial U}{\partial N}\right)_{S,V}, \tag{1.15}$$

therefore have a universal meaning. Moreover, demanding that mixed derivatives coincide leads to the *Maxwell relations*

$$\left(\frac{\partial T}{\partial V}\right)_{S,N} = -\left(\frac{\partial P}{\partial S}\right)_{V,N}, \quad \left(\frac{\partial T}{\partial N}\right)_{S,V} = \left(\frac{\partial \mu}{\partial S}\right)_{V,N},$$
$$\left(\frac{\partial P}{\partial N}\right)_{S,V} = -\left(\frac{\partial \mu}{\partial V}\right)_{S,N}. \tag{1.16}$$

The specific heat at constant volume $C_V$, which we first introduced in Eq. (1.12), can be related to $U$ and $S$ according to

$$C_V = \left(\frac{\partial U}{\partial T}\right)_{V,N} = T\left(\frac{\partial S}{\partial T}\right)_{V,N}. \tag{1.17}$$

Alternatively, one can invert Eq. (1.14) to regard the *entropy* as a thermodynamic potential with natural variables $U$, $V$, and $N$, so that

$$dS = \frac{dU}{T} + \frac{P}{T}dV - \frac{\mu}{T}dN, \tag{1.18}$$

and therefore

$$\frac{1}{T} = \left(\frac{\partial S}{\partial U}\right)_{V,N}, \quad \frac{P}{T} = \left(\frac{\partial S}{\partial V}\right)_{U,N}, \quad \frac{\mu}{T} = -\left(\frac{\partial S}{\partial N}\right)_{U,V}. \tag{1.19}$$

One can now define the *enthalpy*,

$$H = U + PV, \tag{1.20}$$

and Eq. (1.14) implies that

$$dH = TdS + VdP + \mu\, dN, \tag{1.21}$$

so that

$$T = \left(\frac{\partial H}{\partial S}\right)_{P,N}, \quad V = \left(\frac{\partial H}{\partial P}\right)_{S,N}, \quad \mu = \left(\frac{\partial H}{\partial N}\right)_{S,P}. \tag{1.22}$$

The equality of mixed derivatives now implies the additional Maxwell relations

$$\left(\frac{\partial T}{\partial P}\right)_{S,N} = \left(\frac{\partial V}{\partial S}\right)_{P,N}, \quad \left(\frac{\partial T}{\partial N}\right)_{S,P} = \left(\frac{\partial \mu}{\partial S}\right)_{P,N},$$
$$\left(\frac{\partial V}{\partial N}\right)_{S,P} = \left(\frac{\partial \mu}{\partial P}\right)_{S,N}. \tag{1.23}$$

One is also led to define the specific heat at constant pressure $C_P$, according to

$$C_P = \left(\frac{\partial H}{\partial T}\right)_{P,N} = T\left(\frac{\partial S}{\partial T}\right)_{P,N}. \tag{1.24}$$

The *Helmholtz free energy* is then defined as

$$A = U - TS, \tag{1.25}$$

and Eq. (1.14) implies that

$$dA = -S\,dT - P\,dV + \mu\,dN, \tag{1.26}$$

so that

$$S = -\left(\frac{\partial A}{\partial T}\right)_{V,N}, \quad P = -\left(\frac{\partial A}{\partial V}\right)_{T,N}, \quad \mu = \left(\frac{\partial A}{\partial N}\right)_{T,V}. \tag{1.27}$$

The equality of mixed derivatives now implies the additional Maxwell relations

$$\left(\frac{\partial S}{\partial V}\right)_{T,N} = \left(\frac{\partial P}{\partial T}\right)_{V,N}, \quad \left(\frac{\partial S}{\partial N}\right)_{T,V} = -\left(\frac{\partial \mu}{\partial T}\right)_{V,N},$$
$$\left(\frac{\partial P}{\partial N}\right)_{T,V} = -\left(\frac{\partial \mu}{\partial V}\right)_{T,N}. \tag{1.28}$$

The *Gibbs free energy* is defined as

$$G = A + PV, \tag{1.29}$$

and therefore

$$dG = -S\,dT + V\,dP + \mu\,dN, \tag{1.30}$$

so that

$$S = -\left(\frac{\partial G}{\partial T}\right)_{P,N}, \quad V = \left(\frac{\partial G}{\partial P}\right)_{T,N}, \quad \mu = \left(\frac{\partial G}{\partial N}\right)_{T,P}. \tag{1.31}$$

The equality of mixed derivatives now implies one more set of Maxwell relations:

$$\left(\frac{\partial S}{\partial P}\right)_{T,N} = -\left(\frac{\partial V}{\partial T}\right)_{P,N}, \quad \left(\frac{\partial S}{\partial N}\right)_{T,P} = -\left(\frac{\partial \mu}{\partial T}\right)_{P,N},$$
$$\left(\frac{\partial V}{\partial N}\right)_{T,P} = \left(\frac{\partial \mu}{\partial P}\right)_{T,N}. \tag{1.32}$$

There is something interesting here: insofar as $G$ is an extensive function, it should scale proportionally to the number $N$ of particles, but it is naturally a function of $N$ and of the two intensive variables $T$ and $P$. As a result

$$G(T,P,N) = N\,\mu(T,P), \tag{1.33}$$

so that $G$ and the chemical potential $\mu$ are intimately related, and consequently

$$d\mu = -s\,dT + v\,dP, \tag{1.34}$$

with

$$s = \frac{S}{N}, \quad v = \frac{V}{N} \tag{1.35}$$

describing the specific entropy and the specific volume, or the entropy and volume per particle.

Finally, the *Grand Potential* is defined as

$$\Omega = A - \mu N, \tag{1.36}$$

and therefore

$$d\Omega = -SdT - PdV - Nd\mu, \tag{1.37}$$

so that

$$S = -\left(\frac{\partial \Omega}{\partial T}\right)_{V,\mu}, \quad P = -\left(\frac{\partial \Omega}{\partial V}\right)_{T,\mu}, \quad N = -\left(\frac{\partial \Omega}{\partial \mu}\right)_{T,V}. \tag{1.38}$$

The equality of mixed derivatives now implies a final set of Maxwell relations

$$\left(\frac{\partial S}{\partial V}\right)_{T,\mu} = \left(\frac{\partial P}{\partial T}\right)_{V,\mu}, \quad \left(\frac{\partial S}{\partial \mu}\right)_{T,V} = \left(\frac{\partial N}{\partial T}\right)_{V,\mu},$$
$$\left(\frac{\partial P}{\partial \mu}\right)_{T,V} = \left(\frac{\partial N}{\partial V}\right)_{T,\mu}. \tag{1.39}$$

Again, insofar as $\Omega$ is an extensive function, which should depend on $V$ and on the two intensive variables $T$ and $\mu$,

$$\Omega(T, V, \mu) = -P(T, \mu)\, V, \tag{1.40}$$

on account of Eq. (1.37).

Note that an additional step to build $\Omega + PV$ would apparently lead to a thermodynamic potential depending only on the intensive variables $(T, P, \mu)$, which can only vanish by consistency. In this fashion one can recover Eqs. (1.33) and (1.34).

## 1.3 Comparison between $C_P$ and $C_V$

We can now explore some important consequences of thermodynamics, and to begin with we would like to compare the two types of specific heats that we have encountered in Eqs. (1.17) and (1.24). There is a nice trick to do this, which is used at length in the classic text of Landau and Lifshitz. It rests on an important property of Jacobians, which satisfy the chain rule like ordinary derivatives. Jacobians associated to $2 \times 2$ matrices of the type

$$\frac{\partial(A,B)}{\partial(C,D)} = \det \begin{vmatrix} \left(\frac{\partial A}{\partial C}\right)_D & \left(\frac{\partial A}{\partial D}\right)_C \\ \left(\frac{\partial B}{\partial C}\right)_D & \left(\frac{\partial B}{\partial D}\right)_C \end{vmatrix} \tag{1.41}$$

can be used to express $C_V$ as

$$C_V = T\frac{\partial(S,V)}{\partial(T,V)} = T\frac{\partial(S,V)}{\partial(T,P)}\frac{\partial(T,P)}{\partial(T,V)} = T\frac{\partial(S,V)}{\partial(T,P)}\left(\frac{\partial P}{\partial V}\right)_T. \tag{1.42}$$

Expanding the remaining determinant gives

$$C_V = T\left(\frac{\partial S}{\partial T}\right)_P - T\left(\frac{\partial S}{\partial P}\right)_T\left(\frac{\partial P}{\partial V}\right)_T\left(\frac{\partial V}{\partial T}\right)_P, \tag{1.43}$$

and the first term is $C_P$, so that one can conclude that

$$C_V = C_P + T\left[\left(\frac{\partial V}{\partial T}\right)_P\right]^2\left(\frac{\partial P}{\partial V}\right)_T, \tag{1.44}$$

where we have used

$$\left(\frac{\partial P}{\partial V}\right)_T \left(\frac{\partial V}{\partial P}\right)_T = 1 \tag{1.45}$$

and a Maxwell relation in Eq. (1.32). We shall prove shortly that, in general,

$$\left(\frac{\partial P}{\partial V}\right)_T \leq 0, \tag{1.46}$$

so that, under the same circumstances, $C_P \geq C_V$, and for a stable equilibrium, as we are about to see, the strong inequality $C_P > C_V$ holds. In Chapter 6 we shall elaborate on some subtleties that emerge in the presence of first-order phase transitions.

## 1.4 Fluctuations

We can now discuss three types of equilibrium conditions for a pair of subsystems, and their important implications.

1. **Thermal Equilibrium**
   Let us consider two subsystems $\mathcal{S}_1$ and $\mathcal{S}_2$ of a given isolated system $\mathcal{S}$ of fixed volumes, in *thermal contact* with each other and weakly interacting, so that for the entropy one can conclude that $S \simeq S_1 + S_2$. If the energy of subsystem $\mathcal{S}_1$ undergoes a variation $dU_1$, the energy of subsystem $\mathcal{S}_2$ thus undergoes the opposite variation $dU_2 = -dU_1$. Consequently, the total entropy change is

$$dS = \left(\frac{\partial S_1}{\partial U}\right)_V dU_1 + \left(\frac{\partial S_2}{\partial U}\right)_V dU_2 = \left(\frac{1}{T_1} - \frac{1}{T_2}\right) dU_1. \tag{1.47}$$

   We know from the second law of thermodynamics that for an isolated system $dS \geq 0$, and therefore if $dU_1 > 0$ one can conclude that $T_1 < T_2$, and vice versa. In other words, heat flows from warmer bodies to colder ones. If the two systems are in thermal equilibrium, $S$ does not vary and the two temperatures $T_1$ and $T_2$ coincide. This fact is often referred to as the *zeroth law of Thermodynamics*.
2. **Mechanical Equilibrium**
   Let us again consider two subsystems $\mathcal{S}_1$ and $\mathcal{S}_2$, in *thermal* equilibrium and thus at the same temperature $T$, but in *mechanical contact* with each other via a moving partition and again weakly interacting, so that $S \simeq S_1 + S_2$. Hence, if the volume of subsystem $\mathcal{S}_1$ undergoes a variation $dV_1$, the volume of subsystem $\mathcal{S}_2$ undergoes the opposite variation $dV_2 = -dV_1$. The total entropy change is thus

$$dS = \left(\frac{\partial S_1}{\partial V}\right)_T dV_1 + \left(\frac{\partial S_2}{\partial V}\right)_T dV_2 = \frac{1}{T}(P_1 - P_2)\, dV_1. \tag{1.48}$$

   We know from the second law of thermodynamics that for an isolated system $dS \geq 0$, and therefore if $dV_1 > 0$ one can conclude that $P_1 > P_2$, and vice versa. In other words, we have recovered another well-known fact: the partition moves toward the region with lower pressure. In equilibrium, the two pressures $P_1$ and $P_2$ coincide.

3. **Diffusive Equilibrium**
Consider now two systems in equilibrium that can exchange particles, and consider variations $dN_1$ and $dN_2$, in the same spirit of what we did above, for a pair of systems in *thermal* and *mechanical* equilibrium, and thus at the same temperature $T$ and at the same pressure $P$. This leads to

$$dS = \left(-\mu_1 + \mu_2\right)\frac{dN_1}{T}, \tag{1.49}$$

so that demanding that $dS \geq 0$ one can deduce that $\mu_1 \geq \mu_2$ if $dN_1 < 0$, and vice versa. In other words, the particles tend to move into the system with lower chemical potential, and in equilibrium the two chemical potentials $\mu_1$ and $\mu_2$ coincide.

## 1.5 Stability

We can now turn to fluctuations, and to this end let us consider a subsystem 1 of a large system 2, which should be in thermal and mechanical equilibrium within it, thus having the same temperature $T_0$ and the same pressure $P_0$ as the larger system, according the first and second equilibrium conditions given in Section 1.4. Our task is to compute the variation of the internal energy $U$, which depends naturally on the thermodynamic variables that are not fixed a priori, $S$ and $V$. Keeping terms up to second order while leaving implicit, for brevity, the variables that are kept fixed, the result is

$$\begin{aligned} dU_1 &= T_0\, dS_1 - P_0\, dV_1 \\ &+ \frac{1}{2}\left[\frac{\partial^2 U_1}{\partial S^2}\, dS_1^2 + 2\frac{\partial^2 U_1}{\partial S\,\partial V}\, dS_1\, dV_1 + \frac{\partial^2 U_1}{\partial V^2}\, dV_1^2\right]. \end{aligned} \tag{1.50}$$

Since $T_0$ and $P_0$ are fixed, this expression can be recast in the more suggestive form

$$dG = \frac{1}{2}\left[\frac{\partial^2 U_1}{\partial S^2}\, dS_1^2 + 2\frac{\partial^2 U_1}{\partial S\,\partial V}\, dS_1\, dV_1 + \frac{\partial^2 U_1}{\partial V^2}\, dV_1^2\right], \tag{1.51}$$

so that a stable equilibrium demands that

$$\frac{\partial^2 U_1}{\partial S^2}\, dS_1^2 + 2\frac{\partial^2 U_1}{\partial S\,\partial V}\, dS_1\, dV_1 + \frac{\partial^2 U_1}{\partial V^2}\, dV_1^2 > 0. \tag{1.52}$$

This condition translates into the following inequalities for the first coefficient and the discriminant:

$$\begin{aligned} &\frac{\partial^2 U_1}{\partial S^2} > 0, \\ &\frac{\partial^2 U_1}{\partial S^2}\,\frac{\partial^2 U_1}{\partial V^2} - \left(\frac{\partial^2 U_1}{\partial S\,\partial V}\right)^2 > 0, \end{aligned} \tag{1.53}$$

and taking into account that

$$\left(\frac{\partial^2 U_1}{\partial S^2}\right)_V = \left(\frac{\partial T}{\partial S}\right)_V = \frac{T}{C_V}, \tag{1.54}$$

the first condition that one can extract from the inequality is

$$C_V > 0. \tag{1.55}$$

One can then recast the second inequality of Eqs. (1.53) in the form

$$\frac{\partial(T,P)}{\partial(S,V)} \equiv \left(\frac{\partial T}{\partial S}\right)_V \left(\frac{\partial P}{\partial V}\right)_S - \left(\frac{\partial T}{\partial V}\right)_S \left(\frac{\partial P}{\partial S}\right)_V < 0, \tag{1.56}$$

and having identified this expression with a Jacobian, one can apply the chain rule to turn it into

$$\frac{\frac{\partial(T,P)}{\partial(T,V)}}{\frac{\partial(S,V)}{\partial(T,V)}} \equiv \frac{\left(\frac{\partial P}{\partial V}\right)_T}{\left(\frac{\partial S}{\partial T}\right)_V} = \frac{T}{C_V}\left(\frac{\partial P}{\partial V}\right)_T < 0. \tag{1.57}$$

The second condition that follows from stability is therefore

$$\left(\frac{\partial P}{\partial V}\right)_T < 0. \tag{1.58}$$

This was also the missing information required to conclude that $C_P > C_V$ in Section 1.3.

## 1.6 Specific Heat and Compressibility

One can now deduce an important relation between the isothermal and isentropic compressibilities, defined as

$$\kappa_T = -\frac{1}{V}\left(\frac{\partial V}{\partial P}\right)_T, \qquad \kappa_S = -\frac{1}{V}\left(\frac{\partial V}{\partial P}\right)_S, \tag{1.59}$$

and the specific heats. Indeed, the identity

$$\frac{\partial(S,P)}{\partial(T,P)}\frac{\partial(T,V)}{\partial(S,V)} = \frac{\partial(T,V)}{\partial(T,P)}\frac{\partial(S,P)}{\partial(S,V)}, \tag{1.60}$$

which is implied by the chain rule of Jacobians, translates into the relation

$$\frac{\left(\frac{\partial S}{\partial T}\right)_P}{\left(\frac{\partial S}{\partial T}\right)_V} = \frac{\left(\frac{\partial V}{\partial P}\right)_T}{\left(\frac{\partial V}{\partial P}\right)_S}, \tag{1.61}$$

which is equivalent to

$$\frac{C_P}{C_V} = \frac{\kappa_T}{\kappa_S}. \tag{1.62}$$

Note that, as a result of Eqs. (1.44) and (1.46), this implies that for stable systems the isothermal compressibility is normally larger than the isentropic compressibility.

Another interesting relation links the isothermal compressibility to the chemical potential. Let us note that, *at a fixed temperature* (omitting for brevity the corresponding subscript in the next equation),

$$\left(\frac{\partial P}{\partial V}\right)_N = \frac{\partial(P,N)}{\partial(P,V)}\frac{\partial(P,V)}{\partial(V,N)} = -\left(\frac{\partial N}{\partial V}\right)_P \left(\frac{\partial P}{\partial N}\right)_V. \tag{1.63}$$

Now using

$$\left(\frac{\partial N}{\partial V}\right)_{P,T} = n \tag{1.64}$$

and the last Maxwell relation in (1.28), one is led to

$$\left(\frac{\partial P}{\partial V}\right)_{N,T} = n \left(\frac{\partial \mu}{\partial V}\right)_{N,T}, \tag{1.65}$$

or equivalently to

$$-V\left(\frac{\partial P}{\partial V}\right)_{N,T} = n^2 \left(\frac{\partial \mu}{\partial n}\right)_{N,T}, \tag{1.66}$$

since $\mu$, an intensive function, depends on $V$ via the combination $n = \frac{N}{V}$. Therefore, the isothermal compressibility and the chemical potential are related to one another according to

$$\frac{1}{\kappa_T} = n^2 \left(\frac{\partial \mu}{\partial n}\right)_N. \tag{1.67}$$

The stability condition $\kappa_T > 0$ thus guarantees that the chemical grows with the density $n$. Hence, adding a new particle becomes more costly as the gas becomes denser. The effect is naturally less significant for systems with larger $\kappa_T$, which are more easily compressible.

## 1.7 The Ideal Monatomic Gas

This system plays a central role in thermodynamics and statistical physics and will recur in the following chapters. It provides results of universal value for monatomic gases at high temperatures and small densities. The equation of state for the ideal gas is, as is well known,

$$PV = Nk_B T, \tag{1.68}$$

while the specific heat at constant volume is

$$C_V = \frac{3}{2} N k_B, \tag{1.69}$$

and Eq. (1.44) then implies that

$$C_P = \frac{5}{2} N k_B. \tag{1.70}$$

The internal energy for this system is simply

$$U = \frac{3}{2} N k_B T, \tag{1.71}$$

and is independent of $V$ at a given temperature. Consequently

$$dS = \frac{3}{2} N k_B \frac{dT}{T} + \frac{P}{T} dV \tag{1.72}$$

can be integrated to obtain

$$S = \frac{3}{2} N k_B \log\left(\frac{T}{T_0}\right) + N k_B \log\left(\frac{v}{N v_0}\right), \tag{1.73}$$

where the dependence on the specific volume $v$ makes $S$ properly extensive, a point that we shall elaborate upon in Chapter 2. The two reference values $T_0$ and $v_0$ make the arguments of the logarithms properly dimensionless, while also introducing an additive constant in $S$, which is so far arbitrary. In Chapter 2 we shall connect them to microscopic properties of the gas. Moreover, as we shall see in Chapter 8, polyatomic molecules behave in a similar fashion at room temperatures and small densities, but have different values of their specific heats.

## Bibliographical Notes

More details on various aspects of thermodynamics can be found in many books listed in the bibliography, for instance in [16, 36] or in [54].

## Problems

**1.1** Show that

$$dS = \frac{C_V}{T}\, dT + \left(\frac{\partial P}{\partial T}\right)_V dV,$$

$$dU = C_V\, dT + \left[T\left(\frac{\partial P}{\partial T}\right)_V - P\right] dV.$$

**1.2** Compute the efficiency of an ideal Carnot cycle (Figure 1.2) performed by a perfect gas, which is defined, as is well known, by pair of isothermal curves and a pair of adiabatic ones.

**1.3** Compute the efficiency of an ideal Joule cycle (Figure 1.2) performed by a perfect gas, which is defined by a pair of adiabatic curves, together with a pair of isobaric ones. Express it as a function of the constant $\gamma = \frac{C_P}{C_V}$ and of the ratio of the two pressures, $P_2$, and $P_1$, of the isobaric transformations.

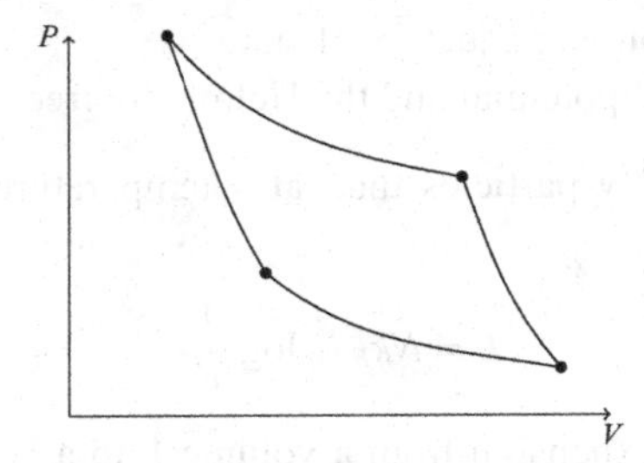


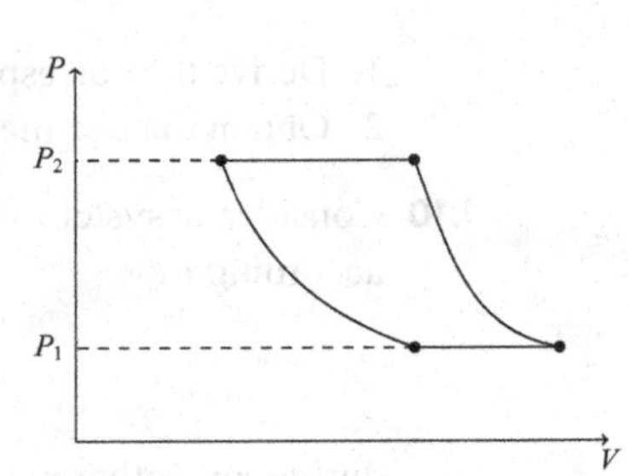


Figure 1.2 Reversible (Left) Carnot and (Right) Joule cycles.

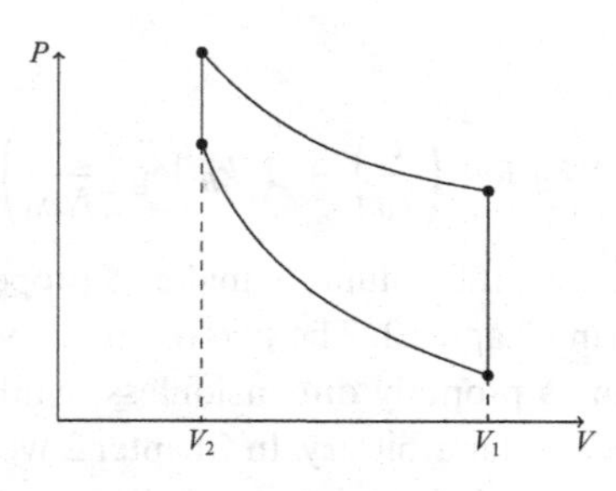

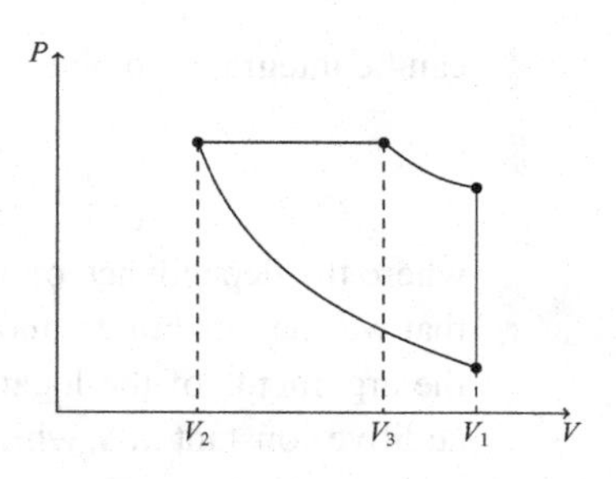


Figure 1.3 Reversible (Left) Otto and (Right) Diesel cycles.

**1.4** Compute the efficiency of an ideal Otto cycle (Figure 1.3) performed by a perfect gas, which is defined by a pair of adiabatic curves and a pair of isochoric ones. Express it as a function of $\gamma = \frac{C_P}{C_V}$ and the compression rate $r = \frac{V_1}{V_2}$.

**1.5** Compute the efficiency of an ideal Diesel cycle (Figure 1.3) performed by a perfect gas, which is defined by a pair of adiabatic curves, together with an isobaric and an isochoric one. Express it as a function of the constant $\gamma = \frac{C_P}{C_V}$ and the compression rates $r = \frac{V_1}{V_2}$ and $\alpha = \frac{V_3}{V_2}$.

**1.6** Prove that two adiabatic curves in the $(P, V)$ plane cannot intersect for any given substance.

**1.7** Compute the variation of the Gibbs free energy along a reversible isothermal compression of a perfect gas of $N$ particles, if the initial and final pressures are related by $P_2 = \sigma P_1$.

**1.8** Consider a system whose chemical potential takes the form

$$\mu = k_B T \log P + g(T),$$

where $g(T)$ is a known function.

1. Derive the corresponding equation of state.
2. Compute the internal energy.

**1.9** A system is described by the grand potential

$$\Omega = -\alpha V T^{\frac{5}{2}} e^{\frac{\mu}{k_B T}},$$

where $\alpha$ is a constant parameter.

1. Derive the corresponding equation of state
2. Obtain the chemical potential and the Helmholtz free energy.

**1.10** Consider a system of $N$ particles that, at a temperature $T_0$, performs work according to

$$L = N k_B T_0 \log \frac{V}{V_0}$$

during an isothermal expansion from a volume $V$ to a volume $V_0$, and whose entropy is

$$S = Nk_B \frac{V_0}{V}\left(\frac{T}{T_0}\right)^{\alpha}.$$

1. Find the equation of state of the system.
2. Find the corresponding Helmholtz free energy.
3. Find the work performed by the system during an isothermal expansion at any given temperature $T$.

**1.11** Consider a system that satisfies the relation $PV = Nk_B T_0$ at a given temperature $T_0$ and whose entropy is given by

$$S = 4Nk_B \left(\frac{T}{T_0}\right)^3 \log\left(\frac{P_0}{P}\right).$$

What is the equation of state for this system? Compute $C_P$ and $C_V$ explicitly, and verify that they are consistent with Eq. (1.44).

**1.12** Consider two ideal monatomic gases with the same number of particles, $N$, initially at temperatures $T_1$ and $T_2$ and in mechanical equilibrium in nearby vessels separated by a moving insulating partition. What is the entropy change of the system when it reaches equilibrium after the partition is removed?

**1.13** Consider two ideal monatomic gases containing the same number of particles, $N$, and held at the same temperature $T$ but initially in different volumes $V_1$ and $V_2$ in nearby vessels. Compute the total entropy variation of the system when it reaches equilibrium after the partition between the two nearby vessels is removed. How can one interpret the result?

**1.14** $N$ molecules of an ideal gas contained in an isolated vessel are compressed adiabatically from a volume $2V$ to a volume $V$. A valve is then opened, so that the system returns to occupy the volume $2V$. Given the initial and final temperatures $T_i$ and $T_f$, compute the total entropy variation of the system and the work performed by the gas. How can one interpret the result?

**1.15** Consider a system of $N$ particles for which

$$P = \frac{Nk_B T}{V} + f(T)\,\frac{N^2}{V^2}.$$

Compute the heat exchange $Q$ and the work performed at constant $T$ when the volume changes reversibly from $V_1$ to $V_2$.

**1.16** Consider a system whose equation of state is

$$V = V_0\, e^{\frac{T}{T_0} - \frac{P}{P_0}}.$$

1. Compute $C_P - C_V$ for the system.
2. Show that in this case $C_V$ is only a function of $T$.

**1.17** The efficiency of a refrigerator is defined as the ratio

$$\xi = \left|\frac{Q_2}{W}\right|$$

between the heat absorbed and the work needed to do it. Compute the efficiency of the Carnot cycle of Figure 1.2 working in reverse as a refrigerator between two temperatures $T_1$ and $T_2 > T_1$.

**1.18** Repeat the preceding analysis for a refrigerator undergoing in a similar fashion the Joule cycle of Figure 1.2.

**1.19** Repeat the preceding analysis for a refrigerator undergoing in a similar fashion the Otto cycle of Figure 1.3.

**1.20** Repeat the preceding analysis for a refrigerator undergoing in a similar fashion the Diesel cycle of Figure 1.3.

# 2 The Classical Ensembles

In this chapter we begin to explore our subject proper. We address the microcanonical ensemble, the canonical ensemble for Boltzmann systems, the role of fluctuations, and the equivalence between the two ensembles in the thermodynamic limit. Their distribution functions are derived by maximizing the von Neumann entropy, and the results applied to two significant examples. In Appendix B we review the more conventional approach, which is based on equilibrium populations.

## 2.1 Time Averages and Ensemble Averages

Consider a large number $N$ of point particles, with masses $m_i$, which are subject to the laws of classical mechanics. Ideally, one might wish to monitor their behavior by solving Newton's equations

$$m_i \frac{d\mathbf{v}_i}{dt} = \sum_{j \neq i} \mathbf{f}_{ji} + \mathbf{F}_i, \tag{2.1}$$

where the $\mathbf{v}_i$ are the particle velocities, $\mathbf{f}_{ji}(= -\mathbf{f}_{ij})$ is the force exerted by particle $j$ on particle $i$, and $\mathbf{F}_i$ is an external force acting on the system. In practice, however, this is technically unfeasible. Additionally, collecting such detailed information is not necessary. Rather, it is more useful to characterize macroscopic equilibrium configurations for systems of this type by using suitable averages, since they play a ubiquitous role in our understanding of nature. Classical thermodynamics tells us that a limited number of variables will suffice. Naively, one could consider time averages of the type

$$\overline{A} = \lim_{T\to\infty} \frac{1}{T} \int_0^T A(\mathbf{p}(t), \mathbf{q}(t))\, dt, \tag{2.2}$$

where $\mathbf{q}(t)$ and $\mathbf{p}(t)$ denote, collectively, positions and momenta for all particles. Computing these time averages, however, would be prohibitively difficult, since it would entail solving Eq. (2.1). However, there is a way out that we now illustrate.

## 2.2 The Microcanonical Ensemble

In order to characterize equilibrium distributions, it is convenient to begin by taking a closer look at classical mechanics and referring to "phase space," where the positions $\mathbf{q}$ and the momenta $\mathbf{p}$ of all particles are recorded. Given this information, which

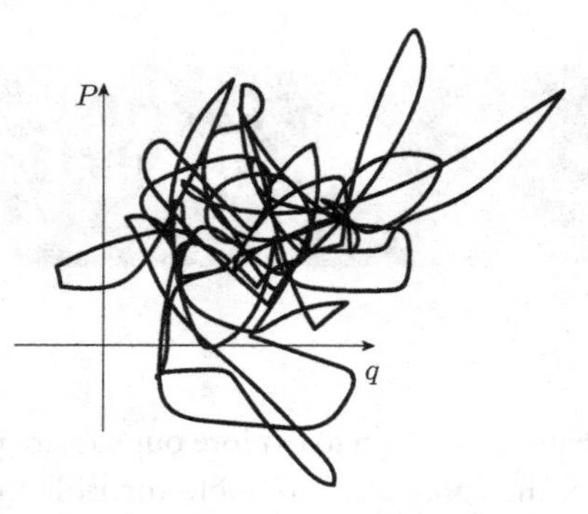


Figure 2.1 A schematic representation of trajectories in phase space.

includes all initial data for Eq. (2.1), one could follow, in principle, all subsequent motions. These would be recorded as curves wandering around phase space as shown, schematically, in Figure 2.1, and the many orders of magnitude that separate microscopic and macroscopic timescales, which we shall estimate shortly, would effectively turn the scribbles into smooth coloring functions $\rho(\mathbf{p},\mathbf{q})$. Time averages thus lead to configuration averages of the type

$$\langle A\rangle = \int d^{3N}\mathbf{p}\, d^{3N}\mathbf{q}\, \rho(\mathbf{p},\mathbf{q})\, A(\mathbf{p},\mathbf{q}). \tag{2.3}$$

It is very important, in this respect, to stress that the phase–space measure in Eq. (2.3) has an invariant meaning along the dynamics. This result is known as Liouville's theorem. It follows from Hamilton's equations, which can be presented schematically as

$$\dot{q} = \frac{\partial H(p,q)}{\partial p}, \qquad \dot{p} = -\frac{\partial H(p,q)}{\partial q}, \tag{2.4}$$

where $H$ is the Hamiltonian, which one may identify as the total energy for these types of systems. The Hamilton equations imply that, after an infinitesimal time interval $\delta t$,

$$\begin{aligned} q(t+\delta t) &\simeq q(t) + \delta t\, \frac{\partial H(p,q)}{\partial p}, \\ p(t+\delta t) &\simeq p(t) - \delta t\, \frac{\partial H(p,q)}{\partial q}, \end{aligned} \tag{2.5}$$

and consequently,

$$\frac{[dp\,dq]\,(t+\delta t)}{[dp\,dq]\,(t)} = \begin{vmatrix} 1 - \frac{\partial^2 H}{\partial p\,\partial q}\,\delta t & -\frac{\partial^2 H}{\partial q^2}\,\delta t \\ -\frac{\partial^2 H}{\partial p^2}\,\delta t & 1 + \frac{\partial^2 H}{\partial p\,\partial q}\,\delta t \end{vmatrix} \simeq 1 + \mathcal{O}(\delta t)^2. \tag{2.6}$$

Since the phase–space measure has an invariant meaning, equilibrium configurations translate into time-independent $\rho(\mathbf{p},\mathbf{q})$.

The key "ergodic hypothesis," which we have tried to justify intuitively, identifies the time averages $\overline{A}$ of interest with far simpler ensemble averages,

$$\overline{A} = \langle A\rangle. \tag{2.7}$$

This correspondence reflects the expectation that, *for an isolated thermodynamic system in equilibrium, macroscopic quantities should not depend on specific choices of initial conditions, but only on its total energy*. If the choice of initial conditions for coordinates and momenta, here denoted collectively as $z$, is immaterial, one can indeed average Eq. (2.2) over all initial conditions corresponding to a given total energy $U$, i.e. belonging to the manifold

$$\Sigma_U = \{z_0; H(z_0) = U\}, \tag{2.8}$$

and taking Liouville's theorem into account:

$$\frac{1}{T}\int_{\Sigma_U}\int_0^T A(z(t;z_0))\,dt\,d\Sigma_U(z_0) = \frac{1}{T}\int_0^T\int_{\Sigma_U} A(z)\,d\Sigma_U(z)\,dt. \tag{2.9}$$

In this fashion, the time integral drops out and one recovers Eq. (2.7). All microstates with total energy $U$ therefore have the same probability of being realized, and equilibrium averages take the form[1]

$$\langle A\rangle = \frac{1}{\Gamma(U)}\int A(z)\,|\nabla H(z)|\,\delta(H(z)-U)\,dz. \tag{2.10}$$

The last integral extends over all phase space, and $\Gamma(U)$ is the total measure of $\Sigma_U$:

$$\Gamma(U) = \int_{\Sigma_U} d\Sigma_U(z) = \int |\nabla H(z)|\,\delta(H(z)-U)\,dz, \tag{2.11}$$

where the Dirac $\delta$ function fixes the total energy. The Dirac $\delta$ function is defined, in the simplest possible one-dimensional setting, by the property

$$\int_{-\infty}^{\infty} dx\,f(x)\,\delta(x) = f(0), \tag{2.12}$$

for any smooth function $f(x)$.

Let us pause to take a closer look at the relevant time scales. For example, one can estimate typical periods for an electron circling around the nucleus in Bohr's atomic model of hydrogen, which we shall discuss in Chapter 3, translating typical distance scales

$$a \simeq 10^{-8}\ \text{cm} \tag{2.13}$$

into corresponding periods

$$T \simeq \frac{2\pi a}{c} \simeq 10^{-18}\ \text{sec}. \tag{2.14}$$

For other systems the estimate can grow, but time scales below $10^{-15}$ sec are typical for atomic or molecular systems, and lie more than 10 orders of magnitude below typical macroscopic values. Therefore, from the vantage point of microscopic dynamics, large systems subject to macroscopic actions move, typically very slowly, from one equilibrium configuration to another. This is the reason behind the disappearance of time in the ordinary description of thermodynamics.

One can now convert Eq. (2.11) into an expression for the entropy of the isolated system of $N$ particles, but we shall actually resort to the simpler expression,

$$\Gamma_N(U) = \int d^{3N}\mathbf{p}\,d^{3N}\mathbf{q}\,\delta\left(\frac{H(p,q)}{U} - 1\right), \tag{2.15}$$

which is equivalent to Eq. (2.11) in the large-$N$ limit and where the argument of Dirac's $\delta$ is properly dimensionless. This function suffices to account for the number

[1] The factor $|\nabla H(z)|$ defines the measure on the surface of fixed total energy, via a counterpart of $\delta(f(x))|f'(x)| = \delta(x-x_0)$, which holds when $f(x)$ has a single zero at $x_0$.

of available configurations, and hence naturally measures the disorder of the system, in a way that increases exponentially with $N$. Following Boltzmann, one can therefore link $\Gamma$ to the entropy $S$ as

$$S = k_B \log \Gamma_N(U), \tag{2.16}$$

where $k_B$ is the Boltzmann constant that was introduced in Eq. (1.9). In computing $S$, one can safely ignore in $\Gamma_N(U)$ the correction factors that do not grow exponentially with $N$, and for this reason we have left out in Eq. (2.15) the measure factor that was present in Eq. (2.11).

Equation (2.15) is essentially correct, except for a few amendments that we shall now introduce. To this end, let us explore its consequences for a perfect gas of $N$ particles of mass $m$ with negligible mutual interactions[2] that are confined to a volume $V$, for which

$$H = \sum_{i=1}^{N} \frac{\mathbf{p}_i^2}{2m}. \tag{2.17}$$

Particle collisions are clearly an essential ingredient for attaining equilibrium via mutual energy transfers, and thus for the validity of the ergodic hypothesis. Nonetheless, as we shall see in detail in Chapter 6, neglecting the contribution to H of mutual interactions is justified for rarefied gases at room temperature.

To begin with, one can compute Eq. (2.15) using polar coordinates in momentum space, for convenience, thus reducing it to

$$\Gamma_N(U) = V^N \, \Omega_{3N} \int_0^\infty dp \, p^{3N-1} \, \delta\left(\frac{p^2}{2\,m\,U} - 1\right), \tag{2.18}$$

where $\Omega_{3N}$ is the counterpart, in $3N$ dimensions, of the complete solid angle of three-dimensional Euclidean space. In general, one can obtain $\Omega_{3N}$ by comparing two ways of computing a Gaussian integral,

$$\left(\int dx\, e^{-x^2}\right)^{3N} = \pi^{\frac{3N}{2}} = \Omega_{3N} \int_0^\infty dp \, p^{3N-1} \, e^{-p^2}, \tag{2.19}$$

in Cartesian or polar coordinates. However, let us supplement this expression with a few details to clarify its meaning, while also preparing the ground for the ensuing discussion. The first detail is the two-dimensional case of this relation, which computes the square of the Gaussian integral in terms of an elementary one, namely

$$\left(\int dx\, e^{-x^2}\right)^{2} = 2\,\pi \int_0^\infty dp \, p \, e^{-p^2} = \pi. \tag{2.20}$$

More generally, we shall often need "shifted" Gaussian integrals of the type

$$\int_{-\infty}^{\infty} dx\, e^{-\alpha\, x^2 + \beta\, x} = \sqrt{\frac{\pi}{\alpha}}\, e^{\frac{\beta^2}{4\alpha}}, \tag{2.21}$$

which can be simply deduced from the previous case completing the square, and their higher-dimensional counterparts. The preceding result extends indeed to integrals over $n$-dimensional Euclidean space of the form

[2] We shall use the terms perfect gas and ideal gas interchangeably to refer to systems where mutual interactions are negligible, even when we take quantum effects into account, as in Chapter 8.

$$\int d^n\mathbf{x}\, e^{-\mathbf{x}^T A\, \mathbf{x} + \mathbf{b}^T \mathbf{x}} = \left(\frac{\pi^n}{\det A}\right)^{\frac{1}{2}} e^{\frac{1}{4}\mathbf{b}^T A^{-1}\mathbf{b}}, \tag{2.22}$$

with $A$ being a positive-definite real symmetric matrix, which will play a role in the ensuing chapters. Let us also introduce the Euler $\Gamma$ function, which can be defined for $\mathrm{Re}(x) > 0$ as

$$\Gamma(x) = \int_0^\infty du\, e^{-u}\, u^{x-1}. \tag{2.23}$$

Integrating by parts, one can see that

$$\Gamma(x+1) = x\,\Gamma(x), \tag{2.24}$$

while the two special cases $\Gamma(1) = 1$ and $\Gamma\left(\frac{1}{2}\right) = \sqrt{\pi}$ follow by direct computation. Consequently, if $n$ is an integer, the two special sets of values

$$\Gamma(n+1) = n!\,, \qquad \Gamma\left(n + \frac{1}{2}\right) = \frac{(2n-1)!\,!}{2^n}\,\sqrt{\pi} \tag{2.25}$$

are expressible in terms of $n!$, the product of all integers up to and including $n$, and of the double factorial $(2n-1)!\,!$, the product of all odd integers up to and including $(2n-1)$. Wrapping up, our first conclusion is that

$$\Omega_N = \frac{2\,\pi^{\frac{N}{2}}}{\Gamma\left(\frac{N}{2}\right)}. \tag{2.26}$$

This general expression includes the familiar results for $N = 2$, $\Omega_2 = 2\pi$, and for $N = 3$, $\Omega_3 = 4\pi$, as special cases.

Letting now

$$y = \frac{p^2}{2m}, \tag{2.27}$$

one can readily turn Eq. (2.18) into

$$\Gamma_N(U) = \frac{V^N (2\pi m)^{\frac{3N}{2}}}{\Gamma\left(\frac{3N}{2}\right)}\, U^{\frac{3N}{2}}, \tag{2.28}$$

and therefore

$$S = N k_B \log[V(2\pi m)^{\frac{3}{2}}] - \log\Gamma\left(\frac{3N}{2}\right) + \frac{3N}{2}\log U. \tag{2.29}$$

It is now interesting to take a closer look at the behavior of Euler's $\Gamma$ for large positive values of its argument. To this end, let us change the integration variable in Eq. (2.23) by letting $u = x\tau$, so that

$$\Gamma(x+1) = x^{x+1}\int_0^\infty d\tau\, e^{-x(\tau - \log\tau)}. \tag{2.30}$$

This way of presenting the integral allows a simple application of a powerful technique known as Laplace's method. For large positive values of $x$, the integral should be dominated by values of $\tau$ close to the minimum of the exponent, which is attained for $\tau = 1$. Keeping only the first two terms in the corresponding Taylor series suffices to capture the dominant behavior

$$\Gamma(x+1) \sim x^{x+1} e^{-x} \int_0^\infty d\tau\, e^{-\frac{x}{2}(\tau-1)^2} = x^{x+\frac{1}{2}} e^{-x} \sqrt{2\pi}, \tag{2.31}$$

where in the last step we have extended the integral to the whole axis, which is possible up to an exponentially small error. This important result, known as Stirling's formula, also implies that for large values of $N$,

$$\log \Gamma(N) \sim N \log N - N, \tag{2.32}$$

and makes it possible to recast Eq. (2.29) in the form

$$S = N k_B \log[V(2\pi m)^{\frac{3}{2}}] + \frac{3 N k_B}{2} \left[\log\left(\frac{2\,U}{3\,N}\right) + 1\right]. \tag{2.33}$$

Stirling's formula is the first term of an asymptotic expansion, a divergent series that can yield, nonetheless, very accurate estimates of $\Gamma(x)$ for large values of $x$. However, in this case Eq. (2.32) also affords a quick alternative derivation, obtained by writing

$$\log N! = \sum_{k=1}^{N} \log k \simeq \int_1^N \log x\, dx \simeq N \log N - N. \tag{2.34}$$

This is an instructive comment, since the replacement of sums by integrals will be a recurrent type of approximation in the following chapters. In general, the methods that we are developing recover classical thermodynamics only asymptotically, as $N \to \infty$, the very limit where $\Gamma$ attains a simple form. At any rate, in typical applications $N = \mathcal{O}(10^{23})$, and therefore the Stirling approximation is very well justified.

The result in Eq. (2.33), however, is still not correct, but we are close to a final fix. The problem is that $V$ is an extensive quantity, and therefore $N \log V$ is not extensive. The resulting behavior of $S$ would lead to the so-called Gibbs paradox: the entropy would increase when two identical gases mix! However, an extensive entropy is obtained if $V$ is replaced by $\frac{V}{N}$ in Eq. (2.33). In view of Eq. (2.32), this can be simply attained if Eq. (2.18) is supplemented with an overall factor $\frac{1}{N!}$, and from now on this factor will be always present when the particles are identical and can invade the whole volume $V$. However, there is one last error in Eq. (2.33): $\Gamma$ should be dimensionless, since after all it counts states and ends up in the argument of a logarithm in Eq. (2.16). Hence, the phase–space measure should be weighted by a quantity of dimension $\left[L^2 M T^{-1}\right]$, raised to the $3N$-th power. Quantum mechanics, as we shall see, requires that this quantity be precisely the Planck's constant $h = 6.626 \times 10^{-34}\,\mathrm{J \cdot s}$.

## Microcanonical Ensemble

The expression for the entropy in the "microcanonical ensemble" that we shall use in the following is

$$S = k_B \log\left\{\left[\frac{1}{N!}\right] \int \frac{d^{3N}\mathbf{p}\, d^{3N}\mathbf{q}}{h^{3N}}\, \delta\left(\frac{H(p,q)}{U} - 1\right)\right\}. \tag{2.35}$$

It must involve a factor $\frac{1}{N!}$ whenever the particles are identical and free to explore the volume $V$. The connection with thermodynamics relies on

$$dS = \frac{dU}{T} + \frac{P}{T}dV - \frac{\mu}{T}dN, \tag{2.36}$$

so that the temperature $T$, the pressure $P$ and the chemical potential $\mu$ are recovered from

$$\frac{1}{T} = \left(\frac{\partial S}{\partial U}\right)_{V,N}, \quad \frac{P}{T} = \left(\frac{\partial S}{\partial V}\right)_{U,N}, \quad \frac{\mu}{T} = -\left(\frac{\partial S}{\partial N}\right)_{U,V}. \tag{2.37}$$

The interpretation of this setting is that a huge number of unobservable "microstates" corresponds to an observable "macrostate," characterized by given values of the few thermodynamic variables $U$, $V$, and $N$ that an observer can adjust. Note that in this fashion, Eq. (2.33) has also become the correct *Sackur–Tetrode entropy* for the perfect gas,

$$S = Nk_B\left[\log\left(\frac{V}{N}\right) + \frac{3}{2}\log\left(\frac{4\pi mU}{3Nh^2}\right) + \frac{5}{2}\right]. \tag{2.38}$$

It is instructive to repeat the construction for a system of harmonic oscillators with distinct equilibrium positions at $\mathbf{q}_{0i}$, so that

$$H = \sum_{i=1}^{N}\frac{\mathbf{p}_i^2}{2m} + \sum_{i=1}^{N}\frac{m\,\omega^2\left(\mathbf{q}_i - \mathbf{q}_{0i}\right)^2}{2}, \tag{2.39}$$

which is also an interesting classical model for the vibrations of a crystal lattice. Now the proper starting point,

$$\Gamma_N(U) = \int\frac{d^{3N}\mathbf{p}\,d^{3N}\mathbf{q}}{h^{3N}}\,\delta\left(\frac{H(p,q)}{U} - 1\right), \tag{2.40}$$

lacks the overall $N!$, since the oscillators are distinguished by their equilibrium positions, and can be turned into a $6N$-dimensional integral of the same form as the one for the ideal gas after rescaling the variables. The end result is

$$\Gamma_N(U) = \left(\frac{U}{\hbar\,\omega}\right)^{3N}\frac{1}{\Gamma(3N)}, \tag{2.41}$$

so that, making use of Eq. (2.32),

$$S = 3Nk_B\left[\log\left(\frac{U}{3N\hbar\,\omega}\right) + 1\right], \tag{2.42}$$

which is a properly extensive quantity. Note the first occurrence, here, of the combination

$$\hbar = \frac{h}{2\pi}, \tag{2.43}$$

which will play a ubiquitous role in the following. While the current setup is correct for all cases of interest here, we shall return in Chapter 8 to some subtleties that quantum mechanics brings along for indistinguishable particles.

Let us now return to thermodynamics, referring to Eqs. (2.37). For the perfect gas, Eq. (1.14) implies that

$$\frac{1}{T} = \left(\frac{\partial S}{\partial U}\right)_V = \frac{3Nk_B}{2\,U}, \tag{2.44}$$

$$P = \left(\frac{\partial S}{\partial V}\right)_U = \frac{Nk_B}{V}, \tag{2.45}$$

or equivalently the law of perfect gases

$$P\,V = N k_B\,T, \tag{2.46}$$

and the main result of the kinetic theory of gases,

$$U = \frac{3}{2}\,N k_B\,T. \tag{2.47}$$

This relation reflects the random collisions of gas particles against a wall of the container orthogonal, say, to the $x$ axis. These transfer to it, surface and per unit time, the momentum

$$(2\,m\,v_x) \times \left(\frac{n}{2}\,v_x\right), \tag{2.48}$$

where $n = \frac{N}{V}$. The second factor counts the number of particles that impinge on the wall per unit time, which are half of those within a distance $v_x$ from it, since the others proceed on average in the opposite direction. Equation (2.47) follows since in equilibrium, symmetry considerations clearly imply that

$$\langle v_x^2 \rangle = \langle v_y^2 \rangle = \langle v_z^2 \rangle = \frac{1}{3}\,\langle \mathbf{v}^2 \rangle, \tag{2.49}$$

and therefore

$$P\,V = \frac{2}{3}\,U. \tag{2.50}$$

When Eq. (2.47), which follows from this result and the equation of state, is combined with the definition of the specific heat at constant volume

$$C_V = \left(\frac{\partial\,U}{\partial\,T}\right)_V, \tag{2.51}$$

for the perfect gas one is led to

$$C_V = \frac{3\,N k_B}{2}. \tag{2.52}$$

This result serves as a first illustration of the *classical equipartition law*: each translational degree of freedom contributes $\frac{1}{2}\,k_B\,T$ to the internal energy $U$.

For the gas of oscillators, proceeding along similar lines, one can deduce from Eq. (2.42) that the pressure vanishes, consistent with the expectation that in this case the particles are confined to neighborhoods of their equilibrium positions, while

$$U = 3\,N k_B\,T, \qquad C_V = 3\,N k_B. \tag{2.53}$$

This result is again compatible with equipartition, since there are also $3N$ vibrational degrees of freedom, each of which contributes an additional $\frac{1}{2}\,k_B\,T$ to the internal energy $U$.

### 2.2.1 Von Neumann Entropy

Let us now elaborate further on the meaning of Eq. (2.35) and on the related concepts of "microstates" and "macrostates." To this end, let us consider a box containing $n_1 + n_2$ marbles, while focusing on its two subsystems of $n_1$ white and $n_2$ black marbles. The probabilities that a marble belongs to either of them are

$$p_1 = \frac{n_1}{n_1 + n_2}, \qquad p_2 = \frac{n_2}{n_1 + n_2}, \tag{2.54}$$

and the entropy of the system,

$$\frac{S}{k_B} = \log(n_1 + n_2), \tag{2.55}$$

can be naturally decomposed as

$$\frac{S}{k_B} = p_1 \log n_1 + p_2 \log n_2 + \sigma(n_1, n_2). \tag{2.56}$$

The first two contributions record the entropies associated to the sets of white and black marbles, weighted by the corresponding probabilities. On the other hand, the last term,

$$\sigma(n_1, n_2) = -p_1 \log p_1 - p_2 \log p_2, \tag{2.57}$$

which is called "von Neumann entropy," records the uncertainty related to whether the marbles are black or white. This simple model captures what one is supposed to count, in the general case, to define the entropy resulting from the effect of many distinct microstates. The quantity of interest is the distribution of the corresponding points in phase space that manifest themselves in the same macrostate, ignoring finer details. The actual rationale behind this choice is that, as we shall see, quantum mechanics would make them inaccessible.

There is a convenient and conceptually important way of recovering Eq. (2.33) that rests on the von Neumann entropy. Let us illustrate the procedure by considering a system with, say, a large set of microstates that corresponds to a given total energy $U$, with probabilities $p_i$. Letting

$$\sigma = -\sum_i p_i \log p_i - \lambda\left(\sum_i p_i - 1\right), \tag{2.58}$$

one can extremize $\sigma$ with respect to the $p_i$, while the Lagrange multiplier $\lambda$ fixes the total probability to one. As a result,

$$\frac{\partial \sigma}{\partial p_i} = -1 - p_i - \lambda = 0. \tag{2.59}$$

Therefore, with a total of $\Gamma_N(U)$ microstates,

$$p_i = \frac{1}{\Gamma_N(U)}, \tag{2.60}$$

which satisfies the constraint enforced by the Lagrange multiplier $\lambda$. Note that we have identified a maximum value for $\sigma$, since

$$\frac{\partial^2 \sigma}{\partial p_i\, \partial p_j} = -\delta_{ij}, \tag{2.61}$$

and

$$\sigma_{\max} = \log \Gamma_N(U) \tag{2.62}$$

is naturally identified with $\frac{S}{k_B}$. This maximum principle thus leads again to the expressions of Section 2.2.

## 2.3 The Canonical Ensemble

Having introduced the von Neumann entropy, we can now generalize the preceding considerations to collections of particles in contact with a heat reservoir. This fixes the temperature $T$ of the system of interest, but introduces energy fluctuations. Only the average total energy is fixed in this "canonical ensemble" and can be taken to define the internal energy $U$ of the system. The proper generalization of Eq. (2.58) is therefore

$$\sigma = -\sum_I p_I \log p_I - \lambda_1 \left( \sum_I p_I - 1 \right) - \lambda_2 \left( \sum_I p_I E_I - U \right), \tag{2.63}$$

where the label $I$ is meant to stress that here the sums include *all states* of the $N$-particle system, with all possible values of the corresponding energies $E_I$.

Setting to zero the derivative with respect to $p_I$ now gives

$$p_I = \frac{1}{Z} e^{-\lambda_2 E_I}, \tag{2.64}$$

$$\frac{S}{k_B} = \log Z + \lambda_2 U, \tag{2.65}$$

which also brings to the forefront the "partition function"

$$Z = \sum_I e^{-\lambda_2 E_I}. \tag{2.66}$$

This quantity lies at the heart of this formulation, and of statistical physics altogether: it determines all thermodynamics quantities, and to begin with

$$U = \sum_I p_I E_I = -\frac{\partial}{\partial \lambda_2} \log Z. \tag{2.67}$$

Our next task is to attach a proper physical interpretation to the Lagrange multiplier $\lambda_2$, and to this end we start again from Eq. (1.14) to recover the temperature $T$ along the lines of Eq. (2.44). In detail, Eq. (2.65) yields

$$\frac{1}{k_B T} \equiv \beta = \frac{1}{k_B} \left( \frac{\partial S}{\partial U} \right)_V = \frac{\partial \log Z}{\partial \lambda_2} \frac{\partial \lambda_2}{\partial U} + U \frac{\partial \lambda_2}{\partial U} + \lambda_2, \tag{2.68}$$

and the first two terms cancel if one takes Eq. (2.67) into account. Note that the energy levels $E_I$ were kept fixed, since they can only depend on the boundary conditions granting the confinement of the system within a box, and thus on the volume $V$. In conclusion, one is led to identify the Lagrange multiplier $\lambda_2$ with

$$\beta = \frac{1}{k_B T}, \tag{2.69}$$

a parameter that is sometimes called the "inverse temperature" or "coldness," and then Eq. (2.65) links the partition function $Z$ to the Helmholtz free energy (1.25) as

$$\beta A = - \log Z. \tag{2.70}$$

Moreover, in equilibrium

$$p_I = \frac{1}{Z} e^{-\beta E_I}. \tag{2.71}$$

## Canonical Ensemble

The key ingredient of the canonical ensemble is the partition function

$$Z = \sum_I e^{-\beta E_I} = e^{-\beta A}, \tag{2.72}$$

which becomes

$$Z = \left[\frac{1}{N!}\right] \int \frac{d^{3N}\mathbf{p}\, d^{3N}\mathbf{q}}{h^{3N}} e^{-\beta H(\mathbf{p},\mathbf{q})} \tag{2.73}$$

in the continuum limit, i.e. when $k_B T$ is much larger than typical energy spacings. As for the microcanonical ensemble, the phase-space measure involves Planck's constant, and we can refer to Section 2.2 for a discussion of the coefficient $\frac{1}{N!}$. From the Helmholtz free energy $A$ one can then obtain

$$S = -\left(\frac{\partial A}{\partial T}\right)_{V,N}, \quad P = -\left(\frac{\partial A}{\partial V}\right)_{T,N}, \quad \mu = \left(\frac{\partial A}{\partial N}\right)_{T,V}, \tag{2.74}$$

and the internal energy as a function of $(T, V, N)$, as

$$U = \left(\frac{\partial\,(\beta A)}{\partial\,\beta}\right)_{V,N}. \tag{2.75}$$

For the ideal gas,

$$Z = \frac{V^N}{h^{3N} N!} \int d^{3N}\mathbf{p}\, e^{-\sum_{i=1}^N \beta \frac{\mathbf{p}_i^2}{2m}} = \frac{V^N}{h^{3N} N!} \left[\int d^3\mathbf{p}\, e^{-\beta\frac{\mathbf{p}^2}{2m}}\right]^N, \tag{2.76}$$

or

$$Z = \frac{Z_1^N}{N!}, \tag{2.77}$$

where

$$Z_1 = \frac{V}{h^3} \int d^3\mathbf{p}\, e^{-\beta\frac{\mathbf{p}^2}{2m}}. \tag{2.78}$$

This step highlights a general simplification resulting from the canonical formalism: when the $N$ constituents have no mutual interactions, $Z$ is determined by the phase–space integral related to a single constituent, possibly up to an overall $\frac{1}{N!}$ as explained in Section 2.2. From Eq. (2.78) one can simply recover the Maxwell distribution in momentum space. Taking into account the measure thus gives

$$f(p) = 4\pi \left(\frac{\beta}{4\pi m}\right)^{\frac{3}{2}} p^2\, e^{-\beta\frac{p^2}{2m}}, \tag{2.79}$$

whose integral is here normalized according to

$$\int_0^\infty f(p)\, dp = 1. \tag{2.80}$$

Computing the integral in Eq. (2.76) leads finally to

$$Z = \frac{V^N}{N!} \left[ \frac{2\,\pi\, m}{h^2\, \beta} \right]^{\frac{3N}{2}}, \tag{2.81}$$

so that, for large values of $N$, making use of Eqs. (2.72) and (2.32),

$$A = -\, N k_B\, T \left\{ 1 + \log \left[ \frac{V}{N} \left( \frac{2\,\pi\, m\, k_B\, T}{h^2} \right)^{\frac{3}{2}} \right] \right\}. \tag{2.82}$$

This result implies a convenient expression for the entropy $S$ as a function of $T$ and $V$,

$$S = N k_B \left\{ \frac{5}{2} + \log \left[ \frac{V}{N} \left( \frac{2\,\pi\, m\, k_B\, T}{h^2} \right)^{\frac{3}{2}} \right] \right\}, \tag{2.83}$$

and an expression for the chemical potential $\mu$ as a function of $T$ and $V$,

$$\mu = -\, k_B\, T \log \left[ \frac{V}{N} \left( \frac{2\,\pi\, m\, k_B\, T}{h^2} \right)^{\frac{3}{2}} \right]. \tag{2.84}$$

From these results one can simply recover the equation of state (2.46) and the equipartition law (2.47). Finally, let us note that Eq. (2.84) is often presented in the form

$$\mu = k_B\, T \log \left( \frac{n}{n_Q(T)} \right), \tag{2.85}$$

where $n = \frac{N}{V}$ is the number density and

$$n_Q(T) = \left( \frac{2\,\pi\, m\, k_B\, T}{h^2} \right)^{\frac{3}{2}} \tag{2.86}$$

is a characteristic "quantum density" that will play an important role in the next chapters.

For increasing temperatures and/or decreasing densities, the chemical potential becomes more and more negative, since in dilute systems the entropy gain favors the introduction of additional particles. Eq. (2.85) is often used in chemistry, in the form

$$n = n_Q(T)\, e^{\frac{\mu}{k_B T}}, \tag{2.87}$$

to define the concentrations in chemical reactions. We shall return to this point in Section 2.4.2.

We can repeat the exercise for the gas of distinguishable harmonic oscillators, for which

$$Z = \int \frac{d^{3N}\mathbf{p}\, d^{3N}\mathbf{q}}{h^{3N}}\, e^{-\sum_{i=1}^N \beta \left( \frac{\mathbf{p}_i^2}{2m} + \frac{m\,\omega^2}{2} (\mathbf{q}_i - \mathbf{q}_{0i})^2 \right)} = \left( \frac{1}{\beta\, \hbar\, \omega} \right)^{3N}, \tag{2.88}$$

so that in this case,

$$A = 3\,Nk_BT\log\left(\frac{\hbar\,\omega}{k_B\,T}\right), \tag{2.89}$$

and therefore

$$S = 3\,Nk_B\left[1 - \log\left(\frac{\hbar\,\omega}{k_B\,T}\right)\right]. \tag{2.90}$$

Moreover,

$$U = 3\,N k_B\,T, \tag{2.91}$$

and consequently

$$C_V = \left(\frac{\partial U}{\partial T}\right)_V = 3\,Nk_B. \tag{2.92}$$

This result underlies the Dulong–Petit law: this simple setup provides a model for crystal vibrations in solids at room temperatures, which links their specific heats to the number of atoms. Note that for this system the chemical potential,

$$\mu = 3\,k_B\,T\log\left(\frac{\hbar\,\omega}{k_B\,T}\right), \tag{2.93}$$

is independent of the density.

Equations. (2.52), (2.53), and (2.91) can be regarded as special cases of a more general result,

$$\left\langle \xi\,\frac{\partial H}{\partial \xi}\right\rangle = \frac{1}{Z}\int\frac{d^{3N}\mathbf{p}\,d^{3N}\mathbf{q}}{[N!\,]\,h^{3N}}\,\xi\,\frac{\partial H}{\partial \xi}\,e^{-\beta H} = k_B\,T, \tag{2.94}$$

where $\xi$ can be a component of the coordinate or momentum vectors. This relation is often called "thermal virial theorem": it can be simply proved by integrating by parts, and it lies at the heart of the equipartition theorems, which follow from it when $H$ depends quadratically on $\xi$.

Since we have presented two distinct methods to link microscopic data to thermodynamics, it is now time to address the equivalence between the microcanonical ensemble and the more convenient canonical ensemble. The starting point for this discussion is provided by Eqs. (2.72) and (2.75), which lead to

$$U = \langle E\rangle = \frac{\sum_I E_I\,e^{-\beta\,E_I}}{\sum_I e^{-\beta\,E_I}} \tag{2.95}$$

for the average internal energy. Differentiating with respect to $\beta$ now gives

$$\frac{\partial U}{\partial \beta} = -\frac{\sum_I E_I^2\,e^{-\beta\,E_I}}{\sum_I e^{-\beta\,E_I}} + \left(\frac{\sum_I E_I\,e^{-\beta\,E_I}}{\sum_I e^{-\beta\,E_I}}\right)^2 \equiv -\,\langle(E - \langle E\rangle)^2\rangle, \tag{2.96}$$

which determines the energy fluctuations around the equilibrium distribution provided by the canonical ensemble. Note that, once more, the derivative with respect to $\beta$ is taken keeping the boundaries fixed, so that the volume $V$ and the available energies $E_I$, which can depend on the boundary conditions, are all constant. On the other hand,

$$\left(\frac{\partial U}{\partial \beta}\right)_V = -\,k_B\,T^2\,C_V, \tag{2.97}$$

and therefore one obtains the important result

$$k_B T^2 C_V = \langle (\Delta E)^2 \rangle \equiv \langle (E - \langle E \rangle)^2 \rangle. \tag{2.98}$$

In conclusion, *the specific heat $C_V$ measures the energy fluctuations in the canonical ensemble*. However, the specific heat is an extensive variable, which scales proportionally to the number $N$ of constituents, and therefore

$$\sqrt{\frac{\langle (\Delta E)^2 \rangle}{\langle E \rangle^2}} \sim \frac{1}{\sqrt{N}}. \tag{2.99}$$

For the large values $N = \mathcal{O}(10^{23})$ that present themselves in macroscopic systems, the microcanonical and canonical ensembles are thus equivalent. In particular, for a perfect gas

$$\langle (\Delta E)^2 \rangle = \frac{3}{2} N (k_B T)^2. \tag{2.100}$$

### 2.3.1 More on Identical Particles

The two examples in the preceding section can be framed in more general terms, which allows us to highlight some universal facts on weakly-interacting systems. To this end, let us consider again the canonical partition function (2.72) for $N$ noninteracting particles with allowed single-particle energies $E_\alpha$. It can be cast in the form

$$Z_N = \sum_{\alpha_1, \ldots, \alpha_N} g_{\alpha_1 \cdots \alpha_N} e^{-\beta(E_{\alpha_1} + \cdots + E_{\alpha_N})}, \tag{2.101}$$

resorting again to discrete labels, where $g_{\alpha_1 \cdots \alpha_N}$ is the number of ways of distributing the particles among the allowed energy levels, which we shall also call *degeneracy* factor.

For distinguishable particles the degeneracy is simply

$$g_{\alpha_1 \cdots \alpha_N} = g_{\alpha_1} g_{\alpha_2} \cdots g_{\alpha_N}, \tag{2.102}$$

in terms of single-particle degeneracies $g_\alpha$, using the notation of Appendix B, and therefore the partition function attains the factorized form

$$Z_N = \sum_{\alpha_1} g_{\alpha_1} e^{-\beta E_{\alpha_1}} \cdots \sum_{\alpha_N} g_{\alpha_N} e^{-\beta E_{\alpha_N}} = Z_1^N, \tag{2.103}$$

where, as in Section 2.3, $Z_1$ is the single-particle contribution. An example of this type is the gas of harmonic oscillators, whose partition function is given in Eq. (2.88).

For indistinguishable particles, when the *a priori* probability that two particles belong to the same state is *negligible*, one can instead obtain a fair estimate of $g_{\alpha_1 \cdots \alpha_N}$ by first treating the particles as distinguishable and then accounting for their identity by an overall $\frac{1}{N!}$:

$$g_{\alpha_1 \cdots \alpha_N} \simeq \frac{g_{\alpha_1} g_{\alpha_2} \cdots g_{\alpha_N}}{N!}. \tag{2.104}$$

As a result, for dilute gases of identical particles Eq. (2.103) should be replaced with

$$Z_N = \frac{Z_1^N}{N!}. \tag{2.105}$$

This modification affects the entropy, so that

$$S_{\text{indist}} = S_{\text{dist}} - k_B \log N!, \tag{2.106}$$

and the last term was crucial to guarantee the extensive nature of the results for the free-particle entropy in Eqs. (2.38) and (2.83).

The ultimately indistinguishable nature of identical particles is, indeed, what lies behind the factors of $\frac{1}{N!}$ that we introduced in Eqs. (2.35) and (2.73). We shall explore this issue in detail in Chapter 8, after taking quantum mechanics into account. For the time being, however, we can provide a simple quantitative estimate characterizing the regime where the approximation (2.104) holds. The counting is reliable when the system is dilute, so that the number, $n_\alpha$, of particles with energy $E_\alpha$ is small compared to the total number of allowed states $g_\alpha$. The nondegeneracy condition is thus

$$\frac{n_\alpha}{g_\alpha} \ll 1, \tag{2.107}$$

and, referring to the single-particle equilibrium populations of Eq. (B.4),

$$n_\alpha = \frac{g_\alpha N}{Z_1} e^{-\beta E_\alpha}, \tag{2.108}$$

one can regard a system as nondegenerate if

$$\eta = \frac{N}{Z_1} \ll 1. \tag{2.109}$$

This nondegeneracy parameter $\eta$ will recur in Chapter 8, and for the ideal gas it affords the suggestive rewriting

$$\eta = \frac{n}{n_Q(T)}, \tag{2.110}$$

as the ratio between the actual density $n$ of the gas and the characteristic density $n_Q(T)$ defined in Eq. (2.86).

## 2.4 Two Examples

We can now discuss two significant applications of the canonical ensemble.

### 2.4.1 Particles in a Constant Gravitational Potential

Let us consider a dilute gas of massive particles confined to a vertical cylinder, with height $\ell$ and two ends of area $L^2$. Taking into account the effect of gravity on the gas, the single-particle Hamiltonian is

$$H_1(\mathbf{p}, \mathbf{q}) = \frac{\mathbf{p}^2}{2m} + mgz, \tag{2.111}$$

where $z$ is the vertical coordinate. The one-particle partition function is thus

$$Z_1 = \int \frac{d^3\mathbf{p}\, d^3\mathbf{q}}{h^3} e^{-\beta H_1(\mathbf{p},\mathbf{q})} = \frac{L^2}{\beta mg}\left(\frac{2\pi m}{\beta h^2}\right)^{3/2} (1 - e^{-\beta mg\ell}), \tag{2.112}$$

and reduces to the standard result for a free gas for $\beta mg\ell \ll 1$, i.e. for a short cylinder or at high temperatures. On the other hand, it becomes

$$Z_1 = \frac{L^2}{\beta mg}\left(\frac{2\pi m}{\beta h^2}\right)^{3/2} \tag{2.113}$$

for $\beta mg\ell \gg 1$, i.e. for a long cylinder or at low temperatures, when the gravitational potential $mgz$ confines the gas far away from the upper lid of the cylinder.

The Helmholtz free energy

$$A = -Nk_BT\log Z_1 \tag{2.114}$$

determines the pressure on the upper end,

$$P(\ell) = -\frac{1}{L^2}\frac{\partial A}{\partial \ell} = \frac{Nmg}{L^2}\frac{1}{e^{\beta mg\ell} - 1}. \tag{2.115}$$

The dominant term for large values of $\ell$ is then

$$P(\ell) = \frac{Nmg}{L^2} e^{-\beta mg\ell}, \tag{2.116}$$

which is the isothermal model of an atmospheric column, while as $\ell \to 0$,

$$P \sim \frac{k_BT}{L^2\,\ell}, \tag{2.117}$$

which the usual ideal gas law. The different behaviors reflect the dependence on $z$ of the gas density within a column of height $\ell$,

$$n(z) = \frac{N\beta mg}{L^2}\frac{e^{-\beta mgz}}{1 - e^{-\beta mg\ell}}, \tag{2.118}$$

where the denominator tends to one for large values of $\ell$ or at low temperatures.

### 2.4.2 Saha Formula and Law of Mass Action

The canonical ensemble can be nicely applied to systems whose elements possess an internal structure. This is the case, for instance, for a gas of atoms, where in the Hamiltonian of a single atom,

$$H = \frac{\mathbf{P}_{\rm cm}^2}{2M} + H_{\rm int}, \tag{2.119}$$

one can distinguish the free motion of the center of mass from a second contribution, which takes into account the interactions of the electrons with the corresponding nucleus. The canonical partition function of a dilute system of this type describing $N$ noninteracting atoms thus takes the form

$$Z_N = \frac{1}{N!} Z_{\rm cm}^N\, Z_{\rm int}^N, \tag{2.120}$$

since the internal and translational degrees of freedom are decoupled in Eq. (2.119).

A remarkable fact, which is also rather puzzling at first sight, is that the internal partition function $Z_{\text{int}}$ is often formally divergent at any finite temperature. This difficulty presents itself, for instance, for a single hydrogen atom, whose energy levels, $E_n$, and degeneracies, $g_n$, are, as we shall see in Chapter 3,

$$E_n = -\frac{\epsilon_0}{n^2}, \qquad g_n = 2n^2, \tag{2.121}$$

where $\epsilon_0 = \frac{me^4}{2\hbar^2} \approx 13.6$ eV is Rydberg's constant. This apparent paradox is resolved by noting that, if arbitrarily excited electron states were to contribute significantly to the partition function, the very notion of atomic gas would break down, since each electron could be very far from the corresponding nucleus. Moreover, in that situation electron–electron interactions, and even the interactions of individual electrons with the walls of the container, would not be negligible. At room temperature, where $\beta\epsilon_0 \approx 5 \cdot 10^2 \gg 1$, a satisfactory way to treat electron–nucleus partition functions is to limit the sum in $Z_{\text{int}}$ to the first few available states. Here we shall actually content ourselves with the ground-state contribution, so that

$$Z_N = \frac{1}{N!}\left(\frac{2\pi m k_B T}{h^2}\right)^{3N/2}\left(g_0\, e^{-\beta E_0}\right)^N. \tag{2.122}$$

We shall return to the physics of atoms and molecules in Chapter 8.

As an example, let us consider a gas comprising an atomic species $X$ in equilibrium, via

$$X \longleftrightarrow X^+ + e^-, \tag{2.123}$$

with its first ionized form $X^+$. Denoting by $N_X$ the number of $X$ atoms, by $N_{X^+}$ the number of ions, and by $N_e$ the number of electrons, the partition function reads

$$Z = \frac{Z_X^{N_X}}{N_X!}\,\frac{Z_{X_+}^{N_{X^+}}}{N_+!}\,\frac{Z_e^{N_e}}{N_e!}. \tag{2.124}$$

However, the three particle numbers $N_X$, $N_{X^+}$, and $N_e$ are not independent, since they must satisfy some stoichiometric relations. First, the total number

$$N = N_X + N_{X^+} \tag{2.125}$$

of $X$ atoms and $X^+$ ions must be a constant, equal to the total number of atoms in the absence of ionization. Moreover, by the overall neutrality of the gas

$$N_e = N_{X^+}. \tag{2.126}$$

Minimizing the Helmholtz free energy, $A = -k_B T \log Z$, with respect to variations of $N_+$ performed at fixed $N$ yields, in the thermodynamic limit,

$$\frac{N_{X^+}^2}{N - N_{X^+}} = \frac{Z_{X^+} Z_e}{Z_X}, \tag{2.127}$$

which is equivalent to the link

$$\mu_X = \mu_{X^+} + \mu_e \tag{2.128}$$

among the chemical potentials. The single-particle partition functions for the neutral and ionized species are

$$Z_X = V\left(\frac{2\pi M k_B T}{h^2}\right)^{3/2} g_0\, e^{-\beta E_0}, \quad Z_{X^+} = V\left(\frac{2\pi M k_B T}{h^2}\right)^{3/2} g_{0+}\, e^{-\beta E_{0+}}, \tag{2.129}$$

where the small mass difference between them has been neglected. On the other hand, for the electrons,

$$Z_e = 2\, V\left(\frac{2\pi m_e k_B T}{h^2}\right)^{3/2}, \tag{2.130}$$

where the factor of two accounts for the spin–$\frac{1}{2}$ degeneracy. The electron has indeed a two-valued internal degree of freedom, as we shall see in Chapter 3 and in Appendix F. Finally, denoting by

$$x = \frac{N_{X^+}}{N} \tag{2.131}$$

the ionization fraction and considering a weakly ionized gas, so that $x \ll 1$, Eq. (2.127) becomes the Saha formula

$$x = \sqrt{\frac{2V}{N}} \left(\frac{2\pi m_e k_B T}{h^2}\right)^{3/4} \sqrt{\frac{g_{0+}}{g_0}}\; e^{-\frac{\beta \Delta E}{2}}. \tag{2.132}$$

The number of ionized atoms is thus governed by the factor $e^{-\frac{1}{2}\beta\Delta E}$, where

$$\Delta E = |E_0| - |E_{0+}| > 0 \tag{2.133}$$

is the first ionization energy. Note also that less dense gases are more likely to become ionized, due to the $\sqrt{\frac{V}{N}}$ factor. This reflects the entropy gain induced by the larger number of states that are available at larger specific volumes.

For a more general reaction

$$a_1 X_1 + \cdots a_I X_I \longleftrightarrow b_1 Y_1 + \cdots + b_J Y_J, \tag{2.134}$$

one would minimize the Helmholtz free energy $A = -k_B T \log Z$, where

$$Z = \frac{Z_{X_1}^{M_1}}{M_1!} \cdots \frac{Z_{X_I}^{M_I}}{M_I!} \frac{Z_{Y_1}^{N_1}}{N_1!} \cdots \frac{Z_{Y_J}^{N_J}}{N_J!}, \tag{2.135}$$

with $M_i$ and $N_i$ the numbers of particles belonging to the various species in the initial and final states, taking into account the stoichiometric constraints that generalize Eqs. (2.125) and (2.126),

$$\frac{M_i}{a_i} + \frac{N_1}{b_1} = \frac{M_i(0)}{a_i}, \qquad \frac{N_j}{b_j} = \frac{N_1}{b_1}. \tag{2.136}$$

Here $i = 1, \ldots, I$ and $j = 1, \ldots, J$, while $M_i(0)$ denotes the initial quantity of the $i$th reagent. This gives finally

$$\left(\frac{Z_{X_1}}{M_1}\right)^{a_1} \cdots \left(\frac{Z_{X_I}}{M_I}\right)^{a_I} = \left(\frac{Z_{Y_1}}{N_1}\right)^{b_1} \cdots \left(\frac{Z_{Y_J}}{N_J}\right)^{b_J}, \tag{2.137}$$

which is known as the law of mass action, or, taking the logarithm,

$$a_1 \mu_{X_1} + \cdots + a_I \mu_{X_I} = b_1 \mu_{Y_1} + \cdots + b_J \mu_{Y_J}, \tag{2.138}$$

where the $\mu_i$ are the chemical potentials of the various species.

## Bibliographical Notes

Pedagogical introductions to the microcanonical and canonical ensembles can be found in many excellent books, including [23, 36, 49, 50, 51, 54].

## Problems

**2.1** Deduce the distribution of single-particle energies for a perfect gas of particles with mass $m$ at a temperature $T$. Keep the number $D$ of spatial dimensions arbitrary.

**2.2** Consider a gas of particles in $d$ dimensions for which the dispersion relation

$$\epsilon(\mathbf{p}) = \alpha |\mathbf{p}|^{\nu}$$

holds, where $\nu$ and $\alpha$ are positive constants.

1. Compute the entropy $S$ in terms of the total energy $U$, the volume $V$, and the number of particles $N$ in the microcanonical ensemble.
2. Repeat the calculation in the canonical ensemble, for a given temperature $T$, and verify that the energy per particle satisfies

$$\frac{U}{N} = \frac{D}{\nu}\, k_B\, T.$$

Appendix D contains a useful result.

3. Compute the ratio $\gamma = \frac{C_P}{C_V}$ for this system.

**2.3** Consider a classical system with a generic Hamiltonian $H(z)$, where we denote collectively canonical positions $q$ and momenta $p$ by $z = (z_0, z_1, \ldots)$.

1. Prove the identity

$$\left\langle z_i \frac{\partial H}{\partial z_j} \right\rangle = \delta_{ij}\, k_B T.$$

2. Use this result to prove that, if for some canonical variable $z_0$ the Hamiltonian takes the form

$$H(z) = z_0^2\, H_0(z_1, z_2, \ldots) + H_1(z_1, z_2, \ldots),$$

where $H_0$, $H_1$ are suitable positive functions that do not depend on $z_0$,

$$\left\langle z_0^2\, H_0(z_1, z_2, \ldots) \right\rangle = \frac{k_B T}{2}.$$

This result is a special case of the equipartition theorem.

3. Repeat the analysis for the Hamiltonian of the preceding problem, thus recovering Eq. (2.139).

**2.4** Consider a system of $N$ massive relativistic particles, for which

$$H = \sqrt{\mathbf{p}^2\, c^2 + m^2\, c^4}.$$

1. Compute the partition function and the energy per particle $\frac{U}{N}$ in the canonical ensemble.
2. Discuss the limits of low and high temperatures, and compare the results with the behaviors of nonrelativistic particles of mass $m$, with

$$H = \frac{\mathbf{p}^2}{2m},$$

and of massless relativistic particles, with

$$H = c\ |\mathbf{p}|.$$

The integral is elementary for $d = 2$, while in general you will be led to consider the modified Bessel functions $K_\nu$ and their asymptotic expansions, which are discussed, for example, in [1].

**2.5** Consider a classical gas of $N$ identical particles not mutually interacting, such that for each of them

$$H = \frac{\mathbf{p}^2}{2m}.$$

The particles are free to move in a *two-dimensional* volume $V_2$ if $H < E_0$, for a given energy scale $E_0$, while they can move in a *four-dimensional* "volume" $V_4 = (V_2)^2$ for $H \geq E_0$.

Compute the canonical partition function $Z$ of the system and discuss the behavior of $C_V$ in the two limits of high and low $T$.

**2.6** Consider a classical gas of $N$ non-relativistic particles of mass $m$, not mutually interacting and confined in an annular region $a < r < b$ of the $(x, y)$ plane, where $r = \sqrt{x^2 + y^2}$ is a radial coordinate. The particles are also subject to the potential

$$V(r) = \kappa \log\left(\frac{r}{a}\right),$$

where $\kappa$ is a positive constant.

1. Compute the canonical partition function of the system and study the dependence on the temperature $T$ of the force per unit length exerted by the gas on the external wall $r = b$, if $\frac{b}{a} \gg 1$. What type of collective behavior does the system manifest?
2. Study the behavior of the specific heat $C_V$ as a function of the temperature $T$.

**2.7** Consider a gas of $N$ noninteracting particles in an area $A$ in two dimensions for which

$$H = \sum_{i=1}^{N} \frac{\mathbf{p}_i^2}{2\,m},$$

if there is a forbidden band of energies between $E$ and $E + \Delta$.

a. Write the partition function of the system.

b. Study the behavior of the specific heat $C_V$ of the system as a function of the temperature $T$.

**2.8** Compute the canonical partition function for a system of $N$ noninteracting particles in two dimensions, confined to a disk of radius $R$ centered at the origin, for which

$$H = \sum_{i=1}^{N} \frac{\mathbf{p}_i^2}{2m} + \sum_{i=1}^{N} \frac{m\,\omega^2\,\left(\mathbf{r}_i^2\right)}{2}.$$

1. Discuss the behavior of the free energy and of the specific heat $C_V$ as the temperature $T$ varies.
2. Discuss the behavior of the pressure on the boundary of the disk as the temperature $T$ varies.

**2.9** Consider a gas $N$ noninteracting atoms of type $A$, of mass $m_A$, confined in a volume $V$, and the reaction

$$A \to B + C,$$

which transforms 30% of them into atoms of species $B$ and $C$, with masses $m_B$ and $m_C$. Compute the corresponding entropy variation.

**2.10** Consider a gas of $N$ noninteracting particles in an area $A$ in two dimensions for which

$$H = \sum_{i=1}^{N} \frac{\mathbf{p}_i^2}{2\,m}.$$

1. Compute the fraction of particles whose energy $E < \frac{k_B T}{2}$.
2. Compute the probability that each of the $N$ particles has energy $E < \frac{k_B T}{2}$ and discuss the behavior of this result in the large-$N$ limit.

**2.11** Consider a gas of $N$ noninteracting particles in an area $A$ in two dimensions for which

$$H = \sum_{i=1}^{N} \frac{\mathbf{p}_i^2}{2\,m} + V(\mathbf{x}),$$

with

$$V(\mathbf{x}) = \begin{cases} \frac{1}{2}\,m\,\omega^2\,a\,|\mathbf{x}| & |\mathbf{x}| < a\,; \\ \frac{1}{2}\,m\,\omega^2\,\,|\mathbf{x}|^2 & |\mathbf{x}| > a. \end{cases}$$

a. Compute the partition function and the Helmholtz free energy of the system;
b. Study the behavior of the specific heat of the system as a function of the temperature $T$.

**2.12** Consider a gas of $2N$ particles in a volume $V$ in three dimensions, bound in pairs by harmonic potentials according to

$$H = \sum_{i=1}^{N} \left( \frac{\mathbf{p}_{2i}^2}{2\,m} + \frac{\mathbf{p}_{2i+1}^2}{2\,m} + \frac{1}{2}\,m\,\omega^2\,|\mathbf{r}_{2i} - \mathbf{r}_{2i+1}|^2 \right).$$

Compute the entropy variation that would accompany the breaking of all mutual harmonic links.

**2.13** Consider a gas of $N$ chains that are not mutually interacting and comprise $\ell$ elements. Each element can occupy two possible states, the state $|0\rangle$ with energy 0 and the state $|1\rangle$ with energy $\epsilon$, subject to the restriction that, if the $(k-1)$th element is in $|0\rangle$, the $k$th element can only be in $|0\rangle$.

1. Compute the partition and the specific heat of the system.
2. Repeat the analysis for chains whose elements can occupy the two states with no restrictions.

**2.14** Consider a system with $N$ nodes. Each node can be in a ground state with energy $E = 0$, or in an excited state with energy $E = \epsilon > 0$.

1. Obtain the relation between the internal energy $U$ and the temperature $T$, working in the microcanonical ensemble.
2. Study the behavior of $T$ and $S$ as $U$ varies from 0 to $N\epsilon$ and interpret the result.
3. Repeat the analysis in the canonical ensemble.

**2.15** Consider a one-dimensional gas confined to the interval $0 \leq x \leq n_M a$, whose $N$ particles are subject to the staircase potential

$$V(x) = n V_0 \qquad \big(na < x < (n+1)a\big),$$

where $n = 0, 1, \ldots, n_M - 1$, and $n_M$ is finite.

1. Compute the partition function of the system.
2. Obtain the pressure by comparing the free energies of system with $n_M$ and $n_{M+1}$ steps, thus building a discrete version of Eq. (1.27).
3. Discuss the limiting behavior for $n_M$ large, $\beta\, n_M V_0$ large, and $\beta\, V_0$ fixed (long staircase).
4. Discuss the limiting behavior for $n_M$ large, $\beta\, n_M V_0$ finite, and $\beta\, V_0$ small (continuum limit of the staircase) and compare with Section 2.4.1.

**2.16** Consider an ideal gas of $N$ particles of mass $m$, not mutually interacting but subject to gravity and contained in a vertical cylinder of radius $R$ and height $H$ that rotates around its axis, aligned with the $z$ axis of a Cartesian coordinate system, with angular velocity $\omega$. We want to characterize the equilibrium properties of the system when the particles follow, on average, the rotation of the cylinder.

1. Identify the relevant transformations of coordinates and momenta to describe particles that follow, on average, the rotating cylinder when the system reaches equilibrium.
2. Show that the resulting single-particle Hamiltonian is

$$H = \frac{1}{2m}\left[(p_x + m\,\omega\, y)^2 + \big(p_y - m\,\omega\, x\big)^2 + p_z^2\right] + m\,g\,z - \frac{1}{2}\,m\,\omega^2\,r^2,$$

with $r = \sqrt{x^2 + y^2}$, while the phase–space measure is invariant under the transformations.
3. Compute the partition function of the system and the pressure exerted on its external surface.

4. Compute the pressure on the upper lid.
5. What is the probability distribution $P(r, z)$ for a particle in this system?

**2.17** Consider an ideal gas of $N$ particles contained in a given volume $V_0$, at a temperature $T$. We would like to characterize the probability distribution that $M < N$ particles of the system occupy a volume $V < V_0$.

1. From the phase–space integral for the partition function, deduce that this probability distribution has the form

$$d\Pi = C\left(\frac{V}{V_0}\right)^M \left[1 - \left(\frac{V}{V_0}\right)\right]^{N-M} d\left(\frac{V}{V_0}\right),$$

and determine the normalization constant $C$.
2. Use this probability distribution to compute $\langle V \rangle$, the variance $\langle (\Delta V)^2 \rangle$, and the relative volume fluctuations, and comment on their scaling properties in the thermodynamic limit $M \gg 1$ and $N \gg 1$, and in particular when $M \ll N$.

**2.18** Consider a gas of $N$ particles of mass $m$, not mutually interacting but confined to the surface of a sphere of radius $R$.

1. Express the Lagrangian

$$\mathcal{L} = \frac{\mathbf{p}^2}{2m}$$

in terms of the polar coordinates $(\theta, \phi)$ and derive the corresponding Hamiltonian.
2. Compute the partition function, the internal energy, and the specific heat $C_V$ for the system.
3. What is the resulting measure for the coordinates q in terms of the two polar angles $\theta$ and $\phi$, after integrating over the momenta?

**2.19** Consider a set of $N$ "classical rotors" located at fixed positions in a three-dimensional lattice and not mutually interacting. The rotors are subject to an external force, so that for any of them

$$H = -\gamma \cos\theta,$$

where $\theta$ denotes the usual polar angle with respect to the $z$ axis of a Cartesian coordinate system. Study the behavior of $\langle \cos\theta \rangle$ and of the corresponding fluctuation as the temperature $T$ varies.

**2.20** Consider the equilibrium conditions for the ionization of a neutral hydrogen gas, according to the reaction

$$H \longleftrightarrow p^+ + e^-.$$

1. Study the dependence on the temperature $T$ of the fraction of ionized atoms.
2. Discuss the physical interpretation of the high-$T$ and low-$T$ limits of the result.

# 3 Aspects of Quantum Mechanics

In this chapter we collect a number of basic facts about quantum mechanics that will be useful in the remainder of the book. These include detailed derivations of some key WKB results on tunneling and energy bands in solids. We also address some basic aspects of functional integrals and the role of instantons in tunneling phenomena. We conclude with a description of the density matrix in quantum mechanics.

## 3.1 Some General Properties

Our starting point is the Schrödinger equation:

$$H|\psi(t)\rangle = i\hbar \frac{\partial}{\partial t}|\psi(t)\rangle. \tag{3.1}$$

We are interested in time-independent Hamiltonians $H$. In this case one can separate variables, looking for energy eigenstates, i.e. solutions of the form

$$|\psi(t)\rangle = e^{-\frac{iE_\alpha t}{\hbar}}\,|\psi_\alpha\rangle, \tag{3.2}$$

which brings about the eigenvalue problem

$$H|\psi_\alpha\rangle = E_\alpha\,|\psi_\alpha\rangle. \tag{3.3}$$

Solving this eigenvalue problem determines complete sets of states $|\psi_\alpha\rangle$. The Hermitian nature of $H$, the main observable of the system, implies the reality of its eigenvalues $E_\alpha$ and the orthogonality of the eigenvectors corresponding to different eigenvalues. For now, let us refer to discrete spectra, which will often characterize the systems of interest here, and thus normalize the states according to

$$\langle\psi_\alpha|\psi_{\alpha'}\rangle = \delta_{\alpha,\alpha'}. \tag{3.4}$$

With continuum spectra, the Dirac $\delta$ function would appear on the right-hand side and, in general, further labels may be needed to fully characterize the states when degeneracies $g_\alpha$ are present. At any rate, the complete set of $|\psi_\alpha\rangle$ allows one to write the general solution of the Schrödinger problem in the form

$$|\psi(t)\rangle = \sum_\alpha C_\alpha\, e^{-\frac{iE_\alpha t}{\hbar}}\,|\psi_\alpha\rangle. \tag{3.5}$$

The interpretation of quantum mechanics and the interaction of particles with detectors are still subject to debate. Here, however, we shall content ourselves with the simplest setup of a quantum system interacting with a large, classical, apparatus, and thus with the standard Copenhagen interpretation. A key postulate states that the

$E_\alpha$ are the only possible outcomes of energy measurements, and the normalization of Eq. (3.5) leads to

$$\sum_\alpha |C_\alpha|^2 = 1, \tag{3.6}$$

which reflects the standard interpretation of $|C_\alpha|^2$ as the probability that an energy measurement effected on the system will yield the value $E_\alpha$. The prediction for the average value of energy measurements is the weighted sum

$$\langle\psi(t)|H|\psi(t)\rangle = \sum_\alpha |C_\alpha|^2 E_\alpha. \tag{3.7}$$

After a measurement takes place at a time $t_0$, with result $E_\beta$, the wavefunction "collapses" to the corresponding energy eigenstate and then evolves as

$$|\psi(t)\rangle = e^{-\frac{iE_\beta(t-t_0)}{\hbar}} |\psi_\beta\rangle. \tag{3.8}$$

Energy conservation manifests itself in the fact that subsequent measurements will continue to yield $E_\beta$ with unit probability, in view of Eqs. (3.5) and (3.7). However, for observables not commuting with $H$, the probability of obtaining different results will evolve in time. Thus, given any observable $\mathcal{O}$, with corresponding eigenvectors and eigenvalues such that

$$\mathcal{O}|O_k\rangle = o_k|O_k\rangle, \tag{3.9}$$

the probability that a measurement of $\mathcal{O}$ at time $t$ yields the value $o_k$,

$$\left|\langle O_k|\psi(t)\rangle\right|^2, \tag{3.10}$$

will generally depend on $t$. If the outcome of a measurement is $o_k$, the wavefunction collapses to $|O_k\rangle$, and the subsequent evolution is determined by expanding $|O_k\rangle$ in energy eigenstates, as in Eq. (3.5).

## 3.2 The Free Particle

Our first example concerns the free particle, for which

$$H = \frac{\mathbf{p}^2}{2m} = -\frac{\hbar^2}{2m}\nabla^2, \tag{3.11}$$

which reflects the relation

$$\langle x|\mathbf{p}|\psi(t)\rangle = \frac{\hbar}{i}\nabla\langle \mathbf{x}|\psi(t)\rangle. \tag{3.12}$$

The last bracket, $\langle\mathbf{x}|\psi(t)\rangle$, is the wavefunction $\psi(\mathbf{x}, t)$.

In the spirit of the preceding sections we shall often consider a particle confined to a finite cubic box of side $L = V^{\frac{1}{3}}$, and we shall also impose *periodic* boundary conditions. The corresponding energy eigenfunctions are labeled by $\mathbf{n} = (n_1, n_2, n_3)$, a triple of integers, and read

$$\psi_\mathbf{n}(\mathbf{r}) = \frac{1}{\sqrt{V}} e^{i\frac{2\pi}{L}\mathbf{n}\cdot\mathbf{r}}. \tag{3.13}$$

They are momentum eigenstates with eigenvalues

$$\mathbf{p_n} = \frac{2\pi\hbar}{L}\,\mathbf{n}, \tag{3.14}$$

and are also energy eigenstates with eigenvalues

$$E_\mathbf{n} = \frac{\mathbf{p_n}^2}{2m}, \tag{3.15}$$

so that the decomposition of a generic wavefunction reads

$$\psi(\mathbf{r},t) = \sum_\mathbf{n} C_\mathbf{n} \frac{1}{\sqrt{V}}\, e^{i\frac{2\pi}{L}\mathbf{n}\cdot\mathbf{r}}\, e^{-\frac{iE_\mathbf{n}t}{\hbar}}. \tag{3.16}$$

The momentum wavefunctions are orthonormal:

$$\int_V d^3\mathbf{x}\, \psi^*_\mathbf{m}(\mathbf{r})\, \psi_\mathbf{n}(\mathbf{r}) = \delta_{\mathbf{m},\mathbf{n}}. \tag{3.17}$$

They are also a complete set, since

$$\frac{1}{V}\sum_\mathbf{n} e^{i\frac{2\pi}{L}\mathbf{n}\cdot(\mathbf{r}-\mathbf{r}')} = \delta(\mathbf{r}-\mathbf{r}'), \tag{3.18}$$

where here $\delta$ denotes a periodic extension of the usual Dirac $\delta$ function.

In the large-volume limit, the momenta in Eq. (3.14) become continuous vectors, and Eq. (3.18) tends to an integral sum. This is best appreciated after recasting it in the form

$$\frac{1}{(2\pi\hbar)^3}\sum_\mathbf{n} e^{i\frac{2\pi}{L}\mathbf{n}\cdot(\mathbf{r}-\mathbf{r}')}\left(\frac{2\pi\hbar}{L}\right)^3 = \delta(\mathbf{r}-\mathbf{r}'), \tag{3.19}$$

since the factor that multiplies the exponentials builds up the measure $d^3\mathbf{p}$ in the large-volume limit. Indeed, we shall often use the replacement

$$\sum_\mathbf{n} \to \frac{V}{(2\pi\hbar)^3}\int d^3\mathbf{p} = \frac{V}{h^3}\int d^3\mathbf{p} \tag{3.20}$$

for the large volumes relevant to our discussion. This result has a direct bearing on statistical mechanics, since it determines the proper phase-space measure, and is ultimately the rationale behind the introduction of $h$ in Eqs. (2.35) and (2.73).

In the large volume limit, one can also directly define the wavefunctions

$$\psi_\mathbf{p}(\mathbf{r},t) = \frac{1}{(2\pi\hbar)^{\frac{3}{2}}}\, e^{i\frac{\mathbf{p}\cdot\mathbf{r}}{\hbar} - \frac{ip^2 t}{2m\hbar}}, \tag{3.21}$$

which are orthonormal in Dirac's sense:

$$\int d^3\mathbf{x}\, \psi^*_\mathbf{p}(\mathbf{r},t)\, \psi_\mathbf{q}(\mathbf{r},t) = \delta\,(\mathbf{p}-\mathbf{q}), \tag{3.22}$$

where $\delta$ is the usual Dirac $\delta$ function.

The evolution of the probability distribution for a particle that is initially well localized, say, around the origin, so that

$$\psi(x,0) = \left(\frac{\alpha}{\pi}\right)^{\frac{1}{4}} e^{-\frac{\alpha x^2}{2}}, \tag{3.23}$$

is an instructive example. Note that for this wavefunction

$$\langle x \rangle = 0, \quad \langle p \rangle = 0, \tag{3.24}$$

and even in quantum mechanics we shall often use this type of shorthand notation to indicate expectations values such as

$$\langle x \rangle = \langle \psi | x | \psi \rangle, \tag{3.25}$$

and similarly for $\langle p \rangle$. The first result holds since $\psi(x,0)$ is an even function of $x$ and the second follows from its reality, and moreover

$$\langle (\Delta x)^2 \rangle = \frac{1}{2\,\alpha}, \quad \langle (\Delta p)^2 \rangle = \frac{\hbar^2\,\alpha}{2}, \tag{3.26}$$

where

$$\langle (\Delta x)^2 \rangle = \langle \psi |\, (x - \langle x \rangle)^2 \,| \psi \rangle, \tag{3.27}$$

and similarly for $\langle (\Delta p)^2 \rangle$.

One can determine the time evolution of this packet by decomposing it into momentum eigenfunctions according to

$$\psi(x,0) = \frac{1}{\sqrt{2\,\pi\,\alpha\,\hbar^2}} \left(\frac{\alpha}{\pi}\right)^{\frac{1}{4}} \int dp\, e^{\frac{ipx}{\hbar}}\, e^{-\frac{p^2}{2\,\alpha\,\hbar^2}}. \tag{3.28}$$

Evolving the momentum eigenfunctions according to Eq. (3.21) then leads to

$$\psi(x,t) = \frac{1}{\sqrt{2\,\pi\,\alpha\,\hbar^2}} \left(\frac{\alpha}{\pi}\right)^{\frac{1}{4}} \int dp\, e^{\frac{ipx}{\hbar}}\, e^{-\frac{p^2}{2\,\alpha\,\hbar^2} - \frac{ip^2\,t}{2\,m\,\hbar}}. \tag{3.29}$$

The effect of the time evolution is thus the replacement of

$$\frac{1}{\alpha} \to \frac{1}{\alpha} + \frac{i\,\hbar\,t}{m} \tag{3.30}$$

in the exponent of Eq. (3.28), and therefore

$$\psi(x,t) = \left(\frac{\alpha}{\pi}\right)^{\frac{1}{4}} \frac{1}{\sqrt{1 + \frac{i\,\alpha\,\hbar\,t}{m}}}\, e^{-\frac{\alpha\,x^2}{2\left[1 + \frac{i\,\alpha\,\hbar\,t}{m}\right]}}, \tag{3.31}$$

so that

$$\left|\psi(x,t)\right|^2 = \left(\frac{\alpha}{\pi}\right)^{\frac{1}{2}} \frac{1}{\sqrt{1 + \frac{\alpha^2\,\hbar^2\,t^2}{m^2}}}\, e^{-\frac{\alpha\,x^2}{\left[1 + \frac{\alpha^2\,\hbar^2\,t^2}{m}\right]}}, \tag{3.32}$$

and a direct comparison with Eq. (3.26) shows that

$$\langle (\Delta x)^2\,(t) \rangle = \langle (\Delta x)^2\,(0) \rangle + \frac{\langle (\Delta p)^2\,(0) \rangle}{m^2}\, t^2. \tag{3.33}$$

This equation quantifies how the time evolution effectively smears the initial wave packet.

It is now simple and instructive to see how classical dynamics can be formally recovered in the limit $\hbar \to 0$. To this end, let us first relax the two conditions in Eq. (3.24), turning the wavefunction into

$$\psi(x,0) = \left(\frac{\alpha}{\pi}\right)^{\frac{1}{4}} e^{-\frac{\alpha\,(x-x_0)^2}{2} + i\frac{p_0 x}{\hbar}}, \tag{3.34}$$

so that for the new initial wave packet

$$\langle x \rangle = x_0, \qquad \langle p \rangle = p_0, \tag{3.35}$$

while the mean square deviations are still as in (3.26). Retracing the preceding derivation, one can see that the resulting wavefunction at time $t$ is

$$\psi(x,t) = \left(\frac{\alpha}{\pi}\right)^{\frac{1}{4}} \frac{1}{\sqrt{1 + \frac{i\alpha\hbar t}{m}}} \, e^{-\frac{\alpha\left(x - x_0 - \frac{p_0}{m}t\right)^2}{2\left[1 + \frac{i\alpha\hbar t}{m}\right]} - \frac{i(p-p_0)x_0}{\hbar} - \frac{itp_0^2}{2\hbar m}}, \tag{3.36}$$

and the probability distribution function reads

$$|\psi(x,t)|^2 = \left(\frac{\alpha}{\pi}\right)^{\frac{1}{2}} \frac{1}{\sqrt{1 + \frac{\alpha^2\hbar^2 t^2}{m^2}}} \, e^{-\frac{\alpha\left(x - x_0 - \frac{p_0}{m}t\right)^2}{\left[1 + \frac{\alpha^2\hbar^2 t^2}{m}\right]}}. \tag{3.37}$$

The classical limit ought to be performed in such a way as to guarantee that, as $\hbar \to 0$, the initial wave packet becomes sharply peaked around $x_0$ in position space and $p_0$ in momentum space. This can be achieved by letting

$$\alpha = \frac{\alpha_0}{\hbar}, \tag{3.38}$$

and formally taking the limit $\hbar \to 0$ for fixed $\alpha_0$, so that both initial uncertainties (3.26) tend to zero in a symmetric fashion. The probability distribution of Eq. (3.37) will then approach a $\delta$ function localized on the classical trajectory

$$|\psi(x,t)|^2 \sim \delta\left(x - x_0 - \frac{p_0 t}{m}\right). \tag{3.39}$$

One may also verify that the corresponding probability distribution in momentum space approaches $\delta(p - p_0)$. When the limit $\hbar \to 0$ is taken in this fashion, one can specify arbitrarily accurate initial conditions $x_0, p_0$ for the particle, which will move at successive instants as predicted by the classical equation of motion.

A couple of comments are now in order. First, a classical interpretation can only be retrieved at the level of probabilities, and not of transition amplitudes since, in the limit that we have identified, the wavefunction becomes singular. Furthermore, the dispersion induced by the free evolution that we have stressed in Eq. (3.33) implies that the classical limit does not commute with the large-$t$ limit, as can be seen directly from Eq. (3.37).

## 3.3 The Harmonic Oscillator

Let us now describe the one-dimensional harmonic oscillator. In this case the dynamics are determined by

$$H = \frac{p^2}{2m} + \frac{1}{2} m\,\omega^2 x^2, \tag{3.40}$$

or, dividing by $\hbar\omega$, which is the typical energy scale of the problem, by

$$\frac{H}{\hbar\omega} = \frac{p^2}{2m\hbar\omega} + \frac{m\omega}{2\hbar} x^2. \tag{3.41}$$

The peculiar form of this Hamiltonian suggests we attempt a factorization, which leads one to define the lowering and raising operators

$$a = \sqrt{\frac{m\omega}{2\hbar}}\, x + i \frac{p}{\sqrt{2m\hbar\omega}}, \quad a^{\dagger} = \sqrt{\frac{m\omega}{2\hbar}}\, x - i \frac{p}{\sqrt{2m\hbar\omega}}, \tag{3.42}$$

or

$$a = \sqrt{\frac{m\omega}{2\hbar}}\, x + \sqrt{\frac{\hbar}{2m\omega}} \frac{d}{dx}, \quad a^{\dagger} = \sqrt{\frac{m\omega}{2\hbar}}\, x - \sqrt{\frac{\hbar}{2m\omega}} \frac{d}{dx}, \tag{3.43}$$

in the coordinate representation (3.12). The canonical commutation relation

$$[x,p] = i\hbar \tag{3.44}$$

then implies that

$$[a, a^{\dagger}] = 1, \tag{3.45}$$

and consequently

$$H = \hbar\,\omega \left( a^{\dagger}\, a + \frac{1}{2} \right). \tag{3.46}$$

This expression implies that all energy eigenvalues are bounded from below by the last term, which is usually called "zero-point energy," since

$$\langle \psi | a^{\dagger}\, a | \psi \rangle = |\, a | \psi \rangle |^2 \geq 0. \tag{3.47}$$

The equality holds if $a$ annihilates the state or, recalling Eq. (3.12), if

$$\frac{d\psi_0(x)}{dx} + \frac{m\,\omega\, x}{\hbar}\, \psi_0(x) = 0. \tag{3.48}$$

This equation identifies the ground state and is solved by a Gaussian wavefunction. With the proper normalization of Eq. (3.4),

$$\psi_0(x) = \left( \frac{m\,\omega}{\pi\,\hbar} \right)^{\frac{1}{4}} e^{-\frac{m\omega}{2\hbar} x^2}, \tag{3.49}$$

and one can verify that

$$\langle \psi_0 | x^2 | \psi_0 \rangle = \frac{\hbar}{2m\omega}, \tag{3.50}$$

using Eq. (2.21) and its derivative with respect to $\alpha$. The wavefunction of Eq. (3.49) peaks at the origin but is slightly spread around it, so that a particle in the ground state is essentially localized at the minimum of the harmonic potential but "explores" nearby regions of higher energy.

Symmetry considerations indicate that the average values of $x$ and $p$ vanish in any energy eigenstate. We have just seen this explicitly for the ground-state wavefunction of Eq. (3.49), and we shall see it soon for the other eigenstates. Assuming momentarily that this is the case, one can turn Eq. (3.40) into a relation involving the uncertainties in $x$ and $p$:

$$\langle H \rangle = \frac{\langle (\Delta p)^2 \rangle}{2m} + \frac{1}{2}\, m\,\omega^2 \, \langle (\Delta x)^2 \rangle. \tag{3.51}$$

Combining this expression with the uncertainty principle

$$\langle (\Delta p)^2 \rangle \langle (\Delta x)^2 \rangle \geq \frac{\hbar^2}{4} \tag{3.52}$$

then implies that

$$\langle H \rangle \geq \frac{\hbar^2}{8m\langle (\Delta x)^2 \rangle} + \frac{1}{2} m\,\omega^2 \langle (\Delta x)^2 \rangle, \tag{3.53}$$

and the lowest value obtains for

$$\langle (\Delta x)^2 \rangle = \frac{\hbar}{2m\omega}, \tag{3.54}$$

so that *the ground state of this system saturates the uncertainty principle.*

We can now determine the whole spectrum, and the relevant observation to this end is that

$$[H, a^\dagger] = \hbar\,\omega [a^\dagger a, a^\dagger] = \hbar\,\omega\, a^\dagger [a, a^\dagger] = \hbar\,\omega\, a^\dagger. \tag{3.55}$$

It implies that, given a state $|s\rangle$ with energy eigenvalue $E_s$, one can construct another state, $a^\dagger |s\rangle$, with energy eigenvalue $E_s + \hbar\omega$. In a similar fashion, applying $a$ to a state lowers the eigenvalue by $\hbar\omega$, and this process terminates, as we have seen, with the ground state, whose state vector we now call $|0\rangle$. All in all, one thus finds the infinite chain of normalized states

$$|n\rangle = \frac{(a^\dagger)^n}{\sqrt{n!}} |0\rangle, \tag{3.56}$$

which correspond to the *equally spaced spectrum*

$$E_n = \hbar\omega \left( n + \frac{1}{2} \right). \tag{3.57}$$

The next issue that we would like to address has to do with what replaces, in this quantum harmonic oscillator, the classical motion

$$x(t) = A \cos(\omega t + \phi), \tag{3.58}$$

which ought to emerge again, somehow, in the $\hbar \to 0$ limit. To begin with, the general evolution of a state

$$|\psi\rangle = \sum_n C_n |n\rangle \tag{3.59}$$

is described for this system by

$$|\psi(t)\rangle = \sum_{n=0}^{\infty} C_n\, e^{-i\left(n+\frac{1}{2}\right)\omega t} \frac{(a^\dagger)^n}{\sqrt{n!}} |0\rangle, \tag{3.60}$$

and this expression is indeed periodic, up to an overall phase $e^{-\frac{i\omega t}{2}}$ that has no effect on measurable quantities. There is a special class of states, the so-called coherent states, which one obtains by choosing

$$C_n = \frac{\lambda^n}{\sqrt{n!}}\, e^{-\frac{|\lambda|^2}{2}} \tag{3.61}$$

for an arbitrary complex number $\lambda$, so that

$$\sum_{n=0}^{\infty} |C_n|^2 = 1. \tag{3.62}$$

Note also that

$$|C_n|^2 = \frac{|\lambda|^{2n}}{n!} \, e^{-|\lambda|^2}, \tag{3.63}$$

so that the probabilities of measuring the different energy eigenvalues are Poisson distributed, as explained in Appendix A. Starting from a coherent state

$$|\psi_\lambda(0)\rangle = e^{-\frac{|\lambda|^2}{2}} e^{\lambda\, a^\dagger} \, |0\rangle, \tag{3.64}$$

which solves the equation

$$a|\psi_\lambda(0)\rangle = \lambda \, |\psi_\lambda(0)\rangle, \tag{3.65}$$

the subsequent time evolution leads to

$$|\psi_\lambda(t)\rangle = e^{-\frac{|\lambda|^2+i\omega t}{2}} \sum_{n=0}^{\infty} \frac{\left(\lambda\, e^{-i\omega t}\, a^\dagger\right)^n}{n!} \, |0\rangle = e^{-\frac{|\lambda|^2+i\omega t}{2}} \, e^{\lambda\, a^\dagger\, e^{-i\omega t}} |0\rangle, \tag{3.66}$$

so that

$$|\psi_\lambda(t)\rangle = e^{-\frac{i\omega t}{2}} \, |\psi_{\lambda\, e^{-i\omega t}}(0)\rangle. \tag{3.67}$$

Therefore, leaving aside the overall phase, the parameter $\lambda$ evolves according to

$$\lambda \;\rightarrow\; \lambda\, e^{-i\omega t}. \tag{3.68}$$

We can get a closer view of the meaning of this result by making use of Eq. (3.65), which translates into the differential equation

$$\left(\sqrt{\frac{m\omega}{2\hbar}}\, x + \sqrt{\frac{\hbar}{2m\omega}}\, \frac{d}{dx}\right) \psi_\lambda(x) \;=\; \lambda\, \psi_\lambda(x), \tag{3.69}$$

whose properly normalized solution reads

$$\psi_\lambda(x) \;=\; \left(\frac{m\omega}{\pi\hbar}\right)^{\frac{1}{4}} \, e^{-\frac{m\omega}{2\hbar}x^2 + \lambda\, x \sqrt{\frac{2m\omega}{\hbar}} - Re(\lambda)^2}. \tag{3.70}$$

The corresponding probability distribution is a Gaussian packet whose center is displaced in a way determined by $Re(\lambda)$:

$$\left|\psi_\lambda(x)\right|^2 \;=\; \sqrt{\frac{m\omega}{\pi\hbar}}\; e^{-\frac{m\omega}{\hbar}\left(x \,-\, Re(\lambda)\sqrt{\frac{2\hbar}{m\omega}}\right)^2}. \tag{3.71}$$

Thus, on account of Eq. (3.68), the peak

$$x_c(t) \;=\; |\lambda| \sqrt{\frac{2\hbar}{m\omega}} \;\cos(\omega t \,+\, \phi) \tag{3.72}$$

moves in time precisely as a classical particle would in the harmonic potential. Furthermore, this superposition is indeed coherent, since its width,

$$\langle(\Delta x)^2\rangle \;=\; \frac{\hbar}{2\, m\, \omega}, \tag{3.73}$$

does not increase in time, in contrast with that observed for the free particle in Eq. (3.37). Note that a semiclassical behavior manifests itself in this case for $\lambda \gg 1$, when the amplitude of the oscillations is much larger than the width (3.73).

Can one do better than this for the width? Yes, part of the time. This leads us to discuss "squeezed states," and for brevity we shall only "squeeze" the ground state wavefunction, although one could do the same for coherent states. We thus confine our attention to states of the form

$$|\mu\rangle = C e^{\frac{\mu}{2}(a^\dagger)^2}|0\rangle = C \sum_{k=0}^{\infty} \left(\frac{\mu}{2}\right)^k \frac{\sqrt{(2k)!}}{k!}|2k\rangle, \tag{3.74}$$

so that the time evolution will transform $\mu$ according to

$$\mu \rightarrow \mu\, e^{-2i\omega t}, \tag{3.75}$$

and we shall see shortly that the construction requires that $|\mu| < 1$.

The definition in Eq (3.74) now implies that

$$\left(a - \mu\, a^\dagger\right)|\mu\rangle = 0, \tag{3.76}$$

which can be turned into the differential equation

$$\frac{d\psi_\mu}{dx} + \frac{m\omega}{\hbar}\,\frac{1-\mu}{1+\mu}\,x\,\psi_\mu(x) = 0 \tag{3.77}$$

using Eqs. (3.12) and (3.42). Properly normalized wavefunctions for these states exist for $|\mu| < 1$, and read

$$\psi_\mu(x) = \left(\frac{m\omega}{\pi\hbar}\right)^{\frac{1}{4}} \left(\frac{1 - |\mu|^2}{|1 + \mu|^2}\right)^{\frac{1}{4}} e^{-\frac{m\omega}{2\hbar}\left(\frac{1-\mu}{1+\mu}\right)x^2}. \tag{3.78}$$

A direct comparison with Eq. (3.50) indicates that for these squeezed states

$$\langle(\Delta x)^2\rangle = \frac{\hbar}{2m\omega}\left(\frac{|1 + \mu|^2}{1 - |\mu|^2}\right), \tag{3.79}$$

which evolves according to

$$\langle(\Delta x)^2\rangle(t) = \frac{\hbar}{2m\omega}\left(\frac{1 + |\mu_0|^2 + 2|\mu_0|\cos(2\omega t + \varphi)}{1 - |\mu_0|^2}\right), \tag{3.80}$$

thus oscillating between a lowest value,

$$\langle(\Delta x)^2\rangle_{min} = \frac{\hbar}{2m\omega}\,\frac{1 - |\mu_0|}{1 + |\mu_0|}, \tag{3.81}$$

and a highest one,

$$\langle(\Delta x)^2\rangle_{max} = \frac{\hbar}{2m\omega}\,\frac{1 + |\mu_0|}{1 - |\mu_0|}. \tag{3.82}$$

As a result, one can indeed improve matters with respect to the width (3.54) of coherent states during part of the evolution, albeit paying a price during the rest. This type of setting has acquired an important role when dealing with very weak signals, and in particular for the detection of gravitational waves.

## 3.4 Evolution Kernels and Path Integrals

One can define the unitary "evolution operator"

$$U = e^{-i\frac{Ht}{\hbar}} = \sum_{\alpha} e^{-\frac{iE_{\alpha}t}{\hbar}} |\psi_{\alpha}\rangle\langle\psi_{\alpha}|, \tag{3.83}$$

which determines $|\psi(t)\rangle$ in terms of the initial value $|\psi(0)\rangle$. Then

$$|\psi(t)\rangle = U|\psi(0)\rangle = \sum_{\alpha} e^{-\frac{iE_{\alpha}t}{\hbar}} |\psi_{\alpha}\rangle\langle\psi_{\alpha}|\psi(0)\rangle, \tag{3.84}$$

which coincides with Eq. (3.5), after taking into account that

$$C_{\alpha} = \langle\psi_{\alpha}|\psi(0)\rangle. \tag{3.85}$$

One can also define the evolution kernel as the matrix element of $U$ between two position eigenstates,

$$K(\mathbf{r},\mathbf{r}';t) = \langle\mathbf{r}|\, e^{-i\frac{Ht}{\hbar}}\, |\mathbf{r}'\rangle = \sum_{\alpha} e^{-\frac{iE_{\alpha}t}{\hbar}} \langle\mathbf{r}|\psi_{\alpha}\rangle\langle\psi_{\alpha}|\mathbf{r}'\rangle. \tag{3.86}$$

It is instructive to compute the evolution kernel $K$ for the free particle. One can begin by considering the free particle in a box of volume $V$ (as in Section 3.2), for which

$$K(\mathbf{r},\mathbf{r}';t) = \frac{1}{V}\sum_{\mathbf{n}} e^{i\frac{2\pi}{L}\mathbf{n}\cdot(\mathbf{r}-\mathbf{r}')}\, e^{-\frac{it}{\hbar}\frac{\mathbf{p}^2}{2m}}, \tag{3.87}$$

with the discrete momenta of Eq. (3.14). Then in the infinite volume limit, using Eq. (3.20),

$$K(\mathbf{r},\mathbf{r}';t) = \frac{1}{h^3}\int d^3\mathbf{p}\; e^{\frac{i}{\hbar}\mathbf{p}\cdot(\mathbf{r}-\mathbf{r}')}\, e^{-\frac{it}{\hbar}\frac{\mathbf{p}^2}{2m}}. \tag{3.88}$$

This "shifted" Gaussian integral can be computed using Eq. (2.21), and the end result,

$$K(\mathbf{r},\mathbf{r}';t) = \left(\frac{m}{2\pi i\hbar\, t}\right)^{\frac{3}{2}} e^{\frac{i}{\hbar}\frac{m}{2t}(\mathbf{r}-\mathbf{r}')^2}, \tag{3.89}$$

is quite suggestive, since *the exponent is the action*

$$\mathcal{S} = \int_0^t d\tau\, \frac{m}{2}\left(\frac{d\mathbf{x}}{d\tau}\right)^2 \tag{3.90}$$

*computed on the classical trajectory*

$$\mathbf{x}(\tau) = (\mathbf{r}-\mathbf{r}')\,\frac{\tau}{t} + \mathbf{r}' \tag{3.91}$$

of a free particle that starts at $\mathbf{r}'$ at $\tau = 0$ and within a time interval $t$ moves from $\mathbf{r}'$ to $\mathbf{r}$. The overall factor has a less transparent meaning, but it grants that

$$K(\mathbf{r},\mathbf{r}';t) \longrightarrow \delta\,(\mathbf{r}-\mathbf{r}') \tag{3.92}$$

as $t \to 0^+$.

There is an important "Euclidean variant,"

$$K_E(\mathbf{r}, \mathbf{r}'; \beta\hbar) = K(\mathbf{r}, \mathbf{r}'; -i\beta\hbar) = \sum_\alpha e^{-\beta E_\alpha} \langle \mathbf{r}|\psi_\alpha\rangle \langle \psi_\alpha|\mathbf{r}'\rangle, \tag{3.93}$$

which is closely related to the partition function. The transition to the imaginary time $t = -i\beta\hbar$ can be understood as a rotation of $-\frac{\pi}{2}$ in the complex plane, and is usually called a "Wick rotation." A direct comparison with Eq. (3.89) shows that, for the free particle,

$$K_E(\mathbf{r}, \mathbf{r}'; \beta\hbar) = \left(\frac{m}{2\pi\hbar^2 \beta}\right)^{\frac{3}{2}} e^{-\frac{m}{2\hbar^2\beta}(\mathbf{r}-\mathbf{r}')^2}. \tag{3.94}$$

Incidentally, Eq. (3.88) also implies that

$$\text{tr}\left(e^{-\beta H}\right) = \int d^3\mathbf{r}\, K(\mathbf{r}, \mathbf{r}; -i\beta\hbar) = \int \frac{d^3\mathbf{r}\, d^3\mathbf{p}}{h^3}\, e^{-\beta \frac{\mathbf{p}^2}{2m}}, \tag{3.95}$$

which recovers the classical partition function of an ideal gas.

These results lead to the path integral representation for $K(\mathbf{r}, \mathbf{r}'; t)$ for a particle subject to a generic potential, for which

$$H = \frac{\mathbf{p}^2}{2m} + U(\mathbf{r}). \tag{3.96}$$

Indeed, for a very small time interval $\Delta t$,

$$e^{-\frac{iH\Delta t}{\hbar}} \simeq e^{-\frac{i\Delta t}{\hbar}\frac{\mathbf{p}^2}{2m}}\, e^{-\frac{iU(\mathbf{r})\Delta t}{\hbar}}, \tag{3.97}$$

up to commutators, which are of order $(\Delta t)^2$. Therefore

$$\langle \mathbf{r}_k|e^{-\frac{iH\Delta t}{\hbar}}|\mathbf{r}_l\rangle \simeq \left(\frac{m}{2\pi i\hbar\,\Delta t}\right)^{\frac{3}{2}} e^{\frac{i}{\hbar}\frac{m}{2\Delta t}(\mathbf{r}_k-\mathbf{r}_l)^2}\, e^{-\frac{iU(\mathbf{r}_l)\Delta t}{\hbar}}, \tag{3.98}$$

and the exponent is again, for very small intervals $\Delta t$, the action $\mathcal{S}$:

$$\mathcal{S} = \int_{\tau_0}^{\tau_0+\Delta t} d\tau \left[\frac{m}{2}\left(\frac{d\mathbf{x}}{d\tau}\right)^2 - U(\mathbf{x}(\tau))\right]. \tag{3.99}$$

Alternatively, one can present Eq. (3.98) in the Hamiltonian form, starting from

$$\begin{aligned}\langle \mathbf{r}_k|e^{-\frac{iH\Delta t}{\hbar}}|\mathbf{r}_l\rangle &= \int d^3\mathbf{p}\, \langle \mathbf{r}_k|e^{-\frac{iH\Delta t}{\hbar}}|\mathbf{p}\rangle\langle \mathbf{p}|\mathbf{r}_l\rangle \\ &\simeq \int \frac{d^3\mathbf{p}}{h^3}\, e^{\frac{i}{\hbar}\mathbf{p}\cdot(\mathbf{r}_k-\mathbf{r}_l)-\frac{i}{\hbar}\Delta t\, H(\mathbf{p},\mathbf{r}_l)},\end{aligned} \tag{3.100}$$

and ordering ambiguities are again of higher order in $\Delta t$.

The preceding results become very useful if one splits the interval $[0, t]$ in $N + 1$ intervals of length $\Delta t = \frac{t}{N+1}$ in the formal limit $N \to \infty$. To begin with, in the Lagrangian form induced by Eq. (3.98), inserting a complete set of position eigenstates gives

$$K(\mathbf{r}, \mathbf{r}'; t) = \int_{-\infty}^{\infty} d^3\mathbf{y}_1\, K\left(\mathbf{r}, \mathbf{y}_1; \frac{t}{2}\right) K\left(\mathbf{y}_1, \mathbf{r}'; \frac{t}{2}\right), \tag{3.101}$$

and thus after $N$ steps, with $\mathbf{y}_0 = \mathbf{r}$ and $\mathbf{y}_{N+1} = \mathbf{r}'$,

$$K(\mathbf{r}, \mathbf{r}'; t) = \prod_{i=1}^{N} \int_{-\infty}^{\infty} d^3\mathbf{y}_i \prod_{j=0}^{N} K\left(\mathbf{y}_{j+1}, \mathbf{y}_j; \frac{t}{N+1}\right) \tag{3.102}$$

$$= \int_{-\infty}^{\infty} \left[\prod_{i=1}^{N} d^3\mathbf{y}_i\right] \left(\frac{m}{2\pi i\hbar\Delta t}\right)^{\frac{3(N+1)}{2}} e^{\frac{i}{\hbar}S}.$$

Note that this expression contains $N + 1$ singular factors, $N$ integrations, and $N + 1$ kernels. This formal limit defines a "path integral"

$$K(\mathbf{r}, \mathbf{r}'; t) = \int [\mathcal{D}\mathbf{x}(\tau)]\, e^{\frac{i}{\hbar}S}, \tag{3.103}$$

as a construct that affords a remarkable physical interpretation: one is indeed summing over all trajectories, while also associating to each of them a weight determined by the exponential of the action along it. The price for this neat result is a singular measure, which is induced by the proper normalization of the $N + 1$ kernels. In this formulation it is the Lagrangian, rather than the Hamiltonian as for the Schrödinger equation, which drives the time evolution. On the other hand, starting from the Hamiltonian form (3.100) one can obtain, in the limit,

$$K(\mathbf{r}, \mathbf{r}'; t) = \int [\mathcal{D}\mathbf{x}(\tau)]\,[\mathcal{D}\mathbf{p}(\tau)]\, e^{\frac{i}{\hbar}\int d\tau[\mathbf{p}\cdot\dot{\mathbf{x}} - H(\mathbf{p},\mathbf{r})]}, \tag{3.104}$$

with the simpler measure

$$\prod_{i=1}^{N}\left[\frac{d^3\mathbf{y}_i\, d^3\mathbf{p}_i}{h^3}\right]\frac{d^3\mathbf{p}}{h^3}. \tag{3.105}$$

When the Lagrangian is quadratic in the velocities, this more general expression reduces to the form of Eq. (3.102).

Let us continue along these lines, considering

$$\mathrm{tr}\left(e^{-\beta H}\right) = \int d^3\mathbf{r}\, \langle \mathbf{r}|e^{-\beta H}|\mathbf{r}\rangle, \tag{3.106}$$

which can also be reduced to a path integral. Proceeding as above one can indeed conclude that

$$\mathrm{tr}\left(e^{-\beta H}\right) = \oint [\mathcal{D}\mathbf{x}(\tau)]\, e^{-\frac{1}{\hbar}S_E}, \tag{3.107}$$

where the "loop path integral" symbol is meant to emphasize the presence of an additional integration over the identical values of initial and final points, and where the "Euclidean action" is

$$S_E = \int_0^{\beta\hbar} d\tau \left[\frac{m}{2}\left(\frac{d\mathbf{x}}{d\tau}\right)^2 + U(\mathbf{x}(\tau))\right]. \tag{3.108}$$

These expressions are indeed a shorthand for a limiting form of a Euclidean path integral (3.102), with an additional integration over initial and final points $\mathbf{y}_{N+1} = \mathbf{y}_0 = \mathbf{r}$:

$$\int_{-\infty}^{\infty} d^3\mathbf{x} \int_{-\infty}^{\infty} \left[\prod_{i=1}^{N} d^3\mathbf{y}_i\right] \left(\frac{m}{2\pi\hbar\Delta t}\right)^{\frac{3(N+1)}{2}} \prod_{i=0}^{N} e^{-\frac{\Delta t}{\hbar}\left[\frac{m(\mathbf{y}_{i+1}-\mathbf{y}_i)^2}{2(\Delta t)^2} + U(\mathbf{y}_i)\right]}. \tag{3.109}$$

As we have already stressed, this expression, where $\Delta t = \frac{\hbar\beta}{N+1}$, rests on the "Euclidean kernel" that can be formally obtained, within each interval, from the standard kernel of quantum mechanics via a Wick rotation.

One can also recover the classical partition functions of the preceding chapter from this type of expression, and to this end it is convenient to redefine the integration variable according to

$$\tau \rightarrow \hbar\,\tau. \tag{3.110}$$

In this fashion

$$K_E(\mathbf{r},\mathbf{r}',\hbar\beta) = \int [\mathcal{D}\mathbf{x}(\tau)]\, e^{-\int_0^\beta \left[\frac{m}{2\hbar^2}\left(\frac{d\mathbf{x}}{d\tau}\right)^2 + U(\mathbf{x}(\tau))\right]d\tau}, \tag{3.111}$$

and the classical limit $\hbar \rightarrow 0$ for fixed $\beta$ is dominated by the free classical trajectory

$$\mathbf{x}_0(\tau) = \mathbf{r}' + \frac{(\mathbf{r}-\mathbf{r}')}{\beta}\,\tau, \tag{3.112}$$

even in the presence of the potential. Now letting

$$\mathbf{x}(\tau) = \mathbf{x}_0(\tau) + \xi(\tau), \tag{3.113}$$

where $\xi$ is a small fluctuation of order $\hbar$ such that $\xi(0) = \xi(\beta) = 0$, the path integral is dominated by

$$K_E(\mathbf{r},\mathbf{r}',\hbar\beta) \sim \left(\frac{m}{2\pi\hbar^2\beta}\right)^{\frac{3}{2}} e^{-\frac{m(\mathbf{r}'-\mathbf{r})^2}{2\hbar^2\beta}}\; e^{-\int_0^\beta d\tau\, U(\mathbf{x}_0(\tau))}. \tag{3.114}$$

Note that the first factor is essentially a Dirac $\delta$ function in the classical limit, and therefore one can also replace the exponent in the second factor by $\beta U(\mathbf{r})$. One can thus recover the classical partition function, writing this expression in the form

$$K_E(\mathbf{r},\mathbf{r}';\beta) \sim \int \frac{d^3\mathbf{p}}{h^3}\, e^{\frac{i}{\hbar}\mathbf{p}\cdot(\mathbf{r}-\mathbf{r}') - \beta\left(\frac{\mathbf{p}^2}{2m} + U(\mathbf{r})\right)}, \tag{3.115}$$

so that

$$\mathrm{tr}\left(e^{-\beta H}\right) \sim \int \frac{d^3\mathbf{p}\, d^3\mathbf{r}}{h^3}\, e^{-\beta\left(\frac{\mathbf{p}^2}{2m} + U(\mathbf{r})\right)}. \tag{3.116}$$

This is the classical phase-space integral that generalizes Eq. (3.95), and its measure is normalized by a power of Planck's constant, as anticipated in the preceding sections. A similar reasoning can be used to proceed along these lines from the Hamiltonian form of the kernel in Eq. (3.104).

Finally, the comparison between Eqs. (3.116) and (3.111) provides an intriguing interpretation to the quantum path integral over closed trajectories that determines the partition function in Eq. (3.111). From these simple examples we see that quantum mechanics extends the limiting classical form of Eq. (3.116) along an additional dimension of size $\beta$, so that, for example, a gas of noninteracting quantum particles behaves somehow as a noninteracting gas of "string-like objects" extended along the "time" direction, with elastic interactions among their points.

## 3.5 The Free Particle on a Circle

The case of a free quantum particle on a circle of *finite* radius $R$ is very instructive. It is described again by the Hamiltonian

$$H = \frac{p^2}{2m}, \tag{3.117}$$

but in this case the energy levels,

$$\epsilon_n = \frac{\hbar^2 n^2}{2mR^2} \qquad n \in \mathbb{Z}, \tag{3.118}$$

correspond to the plane waves

$$\psi_n(x) \equiv \langle x|n\rangle = \frac{1}{\sqrt{2\pi R}}\, e^{\frac{i}{\hbar} p_n x}, \tag{3.119}$$

with momenta quantized according to

$$p_n = \frac{\hbar\, n}{R}, \qquad n \in \mathbb{Z}. \tag{3.120}$$

These wavefunctions satisfy the periodicity condition

$$\psi_n(x + 2\pi R) = \psi_n(x), \tag{3.121}$$

since $\frac{x}{R}$ is an angular variable in this problem.

One may wonder how these results can possibly comply with the uncertainty principle, since the spread of any wavefunction on the circle is clearly at most $2\pi R$. It is thus instructive to retrace the standard derivation, in order to demonstrate how the angular nature of $x$ affects it. To this end, let us assume for simplicity that the normalized wavefunction $\psi_0$ of interest is such that

$$\langle x\rangle = 0, \qquad \langle p\rangle = 0, \tag{3.122}$$

since from a wavefunction of this type, as we have seen, one can simply obtain the most general one as

$$\psi(x) = e^{\frac{i\langle p\rangle x}{\hbar}}\, \psi_0\,(x - \langle x\rangle). \tag{3.123}$$

One can now start, as in the standard case, from the inequality

$$0 \leq \int_0^{2\pi R} dx \left|\alpha\, x\, \psi_0 + \hbar\, \frac{\partial\, \psi_0}{\partial x}\right|^2, \tag{3.124}$$

where $\alpha$ is a real number, which leads to

$$0 \leq \alpha^2 \langle (\Delta x)^2 \rangle + \langle (\Delta p)^2 \rangle + \alpha \int_0^{2\pi R} dx\, x \frac{\partial}{\partial x} \left|\psi_0\right|^2. \tag{3.125}$$

The last term can be directly integrated by parts, but on a circle it also yields a boundary contribution, so that the inequality takes the form

$$0 \leq \alpha^2 \langle (\Delta x)^2 \rangle + \langle (\Delta p)^2 \rangle + \alpha\, \hbar \left( \left[ x \left|\psi_0\right|^2 \right]_0^{2\pi R} - 1 \right). \tag{3.126}$$

In view of the periodic nature of $\psi_0$, the last term is simply $2\pi R \left|\psi_0(0)\right|^2$, so that finally

$$0 \leq \alpha^2 \langle (\Delta x)^2 \rangle + \langle (\Delta p)^2 \rangle + \alpha\, \hbar \left( 2\pi R \left|\psi_0(0)\right|^2 - 1 \right), \tag{3.127}$$

and the resulting condition on the discriminant becomes

$$\langle (\Delta x)^2 \rangle \langle (\Delta p)^2 \rangle \geq \frac{\hbar^2}{4} \left( 2\pi R \left|\psi_0(0)\right|^2 - 1 \right)^2. \tag{3.128}$$

In the standard case of a particle on the real axis, the term involving $\psi(0)$ would be absent. Note that this modified uncertainty principle is nicely compatible with the upper bound of $2\pi R$ on the position uncertainty and with the lack of momentum uncertainty that characterizes the eigenfunctions of Eq. (3.119), since the r.h.s. vanishes in this case.

Let us now consider the single-particle partition function,

$$Z_1 = \sum_{n \in \mathbb{Z}} \langle n | e^{-\beta H} | n \rangle = \sum_{n \in \mathbb{Z}} e^{-\frac{\beta \hbar^2}{2mR^2} n^2}, \tag{3.129}$$

or more generally the Euclidean evolution kernel,

$$K_E(x, x'; \hbar \beta) = \sum_{n \in \mathbb{Z}} \frac{1}{2\pi R}\, e^{i \frac{n}{R}(x - x')}\, e^{-\frac{\beta \hbar^2}{2mR^2} n^2}. \tag{3.130}$$

It is instructive to turn this expression, whose Gaussian sum defines a special case of a Jacobi $\Theta$-function, into its Lagrangian form, in order to compare it with that obtained in Section 3.4 for a particle on the real line. The basic tool to this end is the Poisson summation formula,

$$\sum_{n \in \mathbb{Z}} f(n) = \sum_{k \in \mathbb{Z}} \hat{f}(k), \tag{3.131}$$

which converts an integer-spaced sampling of a function $f$ into a corresponding sampling of its Fourier transform:

$$\hat{f}(k) = \int_{-\infty}^{+\infty} e^{-i 2\pi k t} f(t)\, dt. \tag{3.132}$$

The Poisson formula can be derived starting from

$$F(\tau) = \sum_{n \in \mathbb{Z}} f(\tau + n), \tag{3.133}$$

where $f(\tau)$ is assumed to be suitably decreasing for $\tau \to \pm\infty$. $F(\tau)$ is then clearly periodic, with unit period, and one can expand it in a Fourier series:

$$F(\tau) = \sum_{k\in\mathbb{Z}} e^{i2\pi k\tau} \int_0^1 F(t)\, e^{-i2\pi kt}\, dt. \tag{3.134}$$

However, the form of $F(t)$ in Eq. (3.133) implies that the Fourier coefficients build the Fourier transform of $f(t)$,

$$\int_0^1 F(t)\, e^{-i2\pi kt}\, dt = \sum_{n\in\mathbb{Z}} \int_n^{n+1} f(t)\, e^{-i2\pi kt} dt = \int_{-\infty}^{+\infty} f(t)\, e^{-i2\pi kt} dt, \tag{3.135}$$

so that finally

$$\sum_{n\in\mathbb{Z}} f(\tau + n) = \sum_{k\in\mathbb{Z}} e^{i2\pi k\tau} \hat{f}(k), \tag{3.136}$$

and the special case $\tau = 0$ yields Eq. (3.131). We can now apply this important result to the Euclidean kernel of Eq. (3.130), obtaining

$$K_E(x, x'; \hbar\beta) = \sqrt{\frac{m}{2\pi\hbar^2\beta}} \sum_{k\in\mathbb{Z}} e^{-\frac{m}{2\beta\hbar^2}(x-x'-2\pi kR)^2}. \tag{3.137}$$

Note that the infinitely many contributions correctly account for all possible paths from $x$ to $x'$, since the particle has indeed infinitely many options that involve arbitrary numbers of windings around the circle.

There is an interesting variant of these considerations, which results from the addition of a constant "vector potential" along the circle. This turns the free-particle Lagrangian into

$$L = \frac{m}{2}\dot{x}^2 - \frac{\hbar\,\theta}{2\pi R}\,\dot{x}, \tag{3.138}$$

and in the Hamiltonian language translates into the minimal substitution

$$p = m\dot{x} - \frac{\hbar}{2\pi R}\,\theta, \tag{3.139}$$

so that

$$H = \frac{1}{2m}\left(p + \frac{\hbar\,\theta}{2\pi R}\right)^2. \tag{3.140}$$

The wavefunctions are still periodic,

$$\psi_n(x) = \frac{1}{\sqrt{2\pi R}}\, e^{i\frac{n}{R}x}, \tag{3.141}$$

but in this case the Euclidean kernel is

$$K_E(x, x'; \hbar\beta) = \frac{1}{2\pi R} \sum_{n\in\mathbb{Z}} e^{i\frac{n}{R}(x-x')}\, e^{-\frac{\beta\hbar^2}{2mR^2}\left(n+\frac{\theta}{2\pi}\right)^2}. \tag{3.142}$$

In view of the preceding considerations, the Lagrangian form of the kernel, determined again by Eq. (3.136), reads

$$K_E(x, x'; \hbar\beta) = \sqrt{\frac{m}{2\pi\beta\hbar^2}} \sum_{k\in\mathbb{Z}} e^{-\frac{m}{2\beta\hbar^2}(x-x'-2\pi kR)^2}\, e^{-i\frac{\theta}{2\pi R}(x-x'-2\pi kR)}. \tag{3.143}$$

Note also that the partition function,

$$Z = \sum_{k\in\mathbb{Z}} e^{-\frac{\beta\hbar^2}{2mR^2}\left(k+\frac{\theta}{2\pi}\right)^2} = \sqrt{\frac{2\,\pi\,m\,R^2}{\beta\,\hbar^2}}\sum_{k\in\mathbb{Z}} e^{-\frac{2\pi^2 m R^2 k^2}{\beta\hbar^2}}\, e^{i\,k\,\theta}, \tag{3.144}$$

is periodic in $\theta$ with period $2\pi$, and would be identical if, rather than introducing the "vector potential," one were to work with quasi-periodic wavefunctions such that

$$\psi(x + 2\,\pi\,R) = \psi(x)\,e^{i\,\theta}. \tag{3.145}$$

Phase factors of this type and their generalizations play an important role in a number of contexts, which include the charge-conjugation-parity (CP) problem in quantum chromodynamics, the theory of strong interactions.

## 3.6 The Hydrogen Atom

Rutherford's identification of point-like atomic nuclei resurrected, for the hydrogen atom, the two-body problem in a central force, a setup that had marked the early successes of Newtonian mechanics. The corresponding Lagrangian is

$$\mathcal{L} = \frac{m_1}{2}\,\dot{\mathbf{r}}_1^{\,2} + \frac{m_2}{2}\,\dot{\mathbf{r}}_2^{\,2} - \frac{k}{|\mathbf{r}_1 - \mathbf{r}_2|}, \tag{3.146}$$

where the coupling, which would be $k = G_N\, m_1\, m_2$ in the case of gravity, becomes here

$$k = e^2. \tag{3.147}$$

The gravitational force would still be present, but on atomic scales its contribution is about 40 orders of magnitude below the electric one, as the reader can simply verify, and therefore one can safely ignore it in the following.

Isolating the free motion of the center-of-mass coordinate $\mathbf{R}$

$$\mathbf{R} = \frac{m_1\,\mathbf{r}_1 + m_2\mathbf{r}_2}{m_1 + m_2}, \tag{3.148}$$

the dynamics of the relative coordinate,

$$\mathbf{r} = \mathbf{r}_1 - \mathbf{r}_2, \tag{3.149}$$

is captured by the reduced Lagrangian

$$\mathcal{L} = \frac{\mu}{2}\,\dot{\mathbf{r}}^2 - \frac{k}{|\mathbf{r}|}, \tag{3.150}$$

where the "reduced mass"

$$\mu = \frac{m_e\, m_p}{m_e + m_p} \simeq m_e, \tag{3.151}$$

since the proton mass, $m_p \simeq 1.67 \times 10^{-24}$ g, is about 2000 times larger than the electron mass $m_e \simeq 9.1 \times 10^{-28}$ g.

The spherical symmetry of the potential

$$V = -\frac{k}{r}, \tag{3.152}$$

grants the conservation of angular momentum, so that the "point mass" $\mu$ traces planar orbits around the center of attraction in an effective potential

$$V_{\rm eff} = \frac{\ell^2}{2\mu r^2} - \frac{k}{r}. \qquad (3.153)$$

The two conservation laws of energy $E$ and angular momentum $\mathbf{L} = \hat{z}\,\ell$, orthogonal to the plane containing the orbits,

$$\begin{aligned} \frac{1}{2}\mu\,(\dot{r})^2 + V_{\rm eff} &= E, \\ \mu\, r^2\, \dot{\theta} &= \ell, \end{aligned} \qquad (3.154)$$

reduce the problem to quadratures.

There are three classes of trajectories $r(\theta)$ in this system. If the total energy $E$ is positive or zero they are hyperbolic or parabolic and describe scattering processes. On the other hand, if $E$ is negative the trajectories are elliptical (or circular, if the radius corresponds to the minimum of the effective potential (3.153)) and describe bound states. The point mass traces them in accordance with Kepler's three laws:

1. the attraction center lies at one of the two foci;
2. the radius $\mathbf{r}$ sweeps the area bounded by the trajectories at a constant rate;
3. the ratio $\frac{a^3}{T^2}$ between the cube of the semi-major axis and the square of the corresponding period has the universal value

$$\frac{a^3}{T^2} = \frac{k}{4\,\pi^2\,\mu}.$$

Let us concentrate on the case of bounded orbits, in which Eqs. (3.154) are solved by

$$\frac{1}{r(\theta)} = \frac{\mu k}{\ell^2}\left(1 + \cos\theta\sqrt{1 - \frac{2|E|\ell^2}{\mu k^2}}\right), \qquad (3.155)$$

so that $r(\theta)$ traces indeed an ellipse, as seen from one of its foci. The standard Cartesian presentation of the ellipse,

$$\frac{x^2}{a^2} + \frac{y^2}{b^2} = 1, \qquad (3.156)$$

with $a > b$, can be recovered by demanding that the sum of the distances from the two foci, located at $(\pm\sqrt{a^2 - b^2}, 0)$, be constant and equal to $2a$:

$$\sqrt{\left(x - \sqrt{a^2 - b^2}\right)^2 + y^2} + \sqrt{\left(x + \sqrt{a^2 - b^2}\right)^2 + y^2} = 2\,a. \qquad (3.157)$$

The reader can verify that squaring this expression twice does indeed lead to Eq. (3.156), while letting

$$x = \sqrt{a^2 - b^2} + r\cos\theta, \qquad y = r\sin\theta, \qquad (3.158)$$

leads to an expression along the lines of Eq. (3.155):

$$\frac{1}{r(\theta)} = \frac{a}{b^2}\left(1 + \cos\theta\sqrt{1 - \left(\frac{b}{a}\right)^2}\right). \qquad (3.159)$$

By comparison, one can conclude that

$$a = \frac{k}{2|E|}, \quad b = \frac{\ell}{\sqrt{2\mu|E|}}, \tag{3.160}$$

recovering the first of Kepler's laws above. The areal velocity is

$$\mathbf{A} = \frac{1}{2}\,\mathbf{r}\times\mathbf{v} = \frac{\ell}{2\mu}\,\hat{z}, \tag{3.161}$$

and is constant, since it is proportional to the angular momentum $\ell$. Finally, the third law follows, noting that the period is the ratio between the area $\pi ab$ of the ellipse and the magnitude of **A**:

$$T = \frac{2\pi\mu ab}{\ell} = 2\pi\, a^{\frac{3}{2}}\sqrt{\frac{\mu}{k}}, \tag{3.162}$$

after expressing $b$ in terms of $a$.

The fact that the trajectories of bound orbits are elliptical is strictly true only for the special Newtonian (or Coulomb) potential of Eq. (3.150). For instance, a small additional contribution proportional to $\frac{1}{r^2}$ would induce a precession of the perihelion, a striking effect that general relativity brings along inevitably, and to a detectable extent, for the planet Mercury. This peculiar property of the Coulomb–Newton problem reflects the presence of yet another conserved quantity in addition to the total energy $E$ and the angular momentum **L**: the Runge–Lenz vector

$$\mathbf{R} = \frac{\mathbf{p}\times\mathbf{L}}{\mu} - k\hat{\mathbf{r}}, \tag{3.163}$$

where $\hat{\mathbf{r}}$ is the radial unit vector. Substituting in this expression the solution of Eq. (3.155) and noting that the first of Eqs. (3.154) implies that

$$\dot{r}^2 = \frac{k^2}{\ell^2}\left(1 - \frac{2|E|\ell^2}{\mu k^2}\right)\sin^2\theta, \tag{3.164}$$

one can verify that **R** is indeed a constant vector, which points along the semi-major axis of the elliptical trajectories. As a result, the elliptical orbits do not undergo a precession in the classical Kepler problem.

These considerations underlie our understanding of the universe at large scales, but in the microscopic world they lead readily to a contradiction: the accelerated motion of the electron following an elliptical trajectory around the nucleus would entail a very rapid energy loss, which we now estimate. If this result were true, all matter would be unstable, which is fortunately not the case in nature. For simplicity, let us consider a circular trajectory in hydrogen, for which Eq. (3.153), computed at the minimum, implies the relation

$$E_0 = -\frac{e^2}{2\,a}, \tag{3.165}$$

where $a$ is the orbital radius. Classical electrodynamics indicates that the electron–proton system ought to radiate as a dipole, with a dipole moment of magnitude $p = e\,a$, and thus with a resulting power loss

$$P = \frac{\ddot{p}^2}{3\,c^3} \simeq \frac{e^2}{3\,c^3}\left(\omega^2 a\right)^2 \tag{3.166}$$

for the atom. Indeed, for circular trajectories, which are traced at constant angular velocity $\omega$, $\ddot{p} = -\omega^2 p$. Moreover, from the radial equation,

$$\mu\,\omega^2\,a = \frac{e^2}{a^2}, \tag{3.167}$$

so that the balance condition,

$$\frac{e^2}{2\,a^2}\,\dot{a} = -P, \tag{3.168}$$

yields the differential equation

$$a^2\,\dot{a} \;=\; -\,\frac{2}{3}\,\frac{e^4}{\mu^2\,c^3}, \tag{3.169}$$

and by integrating it one can estimate a decay time

$$\tau \;=\; \frac{\mu^2\,(a\,c)^3}{2\,e^4} \;=\; \frac{a}{8\,c}\left[\frac{\mu\,c^2}{E_0}\right]^2 \simeq 3\times 10^{-11}\ \text{s}, \tag{3.170}$$

which is so short that nature clearly does not work this way.

Bohr modified the rules for circular orbits, adding two postulates that rest on Planck's constant $h$, and thus lie outside the realm of classical physics:

- The magnitude $\ell$ of the angular momentum is quantized:

$$\ell \;=\; n\,\hbar, \qquad n = 0, 1, \ldots .$$

For circular orbits in hydrogen, in particular, this implies that

$$r \;=\; n^2\,a_0, \quad \text{with} \quad a_0 \;=\; \frac{\hbar^2}{\mu\,e^2} \simeq 5\times 10^{-9}\ \text{cm}, \tag{3.171}$$

where $a_0$ is usually called the Bohr radius, and therefore the available energies are discrete:

$$E_n \;=\; -\,\frac{e^2}{2\,a_0\,n^2} \;=\; -\,\frac{\mu\,c^2\,\alpha^2}{2\,n^2}. \tag{3.172}$$

Here,

$$\alpha \;=\; \frac{e^2}{\hbar\,c} \;\simeq\; \frac{1}{137} \tag{3.173}$$

is the fine-structure constant, a dimensionless quantity that will recur throughout this book.

- In a $q \to p$ transition, the system absorbs radiation (if $E_p > E_q$), or emits it (if $E_p < E_q$), with the discrete frequencies

$$\hbar\,\omega \;=\; \left|E_p \,-\, E_q\right|. \tag{3.174}$$

These comply with Rydberg's empirical formula, which has been known since the last decades of the nineteenth century.

Quantum mechanics leads to a more refined description of the hydrogen atom, but the energy levels, up to relativistic corrections, remain those found by Bohr. Let us sketch how this result and the leading-order corrections come about. To begin with, one can separate variables by writing the Schrödinger equation for the bound states,

$$-\frac{\hbar^2}{2\,\mu}\nabla^2\psi - \frac{k}{r}\,\psi = E\,\psi, \tag{3.175}$$

in polar coordinates. In this fashion

$$\nabla^2\psi = \frac{1}{r}\frac{\partial^2(r\psi)}{\partial r^2} + \frac{\mathbf{L}^2\,\psi}{r^2}, \tag{3.176}$$

where $\mathbf{L}^2$ is the square of the angular momentum operator

$$\mathbf{L} = \mathbf{r}\times\frac{\hbar}{i}\,\nabla, \tag{3.177}$$

whose components satisfy

$$\left[L_x, L_y\right] = i\hbar L_z \tag{3.178}$$

and two similar relations obtained by cyclic permutations of $(x, y, z)$, or more concisely,

$$\left[L_i, L_j\right] = i\,\hbar\,\epsilon_{ijk}\,L_k \quad (i,j,k = 1,2,3), \tag{3.179}$$

where $\epsilon_{ijk}$ is the totally antisymmetric Levi–Civita symbol. The spherical harmonics $Y_{\ell m}(\theta,\phi)$ satisfy

$$\begin{aligned} L_z\,Y_{\ell m} &= \hbar\,m\,Y_{\ell m}, \quad m = -\ell, -\ell+1, \ldots, \ell \\ \mathbf{L}^2\,Y_{\ell m} &= \hbar^2\,\ell(\ell+1)\,Y_{\ell m}, \quad \ell = 0, 1, \ldots, \end{aligned} \tag{3.180}$$

and the spherical symmetry of the potential makes it possible to look for solutions of the form

$$\psi(\mathbf{r}) = R_{n\ell}(r)\,Y_{\ell m}(\theta,\phi). \tag{3.181}$$

One is thus led to the ordinary differential equation

$$-\frac{\hbar^2}{2\mu}\frac{1}{r}\frac{d^2(rR_{n\ell})}{dr^2} + V_{\text{eff}}(r)R_{n\ell} = E\,R_{n\ell}, \tag{3.182}$$

where $V_{\text{eff}}(r)$ is the combination defined in Eq. (3.153), up to the replacement of $\ell^2$ with $\ell(\ell+1)$. The Schrödinger equation can be solved using a variety of techniques, and one can show that the allowed values of $E$ for the bound states are precisely those of Eq. (3.172), with $\ell = 0, \ldots, n-1$. Taking into account that every value of $\ell$ brings along a degeneration $2\ell+1$ due to the allowed values of $m$, one is thus led to conclude that the $n$th energy level has degeneracy $n^2$.

By purely algebraic means, one can deduce from Eqs. (3.179) that there exist states $|j,m\rangle$, of positive norm, such that

$$\begin{aligned} L_z|j,m\rangle &= \hbar\,m|j,m\rangle\,, \qquad m = -j, -j+1, \ldots, j \\ L^2|j,m\rangle &= \hbar^2 j(j+1)|j,m\rangle\,, j = 0, 1/2, 1, 3/2, \ldots. \end{aligned} \tag{3.183}$$

The additional solutions with $j = 1/2, 3/2, 5/2, \ldots$ are properly interpreted in terms of multi-component wavefunctions. The two-component wavefunctions of spin-1/2 particles in Appendix F are the simplest example of this type.

It is possible to retrieve the discrete spectrum of Eq. (3.172), together with its degeneracy, via a purely algebraic route that we now illustrate. Before elaborating on this point, and in order to simplify the ensuing formulas, let us introduce atomic units, which amount to measuring lengths in units of the Bohr radius $a_0$ and angular momenta in units of $\hbar$, so that formally

$$\mathbf{r} \to a_0\, r, \qquad \mathbf{p} \to \frac{\hbar}{a_0}\,\mathbf{p}, \qquad \mathbf{L} \to \hbar\,\mathbf{L}, \qquad E \to \frac{e^2}{2a_0}\,E. \tag{3.184}$$

In such units, the Hamiltonian takes the simple form

$$H = \mathbf{p}^2 - \frac{2}{r}, \tag{3.185}$$

while the canonical commutation relations become

$$[r_i, p_j] = i\,\delta_{ij}, \tag{3.186}$$

and imply for the angular momentum operator $\mathbf{L} = \mathbf{r} \times \mathbf{p}$ the rescaled commutators

$$[L_i, r_j] = i\epsilon_{ijk} r_k, \qquad [L_i, p_j] = i\epsilon_{ijk} p_k, \qquad [L_i, L_j] = i\epsilon_{ijk} L_k. \tag{3.187}$$

The invariance of the Hamiltonian under rotations implies that it commutes with the angular momentum operator,

$$[H, \mathbf{L}] = 0. \tag{3.188}$$

However, as we learned from the classical analysis of the Newton and Coulomb problems, the system possesses yet another symmetry, which is responsible for the conservation of the Runge–Lenz vector (3.163). Therefore, one is led to associate to this quantity the Hermitian operator

$$\mathbf{R} = \frac{\mathbf{p} \times \mathbf{L} - \mathbf{L} \times \mathbf{p}}{2} - \hat{\mathbf{r}}, \tag{3.189}$$

where the conversion to atomic units also factored out $e^2$. It is now an instructive exercise to prove that

$$[H, \mathbf{R}] = 0, \tag{3.190}$$

so that the symmetry generated by the classical Runge–Lenz vector does extend to the quantum setup, and that $\mathbf{R}$ also satisfies

$$\mathbf{R}^2 = 1 + H + H\mathbf{L}^2, \qquad \left[R_i, R_j\right] = -\,i H \epsilon_{ijk} L_k. \tag{3.191}$$

Focusing now on the bound-state (negative) spectrum of $H$, one can define the operators

$$\mathbf{J}^{\pm} = \frac{\mathbf{L} \pm (-H)^{\frac{1}{2}}\,\mathbf{R}}{2}, \tag{3.192}$$

which by Eq. (3.191) obey

$$\left[J_i^{\pm}, J_j^{\pm}\right] = i\,\epsilon_{ijk} J_k^{\pm}, \qquad [\mathbf{J}^+, \mathbf{J}^-] = 0, \tag{3.193}$$

and

$$(\mathbf{J}^+)^2 = (\mathbf{J}^-)^2 = -\frac{1}{4}\left(1 + H^{-1}\right). \tag{3.194}$$

The operators $\mathbf{J}^+$ and $\mathbf{J}^-$ are two independent "angular momenta," since they both satisfy Eqs. (3.193). Their eigenvalues, which coincide on account of Eq. (3.194), can be both parametrized as $j(j+1)$, with $2j$ a nonnegative integer. Consequently, Eq. (3.194) determines the bound-state spectrum of $H$,

$$E_n = -\frac{1}{(2j+1)^2} \equiv -\frac{1}{n^2}, \qquad \text{with } n = 1, 2, \ldots, \tag{3.195}$$

and after restoring the standard units this result coincides with Bohr's spectrum of Eq. (3.172). On account of the definition (3.192), the ordinary angular momentum operator is simply

$$\mathbf{L} = \mathbf{J}^+ + \mathbf{J}^-, \tag{3.196}$$

and therefore the eigenspace corresponding to a given value of $j$ contains eigenstates of $J_z^+$ and $J_z^-$ with eigenvalues up to $j$, corresponding to integer values of $\ell$, up to a maximum eigenvalue $2j = n - 1$. One thus retrieves the rule according to which, for a given $n$, the quantum number $\ell$ can range from 0 to $n - 1$, and taking into account the $2\ell + 1$ states present for each choice of $\ell$ leads finally to a total degeneracy $n^2$ for any given value of $n$.

The next ingredient is the electron spin, which requires a pair of wavefunctions, rather than one, as discussed in Appendix F, so that, in the absence of spin-dependent interactions, the degeneracy is $2n^2$, or $4n^2$ if one also takes into account the proton spin. Note also that, as $r \to 0$, the probability density behaves as

$$r^2 |R_{nl}|^2 \sim r^{2\ell}, \tag{3.197}$$

so that it vanishes at the origin except when $l = 0$.

Relativistic effects introduce three types of leading-order perturbative corrections to the Bohr spectrum of Eq. (3.172). The effects of perturbations are determined, to lowest order, by the eigenvalues of the matrices

$$\langle i|\Delta H|j\rangle, \tag{3.198}$$

which involve the perturbation $\Delta H$ of the original Hamiltonian $H$ and are restricted to the originally degenerate subspaces.

1. The proper form of the kinetic energy, when expanded in inverse powers of $c$, yields

$$K = \sqrt{\mathbf{p}^2 c^2 + \mu^2 c^4} = \mu c^2 + \frac{\mathbf{p}^2}{2\mu} - \frac{1}{8}\frac{(\mathbf{p}^2)^2}{\mu^3 c^2} + \cdots . \tag{3.199}$$

The term,

$$\Delta H_1 = -\frac{1}{8}\frac{(\mathbf{p}^2)^2}{\mu^3 c^2}, \tag{3.200}$$

is the leading-order correction, and determines

$$\langle n\ell m|\Delta H_1|n\ell m\rangle = \frac{\mu c^2 \alpha^2}{2n^2}\frac{\alpha^2}{n^2}\left(\frac{3}{4} - \frac{n}{\ell + \frac{1}{2}}\right), \tag{3.201}$$

which is four orders of magnitude smaller than Bohr's term (3.172).

2. The relativistic Dirac equation contains an additional "spin-orbit" term of the same order of magnitude,

$$\Delta H_2 = \frac{1}{2\mu^2 c^2}\frac{e^2}{r^3}\,\mathbf{S}\cdot\mathbf{L}, \tag{3.202}$$

which couples together the orbital angular momentum **L** and the spin angular momentum **S**, whose components

$$S_i = \frac{\hbar}{2}\,\sigma_i, \tag{3.203}$$

mix the two portions of the electron wavefunction, as described in Appendix F. This interaction has a more complicated nature, since there are non-vanishing off-diagonal matrix elements $\langle n\ell m|H_1|n\ell' m'\rangle$. However

$$\mathbf{S}\cdot\mathbf{L} = \frac{1}{2}\left(\mathbf{J}^2 - \mathbf{L}^2 - \mathbf{S}^2\right), \tag{3.204}$$

where

$$\mathbf{J} = \mathbf{L} + \mathbf{S} \tag{3.205}$$

is the total (orbital+spin) angular momentum operator, whose components again satisfy commutation relations as in Eq. (3.178). The states that diagonalize the spin-orbit term are the two sets $|njm\rangle$, where $j = \ell \pm \frac{1}{2}$, which are linear combinations of $|n(\ell \pm \frac{1}{2})\left(m - \frac{1}{2}\right)\rangle|+\rangle$ and $|n(\ell \pm \frac{1}{2})\left(m + \frac{1}{2}\right)\rangle|-\rangle$, with $|\pm\rangle$ the two spin-$\frac{1}{2}$ eigenstates, and one can show that

$$\langle njm|(\Delta H_1 + \Delta H_2)|njm\rangle = \frac{\mu c^2\,\alpha^2}{2\,n^2}\,\frac{\alpha^2}{n^2}\left(\frac{3}{4} - \frac{n}{j+\frac{1}{2}}\right). \tag{3.206}$$

This correction is indeed diagonal in the new basis, so that we can read directly the corrections to Bohr's eigenvalues, and *states with neighboring values of $\ell$ and identical values of j remain degenerate*. Since the interaction Hamiltonian is proportional to **L**, one should consider this contribution only for $l \neq 0$.

3. The relativistic Dirac equation gives rise to an additional interaction of the same order, the so-called Darwin term:

$$\Delta H_3 = \frac{\hbar^2}{8\,\mu^2\,c^2}\,\nabla^2 V = \frac{\pi\,\hbar^2\,e^2}{2\,\mu^2\,c^2}\,\delta(\mathbf{x}). \tag{3.207}$$

This only affects states with $\ell = 0$, due to Eq. (3.197), since they are the only ones that can yield nonvanishing contributions at the origin. All in all, this term simply reinstates the contribution for $\ell = 0$ in Eq. (3.206), which therefore holds for all values of $\ell$.

In conclusion, the corrected energy spectrum is

$$E_{njm} = -\frac{\mu\,c^2\,\alpha^2}{2\,n^2} + \frac{\mu c^2\,\alpha^2}{2\,n^2}\,\frac{\alpha^2}{n^2}\left(\frac{3}{4} - \frac{n}{j+\frac{1}{2}}\right) + \mathcal{O}\left(\alpha^6\right), \tag{3.208}$$

for all values of $\ell = 0, 1, \ldots$, and for $j = \ell \pm \frac{1}{2}$, where the "minus" sign is only available for $\ell > 0$. There is a ground state, with $n = 1, \ell = 0, j = \frac{1}{2}$, usually denoted $1S_{\frac{1}{2}}$, with

$$E\left(1S_{\frac{1}{2}}\right) = -\frac{\mu c^2\alpha^2}{2} - \frac{\mu c^2\alpha^4}{8} + \mathcal{O}\left(\alpha^6\right), \tag{3.209}$$

which is doubly degenerate ($m = \pm\frac{1}{2}$). At the next level, $n = 2$, there are 8 states, and the first four are the degenerate $2S_{\frac{1}{2}}$ and $2P_{\frac{1}{2}}$ (with $\ell = 1$) doublets, which have both $j = \frac{1}{2}$. To this order in $\alpha$ there is thus an accidental degeneracy, and

$$E\left(2S_{\frac{1}{2}}\right) = E\left(2P_{\frac{1}{2}}\right) = -\frac{\mu c^2 \alpha^2}{8} - \frac{5}{128}\,\mu c^2 \alpha^4 + \mathcal{O}\left(\alpha^6\right). \tag{3.210}$$

Finally, there is a quadruplet of states ($m = \pm\frac{1}{2}, \pm\frac{3}{2}$), with $\ell = 1$ and $j = \frac{3}{2}$, and

$$E\left(2P_{\frac{3}{2}}\right) = -\frac{\mu c^2 \alpha^2}{8} - \frac{1}{128}\,\mu c^2 \alpha^4 + \mathcal{O}\left(\alpha^6\right). \tag{3.211}$$

The preceding corrections are usually called "fine-structure" effects. As we have anticipated, there is an additional two-fold degeneracy that we left aside, since the corresponding "hyperfine" splittings, due to the proton spin, are about 2000 times smaller than the dominant fine-structure contributions. We shall elaborate further on the $2S_{\frac{1}{2}}$-$2P_{\frac{1}{2}}$ degeneracy in Section 5.2, and we shall return to the effect of nuclear spins in Section 8.9.

## 3.7 The WKB Approximation

The classical one-dimensional dynamics of particles in the presence of a potential energy $U(x)$ can be reduced to quadratures, and affords a nice intuitive characterization: given the total energy $E$, the particle can only access regions where $U < E$, and its motion reverts at the "turning points," where $U = E$. The WKB approximation puts quantum mechanics, if you will, on similar grounds: it allows explicit, if approximate, solutions, of all one-dimensional problems for

$$-\frac{\hbar^2}{2m}\frac{d^2\psi}{dx^2} + U(x)\,\psi = E\,\psi, \tag{3.212}$$

while also associating an intuitive picture to a number of phenomena of great interest.

The change of variable

$$\psi = e^{\frac{i}{\hbar}\mathcal{S}}, \tag{3.213}$$

suggested by the form of the de Broglie waves for the free particle, turns the Schrödinger equation into

$$\frac{1}{2m}\left(\frac{d\mathcal{S}}{dx}\right)^2 - \frac{i\hbar}{2m}\frac{d^2\mathcal{S}}{dx^2} = E - U(x), \tag{3.214}$$

which leads naturally to a series expansion for $\mathcal{S}$ in powers of $\hbar$. Here we shall content ourselves with the first two terms, writing

$$\mathcal{S} = \mathcal{S}_0 + \hbar\,\mathcal{S}_1, \tag{3.215}$$

and retaining terms up to linear order in $\hbar$ gives

$$\begin{aligned} \frac{1}{2m}\left(\frac{d\mathcal{S}_0}{dx}\right)^2 &= E - U(x), \\ \frac{d\,\mathcal{S}_1}{dx} &= \frac{i}{2}\frac{d}{dx}\log\frac{d\mathcal{S}_0}{dx}. \end{aligned} \tag{3.216}$$

In the classically allowed region, where $U(x) < E$, letting

$$p(x) = \sqrt{2m(E - U(x))}, \tag{3.217}$$

the general solution in the WKB approximation reads, to this order,

$$\psi = \frac{A}{\sqrt{p(x)}} e^{\frac{i}{\hbar}\int_{x_0}^{x} p(y)dy} + \frac{B}{\sqrt{p(x)}} e^{-\frac{i}{\hbar}\int_{x_0}^{x} p(y)dy} . \tag{3.218}$$

Note that the complete exponents,

$$\pm \int p\, dx - E\, t, \tag{3.219}$$

of the solutions in classically allowed regions rest on the action computed for the corresponding classical trajectories. These solutions manifest an oscillatory behavior, with a "local" de Broglie wavelength

$$\lambda(x) = \frac{h}{p(x)}, \tag{3.220}$$

and assign to the traveling waves probabilities proportional to $\frac{1}{p(x)}$. Indeed, it ought to be more likely to find the particle in regions where classically it would spend more time. Quantum mechanics, however, also allows for solutions in the classically forbidden region $U(x) > E$,

$$\psi = \frac{A}{\sqrt{|p(x)|}} e^{\frac{1}{\hbar}\int_{x_0}^{x} |p(y)|dy} + \frac{B}{\sqrt{|p(x)|}} e^{-\frac{1}{\hbar}\int_{x_0}^{x} |p(y)|dy}, \tag{3.221}$$

which contain both a growing mode and a decaying one, with

$$|p(x)| = \sqrt{2m(U(x) - E)}. \tag{3.222}$$

The validity of the WKB approximation clearly rests on the inequality

$$\hbar \left|S_0''\right| \ll (S_0')^2, \tag{3.223}$$

which one can recast in the form

$$\lambda\, |U'| \ll 4\, \pi\, |E - U|, \tag{3.224}$$

so that $U$ should vary little, within a de Broglie wavelength $\lambda = \frac{h}{p}$, in comparison with $|E - U|$. This condition, however, is clearly violated at the turning points of the classical motion, which we now analyze in detail.

### 3.7.1 Linear Connection Formulas

We can now connect the two solutions of Eqs. (3.218) and (3.221), which are clearly singular at the classical turning points, solving exactly the Schrödinger equation for a linear potential. Referring, for definiteness, to the "right turning point" (point B) in Figure 3.1, let us call $x^\star$ the corresponding value of $x$. Let us also assume, for the time being, that close to $x^*$ the difference $E - U(x)$ is well captured by a linear function, so that

$$\frac{2m}{\hbar^2}\,(E - U(x)) \simeq \gamma\,(x^* - x)\,. \tag{3.225}$$

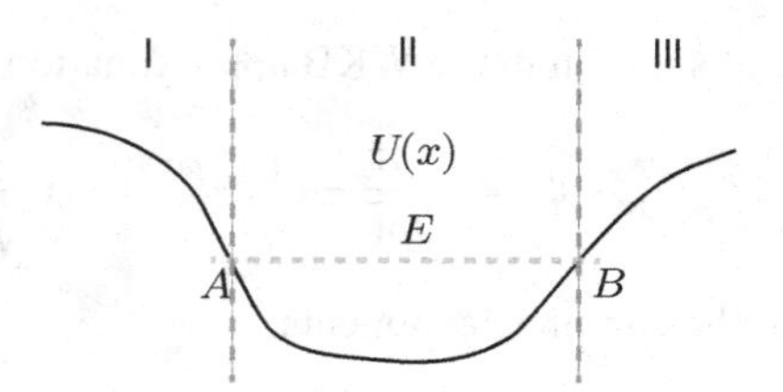


Figure 3.1 A one-dimensional potential $U(x)$ with two turning points $A$ and $B$ (corresponding to $x^\star$ in the text), determined by the total energy $E$. Region II is classically allowed, while regions I and III are classically forbidden.

In compliance with Figure 3.1, let us also assume that $\gamma > 0$, so that a classically allowed region is present for $x < x^\star$. In this fashion, the Schrödinger equation in a neighborhood of the inversion point reduces to

$$\frac{d^2\psi}{dx^2} - \gamma \left(x - x^\star\right) \psi = 0, \tag{3.226}$$

and letting

$$x - x^\star = \gamma^{-\frac{1}{3}}\, \xi, \tag{3.227}$$

leads finally to the Airy equation

$$\frac{d^2\psi}{d\xi^2} - \xi\, \psi = 0. \tag{3.228}$$

The Airy equation can be solved in terms of contour integrals, since letting

$$\psi = \int_\Gamma ds\, e^{s\,\xi} f(s) \tag{3.229}$$

turns it into a first-order differential equation for $f(s)$. Substituting in Eq. (3.228) and integrating by parts gives

$$\int_\Gamma ds\, e^{s\,\xi} \left(s^2 f(s) + f'(s)\right) + \left[e^{s\,\xi} f(s)\right]_{\partial\Gamma} = 0, \tag{3.230}$$

so that

$$\psi = \int_\Gamma ds\, e^{s\,\xi - \frac{s^3}{3}} \tag{3.231}$$

is a solution, provided the contour $\Gamma$ is such that, at its boundary,

$$\left[e^{s\,\xi - \frac{s^3}{3}}\right]_{\partial\Gamma} = 0. \tag{3.232}$$

The contour must be open to build nontrivial solutions, since the integrand in Eq. (3.231) is analytic in the finite plane, and moreover, letting $s = \rho e^{i\theta}$,

$$\left|e^{s\,\xi - \frac{s^3}{3}}\right| = e^{\rho\,\xi\,\cos\theta - \frac{\rho^3}{3}\cos(3\theta)}, \tag{3.233}$$

so that the relevant open contours must tend to infinity within the three wedges

$$-\frac{\pi}{6} + \frac{2k\pi}{3} < \theta < \frac{\pi}{6} + \frac{2k\pi}{3}, \tag{3.234}$$

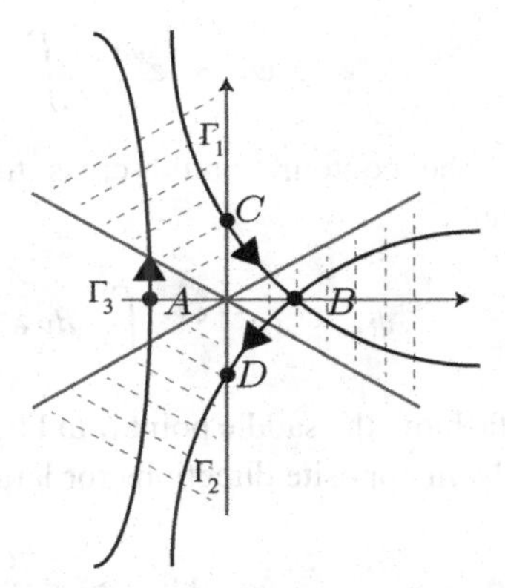


**Figure 3.2** Contours in the complex $s$-plane for the three functions $\psi_i$. $A$ and $B$ are saddle points for $\xi > 0$, while $C$ and $D$ are saddle points for $\xi < 0$.

with $k = 0, 1, 2$. There are apparently three independent contours $\Gamma_k$, with corresponding solutions given by Eq. (3.231). However, there are only two independent solutions of the second-order equation (3.228): the union of the three $\Gamma_k$ deforms smoothly into a closed contour in the finite plane, and consequently the sum of the three $\psi_i$ vanishes, on account of Cauchy's theorem.

Our next task is to obtain asymptotic estimates for the behavior of the three $\psi_k$'s as $\xi \to \pm\infty$, considering the saddle points of the exponent

$$g(s) = s\,\xi - \frac{s^3}{3}, \tag{3.235}$$

which are determined by

$$g'(s) \equiv \xi - s^2 = 0 \longrightarrow \begin{cases} s = \pm\sqrt{\xi} & \text{for } \xi > 0, \\ s = \pm i\sqrt{|\xi|} & \text{for } \xi < 0. \end{cases} \tag{3.236}$$

Consequently, depending on the sign of $\xi$,

$$\begin{aligned} \xi > 0: \quad & g(s)|_{\pm\sqrt{\xi}} \pm \frac{2}{3}\,\xi^{\frac{3}{2}}, \quad g''(s)|_{\pm\sqrt{\xi}} = \mp 2\sqrt{\xi}, \\ \xi < 0: \quad & g(s)|_{\pm i\sqrt{|\xi|}} \mp \frac{2}{3}\,i\,|\xi|^{\frac{3}{2}}, \quad g''(s)|_{\pm i\sqrt{|\xi|}} = \mp 2\,i\sqrt{|\xi|}. \end{aligned} \tag{3.237}$$

In the complex plane, the solutions of $g' = 0$ are saddle points: the real and imaginary parts of an analytic function satisfy the Laplace equation, which constrains their second derivatives with respect to $x$ and $y$ to have opposite signs. The last, crucial step is thus identifying paths of steepest descent through the saddle points, which determine corresponding contour deformations.

To reiterate, and referring to Figure 3.2, in our case the saddle points lie:

1. on the real axis, at $s = \pm\sqrt{\xi}$ (points $A$ and $B$), for $\xi > 0$;
2. on the imaginary axis, at $s = \pm i\sqrt{|\xi|}$ (points $C$ and $D$), for $\xi < 0$.

Let us begin from the point $A$ in Figure 3.2. It concerns $\psi_3$, whose contour $\Gamma_3$ can be deformed to pass through $-\sqrt{\xi}$ in the limit of large positive $\xi$, and in the saddle point approximation

$$\psi_3 \sim e^{\frac{2}{3}\xi^{\frac{3}{2}}} \int_{\Gamma_3} ds\, e^{\sqrt{\xi}\,(s+\sqrt{\xi})^2} . \tag{3.238}$$

As a result, the contour should cross the saddle point vertically, and letting $s = -\sqrt{\xi} + iy$,

$$\psi_3 \sim i\,e^{-\frac{2}{3}\xi^{\frac{3}{2}}} \int_{-\infty}^{\infty} dy\, e^{-\sqrt{\xi}\,y^2} = i\,\frac{\pi^{\frac{1}{2}}}{\xi^{\frac{1}{4}}}\, e^{-\frac{2}{3}\xi^{\frac{3}{2}}} . \tag{3.239}$$

In a similar fashion, the saddle point $B$ in Figure 3.2 concerns $\Gamma_{1,2}$, which are to cross it horizontally, in opposite directions for large positive values of $\xi$, and therefore

$$\psi_{1,2} \sim \pm\,\frac{\pi^{\frac{1}{2}}}{\xi^{\frac{1}{4}}}\, e^{\frac{2}{3}\xi^{\frac{3}{2}}} . \tag{3.240}$$

For $\xi < 0$ there is a novelty: while the contours $\Gamma_{1,2}$ for $\psi_1$ and $\psi_2$ can be deformed to pass, respectively, through the saddle points $C$ and $D$ in Figure 3.2, $\Gamma_3$ can be deformed to pass though both. Let us therefore discuss this last case, since it combines the contributions from the others:

$$\psi_3 \sim e^{i\frac{2}{3}|\xi|^{\frac{3}{2}}} \int_{\Gamma_3} ds\, e^{i\sqrt{|\xi|}\,(s+i\sqrt{|\xi|})^2} + e^{-i\frac{2}{3}|\xi|^{\frac{3}{2}}} \int_{\Gamma_3} ds\, e^{-i\sqrt{|\xi|}\,(s-i\sqrt{|\xi|})^2} . \tag{3.241}$$

The final steps involve the substitutions $s = \mp i\sqrt{|\xi|} + e^{\pm i\frac{\pi}{4}}\,y$ in the two remaining Gaussian integrals, which can be extended to the whole real $y$-axis up to exponentially small errors. The two contributions thus read

$$\psi_3 \sim \frac{\pi^{\frac{1}{2}}}{|\xi|^{\frac{1}{4}}}\, e^{i\frac{\pi}{4}}\, e^{i\frac{2}{3}|\xi|^{\frac{3}{2}}} + \frac{\pi^{\frac{1}{2}}}{|\xi|^{\frac{1}{4}}}\, e^{i\frac{3\pi}{4}}\, e^{-i\frac{2}{3}|\xi|^{\frac{3}{2}}} . \tag{3.242}$$

and add up to

$$\psi_3 \sim 2i\,\frac{\pi^{\frac{1}{2}}}{|\xi|^{\frac{1}{4}}} \cos\left(\frac{2}{3}|\xi|^{\frac{3}{2}} - \frac{\pi}{4}\right). \tag{3.243}$$

Taking into account the two terms in Eq. (3.242) and the way $\Gamma_{1,2}$ pass through $C$ and $D$, one can also conclude that

$$\psi_{1,2} \sim -i\,\frac{\pi^{\frac{1}{2}}}{|\xi|^{\frac{1}{4}}}\, e^{\mp i\left(\frac{2}{3}|\xi|^{\frac{3}{2}} - \frac{\pi}{4}\right)} . \tag{3.244}$$

Summarizing, one is thus led to the connection formulas

$$\begin{aligned} &\mp i\,\frac{\pi^{\frac{1}{2}}}{|\xi|^{\frac{1}{4}}}\, e^{\mp i\left(\frac{2}{3}|\xi|^{\frac{3}{2}} - \frac{\pi}{4}\right)} \longleftrightarrow \frac{\pi^{\frac{1}{2}}}{\xi^{\frac{1}{4}}}\, e^{\frac{2}{3}\xi^{\frac{3}{2}}} , \\ &2\,\frac{\pi^{\frac{1}{2}}}{|\xi|^{\frac{1}{4}}} \cos\left(\frac{2}{3}|\xi|^{\frac{3}{2}} - \frac{\pi}{4}\right) \longleftrightarrow \frac{\pi^{\frac{1}{2}}}{\xi^{\frac{1}{4}}}\, e^{-\frac{2}{3}\xi^{\frac{3}{2}}} . \end{aligned} \tag{3.245}$$

Moreover, for bound states one can always work with real functions, so that one can replace the first couple of links with the real parts, obtaining

$$-\,\frac{\pi^{\frac{1}{2}}}{|\xi|^{\frac{1}{4}}} \sin\left(\frac{2}{3}|\xi|^{\frac{3}{2}} - \frac{\pi}{4}\right) \longleftrightarrow \frac{\pi^{\frac{1}{2}}}{\xi^{\frac{1}{4}}}\, e^{\frac{2}{3}\xi^{\frac{3}{2}}} . \tag{3.246}$$

Referring again to Figure 3.1, one can now express these conditions for the right-most turning point, $B$, in terms of $p(x)$ noting that, for $x > x^\star$ and close to $B$,

$$\frac{1}{\hbar}\int_{x^\star}^{x} dx' \sqrt{2m(U(x')-E)} \simeq \frac{2}{3}\,\gamma^{\frac{1}{2}}(x-x^\star)^{\frac{3}{2}} = \frac{2}{3}\,\xi^{\frac{3}{2}}, \tag{3.247}$$

while for $x < x^\star$ and close to $B$

$$\frac{1}{\hbar}\int_{x}^{x^\star} dx' \sqrt{2m(E-U(x'))} \simeq \frac{2}{3}\,\gamma^{\frac{1}{2}}(x^\star-x)^{\frac{3}{2}} = \frac{2}{3}\,|\xi|^{\frac{3}{2}}. \tag{3.248}$$

The WKB linear connection formulas are finally

$$\begin{aligned} \frac{\mp i}{\sqrt{p(x)}}\, e^{\mp i\left(\frac{1}{\hbar}\int_x^{x^\star} dx'\, p(x') - \frac{\pi}{4}\right)} &\longleftrightarrow \frac{e^{\frac{1}{\hbar}\int_{x^\star}^{x} dx'\, |p(x')|}}{\sqrt{|p(x)|}}, \\ \frac{2}{\sqrt{p(x)}} \cos\left(\frac{1}{\hbar}\int_x^{x^\star} dx'\, p(x') - \frac{\pi}{4}\right) &\longleftrightarrow \frac{e^{-\frac{1}{\hbar}\int_{x^\star}^{x} dx'\, |p(x')|}}{\sqrt{|p(x)|}}, \end{aligned} \tag{3.249}$$

and for bound states

$$-\frac{1}{\sqrt{p(x)}} \sin\left(\frac{1}{\hbar}\int_x^{x^\star} dx'\, p(x') - \frac{\pi}{4}\right) \longleftrightarrow \frac{e^{\frac{1}{\hbar}\int_{x^\star}^{x} dx'\, |p(x')|}}{\sqrt{|p(x)|}}. \tag{3.250}$$

One can repeat these considerations for the left inversion point $A$, and the final answer is that the results are identical, with the endpoints of the integrals properly ordered in all cases. We can now apply these results to some instructive examples.

### 3.7.2 The Bohr–Sommerfeld Rule

Our first example concerns the general solution for the potential well of Figure 3.1. One can build it starting from region I, where

$$\psi_{\mathrm{I}} = \alpha\, \frac{e^{-\frac{1}{\hbar}\int_x^{A} dx'\, |p(x')|}}{\sqrt{|p(x)|}}, \tag{3.251}$$

where $\alpha$ is a constant that will we shall determine shortly by imposing the normalization condition. This choice guarantees that $\psi_{\mathrm{I}}$ vanishes as $x \to -\infty$, and the preceding results imply that

$$\psi_{\mathrm{II}} = \frac{2\,\alpha}{\sqrt{p(x)}} \cos\left(\frac{1}{\hbar}\int_A^{x} dx'\, p(x') - \frac{\pi}{4}\right). \tag{3.252}$$

Before applying the preceding relations at the right-most turning point $B$, however, one must prepare the argument of $\psi_{\mathrm{II}}$ accordingly, writing

$$\frac{1}{\hbar}\int_A^{x} dx'\, p(x') - \frac{\pi}{4} = \gamma - \left(\frac{1}{\hbar}\int_x^{B} dx'\, p(x') - \frac{\pi}{4}\right), \tag{3.253}$$

where

$$\gamma = \frac{1}{\hbar}\int_A^{B} dx'\, p(x') - \frac{\pi}{2}. \tag{3.254}$$

Consequently

$$\psi_{II} = \frac{2\,\alpha}{\sqrt{p(x)}}\left[\cos\gamma\,\cos\left(\frac{1}{\hbar}\int_x^B dx'\,p(x') - \frac{\pi}{4}\right)\right.$$
$$\left. + \sin\gamma\,\sin\left(\frac{1}{\hbar}\int_x^B dx'\,p(x') - \frac{\pi}{4}\right)\right], \tag{3.255}$$

and in order to avoid the growing mode in region III, one must demand that $\gamma = n\,\pi$, with $n = 0, 1, \ldots$. This condition translates into the Bohr–Sommerfeld quantization rule

$$\int_A^B dx\,p(x) = \left(n + \frac{1}{2}\right)\pi\,\hbar, \tag{3.256}$$

which is often written in the form

$$\oint dx\,p(x) = \left(n + \frac{1}{2}\right)h, \tag{3.257}$$

and will play a role in Chapter 9. For any given value of $n$, the complete solution thus reads

$$\psi(x) = \alpha \begin{cases} \dfrac{e^{-\frac{1}{\hbar}\int_x^A dx'\,|p(x')|}}{\sqrt{|p(x)|}} & (x < A), \\ \dfrac{2}{\sqrt{p(x)}}\cos\left(\frac{1}{\hbar}\int_A^x dx'\,p(x') - \frac{\pi}{4}\right) & (A < x < B), \\ (-1)^n\,\dfrac{e^{-\frac{1}{\hbar}\int_B^x dx'\,|p(x')|}}{\sqrt{|p(x)|}} & (x > B). \end{cases} \tag{3.258}$$

The order $n$ of the eigenvalue in Eqs. (3.256) and (3.257) counts the number of nodes of $\psi$ in the classically allowed region between the two turning points $A$ and $B$: no nodes for the ground state ($n = 0$), one node for the first excited state ($n = 1$), and so on. Note also that the result for the energy levels is exact for the harmonic oscillator, since in this case

$$\int_A^B dx\,p(x) = 2\sqrt{2mE}\int_0^{\sqrt{\frac{2E}{m\omega^2}}} dx\,\sqrt{1 - \frac{m\omega^2}{2E}\,x^2} = \frac{\pi E}{\omega}. \tag{3.259}$$

We shall soon be able to appreciate this fact from a different perspective.

The Bohr–Sommerfeld rule contains an important indication regarding the level density. It can be deduced by differentiating both sides of Eq. (3.256) with respect to $E$ while treating $n$, formally, as a continuous variable, since at any rate the approximation will be increasingly reliable for excited states, when the inversion points are further and further apart. This gives

$$\int_A^B \frac{dx}{v(x)} = \pi\hbar\,\frac{dn}{dE}, \tag{3.260}$$

and the integral is one half of the classical oscillation period in the potential well of Figure 3.1, or $\pi/\omega$, so that this relation can be turned into

$$\frac{dE}{dn} = \hbar\,\omega(E). \tag{3.261}$$

This is a neat result: the value of $\omega(E)$ for the classical oscillations of a given total energy $E$ determines the WKB level density around it.

One can simply normalize the wavefunction of Eq. (3.258). Only the classically allowed region contributes, up to exponentially small errors, and

$$4\,|\alpha|^2 \int_A^B \frac{dx}{p(x)} \cos^2\left(\frac{1}{\hbar}\int_A^x dx'\,p(x') - \frac{\pi}{4}\right) \simeq 2\,|\alpha|^2 \int_A^B \frac{dx}{p(x)}, \tag{3.262}$$

since only the average value of the trigonometric function contributes at the high frequencies where the WKB approximation is accurate. The integral is similar to the one discussed for the level density, and one is thus led to the condition

$$1 = |\alpha|^2 \frac{2\pi}{m\omega}, \tag{3.263}$$

so that, up to an overall phase factor, one can choose

$$\alpha = \sqrt{\frac{m\,\omega}{2\pi}}. \tag{3.264}$$

### 3.7.3 Barrier Penetration

Our next example is barrier penetration, and to this end we refer to Figure 3.3. For any energy $E_2$ that exceeds the barrier height there is a phase delay but no reflection in the WKB approximation. On the other hand, for energies such as $E_1$, below the barrier height, one must distinguish, as before, three regions. Assuming that the particle impinges on the barrier from the left, there is only a transmitted wave in region III, and

$$\psi(x) = \alpha \begin{cases} e^{\frac{1}{\hbar}\int_A^B dx'\,|p(x')|}\, \dfrac{e^{i\frac{1}{\hbar}\int_x^A dx'\,|p(x')|} + e^{-i\frac{1}{\hbar}\int_x^A dx'\,|p(x')|}}{\sqrt{p(x)}} & (x < A), \\ \dfrac{e^{\frac{1}{\hbar}\int_x^B |p(x')|}}{\sqrt{|p(x)|}} & (A < x < B), \\ e^{\frac{i\pi}{4}}\, \dfrac{e^{\frac{i}{\hbar}\int_B^x dx'\,p(x')}}{\sqrt{p(x)}} & (x > B). \end{cases} \tag{3.265}$$

This means that the WKB approximation predicts total reflection, up to exponentially small terms. The transmission coefficient, $T$, can be deduced by comparing the amplitudes of transmitted and incident waves, and one thus recovers the famous result

$$|T|^2 = e^{-2\frac{1}{\hbar}\int_A^B dx'\,|p(x')|}. \tag{3.266}$$

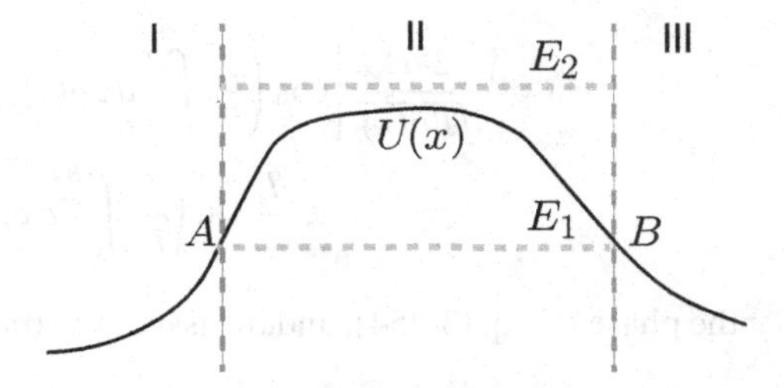


Figure 3.3 Barrier penetration. The WKB approximation yields no reflection for $E = E_2$, and essentially total reflection for $E = E_1$, up to an exponentially small transmission amplitude.

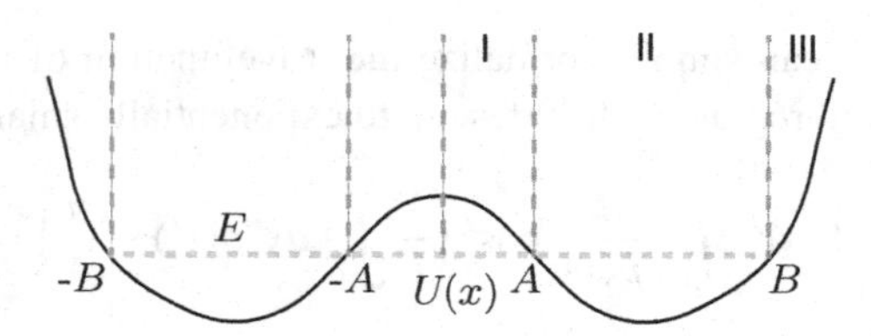


Figure 3.4 Tunneling in the central region of a symmetric double well gives rise to an energy separation between the levels associated to symmetric and antisymmetric wavefunctions.

### 3.7.4 The Symmetric Double Well

Our next example is a symmetric double well. Referring to Figure 3.4, one can work with wavefunctions of definite symmetry under $x \to -x$, since the one-dimensional Schrödinger operator has even parity. Therefore, one can begin in region I with the WKB wavefunctions

$$\psi_{\rm I} \;=\; \frac{\alpha}{\sqrt{|p(x)|}}\left(e^{\frac{1}{\hbar}\int_0^x dx\,|p(x)|} \;\pm\; e^{-\frac{1}{\hbar}\int_0^x dx\,|p(x)|}\right), \tag{3.267}$$

and letting

$$\sqrt{T} \;=\; e^{-\frac{1}{\hbar}\int_0^A dx\,|p(x)|}, \tag{3.268}$$

one can recast $\psi_{\rm I}$ the form

$$\psi_{\rm I} \;=\; \frac{\alpha}{\sqrt{|p(x)|}}\left(\frac{1}{\sqrt{T}}\,e^{-\frac{1}{\hbar}\int_x^A dx\,|p(x)|} \;\pm\; \sqrt{T}\,e^{\frac{1}{\hbar}\int_x^A dx\,|p(x)|}\right), \tag{3.269}$$

as needed for the continuation beyond the turning point at $A$. The connection formulas now imply that

$$\psi_{\rm II} = \frac{2\,\alpha}{\sqrt{T p(x)}}\left[\cos\left(\frac{1}{\hbar}\int_A^x dx\,p(x) - \frac{\pi}{4}\right)\right.$$
$$\left.\mp\,\frac{T}{2}\,\sin\left(\frac{1}{\hbar}\int_A^x dx\,p(x) - \frac{\pi}{4}\right)\right], \tag{3.270}$$

and one more step turns $\psi_{\rm II}$ into an expression that can be directly continued to region III, making use of Eq. (3.254). To begin with, one can rewrite the preceding equation as

$$\psi_{\rm II} = \frac{2\,\alpha}{\sqrt{T p(x)}}\left[\cos\left(\frac{1}{\hbar}\int_x^B dx\,p(x) - \frac{\pi}{4} - \gamma\right)\right.$$
$$\left.\pm\,\frac{T}{2}\,\sin\left(\frac{1}{\hbar}\int_x^B dx\,p(x) - \frac{\pi}{4} - \gamma\right)\right], \tag{3.271}$$

with $\gamma$ the phase of Eq. (3.254), and the issue is to track the coefficient of the function

$$\sin\left(\frac{1}{\hbar}\int_x^B dx\,p(x) \;-\; \frac{\pi}{4}\right), \tag{3.272}$$

which cannot be present for a bound state, since it would translate into a growing mode in the classically forbidden region beyond $B$. This leads to the condition

$$\sin\gamma \pm \frac{T}{2}\cos\gamma \simeq \sin\left(\gamma \pm \frac{T}{2}\right) = 0, \tag{3.273}$$

and thus to

$$\frac{1}{\hbar}\int_A^B dx\,\sqrt{2m(E-U(x))} = \left(n+\frac{1}{2}\right)\pi \mp \frac{T}{2}. \tag{3.274}$$

This modified form of Eq. (3.256) embodies small spectral distortions, so that its solutions are of the form

$$E_{\pm} = E_0 \mp \frac{\Delta}{2}, \tag{3.275}$$

with $E_0$ determined by the standard Bohr–Sommerfeld condition involving the first term above. Consequently, to first order in $\Delta$,

$$\int_A^B dx\sqrt{2m(E_{\mp}-U(x))} = \int_A^B dx\sqrt{2m(E_0-U(x))} \mp \frac{\Delta}{2}\int_A^B \frac{dx}{v}, \tag{3.276}$$

and as we have seen the second integral yields the ratio $\frac{\pi}{\omega(E_0)}$, so that finally

$$\Delta = \frac{\hbar\,\omega(E_0)}{\pi}\,T \tag{3.277}$$

is the level separation. Tunneling thus *lowers the even levels* by $\frac{\Delta}{2}$ and *raises the odd levels* by the same amount. Note, in particular, that the ground-state wavefunction is even, so that the ground-state energy decreases, as pertains to the larger region that is effectively available.

However, close to the minima of a smooth potential such as

$$U(x) = \lambda\left(x^2 - \eta^2\right)^2, \tag{3.278}$$

this result needs to be amended, since the linear approximation and the connection formulas based on Airy functions are not sufficiently accurate there. This potential departs more slowly, quadratically, from the minima in the two wells, and this introduces an important amplification factor, so that

$$\Delta = \frac{\hbar\,\omega}{\pi}\,T\left(\frac{2\pi m^2\omega^3}{\hbar\,\lambda}\right)^{\frac{1}{2}}, \tag{3.279}$$

where $\omega$ is determined by the quadratic approximation to Eq. (3.278) near the bottom of the wells. We shall derive this result in Appendix C.

### 3.7.5 Energy Bands in a One-Dimensional Crystal

Our next example is a one-dimensional model for a crystal, where electrons experience the periodic potential shown in Figure 3.5, together with small mutual interactions. The key idea behind the solution of this problem is that $|\psi|^2$ should be naturally periodic, which requires that $\psi$ be periodic up to a phase factor. This result is usually referred to as Bloch's theorem, according to which

$$\psi(x+a) = e^{\frac{i}{\hbar}pa}\,\psi(x), \tag{3.280}$$

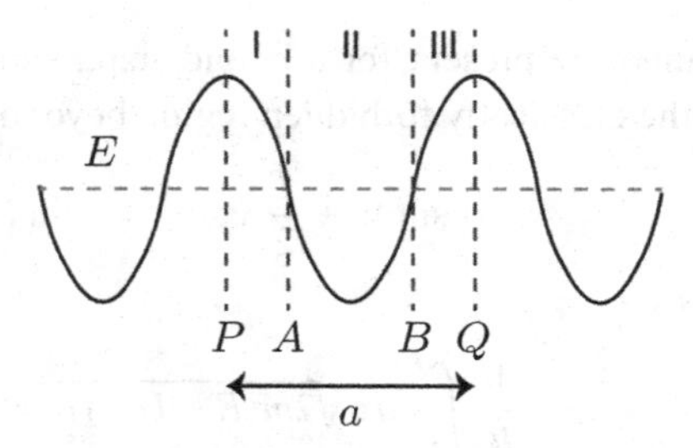


Figure 3.5 Tunneling in a periodic potential splits the energy levels, giving rise to allowed and forbidden bands, with crucial implications for the thermal and electrical behavior of solids.

where $a$ is the "lattice spacing" and $p$ is usually called "quasi-momentum." As usual, we proceed from one region to the next resorting to Eqs. (3.249) and (3.250). We start in region I with a generic wavefunction

$$\psi_{\rm I} = \frac{1}{\sqrt{|p(x)|}} \left( \alpha\, e^{\frac{1}{\hbar}\int_P^x dx'\,|p(x')|} + \beta\, e^{-\frac{1}{\hbar}\int_P^x dx'\,|p(x')|} \right), \tag{3.281}$$

which is readily cast in the form suitable for continuation to region II,

$$\psi_{\rm I} = \frac{1}{\sqrt{|p(x)|}} \left( \frac{\alpha}{\sqrt{T}}\, e^{-\frac{1}{\hbar}\int_x^A dx'\,|p(x')|} + \beta\, \sqrt{T}\, e^{\frac{1}{\hbar}\int_x^A dx'\,|p(x')|} \right), \tag{3.282}$$

where $T$ denotes, as usual, the barrier penetration factor. Consequently,

$$\psi_{\rm II} = \frac{2}{\sqrt{T p(x)}} \left[ \alpha \cos\left( \frac{1}{\hbar} \int_A^x dx'\, p(x') - \frac{\pi}{4} \right) - \frac{\beta\, T}{2} \sin\left( \frac{1}{\hbar} \int_A^x dx'\, p(x') - \frac{\pi}{4} \right) \right]. \tag{3.283}$$

One can now convert this expression into a form suitable for continuation beyond $B$,

$$\psi_{\rm II} = \frac{2}{\sqrt{T p(x)}} \left[ \cos\left( \frac{1}{\hbar} \int_x^B dx'\, p(x') - \frac{\pi}{4} \right) \left( \alpha \cos\gamma - \frac{\beta\, T}{2} \sin\gamma \right) + \sin\left( \frac{1}{\hbar} \int_x^B dx'\, p(x') - \frac{\pi}{4} \right) \left( \alpha \sin\gamma + \frac{\beta\, T}{2} \cos\gamma \right) \right], \tag{3.284}$$

with $\gamma$ the usual phase of Eq. (3.254), and this result implies that

$$\psi_{\rm III} = \frac{1}{\sqrt{|p(x)|}} \left[ \left( \alpha \cos\gamma - \frac{\beta\, T}{2} \sin\gamma \right) e^{-\frac{1}{\hbar}\int_Q^x dx'\,|p(x')|} - e^{\frac{1}{\hbar}\int_Q^x dx'\,|p(x')|} \left( \frac{2\,\alpha}{T} \sin\gamma + \beta \cos\gamma \right) \right]. \tag{3.285}$$

This final expression is to be compared with Eq. (3.281), making use of Eq. (3.280), which leads to the homogeneous system

$$\begin{aligned} \alpha \left( \sin\gamma + \frac{T}{2}\, e^{\frac{i}{\hbar} p a} \right) + \beta\, \frac{T}{2} \cos\gamma &= 0, \\ -\alpha \cos\gamma + \beta \left( e^{\frac{i}{\hbar} p a} + \frac{T}{2} \sin\gamma \right) &= 0. \end{aligned} \tag{3.286}$$

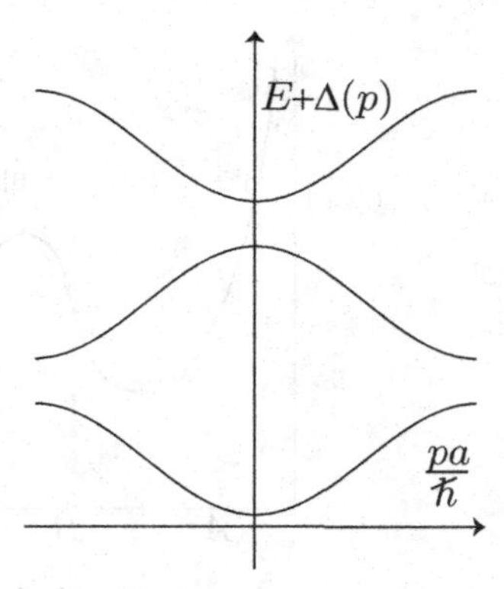


Figure 3.6 Some energy bands of Eq. (3.291).

This linear system can have nontrivial solutions only if

$$\left(\sin\gamma + \frac{T}{2} e^{\frac{i}{\hbar}pa}\right)\left(e^{\frac{i}{\hbar}pa} + \frac{T}{2}\sin\gamma\right) + \frac{T}{2}\cos^2\gamma = 0, \tag{3.287}$$

and keeping terms up to the first order in $T$ leads to the eigenvalue equation

$$T\cos\left(\frac{pa}{\hbar}\right) + \sin\gamma \simeq 0. \tag{3.288}$$

One can solve this equation by recalling that, from the double-well example, the levels of the individual wells ought to split as a result of tunneling. Moreover, as in the previous section (see, in particular, Eq. (3.276)), if $E_n \to E_n + \Delta_n$,

$$\gamma \to n\pi + \frac{\pi\Delta_n}{\hbar\omega}, \tag{3.289}$$

and therefore

$$\sin\left(n\pi + \frac{\pi\Delta}{\omega}\right) \simeq (-1)^n \frac{\pi\Delta_n}{\hbar\omega}, \tag{3.290}$$

so that substituting in Eq. (3.288) gives

$$\Delta_n = (-1)^{n-1}\frac{\hbar\omega}{\pi} T\cos\left(\frac{pa}{\hbar}\right). \tag{3.291}$$

These trigonometric profiles are illustrated in Figure 3.6, and embody a number of facts that are highly relevant in condensed-matter physics, including the concept of "effective mass," defined via the second derivative of the band profiles, which changes sign within a band, giving rise to an exotic behavior for charge carriers.

## 3.7.6 Decay of a Metastable State

As a last example, and referring to Figure 3.7, let us discuss how to compute the lifetime of a metastable state. Now

$$\psi_{\mathrm{I}} = \frac{\alpha}{\sqrt{|p(x)|}} e^{-\frac{1}{\hbar}\int_x^A dx'\,|p(x')|},$$

$$\psi_{\mathrm{II}} = \frac{2\alpha}{\sqrt{p(x)}}\cos\left(\frac{1}{\hbar}\int_A^x dx'\,p(x') - \frac{\pi}{4}\right), \tag{3.292}$$

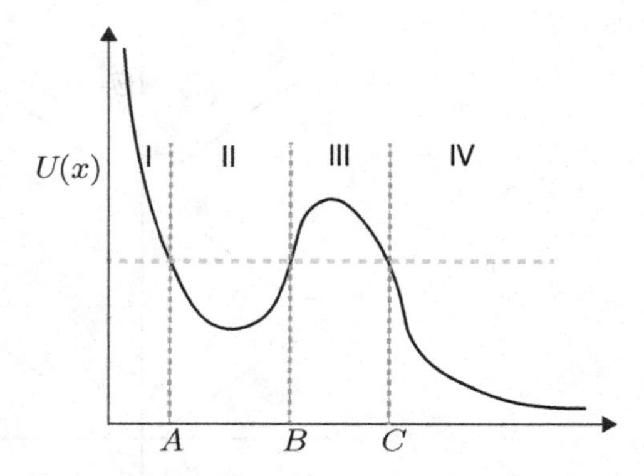


Figure 3.7 The decay of a metastable state can be studied by referring to the potential in the figure and demanding that only outgoing waves be present in the right-most region.

or

$$\psi_{\rm II} = \frac{2\,\alpha}{\sqrt{p(x)}} \left[ \cos\gamma\, \cos\left(\frac{1}{\hbar}\int_x^B dx'\, p(x') - \frac{\pi}{4}\right) + \sin\gamma\, \sin\left(\frac{1}{\hbar}\int_x^B dx'\, p(x') - \frac{\pi}{4}\right)\right], \tag{3.293}$$

and therefore

$$\psi_{\rm III} = \frac{\alpha}{\sqrt{|p(x)|}} \left[ \cos\gamma\; e^{-\frac{1}{\hbar}\int_B^x dx'\, |p(x')|} - 2\, \sin\gamma\; e^{\frac{1}{\hbar}\int_B^x dx'\, |p(x')|}\right], \tag{3.294}$$

or

$$\psi_{\rm III} = \frac{\alpha\, T}{\sqrt{|p(x)|}} \left[ \cos\gamma\; e^{\frac{1}{\hbar}\int_x^C dx'\, |p(x')|} - \frac{2}{T^2}\, \sin\gamma\; e^{-\frac{1}{\hbar}\int_x^C dx'\, |p(x')|}\right]. \tag{3.295}$$

From this last expression one can see that

$$\psi_{\rm IV} = -\frac{\alpha\, T}{\sqrt{p(x)}} \left[ \cos\gamma\, \sin\left(\frac{1}{\hbar}\int_C^x dx'\, p(x') - \frac{\pi}{4}\right) + \frac{4}{T^2}\, \sin\gamma\, \cos\left(\frac{1}{\hbar}\int_C^x dx'\, p(x') - \frac{\pi}{4}\right)\right], \tag{3.296}$$

so that the condition of having only a transmitted wave is

$$\sin\gamma + \frac{i\,T^2}{4}\, \cos\gamma = 0, \tag{3.297}$$

or

$$\gamma = n\,\pi - i\,\frac{T^2}{4}. \tag{3.298}$$

This result translates, as in the previous examples, into a shift of the energy levels

$$\Delta = -\,i\,\frac{T^2}{4}\,\frac{\hbar\,\omega}{\pi}, \tag{3.299}$$

but now it is imaginary. Reinserting the time-dependent factor $e^{-\frac{i}{\hbar}Et}$, we see that the probability of finding the particle in the well decays exponentially in time,

$$|\psi|^2 \sim e^{-\frac{t}{\tau}}, \quad \text{with} \quad \tau = \frac{2\pi}{\omega T^2}, \tag{3.300}$$

which is known as Rutherford's law.

## 3.8 Instantons

We can now reconsider from a functional integral perspective the tunneling effects that we have explored in the preceding sections. This approach has acquired a growing importance over the years, and the ensuing discussion will also provide further insights into the nature of the WKB approximation.

The barrier penetration factor (3.268) rests on the exponent

$$\frac{1}{\hbar}\int_{-A}^{A} dx\,\sqrt{2m(U(x)-E)}, \tag{3.301}$$

where the endpoints identify the classically forbidden region. On the other hand, classically allowed regions lead, as we have seen, to phase factors depending on

$$\frac{1}{\hbar}\int_{x_0}^{x} dx'\,\sqrt{2m(E-U(x'))}, \tag{3.302}$$

which reflect the energy conservation law

$$E = \frac{p^2}{2m} + U(x) \tag{3.303}$$

related to the action

$$\mathcal{S} = \int dt\left[\frac{1}{2}m\left(\frac{dx}{dt}\right)^2 - U(x)\right]. \tag{3.304}$$

It is natural, therefore, to associate eq. (3.301) to the Euclidean action

$$\mathcal{S}_E = \int dt\left[\frac{1}{2}m\left(\frac{dx}{dt}\right)^2 + U(x)\right], \tag{3.305}$$

which is formally identical, for these one-dimensional problems, to Eq. (3.304) with an inverted potential. What we did in previous sections with Airy functions should suggest that the functional counterpart of the WKB approximation rests, in general, on paths in the complex $t$-plane. The references at the end of the chapter discuss detailed developments along this rich research line. Here we shall confine our attention to a relatively simple and well-established case: tunneling effects at the bottom of wells, which can be captured expanding the Euclidean functional integral around *real solutions of the Euclidean equations of motion*. The detailed comparison of the results with the Hamiltonian method, however, will require some additional subtleties discussed in Appendix C, where the connection formulas will be adapted to this case.

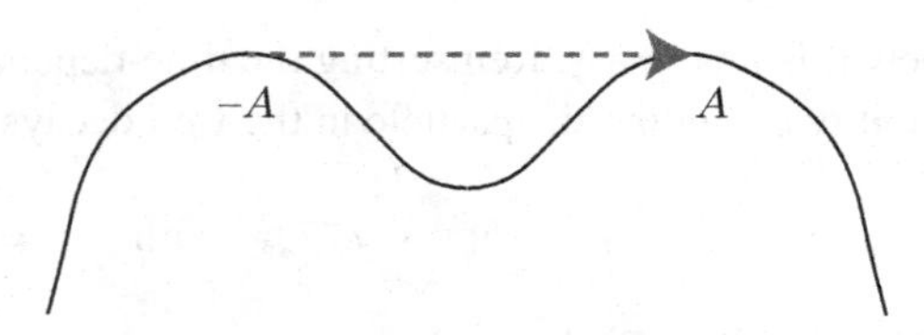


Figure 3.8 The instanton discussed in Section 3.8 is a *real* solution of the Euclidean action that leaves, with infinitesimal speed, the peak at $x = -\eta$ and reaches after an infinite amount of Euclidean time the peak at $x = \eta$. In a similar fashion an anti-instanton would leave $x = \eta$ to reach $x = -\eta$ after an infinite amount of Euclidean time.

In the following, it will be convenient to refer to an explicit form of the double-well potential,

$$U(x) = \lambda \left(\eta^2 - x^2\right)^2 . \tag{3.306}$$

The classical solutions of interest start from one peak of the inverted potential and reach the other peak after an infinite amount of Euclidean time, as shown in Figure 3.8. In this case the Euclidean energy conservation equation reduces to

$$0 = \frac{m}{2}\left(\frac{dx}{dt}\right)^2 - U(x), \tag{3.307}$$

and can be solved using separation of variables, with the end result

$$x_c(t) = \eta \tanh\left[\frac{\omega}{2}(t - t_0)\right]. \tag{3.308}$$

Here

$$\omega^2 = \frac{8\,\lambda\,\eta^2}{m}, \tag{3.309}$$

and the reader should note that $\omega$ is also the frequency of small oscillations at the bottom of the wells. The "instanton" solution in Eq. (3.308) is essentially a step function at $t = t_0$, since $x$ starts at $-\eta$ and then makes a quick transition toward $\eta$ within a time interval around $t_0$ whose width is determined by $\omega$, as in Figure 3.9. There is also an "anti-instanton" solution, simply obtained from Eq. (3.308) by flipping the overall coefficient, which describes a classical Euclidean motion from $\eta$ to $-\eta$.

The Euclidean action computed for the instanton provides the barrier penetration factor $T$ of Eq. (3.268) at the bottom of the well, since for $E = 0$,

$$T = e^{-\frac{1}{\hbar}\mathcal{S}_0}, \qquad \mathcal{S}_0 = \int_{-\eta}^{\eta}\sqrt{2m\,U(x)}\,dx = \int_{-\infty}^{+\infty} m\left(\frac{dx_c}{dt}\right)^2 dt, \tag{3.310}$$

and the Euclidean action for the instanton, as expected, is finite:

$$\mathcal{S}_0 = \frac{m\omega^2\eta^2}{2}\int_{-\infty}^{\infty}\frac{d\tau}{\cosh^4\tau} = \frac{2}{3}\,m\,\omega\,\eta^2. \tag{3.311}$$

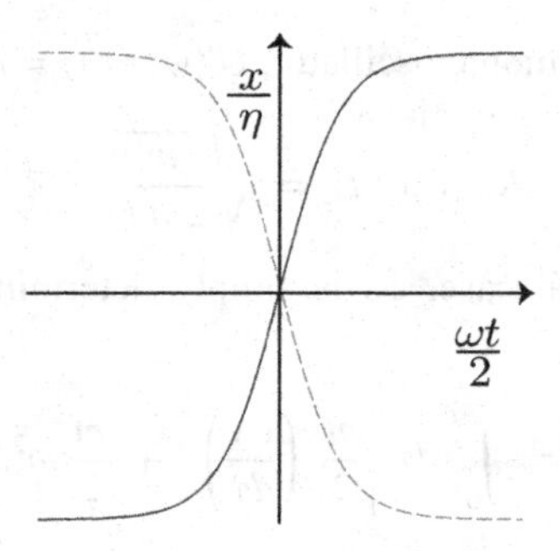


Figure 3.9 Evolution of (solid line) an instanton and (dotted line) an anti-instanton in Euclidean time.

Our next task is to obtain, using these techniques, a WKB estimate of the corrected ground-state energy for the double well: its main ingredients are the two Euclidean evolution kernels:

$$\langle -\eta| e^{-\frac{Ht}{\hbar}} | -\eta\rangle, \qquad \langle \eta| e^{-\frac{Ht}{\hbar}} | -\eta\rangle. \tag{3.312}$$

Let us begin from the first evolution kernel, which is closely linked to the harmonic oscillator, and is simpler. In general, when computing Euclidean kernels,

$$K_E(x,x';t) = \int [\mathcal{D}\mathbf{x}(\tau)]\, e^{-\frac{1}{\hbar}\mathcal{S}_E[x(\tau)]}, \tag{3.313}$$

it is convenient to change variable according to

$$x(\tau) = x_c(\tau) + \xi(\tau), \tag{3.314}$$

where $x_c(\tau)$ is the classical trajectory such that $x_c(0) = x'$ and $x_c(t) = x$, so that $\xi(0) = \xi(t) = 0$. Then

$$\mathcal{S}_E[x_x + \xi] = \mathcal{S}_0 + \mathcal{S}_E^{(2)}[x_c, \xi] + \mathcal{O}\left(\xi^3\right), \tag{3.315}$$

where we have denoted by $\mathcal{S}_0$ the contribution of the classical trajectory, and

$$\mathcal{S}_E^{(2)}[x_c, \xi] = \frac{m}{2}\int_0^t d\tau\, \xi(\tau)\left(-\frac{d^2}{d\tau^2} + \frac{1}{m}\, U''(x_c(\tau))\right)\xi(\tau). \tag{3.316}$$

For any quadratic action, there are no $\mathcal{O}(\xi^3)$ terms in Eq. (3.315), and

$$K_E(x,x';t) = e^{-\frac{\mathcal{S}_0}{\hbar}} \int [\mathcal{D}_{\xi}(\tau)]\, e^{-\frac{1}{\hbar}\mathcal{S}_E^{(2)}[x_c,\xi]}. \tag{3.317}$$

In the general case, this decomposition is still useful as an approximation, since the first two terms in the expansion in $\xi$ capture the two leading WKB contributions, while higher-order ones build up a series of corrections in powers of $\hbar$. Then, the detailed definition of the functional integral in Eq. (3.102) leads to the WKB formula

$$K_E(x,x';t) \simeq \sqrt{\frac{m}{2\,\pi\,\hbar}}\, e^{-\frac{\mathcal{S}_0}{\hbar}} \left[\det\left(-\frac{d^2}{dt^2} + \frac{1}{m}\, U''(x_c(\tau))\right)\right]^{-\frac{1}{2}}, \tag{3.318}$$

which we have associated to a determinant, via a natural infinite-dimensional generalization of Eq. (2.22). We shall return shortly to characterize operator determinants of this type more precisely.

For the harmonic oscillator, $U''(x_c(\tau)) = m\,\omega^2$, and one finds

$$K_E(x, x'; t) = \sqrt{\frac{m}{2\,\pi\,\hbar}}\, e^{-\frac{\mathcal{S}_0}{\hbar}} \left[\det\left(-\frac{d^2}{dt^2} + \omega^2\right)\right]^{-\frac{1}{2}}. \tag{3.319}$$

In this special case, $\mathcal{S}_0$ is simply determined by $x$ and its first derivative at the endpoints,

$$\mathcal{S}_0 = \int_0^t d\tau \left[\frac{m}{2}\left(\frac{dx}{d\tau}\right)^2 + \frac{m}{2}\,\omega^2\,x^2\right] = \left[\frac{m}{2}\,x(\tau)\,\frac{dx}{d\tau}\right]_0^t, \tag{3.320}$$

as can be seen by making use of the classical equation of motion, while the classical solution of interest reads

$$x(\tau) = x' \cosh \omega\tau + \left(\frac{x - x' \cosh \omega t}{\sinh \omega t}\right) \sinh \omega\tau, \tag{3.321}$$

so that finally

$$\frac{\mathcal{S}_0}{\hbar} = \frac{m\omega}{2\hbar \sinh \omega t}\left[\left(x^2 + x'^2\right)\cosh \omega t - 2xx'\right]. \tag{3.322}$$

### 3.8.1 The Gel'fand–Yaglom Equation

Let us now turn to the determinant in Eq. (3.318), which we can relate to the elegant Gel'fand–Yaglom equation. To this end, let us first recall the discretized functional integral over the fluctuations in Eq (3.102), which we adapt to Eq. (3.317):

$$\int_{-\infty}^{\infty} \left[\prod_{i=1}^{N} d\xi_i\right] \left(\frac{m}{2\pi\hbar\,\epsilon}\right)^{\frac{N+1}{2}} \prod_{i=0}^{N} e^{-\frac{\epsilon}{\hbar}\left[\frac{m(\xi_{i+1}-\xi_i)^2}{2\,\epsilon^2} + \frac{1}{2}\,U''(x_{c,i})\,\xi_i^2\right]}. \tag{3.323}$$

Here, $y_0 = y_{N+1} = 0$, and a factor $\frac{1}{\sqrt{\epsilon}} = \sqrt{\frac{N+1}{t}}$ is to be associated to all the $N+1$ kernels that grant infinitesimal time steps in the path integral. Hence, using Eq. (2.22), one can verify that Eq. (3.323) takes the form

$$\sqrt{\frac{m}{2\,\pi\,\hbar}} \lim_{N\to\infty, \epsilon\to 0} \left[D_N\right]^{-\frac{1}{2}}, \tag{3.324}$$

where

$$D_N = \epsilon \det \begin{pmatrix} \epsilon^2\,\omega_N^2 + 2 & -1 & 0 & \dots & 0 \\ -1 & \epsilon^2\,\omega_{N-1}^2 + 2 & -1 & \dots & 0 \\ 0 & -1 & \epsilon^2\,\omega_{N-2}^2 + 2 & \dots & 0 \\ \vdots & \vdots & \vdots & \ddots & \end{pmatrix}, \tag{3.325}$$

with Dirichlet boundary conditions, as pertains to $\xi(\tau)$, and with

$$\omega_i^2 = \frac{1}{m}\,U''(x_{c,i}), \tag{3.326}$$

since with discrete time intervals

$$\frac{d^2}{dt^2} f(t) \to \frac{f(t+\epsilon) + f(t-\epsilon) - 2f(t)}{\epsilon^2}. \tag{3.327}$$

Note that the presence of $N+1$ kernels and only $N$ integrations implies that the proper definition of $D_N$ includes an overall factor of $\epsilon$.

Expanding the determinant $D_N$ from the first row, one can now obtain the recursion relation

$$D_N = \left(\epsilon^2 \omega_N^2 + 2\right) D_{N-1} - D_{N-2}, \tag{3.328}$$

between determinants of different orders, or

$$\frac{D_N + D_{N-2} - 2 D_{N-1}}{\epsilon^2} = \omega_N^2 D_{N-1}. \tag{3.329}$$

In the continuum limit this becomes the differential equation

$$\frac{d^2 D(\tau)}{d\tau^2} - \omega^2(\tau) D(\tau) = 0, \tag{3.330}$$

and moreover

$$\begin{aligned} D_1 &= \epsilon \left(\epsilon^2 \omega_1^2 + 2\right) \\ D_2 &= \epsilon \left[\left(\epsilon^2 \omega_2^2 + 2\right)\left(\epsilon^2 \omega_1^2 + 2\right) - 1\right], \end{aligned} \tag{3.331}$$

so that

$$D_2 - D_1 = \epsilon \left[1 + \epsilon^2 \omega_1^2 + 2\,\epsilon^2 \omega_2^2 + \epsilon^2 \omega_1^2 \omega_2^2\right]. \tag{3.332}$$

In the continuum limit, one is thus left with the initial conditions

$$D(0) = 0, \qquad \left.\frac{dD}{dt}\right|_0 = 1. \tag{3.333}$$

In particular, solving Eq. (3.330) for the harmonic oscillator with these initial conditions gives

$$D(t) = \frac{\sinh \omega t}{\omega} \sim_{t\to+\infty} \frac{1}{2\omega}\, e^{\omega t}. \tag{3.334}$$

One can be more explicit in presenting the relevant solution of Eq. (3.330) with the proper initial conditions in the general case. Indeed, for any two independent solutions $D_1(\tau)$ and $D_2(\tau)$ of Eq. (3.330), the Wronskian,

$$D_1(\tau)\frac{dD_2(\tau)}{d\tau} - D_2(\tau)\frac{dD_1(\tau)}{d\tau} = 1, \tag{3.335}$$

is a constant, here set to one. This equation determines the time derivative of the ratio of $D_2$ and $D_1$, and therefore, if $D_1(\tau)$ is somehow known, a second independent solution is

$$D_2(t) = C D_1(t) + \int_{t_0}^{t} d\tau \, \frac{D_1(t)}{D_1^2(\tau)}, \tag{3.336}$$

where $C$ is an arbitrary integration constant. One can now simply verify that

$$D(t) = D_1(0)\, D_1(t) \int_0^t \frac{d\tau}{D_1^2(\tau)} \tag{3.337}$$

solves the Gel'fand–Yaglom equation and satisfies the initial conditions (3.333). Therefore, it is the determinant of the corresponding differential operator.

### 3.8.2 Kernels for the Double Well

Returning to our original problem, to leading order one can write

$$K_E(-\eta,-\eta;t) = K_E(\eta,\eta;t) = \sqrt{\frac{m\omega}{2\pi\hbar\sinh\omega t}} \sim \sqrt{\frac{m\omega}{\pi\hbar}}\, e^{-\frac{\omega t}{2}}, \tag{3.338}$$

since in this case $x_c = \mp\eta$ and the result is dominated by harmonic oscillations around the two minima. Comparing this expression with Eq. (3.93), one can simply deduce what the ground state energy and the normalization factor of the ground-state wavefunction would be in the absence of tunneling.

Naively, one could also conclude that

$$K_E(\eta,-\eta;t) = \sqrt{\frac{m}{2\pi\hbar}}\, e^{-\frac{1}{\hbar}\mathcal{S}_0}\left[\det\left(-\frac{d^2}{dt^2}+\frac{1}{m}U''\left(x_c(t)\right)\right)\right]^{-\frac{1}{2}}, \tag{3.339}$$

where $\mathcal{S}_0$ is given in Eq. (3.311), since we shall shortly identify $x_c(t)$ with the instanton solution. However, this expression is not correct, since the operator has a "zero mode," an eigenfunction corresponding to a vanishing eigenvalue, which will prove instrumental to endow $K_E(\eta,-\eta;T)$ with a proper dependence on the (large) transition time, as in Eq. (3.338). Indeed, the Euclidean action (3.305) implies that

$$m\frac{d^2x_c}{dt^2} - U'(x_c(t)) = 0, \tag{3.340}$$

and differentiating this equation with respect to $t$ leads to

$$m\frac{d^2}{dt^2}\left(\frac{dx_c}{dt}\right) - U''(x_c(t))\frac{dx_c}{dt} = 0, \tag{3.341}$$

which proves that the time derivative of the instanton solution is a zero mode of the differential operator in Eq. (3.339), which also satisfies the proper boundary conditions since it approaches zero as $t \to \pm\infty$.

In order to deal with the zero mode, one can pretend momentarily that all eigenvalues are nonzero, thus writing for the Gel'fand–Yaglom determinant:

$$D_N^{-\frac{1}{2}} = \frac{1}{\sqrt{\epsilon}}\prod_n \frac{1}{\sqrt{\epsilon^2\,\lambda_n}} = \frac{1}{\sqrt{\epsilon}}\int\prod_n\left(dc_n\sqrt{\frac{m}{2\pi\hbar\,\epsilon^2}}\right)e^{-\frac{m}{2\hbar}\sum_n c_n^2\lambda_n}, \tag{3.342}$$

where $\epsilon^2\lambda_n$ are the eigenvalues of the matrix entering Eq. (3.325), without the overall $\epsilon$. One can now change variable from $c_0$, the zero-mode coefficient, to $t_0$, the transition time of the instanton, which is essentially a step function at $t_0$, according to

$$\delta\xi = \left|\psi_0\right| dc_0 = \left|\frac{dx_c}{dt_0}\right| dt_0, \tag{3.343}$$

where $\psi_0$ is the normalized zero-mode for the fluctuations around the instanton. Moreover, Eq. (3.305) determines the normalization of $\psi_0$, and therefore

$$\psi_0 = \sqrt{\frac{m}{\mathcal{S}_0}}\frac{dx_c}{dt_0}, \tag{3.344}$$

so that finally

$$dc_0 \to \sqrt{\frac{\mathcal{S}_0}{m}}\, dt_0. \tag{3.345}$$

One can thus write, integrating over $t_0$ in a large interval $T$ and comparing with the large-$T$ oscillator kernel of Eq. (3.338),

$$K_E(\eta, -\eta; T) = \sqrt{\frac{m\omega}{\pi\hbar}}\, e^{-\frac{\omega T}{2}}\, \mathcal{K}\, T\, e^{-\frac{1}{\hbar}S_0}, \tag{3.346}$$

where

$$\mathcal{K} = \sqrt{\frac{S_0}{2\pi\hbar}} \left[ \frac{\det\left(-\frac{d^2}{dt^2} + \omega^2\right)}{\det'\left[-\frac{d^2}{dt^2} + \frac{1}{m}\, U''\left(x_c(t)\right)\right]} \right]^{\frac{1}{2}} \tag{3.347}$$

with det′ the determinant without $\lambda_0$ but containing all other factors. Note that an anti-instanton would lead to the same result, since $U(x)$ is an even function of $x$.

Leaving aside momentarily the actual value of $\mathcal{K}$, the step-like form of the instanton solution suggests to build ordered sums (which introduces a factor $\frac{1}{n!}$) with equal numbers of instantons and anti-instantons to get $K_E(-\eta, -\eta; T)$, and with one instanton in excess to get $K_E(\eta, -\eta; T)$. Hence

$$\begin{aligned} K_E(\eta, -\eta; T) &= \sqrt{\frac{m\omega}{\pi\hbar}}\, e^{-\frac{\omega T}{2}} \sum_{n \text{ odd}} \frac{\left(\mathcal{K}\, T e^{-\frac{1}{\hbar}S_0}\right)^n}{n!} \\ &= \frac{1}{2}\sqrt{\frac{m\omega}{\pi\hbar}} \left[ e^{-\frac{T}{\hbar}\left(\frac{\hbar\omega}{2} - \frac{\Delta}{2}\right)} - e^{-\frac{T}{\hbar}\left(\frac{\hbar\omega}{2} + \frac{\Delta}{2}\right)} \right], \end{aligned} \tag{3.348}$$

with

$$\Delta = 2\hbar\, \mathcal{K}\, e^{-\frac{1}{\hbar}S_0}, \tag{3.349}$$

after recognizing the Taylor series for a sinh function. The coefficients of the two resulting exponentials have a neat interpretation, as we have anticipated, in terms of the two naive ground-state wavefunctions localized around the bottoms of the two wells. In particular, the second, negative contribution, reflects the negative overlap between their odd combinations, here computed at $\pm\eta$, where they assume opposite values. In a similar fashion,

$$\begin{aligned} K_E(-\eta, -\eta; T) &= \sqrt{\frac{m\omega}{\pi\hbar}}\, e^{-\frac{\omega T}{2}} \sum_{n \text{ even}} \frac{\left(\mathcal{K}\, T e^{-\frac{1}{\hbar}S_0}\right)^n}{n!} \\ &= \frac{1}{2}\sqrt{\frac{m\omega}{\pi\hbar}} \left[ e^{-\frac{T}{\hbar}\left(\frac{\hbar\omega}{2} - \frac{\Delta}{2}\right)} + e^{-\frac{T}{\hbar}\left(\frac{\hbar\omega}{2} + \frac{\Delta}{2}\right)} \right], \end{aligned} \tag{3.350}$$

where the second term is properly positive, since now the overlap involves odd combinations computed at the same point.

All in all, these expressions reveal that, as a result of tunneling, the ground state of the double well splits into a pair of nearby energy levels, associated to symmetric and antisymmetric combinations of wavefunctions localized around the bottoms of the two wells. The corresponding energy eigenvalues are

$$E_{\pm} = \frac{\hbar\omega}{2} \mp \frac{\Delta}{2}, \tag{3.351}$$

where the upper sign applies to the symmetric combination, which is the actual ground state.

### 3.8.3 The Correction Factor $\mathcal{K}$

We can now conclude this discussion by determining the actual value of $\mathcal{K}$. To begin with, let us characterize the behavior of

$$D_0\left(\frac{T}{2}\right) = \det\left[-\frac{d^2}{dt^2} + \frac{1}{m}\,U''\,(x_c(t))\right]_{t=\frac{T}{2}} \tag{3.352}$$

for large values of $T$, starting from the Gel'fand–Yaglom initial conditions

$$D_0\left(-\frac{T}{2}\right) = 0, \quad D_0'\left(-\frac{T}{2}\right) = 1. \tag{3.353}$$

We know already a normalized solution of

$$\frac{d^2\xi}{dt^2} - \frac{1}{m}\,U''\,(x_c(t))\;\xi = 0 \tag{3.354}$$

is

$$\xi_1 = \sqrt{\frac{m}{\mathcal{S}_0}}\,\frac{\eta\,\omega}{2}\,\frac{1}{\cosh^2\left(\frac{\omega t}{2}\right)}, \tag{3.355}$$

which is *even* and whose asymptotic behavior for large $t$ is

$$\xi_1 \sim_{|t|\to\infty} \alpha\, e^{-\omega|t|}, \tag{3.356}$$

with

$$\alpha = \sqrt{\frac{m}{\mathcal{S}_0}}\,2\,\eta\,\omega. \tag{3.357}$$

Moreover, since $U''$ is even, as can be seen from Eq. (3.306), there is a second independent *odd* solution. Since their Wronskian

$$\mathrm{w} = \xi_1(t)\,\frac{d\xi_2(t)}{dt} - \xi_2(t)\,\frac{d\xi_1(t)}{dt} \tag{3.358}$$

is a constant for Eq. (3.354), one can also see that the second independent solution grows exponentially for large $|t|$. Hence, it can normalized so as to have the asymptotic behavior

$$\xi_2 \sim_{t\to\pm\infty} \pm\,\alpha\, e^{\omega|t|}, \tag{3.359}$$

and then

$$\mathrm{w} = 2\,\alpha^2\,\omega. \tag{3.360}$$

These results imply a corresponding expression for the determinant,

$$D_0(t) = \frac{1}{2\,\omega\,\alpha}\left(e^{\frac{\omega T}{2}}\,\xi_1(t) + e^{-\frac{\omega T}{2}}\,\xi_2(t)\right), \tag{3.361}$$

and consequently

$$D_0\left(\frac{T}{2}\right) \sim_{T\to\infty} \frac{1}{\omega}. \tag{3.362}$$

Note that the presence of a zero mode reflects itself in the lack of the exponential growth with $T$ present in the more conventional case of Eq. (3.334).

The next step needed to determine the correction factor $\mathcal{K}$ of Eq. (3.347) is the identification of the lowest eigenvalue, which is to be removed from $D_0(T)$. This eigenvalue is determined from a variant of Eq. (3.354),

$$-\frac{d^2Y}{dt^2} + \frac{1}{m}U''(x_c(t))\, Y = \lambda\, Y, \tag{3.363}$$

which can be conveniently formulated as the integral equation

$$Y(t) = D_0(t) - \lambda \int_{-\frac{T}{2}}^{\frac{T}{2}} d\tau\, G(t,\tau)\, Y(\tau), \tag{3.364}$$

where the retarded Green function $G(t,\tau)$ satisfies

$$\frac{d^2G(t,\tau)}{dt^2} - \frac{1}{m}U''(x_c(t))\, G(t,\tau) = \delta(t-\tau), \tag{3.365}$$

and $G(t,\tau) = 0$ for $t < \tau$. One can express $G$ in terms of $\xi_{1,2}$ and their Wronskian, according to

$$G(t,\tau) = \frac{\theta(t-\tau)}{\mathrm{w}}\big(\xi_2(t)\,\xi_1(\tau) - \xi_1(t)\,\xi_2(\tau)\big), \tag{3.366}$$

where $\theta$ is the Heaviside step function, usually defined as

$$\theta(x) = \begin{cases} 0 & x < 0 \\ \frac{1}{2} & x = 0 \\ 1 & x > 0. \end{cases} \tag{3.367}$$

In this fashion, $Y(t)$ satisfies the Gel'fand–Yaglom initial conditions, as pertains to the $\lambda = 0$ limit that one is exploring here, and to first order in $\lambda$,

$$Y\left(\frac{T}{2}\right) \simeq \frac{1}{\omega} - \frac{\lambda\,\alpha}{\mathrm{w}}\int_{-\frac{T}{2}}^{\frac{T}{2}} d\tau\,\left(e^{\omega\frac{T}{2}}\,\xi_1(\tau) - e^{-\omega\frac{T}{2}}\,\xi_2(\tau)\right) D_0(\tau), \tag{3.368}$$

and making use of Eq. (3.361),

$$Y\left(\frac{T}{2}\right) \simeq \frac{1}{\omega} - \frac{\lambda}{2\,\omega\,\mathrm{w}}\int_{-\frac{T}{2}}^{\frac{T}{2}} d\tau\,\left(e^{\omega T}\,\xi_1^2(\tau) - e^{-\omega T}\,\xi_2^2(\tau)\right). \tag{3.369}$$

The first term dominates and $\xi_1$ is normalized so that, finally, demanding that $Y\left(\frac{T}{2}\right) \simeq 0$ for large values of $T$, as pertains to a zero mode,

$$\lambda \simeq 2\,\mathrm{w}\, e^{-\omega T} = 4\,\omega\,\alpha^2\, e^{-\omega T}, \tag{3.370}$$

where we have used Eq. (3.358).

This result and Eq. (3.362) translate into

$$\det{}'\left[-\frac{d^2}{dt^2} + \frac{1}{m}U''(x_c(t))\right] = \frac{1}{4\,\omega^2\,\alpha^2}\, e^{\omega T}, \tag{3.371}$$

and using Eq. (3.357) finally leads to

$$\mathcal{K} = \sqrt{\frac{S_0}{2\pi\hbar}}\,\sqrt{2\,\omega}\,\alpha = 2\sqrt{\frac{\eta^2\,\omega^3\,m}{\pi\,\hbar}}. \tag{3.372}$$

The final form for the level splitting of Eq. (3.349) is therefore

$$\Delta = \frac{\hbar\,\omega}{\pi} e^{-\frac{1}{\hbar} S_0} \sqrt{\frac{16\,\pi\,\omega\,m\,\eta^2}{\hbar}}, \tag{3.373}$$

or

$$\Delta = \frac{\hbar\,\omega}{\pi} e^{-\frac{1}{\hbar} S_0} \sqrt{\frac{2\,\pi\,\omega^3\,m^2}{\hbar\,\lambda}}, \tag{3.374}$$

after using Eq. (3.309). This is the result that was announced in Eq. (3.279), and contains an amplification factor with respect to Eq. (3.277) that reflects the slow growth of the barrier near the bottom of the wells. A detailed derivation of the quadratic connection formulas that are needed to obtain this result in the conventional WKB approach is presented in Appendix C.

### 3.8.4 Energy Bands and Metastable States

In a periodic potential instantons (and anti-instantons) can join any pair of nearby peaks of the inverted potential, as shown in Figure 3.10, so that one is generally interested in transition amplitudes between arbitrary pairs of wells,

$$\langle k | e^{-\frac{1}{\hbar} HT} | j\rangle = \sqrt{\frac{m\omega}{\pi\hbar}}\, e^{-\frac{\omega T}{2}} \sum_{n,\overline{n}} \frac{\left(\mathcal{K} T e^{-\frac{1}{\hbar} S_0}\right)^{n+\overline{n}}}{n!\ \overline{n}!}\, \delta_{n-\overline{n}+j-k}. \tag{3.375}$$

The integral representation

$$\delta_{n-\overline{n}+j-k} = \int_{-\pi}^{\pi} \frac{d\theta}{2\pi}\, e^{i\,\theta(n-\overline{n}+j-k)} \tag{3.376}$$

can disentangle the two sums, leading finally to

$$\langle k | e^{-\frac{1}{\hbar} HT} | j\rangle = \sqrt{\frac{m\omega}{\pi\hbar}} \int_{-\pi}^{\pi} \frac{d\theta}{2\pi}\, e^{i\,\theta(j-k) - \frac{1}{\hbar} E(\theta)\, T}. \tag{3.377}$$

Here,

$$E(\theta) = \frac{\hbar\,\omega}{2} - 2\,\mathcal{K}\,\hbar\, e^{-\frac{1}{\hbar} S_0}\, \cos\theta, \tag{3.378}$$

so that, using the explicit expression for the coefficient in the last term given in Eq. (3.374),

$$E(\theta) = \frac{\hbar\,\omega}{2} - \frac{\hbar\,\omega}{\pi}\, e^{-\frac{1}{\hbar} S_0} \sqrt{\frac{2\,\pi\,\omega^3\,m^2}{\hbar\,\lambda}}\, \cos\theta. \tag{3.379}$$

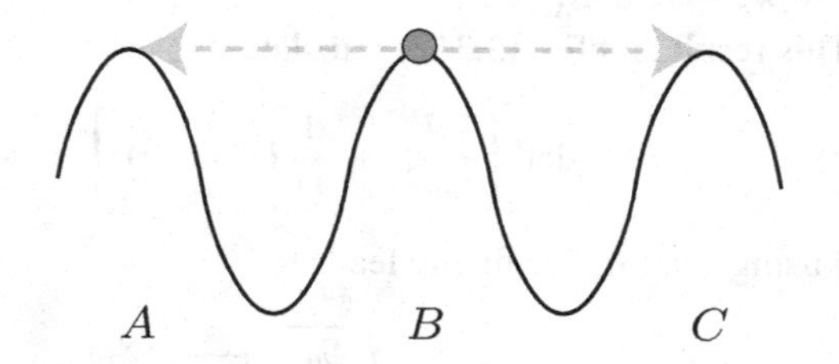


Figure 3.10 In the periodic case (anti)instantons can join any two nearby peaks of the inverted potential.

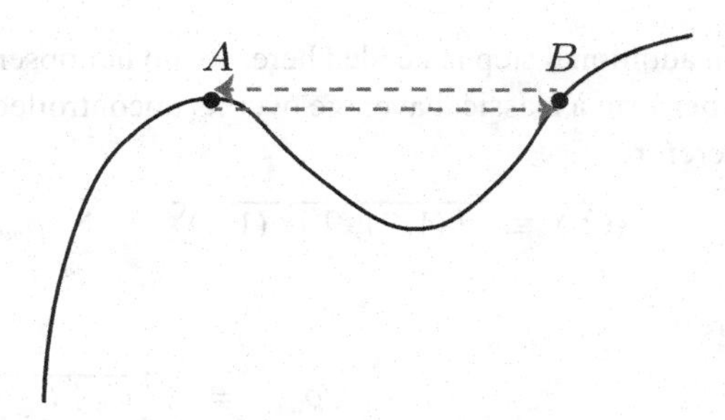


Figure 3.11 The "bounce" solution that corresponds to the decay of a metastable state.

This expression captures the energy splitting of the ground state that builds the lowest energy band in the one-dimensional crystal, and is to be compared with the $n = 0$ case of Eq. (3.291). Their difference resides, once more, in the need to use the quadratic connection formulas of Appendix C near the bottom of the wells.

One can also describe metastable states in terms of instantons, and in this case the basic classical solution, the "bounce," describes a Euclidean motion starting from A, the top of the hill of the inverted potential in Figure 3.11, moving toward B and returning to A after a reflection. The first lesson is that the resulting Euclidean action is twice as large as the one corresponding to the motion from A to B, so that the correction to the ground-state energy involves the square of the coefficient entering the previous cases, consistent with what we saw in the Hamiltonian WKB treatment. The second indication is that the wavefunction $\dot{x}_c$ corresponding to the vanishing eigenvalue, which is to be removed as in the other cases, has a node induced by the reflection, and therefore cannot correspond to the ground state. Consequently, there is a negative eigenvalue, whose presence makes $\mathcal{K}$ imaginary. An imaginary shift to the original energy levels in the absence of tunneling yields a decaying probability to find the particle in the well, consistent with the metastable nature of the ground state.

## 3.9 The Density Matrix

Let us conclude this brief overview of quantum mechanics with a discussion of the density matrix, which we denote by $\rho$ in the following. We can motivate its role considering the system under study (system 1), together with a thermal bath (system 2) to which it is weakly coupled, so that

$$H(1,2) \simeq H(1) + H(2). \tag{3.380}$$

The states decompose accordingly as

$$|\psi(1,2)\rangle = \sum_{m,n} C_{m,n} |m\rangle_1 |n\rangle_2, \tag{3.381}$$

where $|m\rangle_1$ and $|n\rangle_2$ form complete sets for the two systems 1 and 2, and the mean value of an observable $\mathcal{O}_1$ of system 1 is consequently

$$\langle\psi(1,2)|\mathcal{O}_1|\psi(1,2)\rangle = \sum_{m,m',n} C_{m',n} C^*_{m,n} \, {}_1\langle m|\mathcal{O}_1|m'\rangle_1, \tag{3.382}$$

where we have made use of the orthonormality of the states in system 2.

An additional step is needed here: we do not observe system 2, and thus we should also perform a classical average over its uncontrolled details. The quantity of interest is therefore

$$\overline{\langle \mathcal{O}_1 \rangle} \equiv \overline{\langle \psi(1,2)|\mathcal{O}_1|\psi(1,2)\rangle} = \sum_{m,m'} \rho_{mm'} \langle m|\mathcal{O}_1|m'\rangle_1, \tag{3.383}$$

where

$$\rho_{mm'} = \sum_n \overline{C_{m,n}\, C^*_{m',n}}, \tag{3.384}$$

and the overstrike denotes a classical average over fluctuations of the thermal bath. This leads one to define the "density operator," or "density matrix,"

$$\rho = \sum_{m,m'} \rho_{mm'}\, |m\rangle_1\, {}_1\langle m'|, \tag{3.385}$$

in terms of which Eq. (3.383) can be cast in the form

$$\overline{\langle \psi(1,2)|\mathcal{O}_1|\psi(1,2)\rangle} = \mathrm{tr}\left(\rho\, \mathcal{O}_1\right). \tag{3.386}$$

Note that $\rho$ is Hermitian, since Eq. (3.384) implies that

$$\rho_{mm'} = \left(\rho_{m'm}\right)^*, \tag{3.387}$$

and has nonnegative diagonal elements. One can therefore diagonalize it, so that from now on we shall write

$$\rho = \sum_m w_m\, |w_m\rangle \langle w_m|, \tag{3.388}$$

with weights $w_m \geq 0$. The $w_m$ are actually classical probabilities, since in general

$$\overline{\langle \mathcal{O} \rangle} \equiv \mathrm{tr}\left(\rho\, \mathcal{O}\right) = \sum_m w_m \langle w_m|\mathcal{O}|w_m\rangle, \tag{3.389}$$

and therefore the average of the identity operator implies that

$$\mathrm{tr}\left(\rho\right) = \sum_m w_m = 1. \tag{3.390}$$

Furthermore

$$\mathrm{tr}\left(\rho^2\right) = \sum_m w_m^2 \leq 1, \tag{3.391}$$

and the equality holds only for *pure states*, the standard ingredients of quantum mechanics for isolated systems, for which only one of the $w$'s is not zero. In contrast, states are called *mixed* when the strict inequality holds.

Let us now discuss the time evolution of $\rho$. From the behavior of the states $|w_m\rangle$ one can deduce that, in our Schrödinger picture,

$$\rho \to e^{-\frac{i}{\hbar}Ht}\, \rho\, e^{\frac{i}{\hbar}Ht}, \tag{3.392}$$

which translates into the differential equation

$$i\hbar \frac{\partial \rho}{\partial t} = \left[H \rho\right]. \tag{3.393}$$

This shows that, in equilibrium, $H$ and $\rho$ commute, so that the energy eigenstates $|\alpha\rangle$ provide a natural choice for the $|w_m\rangle$, and therefore from now on we shall refer to the decomposition

$$\rho = \sum_\alpha w_\alpha \, |\alpha\rangle \langle\alpha|. \tag{3.394}$$

The density matrix is the key ingredient for the construction of the canonical ensemble in quantum mechanics. Starting from a generic $\rho$ and imposing the appropriate constraints via a pair of Lagrange multipliers, one can build the von Neumann entropy

$$\sigma = -\operatorname{tr}\left(\rho \log \rho\right) - \lambda_1 \left[\operatorname{tr}\left(\rho\right) - 1\right] - \lambda_2 \left[\operatorname{tr}\left(\rho H\right) - U\right], \tag{3.395}$$

along the lines of Section 2.3. Note that this quantity vanishes, as it should, for a pure state, while familiar steps lead again to the partition function, which now becomes

$$Z = \operatorname{tr}\left(e^{-\beta H}\right), \tag{3.396}$$

and to the Helmholtz free energy.

We can conclude this section with the quantum counterpart of classical conservation laws. To this end, let us consider the thermal average (3.389) of a Schrödinger picture operator $\mathcal{O}$ that does not depend explicitly on time, and whose Heisenberg picture counterpart is conserved, so that

$$[\mathcal{O}, H] = 0. \tag{3.397}$$

The thermal average is then constant in time since, using Eq. (3.393) and the cyclic property of the trace,

$$i\hbar \frac{\partial}{\partial t} \overline{\langle \mathcal{O} \rangle} \equiv i\hbar \frac{\partial}{\partial t} tr\left(\rho\, \mathcal{O}\right) = tr\left(\rho\, [\mathcal{O}, H]\right), \tag{3.398}$$

and the last term vanishes on account of Eq. (3.397).

## Bibliographical Notes

Detailed introductions to quantum mechanics can be found, for example, in [14, 35, 52, 53]. Introductions to path integrals in quantum mechanics and quantum field theory are available in [30, 57, 60]. The basic properties of instantons are described in [11, 47, 45, 63]. More details on the WKB approximation, including higher orders and instanton analyses for the double well and other quantum mechanical problems, can be found in J. Zinn-Justin and U. D. Jentschura, Ann. Phys. **313** (2004) 197; 269; **326** (2011) 2186; U. D. Jentschura, A. Surzhykov and J. Zinn-Justin, Ann. Phys. **325** (2010) 1135, and references therein. This is still an active research area in mathematical physics. The dominant instanton contributions in nonsymmetric wells and in the presence of gravity were neatly discussed in S. R. Coleman, Phys. Rev. D **15** 2929 (1977) [erratum: Phys. Rev. D**16** 1248 (1977)], C. G. Callan, Jr. and S. R. Coleman, Phys. Rev. D**16** 1762 (1977), S. R. Coleman and F. De Luccia, Phys. Rev. D**21** 3305 (1980). The special functions that play a role in this chapter are discussed in detail in [1] and [62].

## Problems

**3.1** Consider the vertical motion of a quantum particle of mass $m$, subject to a constant gravitational field near the surface of the Earth, for which therefore

$$H = \frac{p_z^2}{2m} + m g z, \quad z \geq 0.$$

1. Compute the eigenvalues of $H$ in the WKB approximation, and explain the difference with respect to the standard Bohr–Sommerfeld formula.
2. Using this result, write the partition function of a system of noninteracting particles of this type and the corresponding expression for the specific heat $C_V$ of the system.
3. Compute the partition function starting from the corresponding classical phase space and verify that the results coincide in the high–temperature limit.

**3.2** Consider a quantum particle of mass $m$ in one dimension subject to the attractive potential

$$V(x) = -\lambda\, \delta(x),$$

where $\lambda > 0$.

1. Discuss the solutions of the Schrödinger equation for negative energy eigenvalues.
2. Repeat the preceding analysis for positive energy eigenvalues. To this end, build the solutions starting from a plane wave impinging from the left on the potential, together with the corresponding reflected and transmitted waves in the form

$$\psi(x) = \begin{cases} e^{ikx} + R(k)\, e^{-ikx} & (x < 0)\,; \\ T(k)\, e^{ikx} & (x > 0). \end{cases}$$

Compute the reflection and transmission coefficients and show that they are singular for a complex momentum corresponding to the previous negative–energy eigenstate. This is perhaps the simplest manifestation of a resonance phenomenon in quantum mechanics.

**3.3** Consider a quantum particle of mass $m$ in one dimension subject to the attractive potential

$$V(x) = -\lambda \left[\delta(x-a) + \delta(x+a)\right],$$

where $\lambda > 0$.

1. Study the possible bound states of the system corresponding to even wavefunctions, such that $\psi(x) = \psi(-x)$.
2. Study the possible bound states of the system corresponding to odd wavefunctions, such that $\psi(x) = -\psi(-x)$.

**3.4** Consider a quantum particle of mass $m$ in one dimension subject to the attractive potential

$$V(x) = -\lambda \sum_{k=-\infty}^{+\infty} \delta(x - ka),$$

where $\lambda > 0$. Study the bound states of the system.

**3.5** Consider a quantum particle of mass $m$ in two dimensions for which

$$H = \frac{p_x^2 + p_y^2}{2m} + \frac{1}{2} m\, \omega^2 \left(x^2 + y^2 - 2xy\right).$$

1. Compute the eigenvalues and the eigenfunctions of $H$.
2. Compute the partition function of a system of $N$ not mutually interacting particles of this type and the corresponding specific heat $C_P$.

**3.6** Compute the evolution kernel for a harmonic oscillator subject to an external force $F(t)$, for which

$$H = \frac{p^2}{2m} + \frac{1}{2} m\, \omega^2 x^2 - F(t)\, x.$$

**3.7** Consider a quantum particle of mass $m$ in two dimensions for which

$$H = \frac{p_x^2 + p_y^2}{2m} + \frac{1}{2} m\, \omega^2 \left(x^2 + y^2\right).$$

1. Study the spectrum of this system and the corresponding degeneracies.
2. Use the preceding results to compute the partition function for $N$ nonmutually interacting particles of this type and the corresponding free energy.
3. Repeat the analysis considering $2N$ particles in one dimension for which

$$H = \frac{p^2}{2m} + \frac{1}{2} m\, \omega^2 x^2.$$

**3.8** Consider a particle of mass $m$ confined to the region $x \geq 0$ and subject there to the potential

$$V(x) = \frac{1}{2} m\, \omega^2 x^2.$$

1. Study the exact spectrum of the system.
2. Study the spectrum in the WKB approximation.
3. Compute the partition function of the system and the corresponding specific heat $C_V$.

**3.9** Consider a particle in one dimension subject to the potential

$$V(x) = \begin{cases} \frac{1}{2}\, \omega_1^2 x^2 & (x < 0)\,; \\ \frac{1}{2}\, \omega_2^2 x^2 & (x > 0). \end{cases}$$

1. Study the spectrum of the system in the WKB approximation.
2. Compute the partition function and the specific heat $C_V$ of the system.

**3.10** Consider a system of $N$ noninteracting particles, such that the Hamiltonian for each of them reads

$$H = \epsilon \begin{pmatrix} 2 & 1 \\ 1 & 2 \end{pmatrix}.$$

a. Compute the canonical density matrix.
b. Compute the corresponding von Neumann entropy.

**3.11** Consider a two–state system whose density matrix is

$$\rho = \begin{pmatrix} a & b \\ b^* & 1-a \end{pmatrix},$$

with $a$ real.

1. Determine the condition on $a$ and $b$ that grants that $\rho$ is a density matrix.
2. If the spin operators are

$$S_x = \frac{\hbar}{2}\begin{pmatrix} 0 & 1 \\ 1 & 0 \end{pmatrix}, \; S_y = \frac{\hbar}{2}\begin{pmatrix} 0 & -i \\ i & 0 \end{pmatrix}, \; S_z = \frac{\hbar}{2}\begin{pmatrix} 1 & 0 \\ 0 & -1 \end{pmatrix},$$

determine the values of $a$ and $b$ if

$$\langle S_x \rangle = \frac{\hbar}{4}, \qquad \langle S_y \rangle = \langle S_z \rangle = 0.$$

3. Compute the von Neumann entropy associated to $\rho$.

**3.12** The quantum partition function, as we have seen, approaches the classical phase–space integral in the $\hbar \to 0$ limit. Now we want to explore, in one dimension, the first correction to this result.

1. Starting from eq. (3.111), evaluate the Euclidean kernel $K_E(x, x, \hbar\beta)$, retaining also the first nontrivial correction in $\hbar$ to the result in eq. (3.114), and deduce the corresponding modification of the partition function.
2. Repeat the analysis in the Hamiltonian formalism, and to this end take into account that

$$e^{A+B} \simeq e^A\, e^B \left(1 - \frac{1}{2}[A,B] + \frac{1}{8}[A,B]^2 + \frac{1}{6}[A+2B,[A,B]]\right),$$

up to higher–order terms in the commutators.
3. Show that the result is, in both cases

$$Z = \int \frac{dx\, dp}{h} e^{-\beta\left(\frac{p^2}{2m}+U(x)\right)} \left[1 - \frac{\hbar^2\beta^3}{24\, m}\left(U'(x)\right)^2\right] + \cdots .$$

**3.13** Consider a harmonic oscillator whose frequency changes in time, so that in the far past it equals $\omega_1$ and far in the future it attains a different value, $\omega_2$.

1. Express the creation and annihilation operators $(a_2, a_2^\dagger)$ in the far future in terms of $(a_1, a_1^\dagger)$, the creation and annihilation operators in the far past. This type of link is usually called a Bogolyubov transformation.

2. Letting $|0_1\rangle$ denote the ground state in the far past, so that

$$a_1 |0_1\rangle = 0,$$

evaluate the average of the number operator in the far future $a_2^\dagger a_2$ in this state.

**3.14** Consider a particle of mass $m$ and charge $q$, which is subject to the combined effects of an electric field **E** and of a magnetic field **B**.

1. Show that the Lorentz force law follows from the Lagrangian

$$\mathcal{L} = \frac{m}{2}\dot{\mathbf{x}}^2 + \frac{q}{c}\mathbf{A}(\mathbf{x}) \cdot \dot{\mathbf{x}} - q\,V(\mathbf{x}),$$

where **A** and $V$ are the vector and scalar potentials.

2. Show that the corresponding Hamiltonian is

$$H = \frac{1}{2m}\left(\mathbf{p} - \frac{q}{c}\mathbf{A}(\mathbf{x})\right)^2 + q\,V(\mathbf{x}).$$

3. Taking into account that a magnetic field thus introduces in the path integral a "nonintegrable phase factor"

$$e^{\frac{iq}{\hbar c}\int \mathbf{A}\cdot d\mathbf{x}},$$

consider the interaction between the particle and a magnetic monopole of magnetic charge $q_m$ at the origin, whose magnetic field satisfies

$$\nabla \cdot \mathbf{B} = 4\,\pi\, q_m\, \delta\,(\mathbf{r}).$$

Demanding that the nonintegrable phase factor along a path surrounding the monopole be well defined, derive Dirac's quantization condition

$$q\, q_m = \frac{n}{2}\hbar\, c,$$

where $n$ is an integer number.

**3.15** Consider a particle on a circle of radius $R$, as in Section 3.5, which is threaded by a magnetic field described by a potential $A$ that is constant along it.

1. Solve the Schrödinger equation for the system, demanding that the wavefunctions be periodic, so that $\psi\,(x + 2\pi R) = \psi(x)$, and compute the corresponding partition function.
2. Show that the preceding solutions are equivalent to those obtained removing the potential $A$ and demanding that the wavefunctions be quasi–periodic, so that

$$\psi(x + 2\pi R) = e^{i\,\alpha}\,\psi(x),$$

and relate $\alpha$ and $A$.

**3.16** The variational method is a most powerful technique to estimate the ground–state energy of a system in quantum mechanics.

1. For a generic system, show that, for any choice of $|\psi\rangle$ satisfying the proper boundary conditions, the estimate

$$E(\psi) = \frac{\langle\psi|H|\psi\rangle}{\langle\psi|\psi\rangle} \geq E_0$$

holds, where $E_0$ denotes the true ground–state energy.

2. Consider the He atom, for which

$$H = \frac{\mathbf{p}_1^2 + \mathbf{p}_2^2}{2m} - \frac{2\,e^2}{r_1} - \frac{2\,e^2}{r_2} + \frac{e^2}{|\mathbf{r}_1 - \mathbf{r}_2|},$$

where the position vectors $\mathbf{r}_{1,2}$ and the momenta $\mathbf{p}_{1,2}$ refer to the two electrons, and estimate the ground–state energy considering the trial wave-function

$$\psi(\mathbf{r}_1, \mathbf{r}_2) = C\,e^{-\alpha(r_1+r_2)},$$

and minimizing the result with respect to $\alpha$. Interpret physically the end result found for $\alpha$. Note that the resulting estimate is only off by about 2% from the true ground state energy of *He*.

**3.17** The WKB approximation leads to an elegant expression that determines a potential $V(x)$ given the corresponding spectrum. This relation can be regarded as the inverse of the Bohr–Sommerfeld rule, and more specifically of its consequence in eq. (3.260).

1. Specialize the expression for the density of states of eq. (3.260) to an even potential $V(x)$ that is also a growing function for $x > 0$, and recast it into a form that only involves the positive real axis.
2. After dividing both sides of the resulting relation by $\sqrt{\alpha - E}$, integrate them in $E$ and show that, with a suitable choice of integration range, this step leads to

$$x(\alpha) = \frac{\hbar}{\sqrt{2m}} \int_0^{\alpha} \frac{dn(E)}{dE} \frac{dE}{\sqrt{\alpha - E}}.$$

Comment on how this result can determine the potential $V(x)$ in terms of a given spectrum $E(n)$.

**3.18** In classical mechanics, the virial theorem can be used to connect the average kinetic energy to the potential energy in periodic motions, thus providing a powerful method to estimate both. Let us therefore consider, for a system of $N$ particles in classical mechanics, the quantity

$$G = \sum_{i=1}^{N} \mathbf{r}_i \cdot \mathbf{p}_i.$$

1. Show that

$$\dot{G} = 2T + \sum_{i=1}^{N} \mathbf{r}_i \cdot \mathbf{F}_i,$$

where $T$ denotes the kinetic energy of the system.

2. For a periodic motion, deduce from the preceding result the virial theorem

$$\langle T\rangle = -\frac{1}{2}\sum_{i=1}^{N}\langle \mathbf{r}_i\cdot\mathbf{F}_i\rangle .$$

In particular, for a particle in a central potential

$$V(r) = -\frac{k}{r^n},$$

deduce that

$$\langle T\rangle = -\frac{n}{2}\langle V\rangle .$$

3. Rephrase the preceding arguments in terms of a Hermitian operator $G$ in the Heisenberg picture of quantum mechanics and compute the average kinetic energy of the electron in the ground state of Hydrogen.

**3.19** We want to estimate the thermal noise in a resistor, and to this end:

1. Identifying the power dissipated in a resistor,

$$P = \frac{V^2}{R},$$

where $V$ is the voltage at its ends, as the corresponding energy per unit frequency band, introduce a natural Gaussian measure over the voltage fluctuations within a frequency band $\Delta\omega$ compatible with $\langle V\rangle = 0$, and show that

$$\langle(\Delta V)^2\rangle = \frac{1}{2}\,k_B T\,\Delta\,\omega .$$

2. Use this result to obtain an estimate for the mean square noise current in a resistor at temperature $T$.

**3.20** The Morse potential is a nice model of diatomic molecules. Simplifying matters a bit, let us consider a similar potential on the real axis,

$$V(x) = V_0\left[\left(1 - e^{-\frac{x}{\rho}}\right)^2 - 1\right]$$

and address its bound–state spectrum.

1. Show that the change of variable

$$z = e^{-\frac{x}{\rho}}$$

reduces the Schrödinger equation for the bound–state spectrum to

$$\frac{d^2\psi}{dz^2} + \frac{1}{z}\frac{d\psi}{dz} - \gamma^2\left[1 - \frac{2}{z} + \frac{\alpha^2}{z^2}\right]\psi = 0,$$

where

$$\gamma^2 = \frac{2\,m\,\rho^2\,V_0}{\hbar^2}, \qquad \alpha^2 = \frac{|E|}{V_0}.$$

2. Show that a bound-state wavefunction for this system must have the dominant behaviors

$$\psi \sim z^{\alpha\gamma}, \qquad \psi \sim e^{-\gamma z},$$

as $z \to 0$ and $z \to \infty$.

3. Let

$$\psi = z^{\alpha\gamma}\, e^{-\gamma z}\, \phi(z),$$

and look for a power series expansion

$$\phi(z) = \sum_{n=0}^{\infty} c_n z^n.$$

Analyzing the recursion relation for the coefficients, show that for bound states the series must be a polynomial, which determines the eigenvalue spectrum

$$E_n = V_0 \left[ -1 + \frac{2n+1}{\gamma} - \left( \frac{n+\frac{1}{2}}{\gamma} \right)^2 \right],$$

with $n$ an integer such that $n \geq 0$ and $2n+1 < \gamma$. Note that the level density decreases with $n$, as expected from the WKB analysis, since the classical trajectories become longer as the energy increases.

4. Compare the low–lying modes with the harmonic–oscillator spectrum around the bottom of the potential well.

# 4 Systems of Quantum Oscillators

In this chapter we take a first step toward the description of quantum statistical systems. We consider, to begin with, a gas of $N$ three-dimensional distinguishable quantum oscillators characterized by a given frequency $\omega$, and then extend the discussion to systems with continuous frequency spectra. We thus describe Planck's theory of blackbody radiation, which opened the way to quantum mechanics and remains a result of utmost importance, together with some crucial refinements due to Einstein. We conclude with Debye's model of the vibrational specific heat of solids. As in the original treatments, we leave aside the subtle effects introduced by the zero-point energy. We shall return to them in the next chapter.

## 4.1 The Effect of an Energy Gap

The partition function (3.396) for a gas of $N$ distinguishable isotropic oscillators of frequency $\omega$ in three dimensions coincides with the corresponding expression for $3N$ oscillators in one dimension,

$$Z = \left[\mathrm{tr}\, e^{-\beta H}\right]^{3N}, \tag{4.1}$$

where the trace is over the state space of a single oscillator. According to Eq. (3.57), the single-oscillator contribution is the geometric series

$$\mathrm{tr}\, e^{-\beta H} = \sum_{n=0}^{\infty} e^{-\beta\hbar\omega\left(n+\frac{1}{2}\right)} = e^{-\frac{\beta\hbar\omega}{2}} \frac{1}{1 - e^{-\beta\hbar\omega}}, \tag{4.2}$$

and combining this result with Eq. (4.1) one can simply deduce that, for a gas of $3N$ such oscillators,

$$\beta A = \frac{3N\beta\hbar\,\omega}{2} + 3N \log\left(1 - e^{-\beta\hbar\omega}\right). \tag{4.3}$$

Therefore,

$$U = \frac{\partial(\beta A)}{\partial\beta} = \frac{3N\hbar\omega}{2} + \frac{3N\hbar\omega}{e^{\beta\hbar\omega} - 1}, \tag{4.4}$$

where the first term originates from the zero-point energy. From $U$ one can also deduce the specific heat

$$C_V = 3Nk_B \left[\frac{\frac{\hbar\omega}{2k_BT}}{\sinh\left(\frac{\hbar\omega}{2k_BT}\right)}\right]^2, \tag{4.5}$$

and this result contains two important lessons:

1. at low temperatures, where the argument of the hyperbolic function is large,

$$C_V \sim Nk_B \left(\frac{\hbar\omega}{k_B T}\right)^2 e^{-\frac{\hbar\omega}{k_B T}}, \tag{4.6}$$

which tends to zero as $T \to 0$, compatibly with Nerst's theorem. It is proportional to $e^{-\frac{\Delta}{k_B T}}$, where $\Delta$ is the "energy gap" ($\hbar\omega$ in this case) that separates the ground state from the first excited state;

2. at high temperatures, the accessible modes form effectively a continuum, the function within brackets in Eq. (4.5) tends to one, and

$$C_V \sim 3Nk_B. \tag{4.7}$$

This is the specific heat obtained in Sections 2.2 and 2.3 for a gas of classical oscillators.

These results are special manifestations of a general fact. If indeed the single-particle partition function is

$$Z = \sum_\alpha g_\alpha \, e^{-\beta E_\alpha}, \tag{4.8}$$

at low temperatures the first two terms dominate and

$$Z \simeq g_0 \, e^{-\beta E_0} + g_1 \, e^{-\beta E_1}. \tag{4.9}$$

Therefore

$$\beta A \simeq N\beta E_0 - N\frac{g_1}{g_0} \, e^{-\beta\Delta}, \tag{4.10}$$

and

$$C_V \simeq Nk_B \frac{g_1}{g_0}\left(\frac{\Delta}{k_B T}\right)^2 e^{-\frac{\Delta}{k_B T}}, \tag{4.11}$$

so that $C_V$ tends to zero exponentially as determined by the "energy gap"

$$\Delta = E_1 - E_0. \tag{4.12}$$

On the other hand, at high temperatures many states contribute significantly to the partition function, and the classical phase-space integrals provide close approximations to quantum partition sums. Consequently, replacing the sum over $n$ in Eq. (4.2) with an integral one can recover the constant high-$T$ behavior for $C_V$ in Eq. (4.7), as expected from the equipartition law.

This example is often called "Einstein model of specific heats," and reproduces at high temperatures the Dulong–Petit law. At low temperatures, however, it yields an exponential decay for $C_V$. This result is only qualitatively correct: in Section 4.3 we shall see that a more refined model for the vibrational modes of a crystal lattice yields a milder low-$T$ behavior.

## 4.2 Blackbody Radiation

We can now apply the preceding considerations to the important problem of blackbody radiation. Sure enough, the original work where Max Planck derived its spectrum marked the birth of quantum mechanics, but blackbody radiation remains of utmost importance today, and has manifold applications. For example, the Universe hosts the cosmic microwave background (CMB), a relic radiation from its early epochs that has a blackbody-like spectrum with a temperature of about 3 K.

A basic motivation to address this problem can be drawn from a simple phenomenon of everyday life: a piece of wood in the fireplace becomes red as it warms up, and then, as the temperature increases further, its warmest portions turn white with blue traits. This is a nonequilibrium phenomenon, but one can idealize matters and conceive a thermodynamic equilibrium that mimics it. To this end, one can consider a cavity at a temperature $T$, whose walls emit and absorb electromagnetic radiation continuously. By piercing a small hole in one of the walls, an observer would be able to record the properties of the radiation from the small amount emerging from it without significantly perturbing the equilibrium.

To begin with, let us see how far one can get with thermodynamics. The two inputs, to this effect, are the extensive nature of the internal energy, which implies that

$$U = V u(T), \tag{4.13}$$

and the link between the radiation pressure $P$ and $u(T)$,

$$P = \frac{1}{3} u. \tag{4.14}$$

It is sufficient to say, for the time being, that this result reflects the presence of three independent directions along which electromagnetic waves can propagate, and that it can be deduced from the Maxwell equations. As we shall better appreciate in the following, Eq. (4.14) also reflects the dispersion relation for plane waves,

$$k = |\mathbf{k}| = \frac{\omega}{c}, \tag{4.15}$$

where $\mathbf{k}$ is a wave vector, $\omega$ is the frequency and $c \simeq 3\times 10^8$ m/s is the speed of light.

Starting from the first and second laws of thermodynamics, written in the form

$$dS = \frac{dU}{T} + \frac{P}{T} dV, \tag{4.16}$$

and making use of Eqs. (4.13) and (4.14) leads to

$$dS = \frac{4}{3}\frac{u(T)}{T} dV + V\frac{du}{dT} dT. \tag{4.17}$$

This ought to be an exact differential, and equating mixed derivatives leads to

$$\frac{\partial}{\partial T}\left(\frac{4}{3}\frac{u(T)}{T}\right) = \frac{\partial}{\partial V}\left(V\frac{du(T)}{dT}\right), \tag{4.18}$$

or

$$\frac{du}{dT} = 4\frac{u}{T}, \tag{4.19}$$

whose solution is the Stefan–Boltzmann law

$$u(T) = C\,T^4. \tag{4.20}$$

This result is of fundamental importance, and it remains valid in Planck's theory, which will also determine $C$ in terms of fundamental physical constants. On the other hand, thermodynamics alone cannot determine $C$, which emerged here as an integration constant.

Let us now see why the attempts to derive $C$ from microscopic considerations based on classical electrodynamics ran into trouble. To begin with, each of the two independent polarizations of a plane wave in our box of side $L$ behaves as a harmonic oscillator. This can be seen by inserting a plane wave

$$\mathbf{E}(\mathbf{r},t) = e^{i\mathbf{k}\cdot\mathbf{r}}\,\mathbf{e}(t), \tag{4.21}$$

where periodicity demands that

$$\mathbf{k} = \frac{2\,\pi\,\mathbf{n}}{L}, \tag{4.22}$$

in the wave equation

$$\nabla^2\,\mathbf{E} = \frac{1}{c^2}\,\frac{\partial^2\,\mathbf{E}}{\partial\,t^2} \tag{4.23}$$

implied by Maxwell's theory. This step reduces it to

$$\frac{d^2\mathbf{e}(t)}{dt^2} + c^2\,k^2\,\mathbf{e}(t) = 0 \tag{4.24}$$

and implies the dispersion relation (4.15).

We know already that each one-dimensional *classic* harmonic oscillator contributes $k_BT$ to the internal energy, and there are two such oscillators for each plane wave. All in all, the resulting prediction for $U$ would then be

$$U = 2\sum_{\mathbf{n}} k_BT = \frac{2V}{(2\pi)^3}\int d^3\mathbf{k}\,k_BT = \frac{k_BTV}{\pi^2\,c^3}\int_0^\infty d\omega\,\omega^2, \tag{4.25}$$

where we have used Eq. (3.20). This expression is clearly divergent, since the integral extends to arbitrarily high frequencies.

In modern language, Planck replaced the equipartite energies $k_BT$ of the individual modes with the expression in Eq. (4.4), so that

$$U = \frac{2V}{(2\pi)^3}\int d^3\mathbf{k}\left[\frac{\hbar\omega}{2} + \frac{\hbar\omega}{e^{\beta\hbar\omega} - 1}\right]. \tag{4.26}$$

Actually, and perhaps fortunately, Planck was unaware of the zero-point energy, so that he only considered the second term,

$$U = \frac{2V}{(2\pi)^3}\int d^3\mathbf{k}\,\frac{\hbar\omega}{e^{\beta\hbar\omega} - 1} = \frac{V}{\pi^2\,c^3}\int_0^\infty d\omega\,\frac{\hbar\,\omega^3}{e^{\beta\hbar\omega} - 1}, \tag{4.27}$$

whose integral is finite. Let us concentrate on it momentarily, before returning to the zero-point contribution in the following chapter.

Two lessons can be readily drawn from the key result of Eq. (4.27). Firstly, one can recover the Stefan–Boltzmann law, while also obtaining a prediction for $C$ in terms of fundamental physical constants. The change of integration variable to $x = \beta\hbar\omega$ yields indeed an energy density

$$u(T) = \frac{(k_B T)^4}{\pi^2 (\hbar c)^3} \int_0^\infty dx \, \frac{x^3}{e^x - 1}, \tag{4.28}$$

which is consistent with Eq. (4.20). The integral, which is clearly finite, is evaluated in Appendix D, and leads to

$$u(T) = \frac{\pi^2}{15 \hbar^3 c^3} (k_B T)^4. \tag{4.29}$$

Note that this expression diverges, as expected, if $\hbar \to 0$. As we have seen, Planck's constant in SI units is $h = 6.626 \times 10^{-34}$ J · s, but it is perhaps convenient to keep in mind that $\hbar c \simeq 200$ MeV fm, where the size of a proton is about 1 fm or $10^{-15}$ m and its rest mass is about 1 GeV/$c^2$ or 1000 MeV/$c^2$. Then, taking into account that $k_B T_0 \simeq \frac{1}{40}$ eV for $T_0 \simeq 300$ K, this result translates into

$$u(T) \simeq 4 \times 10^{-6} \left(\frac{T}{T_0}\right)^4 \text{J/m}^3. \tag{4.30}$$

The second lesson of Eq. (4.27) has to do with the integrand, which defines the spectral density of blackbody radiation,

$$u_\omega = \frac{\hbar}{\pi^2 c^3} \frac{\omega^3}{e^{\frac{\hbar\omega}{k_B T}} - 1}. \tag{4.31}$$

This expression has an important physical meaning: it describes the spectrum that could be observed if a filter were placed on the little hole of the blackbody cavity, and its product with $\frac{c}{4}$ would yield, as we shall see, the energy density flowing through it per unit time. The integral of $u_\omega$ over all frequencies yields Eq. (4.29), but its behavior at low and high frequencies is markedly different, with a transition for $\hbar\omega \simeq k_B T$. In detail:

- At low frequencies, in the so-called Rayleigh–Jeans region, it increases according to

$$u_\omega \simeq \frac{k_B T \omega^2}{\pi^2 c^3}. \tag{4.32}$$

- At high frequencies, in the so-called Wien region, it decreases according to

$$u_\omega \simeq \frac{\hbar \omega^3}{\pi^2 c^3} e^{-\frac{\hbar\omega}{k_B T}}. \tag{4.33}$$

- One can verify that the derivative of Eq. (4.31) vanishes for

$$\hbar\omega \simeq 2.8 \, k_B T, \tag{4.34}$$

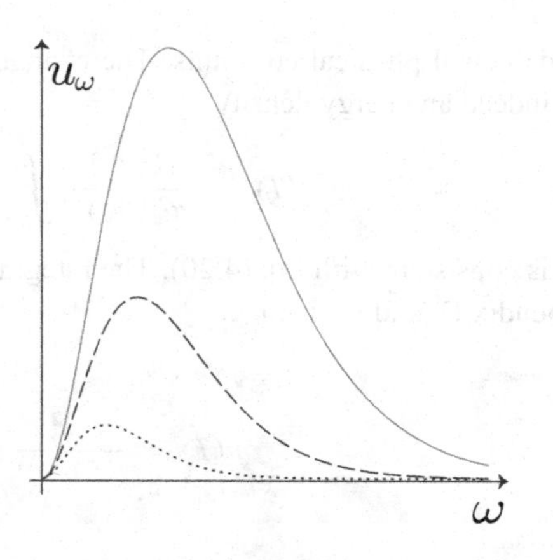


Figure 4.1 Some typical blackbody spectra, for three increasing values of the temperature $T$ (dotted, dashed and solid curves).

which identifies the maximum of $u_\omega$. Consequently, a few thousand Kelvin translate into peak frequencies in the visible spectrum. Some typical curves for $u_\omega$ are shown in Figure 4.1.

The two limiting behaviors in the Rayleigh-Jeans and Wien regions were known to Planck from ongoing experiments, and he actually started from them in his construction of the blackbody spectrum (4.31). He noticed that they can be regarded as expressions of the same type,

$$u_\omega \;=\; \frac{\omega^2}{\pi^2\, c^3}\,\langle E\rangle, \tag{4.35}$$

which rest however on different forms of the average energy $\langle E\rangle$ associated to a single mode. In detail, he was aware that in the Rayleigh–Jeans region

$$\langle E\rangle \;=\; k_B T, \tag{4.36}$$

while in the Wien region

$$\langle E\rangle \;=\; A\,\omega\, e^{-\beta\,\omega\,\gamma}, \tag{4.37}$$

with $A$ and $\gamma$ two constants. A basic relation of thermodynamics, as we have seen, links energy and entropy densities according to

$$\frac{ds}{d\langle E\rangle} \;=\; \frac{1}{T}, \tag{4.38}$$

and therefore the two limiting behaviors translate into

$$\frac{ds}{d\langle E\rangle} \;=\; \begin{cases} \frac{k_B}{\langle E\rangle}\ ; \\ \frac{k_B}{\omega\,\gamma}\,\log\left(\frac{A\,\omega}{\langle E\rangle}\right). \end{cases} \tag{4.39}$$

These two expressions become more similar after taking one more derivative, which leads to

$$\frac{d^2 s}{d\langle E\rangle^2} \;=\; \begin{cases} -\,\frac{k_B}{\langle E\rangle^2}\ ; \\ -\,\frac{k_B}{\omega\,\gamma}\,\frac{1}{\langle E\rangle}, \end{cases} \tag{4.40}$$

which suggested to Planck the interpolating formula

$$\frac{d^2 s}{d\langle E\rangle^2} = -\frac{k_B}{\langle E\rangle^2 + \omega\,\gamma\,\langle E\rangle}. \tag{4.41}$$

This integrates to

$$\frac{1}{T} = \frac{ds}{d\langle E\rangle} = \frac{k_B}{\omega\,\gamma}\log\left[1 + \frac{\omega\,\gamma}{\langle E\rangle}\right], \tag{4.42}$$

consistent with the two limiting behaviors of Eqs. (4.39). Inverting this relation gives

$$\langle E\rangle = \frac{\omega\,\gamma}{e^{\beta\,\omega\,\gamma} - 1}, \tag{4.43}$$

and thus

$$u_\omega = \frac{\gamma\,\omega^3}{\pi^2\,c^3}\,\frac{1}{e^{\beta\,\omega\,\gamma} - 1}. \tag{4.44}$$

Consequently

$$u = \frac{\pi^2}{15\,\gamma^3\,c^3}\,(k_B\,T)^4, \tag{4.45}$$

where, as we now know, $\gamma = \hbar$.

This expression for $u_\omega$ begged to be connected to a discrete partition sum of the type

$$Z = \sum_{n=0}^{\infty} e^{-n\,\beta\,\omega\,\gamma} = \frac{1}{1 - e^{-\beta\,\omega\,\gamma}}, \tag{4.46}$$

and this is the way $E_n = n\hbar\omega$ entered physics in 1900. Comparison of data against the Stefan–Boltzmann law determined $\hbar$, while the peak frequency of Eq. (4.34) provided a precise determination of Boltzmann's constant $k_B$. In this fashion, Eq. (4.43) becomes

$$\langle E\rangle = \frac{\hbar\,\omega}{e^{\frac{\hbar\omega}{k_B T}} - 1}, \tag{4.47}$$

and

$$\langle E\rangle \simeq k_B T, \qquad \langle E\rangle \simeq \hbar\omega\, e^{-\frac{\hbar\omega}{k_B T}} \tag{4.48}$$

in the Rayleigh–Jeans and Wien regions, respectively.

Starting from the partition sum of Eq. (4.46), one can derive the free energy taking into account the different field oscillators, and

$$A = \frac{V(k_B T)^4}{\pi^2(\hbar c)^3}\int_0^\infty x^2\,dx\,\log\left(1 - e^{-x}\right)$$

$$= -\frac{V(k_B T)^4}{3\,\pi^2(\hbar c)^3}\int_0^\infty dx\,\frac{x^3}{e^x - 1} = -\frac{V(k_B T)^4}{45\,\pi^2(\hbar c)^3}, \tag{4.49}$$

resorting again to Appendix D. From this result one can also deduce that

$$S = k_B\,\frac{4\,V(k_B T)^3}{45\,\pi^2(\hbar c)^3}, \text{ and} \tag{4.50}$$

$$P = \frac{(k_B T)^4}{45\,\pi^2(\hbar c)^3}, \tag{4.51}$$

so that the comparison with Eq. (4.29) shows that, indeed,

$$P = \frac{1}{3}\, u \tag{4.52}$$

for blackbody radiation. Note also that, for this system, the entropy $S$ tends to zero as $T$ tends to zero, in contrast with the behavior of the Sackur–Tetrode equation (2.83). This is as it should be: the system ought to tend to an orderly state at a temperature of absolute zero, and blackbody radiation does so, in agreement with the third law of thermodynamics. In the following chapters we shall see why and how the results for matter particles will have to be amended to comply with this type of behavior.

Returning to blackbody radiation, Einstein went much further. He clarified the physical meaning of Planck's quantization, associating the energy quanta $\hbar\omega$ to particles that build up the electromagnetic radiation. It is instructive to retrace his steps, starting from the limiting behavior of Planck's law in the Wien region, which leads us to define a spectral entropy density via

$$\frac{ds_\omega}{du_\omega} = \frac{1}{T} = \frac{k_B}{\hbar\omega}\log\left(\frac{\hbar\,\omega^3}{\pi^2\,c^3}\,\frac{1}{u_\omega}\right). \tag{4.53}$$

Integrating this equation gives

$$s_\omega = \frac{k_B\,u_\omega}{\hbar\omega}\left[1 + \log\left(\frac{\hbar\,\omega^3}{\pi^2\,c^3}\,\frac{1}{u_\omega}\right)\right], \tag{4.54}$$

and therefore, multiplying both sides by the volume

$$S_\omega = \frac{k_B\,U_\omega}{\hbar\omega}\left[1 + \log\left(\frac{\hbar\,\omega^3}{\pi^2\,c^3}\,\frac{V}{U_\omega}\right)\right]. \tag{4.55}$$

Einstein noticed a striking, if not evident, correspondence between this expression and the Sackur–Tetrode formula of Eq. (2.38). In particular, if $V$ varied while maintaining $U_\omega$ fixed, as happens for an ideal gas of particles in isothermal transformations, the resulting entropy change,

$$\Delta S_\omega = \frac{k_B\,U_\omega}{\hbar\omega}\log\left(\frac{V_2}{V_1}\right), \tag{4.56}$$

would match the result determined by Eq. (2.38), up to the identification

$$U_\omega = N\hbar\,\omega. \tag{4.57}$$

His conclusion was that, in the Wien region, blackbody radiation behaves like a gas of particles, of varying numbers, which carry the energy quanta $\hbar\omega$ and are now called photons.

One can compute the photon density in blackbody radiation as

$$\frac{N}{V} = \pi^2\,c^3\int_0^\infty d\omega\,\frac{\omega^2}{e^{\beta\hbar\omega}-1} = \frac{2\,\zeta(3)\,\left(k_B\,T\right)^3}{\pi^2\,(\hbar\,c)^3}, \tag{4.58}$$

using some results in Appendix D, where $\zeta(3) \simeq 1.2$, and the result is about $5 \times 10^{14}$ photons/m$^3$ at room temperature. Moreover, a minor addition to the preceding arguments determines the actual energy or photon flux across a small aperture starting from Eqs. (4.28) and (4.58). These are simply determined by the energy or photon densities, up to a factor $c$ and a geometrical factor

$$\frac{1}{4\pi} \int_{\langle 2\pi \rangle} \cos\theta \, d\Omega = \frac{1}{2} \int_0^{\frac{\pi}{2}} \sin\theta \cos\theta \, d\theta = \frac{1}{4}, \tag{4.59}$$

since the propagation occurs at the speed of light and the outgoing flux that one detects originates from half of the solid angle, in the direction orthogonal to the aperture. Hence the final results for the energy and photon fluxes across the aperture are

$$J_E = \frac{\pi^2}{60\, \gamma^3\, c^2} (k_B T)^4, \qquad J_\gamma = \frac{\zeta(3)\, (k_B T)^3}{1/4\, \pi^2\, \hbar^3\, c^2}. \tag{4.60}$$

Note also that, starting from Eq. (4.47), one can compute the fluctuations of blackbody radiation, taking into account Eq. (2.98), and the result is the very interesting formula

$$\langle (\Delta E)^2 \rangle = \hbar\, \omega \langle E \rangle + \langle E \rangle^2. \tag{4.61}$$

The two terms reflect the two limiting forms in Eq. (4.48). In other words, blackbody radiation behaves as an admixture of two components: waves and the energy quanta unveiled by the preceding argument. This was a striking, early manifestation of particle–wave duality.

Years later, Einstein exposed the three leading processes that underlie the dynamical equilibrium between matter and radiation, and thus the blackbody spectrum. In order to illustrate his argument, let us begin by recalling Bohr's formula,

$$\Delta E = \hbar\, \omega, \tag{4.62}$$

which links the energy separation $\Delta E$ between a pair of levels to the angular frequency $\omega$ of the radiation absorbed or emitted in the corresponding transitions. Taking this into account, Einstein recast the equilibrium energy density of Eq. (4.31) in the equivalent, but very suggestive, form

$$u_\omega = \frac{\hbar \omega^3}{\pi^2\, c^3}\, e^{-\beta \hbar \omega} + u_\omega\, e^{-\beta \hbar \omega} \tag{4.63}$$

Since the Boltzmann factor determines the relative populations of the two levels, he identified *three* distinct mechanisms at work in the dynamical equilibrium between matter and radiation. To begin with, the l.h.s. of Eq. (4.63) is naturally associated to a transition to the upper level resulting from the *absorption* of a quantum and the consequent excitation of matter. However, the same reasoning revealed that *two* distinct processes underlie transitions from the upper level, with the consequent *emission* of a photon: the term independent of $u(T)$ reflects a *spontaneous emission*, while the one proportional to $u(T)$ reflects a *stimulated emission*, a coherent phenomenon would eventually lie at the heart of laser systems.

## 4.3 Debye Theory of Specific Heats

The result of Eq. (4.47) leads to an interesting theory of specific heats that accounts for crystal vibrations in a more realistic fashion, and consequently extends the Dulong–Petit result of Eq. (2.92) to low temperatures,. The starting point is provided in this case by the two relations

$$U = 3 \sum_{\mathbf{k}}^{\mathbf{k}_{max}} \frac{\hbar\omega}{e^{\frac{\hbar\omega}{k_B T}} - 1}, \tag{4.64}$$

$$3\,N = 3 \sum_{\mathbf{k}}^{\mathbf{k}_{max}} 1, \tag{4.65}$$

which link these effects to *phonons*, the quanta of crystal vibrations. These correspond to sound waves, which have three polarizations (two transverse and one longitudinal). The total number of independent sound waves is *finite*, and equals the independent vibration modes of the crystal in three dimensions, which are indeed $3N$ for $N$ atoms.

In the large-volume limit these expressions become

$$U = 3\,\frac{V}{(2\pi)^3} \int^{|\mathbf{k}_{max}|} d^3\mathbf{k}\, \frac{\hbar\omega}{e^{\frac{\hbar\omega}{k_B T}} - 1}, \tag{4.66}$$

$$3\,N = 3\,\frac{V}{(2\pi)^3} \int^{|\mathbf{k}_{max}|} d^3\mathbf{k} \;=\; \frac{V}{2\pi^2}\, k_{max}^3, \tag{4.67}$$

so that the second implies the important relation

$$|\mathbf{k}_{max}| \;=\; \left(\frac{6\pi^2 N}{V}\right)^{\frac{1}{3}} . \tag{4.68}$$

Moreover, for sound waves,

$$\omega \;=\; |\mathbf{k}|\, v_s, \tag{4.69}$$

with $v_s$ the speed of sound, which we assume for simplicity to be the same for transverse and longitudinal waves, and the expression for the internal energy finally becomes

$$U \;=\; \frac{3\,V(k_B T)^4}{2\pi^2(\hbar v_s)^3} \int_0^{\frac{\Theta}{T}} \frac{x^3}{e^x - 1}, \tag{4.70}$$

where $\Theta$, the Debye temperature, is

$$\Theta \;=\; \frac{\hbar\, v_s}{k_B} \left(\frac{6\pi^2 N}{V}\right)^{\frac{1}{3}} . \tag{4.71}$$

Around the Debye temperature the behavior of the specific heat undergoes a transition. For a typical crystal the last factor in the definition of $\Theta$ is of order $1\text{Å}^{-1}$, and with $v_s$ in the range 100–1000 m/s one obtains Debye temperatures in the range 10–100 K. However, there are also materials, such as Beryllium, with Debye temperatures of a few thousand K.

We conclude this chapter taking a closer look at the limiting behavior of the internal energy of the Debye gas for large and small $T$. In the first case the upper bound on the integral is small, and retaining only the leading term of the integrand for small values of $x$ gives

$$U \simeq \frac{V(k_B T)^4}{2\pi^2(\hbar v_s)^3}\left(\frac{\Theta}{T}\right)^3 = 3Nk_B T, \tag{4.72}$$

so that

$$C_V = 3Nk_B, \tag{4.73}$$

which is the Dulong–Petit law. On the other hand, for small $T$ one can extend the upper limit to infinity, and the integral thus approaches $\frac{\pi^4}{15}$, which is the result of Appendix D that we used to obtain Eq. (4.29), so that

$$U = \frac{\pi^2\, V(k_B T)^4}{10\,(\hbar v_s)^3} \tag{4.74}$$

and

$$C_V = k_B \frac{2\,\pi^2\, V(k_B T)^3}{5\,(\hbar v_s)^3}, \tag{4.75}$$

which tends to zero proportionally to $T^3$ at low temperatures.

## Bibliographical Notes

Discussions of blackbody radiation can be found in many books on statistical mechanics, including [36, 49, 50, 51, 54]. These ideas have manifold applications and permeate various aspects of physics. See for example [42, 59, 61] or [38].

## Problems

**4.1** Compute the first quantum correction for the harmonic oscillator using the result from problem 12 in Chapter 3, and compare with Eq. (4.2).

**4.2** Consider an ensemble of non-interacting massless particles, so that

$$H = |p|\,c,$$

which live in one dimension for energies below a certain threshold $E_0$ and in two dimensions for energies above $E_0$. Compute the partition function, the internal energy, and the specific heat $C_V$. Discuss the low-$T$ and high-$T$ limits.

**4.3** Show that, if

$$H = H_0 + \epsilon\, V,$$

with $\epsilon \ll 1$, to first order in $\epsilon$

$$A = A_0 + \epsilon N\langle V\rangle,$$

where $A_0$ is the free energy determined by $H_0$, and

$$\langle V\rangle = \frac{tr\left(e^{-\beta H_0} V\right)}{tr\left(e^{-\beta H_0}\right)}.$$

Assume we have a gas of $N$ noninteracting oscillators in one dimension with a small anharmonic correction to the potential, so that

$$H = \frac{p^2}{2m} + \frac{1}{2} m\,\omega^2 x^2 + \epsilon \left(\frac{x}{x_0}\right)^4,$$

where $x_0$ is a constant. Compute the Helmholtz free energy and study the behavior at fixed length of the specific heat $C_L$ to first order in $\epsilon$.

**4.4** Consider an ensemble of particles in three dimensions, not mutually interacting, and with

$$H = \frac{p^2}{2m} + \epsilon \left(\frac{p}{p_0}\right)^4.$$

Compute the Helmholtz free energy, and study the behavior of the specific heat $C_V$ to first order in $\epsilon$.

**4.5** Consider a gas of $N$ "atoms" of mass $M$, not mutually interacting, confined in a volume $V$ and kept at a temperature $T$. The "atoms" are all of the same type, and have internal excitation levels of energies

$$E_n = (n+1)\,\epsilon \quad (n = 0, 1, \ldots),$$

with degeneracies

$$d_n = n + 1.$$

1. Compute the canonical partition function $Z$ of the gas and its internal energy $U$. Compute also the specific heat at constant volume $C_V$, discussing in detail its limiting behaviors at high and low $T$.
2. Estimate the leading behavior of $C_V$, at low and high $T$, if

$$d_n = (n+1)^k,$$

where $k$ is a generic positive integer.

**4.6** Consider a gas of $N$ "molecules" contained in a volume, $V$, at a temperature $T$. The Hamiltonian for each molecule is

$$H = \frac{\mathbf{p}^2}{2m} + H_1,$$

where $H_1$ models the internal dynamics and reads

$$H_1 = \begin{pmatrix} a & b \\ b^* & -a \end{pmatrix},$$

with $a$ a real number, $b$ a complex number, and $b^*$ the complex conjugate of $b$.

1. Compute the partition function, the internal energy, and the entropy of the system.
2. Compute the specific heat at constant volume $C_V$ and study its behavior as the temperature $T$ varies.

**4.7** Consider a system of noninteracting particles kept at a temperature $T$. The Hamiltonian of a single particle has eigenvalues

$$E_n = \epsilon \log n \qquad n = 1, 2, \ldots,$$

with degeneracy

$$d_n = n + 1.$$

1. Compute the internal energy and the specific heat at constant pressure $C_P$ for the system.
2. Study the limiting behavior of $C_V$ at low temperatures.
3. What is the limiting temperature $T_\ell$ for this system? What happens to $C_V$ as $T$ approaches $T_\ell$ from below?

**4.8** Consider a "polymer" in two dimensions, which one can model as a chain of $N+1$ vectors of fixed length. The angle between two consecutive vectors can be zero (if the chain proceeds in a given direction), and the corresponding energy is then $E = 0$, or it can be $\pm 90^\circ$ (if the chain undergoes a left or right turn), and the corresponding energy is then $E = \epsilon$.

1. Compute the entropy of the system using the microcanonical ensemble, for a given internal energy $U$.
2. Repeat the analysis using the canonical ensemble.
3. Compare the results.

**4.9** Study the internal energy, the specific heat $C_V$, and the entropy for a system analogous to a blackbody, so that

$$H = \sum_{i=1}^{N} c\,|\mathbf{p}_i|,$$

except for a range of forbidden single-particle energies $0 \leq E \leq E_0$.

**4.10** Consider a gas of $N$ noninteracting chains. Each chain has $\ell$ sites $i = 1, \ldots, \ell$, which may be empty, with energy $E = 0$, or full, with energy $E = \epsilon > 0$. However, the $i$th site can be full only if the $(i-1)$th site is also full.

a. Compute the partition function of the system.

b. Study how the internal energy and the specific heat, $C_V$, vary with the temperature $T$ of the system.

**4.11** Assuming that the Sun and the Earth behave like blackbodies at temperatures $T_S$ and $T_E$, provide an estimate of the temperature of the Earth in terms of the radius $R_S$ of the Sun and of the distance $R_{TS}$ between Earth and Sun, making use of the following data

$$T_S \simeq 6 \times 10^3\ \text{K}, \qquad R_S \simeq 7 \times 10^5\ \text{km}, \qquad R_{TS} \simeq 1.5 \times 10^8\ \text{km}.$$

**4.12** Consider the quantization of modes in a one-dimensional segment of finite length $L$, as a toy model for Planck radiation at a temperature $T$, but for a finite value of the length $L$.

1. Find the exact relation between the force exerted by the gas on the endpoints of the segment and the total energy of the gas.
2. Evaluate this force in the high- and low-temperature regimes.
3. If one divides the interval into two portions of different lengths by a partition identical to the ones at the ends, in which direction will the gas tend to push it?

**4.13** Consider a gas of quantum rotators, not mutually interacting, so that for a single component

$$H = \frac{\mathbf{L}^2}{2I}.$$

1. Write the partition function for the system.
2. Study the behavior of the internal energy of the system $U$ and of the corresponding specific heat $C_V$ as the temperature $T$ varies.

**4.14** Study the thermodynamics of a gas of one-dimensional quantum oscillators with distinct equilibrium positions in the microcanonical ensemble.

**4.15** Consider a set of $N$ noninteracting particles of mass $m$ in one dimension, contained in an infinite potential well in the region $0 < x < L$.

1. Compute the energy levels and the partition function of the system.
2. Study the behavior of the internal energy and of the specific heat as the temperature $T$ varies.
3. Study the behavior of the force exerted by the gas on one end of the interval at $x = L$ and compare with the ideal gas law. Notice the residual contribution as $T \to 0$.

**4.16** We want to elaborate in more detail on Planck's blackbody formula, extending the discussion in the text. Consider a system with two energy levels, of energies $E_1$ and $E_2 = E_1 + \hbar\omega$ and degeneracies $g_1$ and $g_2$, in a radiation bath with energy density $\rho(\omega)$. Einstein argued that three key processes underlie the dynamical equilibrium:

1. spontaneous emission of a photon, with probability per unit time $A_{2\to1}$;
2. stimulated emission of a photon, with probability per unit time $B_{2\to1}\,\rho(\omega)$;
3. absorption of a photon, with probability per unit time $B_{2\to1}\,\rho(\omega)$.

These result in the following evolution equations for the densities of components in the two levels, $n_1$ and $n_2$:

$$\frac{dn_1}{dt} = A_{2\to1}\,n_2 - B_{1\to2}\,n_1\,\rho(\omega) + B_{2\to1}\,n_2\,\rho(\omega),$$

$$\frac{dn_2}{dt} = -A_{2\to1}\,n_2 + B_{1\to2}\,n_1\,\rho(\omega) - B_{2\to1}\,n_2\,\rho(\omega).$$

Demanding that Planck's radiation formula $u_\omega$, of Eq. (4.31), describe the equilibrium state of the system, and extract the available information on the three coefficients.

**4.17** Consider a gas of photons that, for frequencies below a given value $\omega_0$, are free to move in three dimensions, while for frequencies above $\omega_0$ can also explore an additional dimension.

1. Obtain expressions for the Helmholtz free energy $A$, the internal energy $U$, the entropy $S$, and the pressure $P$.
2. Study qualitatively how the Stefan–Boltzmann law is modified, and the resulting shape of the blackbody spectrum.

**4.18** Consider a gas of photons in three dimensions, for which there is a forbidden band of frequencies between $\omega_0$ and $2\,\omega_0$.

1. Obtain expressions for the Helmholtz free energy $A$, the internal energy $U$, the entropy $S$, and the pressure $P$.
2. Study qualitatively how the Stefan–Boltzmann law is modified, and the resulting shape of the blackbody spectrum.

**4.19** Discuss how Planck's blackbody spectrum and the Stefan–Boltzmann law would be affected if the dispersion relation for photons were replaced by the nonrelativistic law

$$E = \frac{\mathbf{p}^2}{2\,m} .$$

**4.20** Discuss how Planck's blackbody spectrum and the Stefan–Boltzmann law would be affected if the dispersion relation for photons were replaced by the massive relativistic law

$$E = \sqrt{(\mathbf{p}\,c)^2 + m^2\,c^4} - m\,c^2 ,$$

where we have removed the rest–mass contribution, consistent with the nonrelativistic limit.

# 5 Vacuum Fluctuations

In our treatment of quantum oscillators we left out the zero-point contribution,

$$\langle E \rangle_0 = 2 \sum_{\mathbf{k}} \frac{\hbar\,\omega}{2}, \tag{5.1}$$

following the work of the founding fathers of quantum theory. However, we now know from quantum mechanics that this term should be present. This chapter takes a closer look at this loose end of the original treatment. It concerns zero-temperature effects, and thus is more relevant to quantum mechanics than statistical mechanics proper. However, we cannot resist dwelling on these matters for two reasons. Firstly, because it is intriguing to speculate on how Planck would have reacted had he known of this contribution, which is naively badly divergent. Moreover, these terms underlie important physical effects and are still linked to one of the deepest mysteries of our universe.

One might be tempted to argue that an additive constant should have no bearing on the proper meaning of energy, but there are actually some instances where this is not so, and one simply cannot ignore it. This happens when the contribution in Eq. (5.1), which depends on geometrical data as one can already appreciate from Eq. (3.20), varies because the geometry itself is subject to changes. A setting of this type can be realized in the laboratory, in the so-called Casimir effect. With proper care, as we shall see, one can then obtain striking predictions that were confirmed in detailed experimental tests only at the end of the last Century. Moreover, phenomena of this type were instrumental for the eventual success of quantum field theory in accounting for the fundamental interactions. The Lamb shift (is discussed in Section 5.2) a key example in this respect, but this setting still entails a baffling puzzle related to the vacuum energy of the Universe, the "cosmological constant problem." The tools that we have developed so far suffice to take a first look at these phenomena.

## 5.1 The Casimir Effect

Let us now consider the geometry in Figure 5.1, and try to compute the vacuum energy density in the region between two conducting plates. To this end, one must count the plane waves that are allowed there by the condition that the electric field **E** be orthogonal to the plates. The available contributions depend on the value of $k_z$, and in detail:

- if $k_z = 0$, the waves propagate in directions parallel to the plates, and only the polarization orthogonal to them is allowed, so that

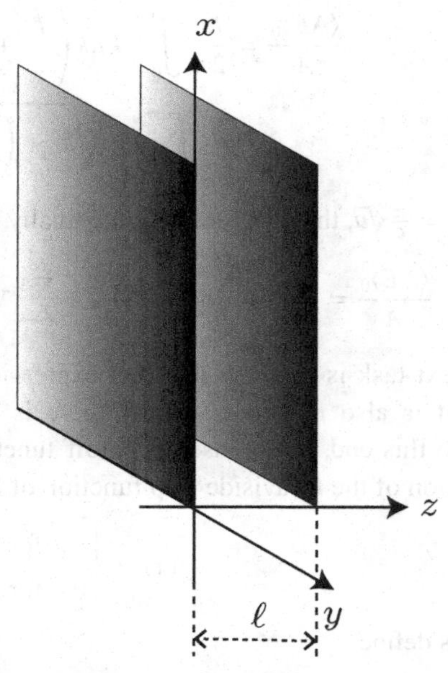


Figure 5.1 A weak attraction develops between two parallel conducting plates as a result of vacuum fluctuations.

$$\mathbf{E} = \hat{z} \sum_{k_x,k_y} C_{k_x,k_y}\, e^{i(k_x x + k_y y)} \; ; \tag{5.2}$$

- if $k_z \neq 0$, two polarizations are allowed, so that

$$\mathbf{E}_i = \sum_{n=0}^{\infty} \sum_{k_x,k_y} \mathbf{e}_i\, C_{k_x,k_y,n}\, e^{i(k_x x + k_y y)} \sin\left(\frac{n\pi z}{\ell}\right). \tag{5.3}$$

  where $i = 1, 2$ and the two unit vectors $\mathbf{e}_i$ are orthogonal to $\mathbf{k}$. Indeed, for any integer $n \neq 0$, the sine function eliminates both polarizations at the two ends $z = 0$ and $z = \ell$, which would contain contributions parallel to the conducting planes.

One is thus led to the zero-point contribution to the internal energy

$$\langle E \rangle_0 = \sum_{k_x,k_y} \left( \frac{\hbar c}{2} \sqrt{k_x^2 + k_y^2} + \hbar c \sum_{n=1}^{\infty} \sqrt{k_x^2 + k_y^2 + \left(\frac{n\pi}{\ell}\right)^2} \right), \tag{5.4}$$

and for macroscopic plates of area $A$ one can replace the sum over $k_x$ and $k_y$ by a double integral, adapting Eq. (3.20) to the case of two dimensions. As we had anticipated, however, the end result

$$\frac{\langle E \rangle_0}{A} = \frac{\hbar c}{2\pi} \int_0^{\infty} k\,dk \left( \frac{k}{2} + \sum_{n=1}^{\infty} \sqrt{k^2 + \left(\frac{n\pi}{\ell}\right)^2} \right), \tag{5.5}$$

is badly divergent. However, one is actually interested in the comparison between this expression and the value that would obtain if the plates were far apart, i.e. in the large-$\ell$ limit, so that the sum over $n$ would also become an integral:

$$\frac{\langle \Delta E \rangle_0}{A} = \frac{\hbar c}{2\pi} \int_0^\infty k\, dk \left( \frac{k}{2} + \sum_{n=1}^{\infty} \sqrt{k^2 + \left(\frac{n\pi}{\ell}\right)^2} - \int_0^\infty dn \sqrt{k^2 + \left(\frac{n\pi}{\ell}\right)^2} \right). \tag{5.6}$$

Letting $k = \frac{\pi}{\ell}\sqrt{u}$, the difference can be finally cast in the form

$$\frac{\langle \Delta E \rangle_0}{A} = \frac{\hbar c\, \pi^2}{4\, \ell^3} \int_0^\infty du \left( \frac{\sqrt{u}}{2} + \sum_{n=1}^{\infty} \sqrt{u + n^2} - \int_0^\infty dn\, \sqrt{u + n^2} \right). \tag{5.7}$$

Our next task is to show that this expression is actually finite. As we shall see shortly, it is also negative, so that there is an attractive force between the two plates. To this end, let us insert a cutoff function $f\left(\sqrt{u + n^2}\right)$, where $f$ is a smooth deformation of the Heaviside step function of Eq. (3.367),

$$f(x) \simeq \begin{cases} 1 & \text{for } x < \Lambda \\ 0 & \text{for } x > \Lambda, \end{cases} \tag{5.8}$$

and let us define

$$F(n) = \int_0^\infty du\, f\left(\sqrt{u + n^2}\right) \sqrt{u + n^2} = \int_{n^2}^\infty dx\, \sqrt{x} f(\sqrt{x}). \tag{5.9}$$

The introduction of $f$ is meant to cut off very high frequencies to which the plates would not respond, without affecting the contributions at low frequencies. All in all, one is thus led to

$$\frac{\langle \Delta E \rangle_0}{A} = \frac{\hbar c\, \pi^2}{4\, \ell^3} \left( \frac{1}{2} F(0) + \sum_{n=1}^{\infty} F(n) - \int_0^\infty dn\, F(n) \right), \tag{5.10}$$

and our next task is to compute this expression. This is done in Appendix D via the Euler–MacLaurin formula, and also via the Abel–Plana formula. The end result is the same, is independent of the cutoff function $f$, and reads

$$\frac{\langle \Delta E \rangle_0}{A} = -\frac{\hbar\, c\, \pi^2}{720\, \ell^3}. \tag{5.11}$$

Note that this result translates into a very small energy density for macroscopic plates,

$$|\langle \Delta E \rangle_0(\text{eV})| = 2.7 \times 10^{-7} \frac{A(\text{cm}^2)}{[\ell(\text{cm})]^3}, \tag{5.12}$$

and indeed the corresponding force was first detected in a convincing way only in the late 1990s (see the Bibliographical Notes at the end of this chapter).

## 5.2 The Lamb Shift

As we saw in Section 3.6, the discrete energy levels of the hydrogen atom can be deduced exactly from the Schrödinger Hamiltonian

$$H = \frac{\mathbf{p}^2}{2m} - \frac{e^2}{r}. \tag{5.13}$$

In Section 3.6, we also reviewed how the inclusion of first-order relativistic corrections modifies the spectrum, leaving behind a degeneracy between the first excited $2S_{\frac{1}{2}}$ and $2P_{\frac{1}{2}}$ levels.

The Lamb shift is a small energy shift that is experienced by the $\ell = 0$ levels of hydrogen, which lifts the $2S_{\frac{1}{2}}$ level slightly above the neighboring $2P_{\frac{1}{2}}$ level. It escaped the original analysis since it is subtly related to vacuum fluctuations of the electromagnetic fields around the atom. The late 1940s witnessed a major breakthrough when quantum electrodynamics, the quantum theory of electron and photon fields, could properly account for this effect. This problem also served as a motivation to develop methods that have paved the way, in more recent decades, to our current understanding of the fundamental interactions. However, following an argument due to Bethe, one can relatively simply obtain a qualitative picture of the phenomenon, treating it as a peculiar perturbation of Schrödinger theory. We can now review this approach, which preceded and stimulated the crucial developments that we have alluded to.

To begin with, one can model the effects of vacuum fluctuations on an electron in an energy eigenstate $|n\rangle$, adding to its position $\mathbf{r}$ a random deviation, according to

$$\mathbf{r} \rightarrow \mathbf{r} + \delta\,\mathbf{r}. \tag{5.14}$$

This affects the potential energy, which can be Taylor expanded around $\mathbf{r}$, and translates naturally into an energy shift

$$\Delta E_n = \frac{1}{6}\langle(\delta\,\mathbf{r})^2\rangle\,\langle n|\nabla^2\mathcal{U}|n\rangle, \tag{5.15}$$

where the angular brackets denote averages over these fluctuations. Rotational symmetry indeed demands that

$$\langle\delta\,\mathbf{r}_i\rangle = 0, \qquad \langle\delta\,\mathbf{r}_i\,\delta\,\mathbf{r}_j\rangle = \frac{1}{3}\,\delta_{ij}\,\langle(\delta\,\mathbf{r})^2\rangle. \tag{5.16}$$

The potential energy satisfies the Poisson equation

$$\nabla^2\,\mathcal{U} = 4\,\pi\,e^2\,\delta(\mathbf{r}), \tag{5.17}$$

which turns Eq. (5.15) into

$$\Delta E_n = \frac{2\,\pi\,e^2}{3}\,\langle(\delta\,\mathbf{r})^2\rangle\,|\psi_n(0)|^2. \tag{5.18}$$

Therefore, only $S$ states, which have $\ell = 0$ and whose wavefunctions do not vanish at the origin, can be affected, and are lifted. As we have stressed, the $2S_{\frac{1}{2}}$ and $2P_{\frac{1}{2}}$ states are degenerate to begin with, so that, in principle, one should diagonalize a $4\times4$ matrix to determine the corrected energy levels. However, the $2S_{\frac{1}{2}}$ states can be treated separately, precisely because the $2P_{\frac{1}{2}}$ states have wavefunctions that vanish at the origin.

Our next task is to relate $\langle(\delta\,\mathbf{r})^2\rangle$ to the vacuum fluctuations of the electromagnetic field. The starting point is the classical equation of motion for the electron fluctuation, subject to what we shall shortly identify as a vacuum electric field fluctuation,

$$m\frac{d^2\delta\mathbf{r}}{dt^2} \simeq -\,e\,\mathbf{E}, \tag{5.19}$$

where we have left out the Lorentz term, which is subdominant for the hydrogen atom, since $\frac{\langle v \rangle}{c} \sim \alpha$, with $\alpha$ the fine-structure constant. The harmonic components satisfy

$$m\,\omega^2\,\delta\,\mathbf{r}_\omega \simeq e\,\mathbf{E}_\omega, \tag{5.20}$$

and they are uncorrelated, so that

$$\langle (\delta\,\mathbf{r})^2 \rangle = \int \frac{d\omega}{2\,\pi} \langle |\delta\,\mathbf{r}_\omega|^2 \rangle = \left(\frac{e}{m}\right)^2 \int \frac{d\omega}{2\,\pi} \frac{\langle |\mathbf{E}_\omega|^2 \rangle}{\omega^4}. \tag{5.21}$$

One can now relate $\langle |\mathbf{E}_\omega|^2 \rangle$ to the energy density in the radiation field, recalling that

$$\langle \mathcal{E} \rangle = \frac{1}{8\,\pi} \int d^3\mathbf{x}\,\langle \mathbf{E}^2 + \mathbf{B}^2 \rangle = \frac{V}{4\,\pi} \int \frac{d\,\omega}{2\,\pi} \langle |\mathbf{E}_\omega|^2 \rangle, \tag{5.22}$$

since the electric and magnetic contributions coincide for radiation. The average energy $\langle \mathcal{E} \rangle$ can then be related to the zero-point fluctuations according to

$$\langle \mathcal{E} \rangle = 2 \sum \frac{\hbar\,\omega}{2} = \frac{2\,V}{(2\pi)^3} \int d^3\mathbf{k}\,\frac{\hbar\,\omega}{2} \to \frac{V\hbar}{\pi\,c^3} \int_{\omega_{\rm min}}^{\omega_{\rm max}} \frac{d\,\omega}{2\pi}\,\omega^3, \tag{5.23}$$

where the integral is naturally restricted to the frequency range that is perturbed by the presence of the hydrogen atom, in analogy with what we did in the discussion of the Casimir effect. The comparison between Eqs. (5.22) and (5.23) gives

$$|\mathbf{E}_\omega|^2 = \frac{4\,\hbar}{c^3}\,\omega^3, \tag{5.24}$$

and substituting in Eq. (5.21),

$$\langle (\delta\,\mathbf{r})^2 \rangle = \frac{2\,\hbar\,e^2}{\pi\,m^2\,c^3} \int_{\omega_{\rm min}}^{\omega_{\rm max}} \frac{d\omega}{\omega} = \frac{2\,\hbar\,e^2}{\pi\,m^2\,c^3} \log\left(\frac{\omega_{\rm max}}{\omega_{\rm min}}\right). \tag{5.25}$$

Note that this expression has a mild dependence on the frequency range, and the natural choice is

$$\frac{\omega_{\rm max}}{\omega_{\rm min}} \sim \frac{a_0}{\lambda_{\rm Compton}} = \frac{\frac{\hbar^2}{m\,e^2}}{\frac{\hbar}{m\,c}} = \frac{\hbar\,c}{e^2} \equiv \frac{1}{\alpha} \simeq 137, \tag{5.26}$$

since $a_0$ is the typical size of the atom, while below $\lambda_{\rm Compton}$, particle production becomes relevant. Note also, once more, the emergence of the fine-structure constant $\alpha$. Taking everything into account, one can write

$$\langle (\delta\,\mathbf{r})^2 \rangle = \frac{2\,\alpha}{\pi}\,\lambda^2{}_{\rm Compton} \log\left(\frac{1}{\alpha}\right), \tag{5.27}$$

and since for the $2S_{\frac{1}{2}}$ states

$$\left|\psi_2(0)\right|^2 = \frac{1}{8\,\pi\,a_0^3}, \tag{5.28}$$

one can finally conclude that

$$\Delta E_2 \simeq \frac{e^2}{2\,a_0}\,\frac{\alpha_e^3}{3\,\pi} \log\left(\frac{1}{\alpha}\right). \tag{5.29}$$

This crude estimate gives a frequency shift of about 700 MHz for the $2S_{\frac{1}{2}}$ level, to be compared with the result of full-fledged quantum electrodynamics, which is about 1000 MHz.

## 5.3 The Cosmological-Constant Problem

There is another, more fundamental, context in which zero-point energies play a role. This occurs when gravity is taken into account, since the vacuum energy then translates into an average curvature of the universe (or, more precisely, into a cosmological constant), whose microscopic derivation along the lines of the preceding section results in a puzzle.

Our first step is to adapt the preceding derivation (see Eq. (5.23)) to a large three-dimensional volume, which would yield

$$\frac{\langle \Delta E \rangle_0}{V} = \frac{1}{4\,\pi^2 (\hbar\, c)^3} \sum_i n_i\, (-1)^{F_i} \int_0^{(pc)_{max}} u^2 du\, \sqrt{u^2 + (m_i\, c^2)^2} \tag{5.30}$$

for a collection of fields (or types of particles), $n_i$ species of which have mass $m_i$. Here, $(-1)^{F_i}$ equals 1 for Bose particles and $-1$ for Fermi particles, which contribute to the vacuum energy with opposite signs. We shall soon become more familiar with these two types of particles. To begin with, it is sufficient to say that photons and electrons are well-known representatives of the two families.

Equation (5.30) can be justified by noting, to begin with, that particle masses enter important counterparts of the wave equation for electric and magnetic fields, so that, for example, the wave equation for a scalar field $\Phi$ of mass $m$ is

$$\frac{1}{c^2}\frac{\partial^2}{\partial t^2}\Phi - \nabla^2\,\Phi + \left(\frac{m\,c}{\hbar}\right)^2 \Phi = 0, \tag{5.31}$$

which is known as the Klein–Gordon equation. Note, once more, that the amplitude $f(t)$ of a generic plane wave

$$\Phi = e^{\frac{i}{\hbar}\mathbf{p}\cdot\mathbf{x}}\, f(t) \tag{5.32}$$

satisfies the equation of motion of a harmonic oscillator:

$$\frac{d^2 f(t)}{dt^2} + \omega^2 f(t) = 0. \tag{5.33}$$

However, now the frequency is determined by

$$\omega^2 = \frac{\mathbf{p}^2\, c^2 + m^2\, c^4}{\hbar^2}, \tag{5.34}$$

which connects it to the relativistic energy of the massive particles associated to $\Phi$,

$$E = \sqrt{\mathbf{p}^2\, c^2 + m^2\, c^4}, \tag{5.35}$$

in such a way that

$$E = \pm\,\hbar\,\omega. \tag{5.36}$$

Up to the sign ambiguity, which reflects the existence of anti-particles, the "quanta" of such a field $\Phi$ are massive particles, in the same way photons are massless ones emerging from Maxwell's theory.

The quantum counterparts of the harmonic oscillators in Eq. (5.33) were considered in Section 3.3, and are described by the Hamiltonians

$$H = \frac{\hbar\omega}{2}\left(a^\dagger a + a a^\dagger\right), \tag{5.37}$$

with

$$\left[a, a^\dagger\right] \equiv a a^\dagger - a^\dagger a = 1. \tag{5.38}$$

The corresponding equally spaced spectra,

$$E_n = \hbar\omega\left(n + \frac{1}{2}\right), \tag{5.39}$$

are precisely tailored to describe the vacuum and the addition to it of $1, 2, \ldots$ Bose particles with given momenta (and different polarizations, if this attribute is present, as for photons). For this very reason particles are usually called field quanta.

Fermi particles, or fermions, constitute a second class of fundamental ingredients in nature, and the electron is the most familiar of them. We shall return to these issues in Chapter 8, but for the time being it is sufficient to say that Fermi particles obey the Pauli principle, which states that any two identical Fermi particles can never be in the same state. Fermi fields satisfy more complicated wave equations, but in essence, one can associate to each of their waves and polarizations a Hamiltonian of the form

$$H = \frac{\hbar\omega}{2}\left(a^\dagger a - a a^\dagger\right), \tag{5.40}$$

where the corresponding creation and annihilation operators satisfy the relations

$$a^2 = 0, \qquad (a^\dagger)^2 = 0, \tag{5.41}$$

and, moreover,

$$\left\{a, a^\dagger\right\} \equiv a a^\dagger + a^\dagger a = 1. \tag{5.42}$$

These conditions imply that only two states exist for such a Fermi oscillator, in contrast with the infinite sequence of states that exist for a Bose oscillator. The two Fermi states are the vacuum, or empty state, $|0\rangle$, and the filled state $|1\rangle$, where

$$\begin{aligned} a^\dagger|0\rangle = |1\rangle, &\qquad a|1\rangle = |0\rangle, \\ a^\dagger|1\rangle = 0, &\qquad a|0\rangle = 0. \end{aligned} \tag{5.43}$$

Finally, using Eq. (5.42) the Hamiltonian of Eq. (5.40) can be recast in the form

$$H = \hbar\omega\left(a^\dagger a - \frac{1}{2}\right), \tag{5.44}$$

with a *negative* zero-point energy, in agreement with our previous claim.

Let us now discuss the origin of the Planck scale, which sets a natural upper bound on the integral in Eq. (5.30) if the ubiquitous gravitational force is taken somehow into account. To this end, let us begin by comparing the Coulomb force between two identical static point masses $m$ carrying the electron charge $e$,

$$F_C = \frac{e^2}{r^2}, \tag{5.45}$$

with their mutual gravitational force

$$F_C = \frac{G_N m^2}{r^2}. \tag{5.46}$$

The analogy with the first expression, which leads to the fine-structure constant

$$\alpha = \frac{e^2}{\hbar c} \simeq \frac{1}{137}, \tag{5.47}$$

and the mass–energy relation suggests that a gravitational "fine-structure function,"

$$\alpha_G(E) = \frac{G_N E^2}{\hbar c^5}, \tag{5.48}$$

ought somehow to play a role in gravitational interactions among field quanta at an energy scale $E$. However, $\alpha_G(E)$ grows quadratically with the energy, and becomes one at

$$E_{Pl} = c^2 \sqrt{\frac{\hbar c}{G_N}} \simeq 10^{19}\text{GeV}. \tag{5.49}$$

This huge value, to be compared with the rest energy of a proton, which is about 1 GeV, is usually called "Planck energy." Two related quantities are the "Planck mass,"

$$m_{Pl} = \sqrt{\frac{\hbar c}{G_N}}, \tag{5.50}$$

and the "Planck length,"

$$\ell_{Pl} = \frac{\hbar c}{E_{Pl}} \simeq 10^{-35} \text{ m}. \tag{5.51}$$

At length scales of the order of $\ell_{Pl}$, spacetime becomes a fuzzy concept, whose nature still awaits a proper clarification. One can get a flavor of these difficulties by noticing that the Compton wavelength of a particle of mass $m_{Pl}$,

$$\lambda_{\text{Compton}} = \frac{\hbar}{m_{Pl} c}, \tag{5.52}$$

which characterizes typical uncertainties of its motion, would be comparable with the Schwarzschild radius

$$r_S = \frac{2Gm}{c^2} \tag{5.53}$$

of a corresponding black hole, which delimits the inner region inaccessible to an external classical observer. Therefore, we should keep in mind, in the following, that one is not quite sure whether our standard continuum ideas desumed from Minkowski space make sense at the Planck scale. These considerations suggest nonetheless that the integrals in Eq. (5.30) be restricted to the range $k < k_{lim}$, with

$$k_{lim} \sim \frac{1}{\ell_{Pl}}. \tag{5.54}$$

The dominant contribution in $k_{lim}$ in Eq. (5.30) is then

$$\frac{\langle \Delta E \rangle_0}{V} \sim \frac{\hbar c\, k_{lim}^4}{16\,\pi^2} \sum_i n_i\,(-1)^{F_i}, \tag{5.55}$$

and, barring cancelations, it translates into a fraction of the Planck energy per Planck volume, since

$$\frac{\langle \Delta E \rangle_0}{V} \sim \frac{E_{Pl}}{(\ell_{Pl})^3} \sim \frac{(E_{Pl})^4}{(\hbar\, c)^3}. \tag{5.56}$$

Barring cancelations, however, since if the particle species were such that

$$\sum_i n_i \, (-1)^{F_i} = 0, \tag{5.57}$$

i.e. if there were equal numbers of Bose and Fermi species in nature, one would be left with the next-to-leading term,

$$\frac{\langle \Delta E \rangle_0}{V} \sim \frac{E_{Pl}^2}{16\,\pi^2\,(\hbar c)^3} \sum_i n_i \left(m_i c^2\right)^2 (-1)^{F_i}, \tag{5.58}$$

obtained by expanding the square root, which is already lower than the previous estimate by the ratio $(m/m_{Pl})^2$, with $m$ a typical particle mass. Moreover, if the particle masses were nicely arranged in nature so that

$$\sum_i n_i \, m_i^2 \, (-1)^{F_i} = 0, \tag{5.59}$$

the dominant contribution would originate from the following term in the expansion, and would be depressed by an overall factor $(m/m_{Pl})^4$ with respect to the naive estimate. We know the species of elementary particles that exist in nature up to the currently explored energies, which lie however about 15 orders or magnitude below the Planck scale, and we know that Eqs. (5.57) and (5.59) do not hold for them.

On the other hand, the cosmological evolution provides one time scale, the age of the universe $t_0$ (about 14 billion years, or the Hubble parameter $H_0 \sim \frac{1}{t_0}$). Given the average mass density $\rho$ in the universe, the quantity $\rho\, G_N$ has dimension $[T]^{-2}$, and therefore cosmology suggests the identification

$$\rho\, G_N \sim \left(\frac{1}{t_0}\right)^2 \sim H_0^2, \tag{5.60}$$

which translates into a macroscopic energy density

$$\frac{\langle E_{macro} \rangle}{V} \sim \frac{(H_0\, c)^2}{G_N}. \tag{5.61}$$

The comparison with the preceding microscopic estimate thus leads to

$$\frac{\langle \Delta E \rangle_0}{E_{macro}} \sim \frac{(E_{Pl})^4}{(\hbar\, c)^3}\,\frac{G_N}{(H_0\, c)^2} = \left(\frac{c}{H_0\, \ell_{Pl}}\right)^2 \sim 10^{120}. \tag{5.62}$$

This is the squared ratio between the distance that a beam of light would have covered since the origin of the universe and the Planck length, and is unthinkably large. Many physicists regard this as the twentyfirst-century counterpart of Planck's conundrum.

As we have already stressed, the urge to bring the preceding scales closer to one another points to Bose–Fermi cancellations. The theory of the fundamental interactions could, in principle, accommodate an exact Bose–Fermi symmetry called supersymmetry, which is deeply related to gravity and would eliminate all terms in Eq. (5.30). An exact supersymmetry would require the presence of Bose and Fermi particle species in equal numbers and with identical masses, which is not the case in nature within the range that was explored so far. At any rate, this lies far below the Planck scale, and a hidden form of supersymmetry, somehow vaguely along the lines of what we shall see for the symmetries of ferromagnets, is not excluded, and could still render the clash far less severe. Still, even possible cancellations beyond the currently observable energies appear unlikely to lower the huge ratio in Eq. (5.62) below $10^{60}$. Hopefully, time and more work will help to shed some light on all this.

## Bibliographical Notes

For a detailed discussion of the Casimir effect, see for instance M. Bordag, U. Mohideen and V. M. Mostepanenko, "New developments in the Casimir effect," Phys. Rept. **353** (2001), 1–205 [arXiv:quant-ph/0106045 [quant-ph]]. The discussion of the Lamb shift follows H. A. Bethe, "The Electromagnetic Shift of Energy Levels," Phys. Rev. **72** (1947) 339. This subject lies at the heart of quantum electrodynamics, and a collection of original papers on the subject can be found in J. Schwinger (ed.), "Selected Papers on Quantum Electrodynamics" (Dover Publications, New York, 1958). Discussions of the cosmological constant problem can be found in many classical books on quantum field theory. For the original paper see Y. B. Zel'dovich, "The cosmological constant and the theory of elementary particles," Sov. Phys. Usp. **11** (1968) 381, reprinted in Gen. Rel. and Gravitation **40** (2008) 1557. We recently came to know that W. Pauli and O. Stern had also considered the problem, in unpublished work, before the Second World War.

# 6 The van der Waals Theory

As we have seen, kinetic energies suffice to recover the ideal gas law. Here, we show how a crude account of short-distance repulsions and longer-range dipole attractions among molecules can capture some universal effects encoded in the van der Waals equation of state. This amendment of the ideal gas law lends itself naturally to a first characterization of phase equilibria.

## 6.1 The Role of Interactions

We can now return to the ideal gas and take a closer look at the issue of mutual interactions, which have been neglected so far. In other words, we would like to justify how a naive description based on the simple Hamiltonian,

$$H = \sum_i \frac{\mathbf{p}_i^2}{2\,m}, \tag{6.1}$$

could yield results that apply convincingly to ordinary gases at room temperature. Interactions are clearly essential to drive these systems toward equilibrium, and yet they ought to give small corrections to Eq. (6.1), since the derivations in Chapter 2 led directly to the ideal gas law.

We have already seen how classical phase space emerges in the limit of large volumes, and now we would like to quantify the limitations of the present setting, in order to prepare the ground for the following sections. To begin with, let us consider the equipartition law (2.47). It implies the important relation

$$\sqrt{\langle \mathbf{p}^2 \rangle} \;=\; \sqrt{3\,m\,k_B\,T}, \tag{6.2}$$

which defines a "thermal de Broglie wavelength"

$$\lambda_T \sim \frac{2\pi\hbar}{\sqrt{3\,m_0\,k_B\,T_0}} \left(\frac{m_0}{m}\right)^{\frac{1}{2}} \left(\frac{T_0}{T}\right)^{\frac{1}{2}} \sim 0.7\,\text{Å} \left(\frac{m_0}{m}\right)^{\frac{1}{2}} \left(\frac{T_0}{T}\right)^{\frac{1}{2}}, \tag{6.3}$$

where the reference values are the mass of $^4$He atoms, for which $m_0\,c^2 \simeq 4$ GeV, and a room temperature of about 300 K, for which $k_B T_0 \simeq \frac{1}{40}$ eV. These considerations indicate that the typical values of $\lambda_T$ at room temperatures are not larger than atomic or molecular sizes, which lie around 1 Å or beyond. We can now estimate the typical inter-molecular distance from Avogadro's number, which associates $N_{\text{Av}} = 6 \times 10^{23}$ gas molecules to a volume $V \simeq 22$ dm$^3$ in ordinary conditions, as

$$d \simeq \left(\frac{V}{N_{Av}}\right)^{\frac{1}{3}} \simeq 30\,\text{Å}, \tag{6.4}$$

which is about an order of magnitude above the typical values of $\lambda_T$ for room temperatures. Therefore, gas molecules behave essentially as classical particles in ordinary conditions. However, we shall see that at low temperatures, where $\lambda_T$ increases, quantum mechanics brings along important new phenomena.

Let us now turn to characterize typical inter-molecular interactions. These combine a short-distance hard-sphere repulsion, which sets in naturally when a pair of molecules are brought close enough to try to overlap, and a weak attraction resulting from (induced) dipole–dipole interactions. The dependence of these dipole interactions on the mutual distance between a pair of molecules can be simply estimated from the typical energy of a dipole **p** in an electric field **E**,

$$\mathcal{E} \sim \mathbf{p} \cdot \mathbf{E}. \tag{6.5}$$

If the dipole is induced by the electric field, this expression is proportional to $|\mathbf{E}|^2$, and for an electric dipole $|\mathbf{E}| \sim \frac{1}{r^3}$. In conclusion, this van der Waals attractive interaction between a pair of molecules decreases proportionally to $\frac{1}{r^6}$. A simple model for the interaction energy between a pair of molecules that combines this attraction with a repulsion proportional to $\frac{1}{r^{12}}$, usually called the Lennard-Jones potential,

$$\mathcal{V} = \mathcal{V}_0 \left[ \left(\frac{r_0}{r}\right)^{12} - \left(\frac{r_0}{r}\right)^6 \right], \tag{6.6}$$

is displayed in Figure 6.1. The same figure also illustrates the "hard-sphere approximation," whereby the repulsive potential is modeled by a lower bound on the minimal mutual distance between molecules, which will play a role in the ensuing discussion. Before proceeding to calculate the partition function for a gas of molecules subject to pair-wise interactions as in Eq. (6.6), let us estimate their typical mean free path $\ell_F$. For any given molecule, the mean free path is determined by the ratio between any length $L$ that it traverses and the number of hard-sphere repulsions that it experiences

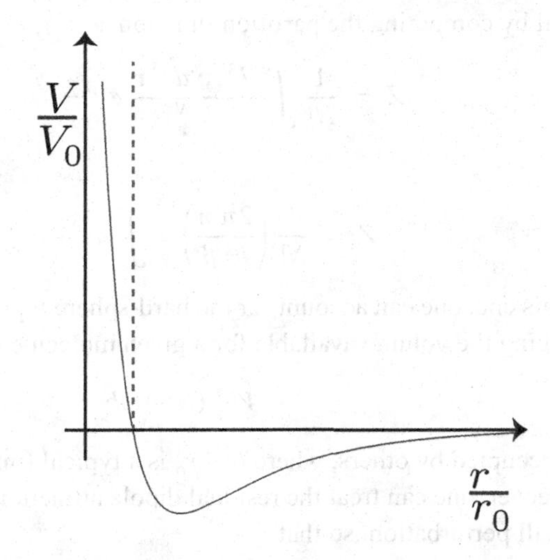


**Figure 6.1** The typical inter-molecular potential of Eq. (6.6), which combines a short-distance repulsive contribution with (continuous line) a weak dipole–dipole attraction, and (dashed line) the short-distance hard-sphere approximation.

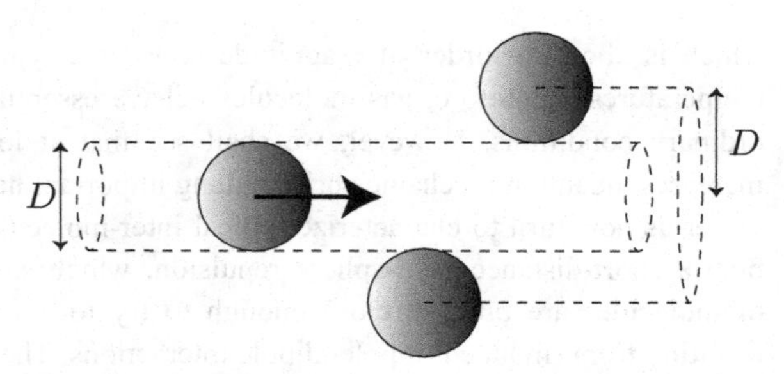


Figure 6.2 An estimate of the "mean free path" in the presence of hard-core interactions. A molecule is affected by all others that lie inside a tube whose axis is determined by its velocity and whose radius equals the molecule's diameter $D$.

on average along it. The latter equals the number density $n$ times the volume of a tube of length $L$ and radius equal to the molecule's diameter, $D = 2r_0$, as can be seen in Figure 6.2. In detail

$$\ell_F = \frac{L}{n\,\pi\,(2r_0)^2\,L} = \frac{1}{4\,\pi\,n\,r_0^2} \sim \left(\frac{d}{r_0}\right)^2 \frac{d}{4\,\pi} \sim 2000\,\text{Å}. \tag{6.7}$$

All in all, the ideal gas model is successful since, in ordinary circumstances, gas molecules are roughly hard classical balls with feeble mutual interactions.

## 6.2 The Partition Function

Our estimates indicate that neglecting mutual interactions is fully justified in ordinary conditions. However, striking results emerge if these mutual interactions are taken into account to lowest order in a low-density approximation, as we can now see in detail by computing the partition function:

$$Z = \frac{1}{N!}\int \frac{d^{3N}\mathbf{p}\,d^{3N}\mathbf{r}}{h^{3N}}\, e^{-\beta\sum_i \frac{\mathbf{p}_i^2}{2m}}\, e^{-\beta\sum_{i<j}\mathcal{V}(|\mathbf{r}_i-\mathbf{r}_j|)}, \tag{6.8}$$

or

$$Z = \frac{1}{N!}\left(\frac{2\pi m}{h^2\,\beta}\right)^{\frac{3N}{2}} \int d^{3N}\mathbf{r}\, e^{-\beta\sum_{i<j}\mathcal{V}(|\mathbf{r}_i-\mathbf{r}_j|)}. \tag{6.9}$$

To this end, one can account for the hard-sphere repulsion via a geometric constraint, reducing the volume available for a given molecule to the portion

$$V-(N-1)b \simeq V-Nb \tag{6.10}$$

not occupied by others, where $b \sim r_0^3$ is a typical (microscopic) hard-sphere volume. Moreover, one can treat the residual dipole attraction, which we shall denote by $v$, as a small perturbation, so that

$$Z \simeq \frac{1}{N!}\left(\frac{2\pi m}{h^2\,\beta}\right)^{\frac{3N}{2}} \int d^{3N}\mathbf{r}\left[1-\beta\sum_{i<j} v\left(\left|\mathbf{r}_i-\mathbf{r}_j\right|\right)\right]. \tag{6.11}$$

Hence, after a change of variables, one can characterize the interaction by a parameter $a$, letting

$$\int d^3\mathbf{x}\, d^3\mathbf{y}\, v(|\mathbf{x}-\mathbf{y}|) = (V-Nb)\int d^3\mathbf{r}\, v(|\mathbf{r}|) \simeq -2\,a\,V, \qquad (6.12)$$

where $a > 0$ for the attractive dipole–dipole interaction, for which $v < 0$ (see Figure 6.1). This finally leads to

$$Z \simeq \frac{1}{N!}\left(\frac{2\pi m}{h^2\,\beta}\right)^{\frac{3N}{2}} (V-Nb)^N \left[1 + \beta\, a\frac{N^2}{V}\right], \qquad (6.13)$$

since the sum in Eq. (6.11) contains $\frac{N(N-1)}{2} \simeq \frac{N^2}{2}$ terms.

## 6.3 The van der Waals Equation of State

For the Helmholtz free energy, by making use of Eq. (6.13) one obtains

$$\beta A = -\frac{3N}{2}\log\left(\frac{2\pi m}{h^2\,\beta}\right) - N\log\left(\frac{V-Nb}{N}\right) - N - \frac{\beta\, a N^2}{V}, \qquad (6.14)$$

again to first order in $a$, and consequently the internal energy,

$$U = \frac{3}{2}Nk_B T - \frac{a\,N^2}{V}, \qquad (6.15)$$

acquires a contribution that reflects the mutual dipole attraction and depends on the volume $V$. Note also that

$$C_V = \frac{3}{2}Nk_B, \qquad (6.16)$$

as for the ideal gas. From the Helmholtz free energy, one can also derive the van der Waals equation of state

$$\left(P + a\frac{N^2}{V^2}\right)(V - Nb) = Nk_B T. \qquad (6.17)$$

Note that this is a cubic equation for $\frac{N}{V}$ as a function of $(P, T)$.
For large volumes,

$$P \simeq \frac{Nk_B T}{V} + \left(\frac{N}{V}\right)^2 (b\,k_B\,T - a), \qquad (6.18)$$

and Eq. (6.17) thus entails a first correction to the ideal gas law that involves, in different ways, the two parameters $a$ and $b$.

Let us now discuss in detail the isothermal curves for the van der Waals equation of state (6.17). To begin with, for high temperatures or low densities the corrections are negligible, and the isothermal curves approach those determined by the ideal gas law. As $T$ is lowered, however, one reaches a "critical" temperature $T_c$ where the isothermal curve has a horizontal inflection point. The two conditions that determine the corresponding values of $(P, V, T)$ read

$$\left.\frac{\partial P}{\partial V}\right|_{\text{crit}} = -\frac{Nk_B T_c}{(V_c - Nb)^2} + \frac{2aN^2}{V_c^3} = 0,$$
$$\left.\frac{\partial^2 P}{\partial V^2}\right|_{\text{crit}} = \frac{2Nk_B T_c}{(V_c - Nb)^3} - \frac{6aN^2}{V_c^4} = 0, \tag{6.19}$$

and imply the two relations

$$Nb = \frac{V_c}{3}, \qquad a\,\frac{N^2}{V_c^2} = \frac{9}{8}\,\frac{N k_B T_c}{V_c}. \tag{6.20}$$

Substituting these results into the van der Waals equation (6.17) turns it into

$$P_c\, V_c = \frac{3}{8} N k_B\, T_c, \tag{6.21}$$

which determines the critical pressure. Finally, in terms of the three relative quantities

$$\overline{P} = \frac{P}{P_c}, \quad \overline{T} = \frac{T}{T_c}, \quad \overline{V} = \frac{V}{V_c}, \tag{6.22}$$

Eq. (6.17) then takes the form

$$\left(\overline{P} + \frac{3}{\overline{V}^2}\right)\left(\overline{V} - \frac{1}{3}\right) = \frac{8}{3}\,\overline{T}. \tag{6.23}$$

Note that this result bears no reference to the intrinsic properties of the gas, and hence possesses a universal meaning within its region of validity.

Let us comment briefly on these facts. The isothermal curves for $T < T_c$, so as the one displayed in Figure 6.3, appear to violate the thermodynamic inequality (1.58), since at a given temperature the pressure should decrease for increasing volume. This occurs because in certain regions the van der Waals equation, which is cubic in $\frac{N}{V}$, admits three real solutions for a given pressure. Physically, however, in that temperature range the system should be trying to capture the liquid–vapor transition. This will require a proper amendment of the isothermal curves, to which we shall return shortly. Moreover, the pressure in Eq. (6.17) becomes negative, in some regions, below about 0.8 $T_c$, which sets a lower limit of applicability for the van der Waals equation, consistent with the approximations that led to it.

## 6.4 Phase Equilibria and the van der Waals Gas

Let us consider a system comprising two different phases in equilibrium. In Chapter 1 we provided an argument to the effect that their two chemical potentials must be equal,

$$\mu_1(P, T) = \mu_2(P, T), \tag{6.24}$$

a condition that must hold on their whole coexistence curve. One can thus compute the differential of both sides, which leads to

$$v_1\, dP - s_1\, dT = v_2\, dP - s_2\, dT, \tag{6.25}$$

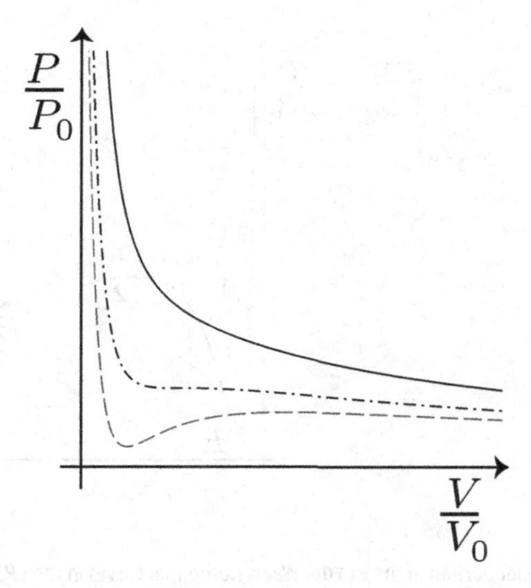


Figure 6.3 Typical isothermal curves in the $(P, T)$ plane for the van der Waals equation of state (6.17), for (continuous line) $T > T_c$, (dash-dotted line) $T = T_c$, and (dashed line) $T < T_c$ (dashed line). Note the strange behavior of the last curve, which violates the thermodynamic inequality (1.58).

since the two phases in equilibrium are characterized by identical values of $P$ and $T$. These considerations lead to the Clausius–Clapeyron equation

$$\frac{dP}{dT} = \frac{s_2 - s_1}{v_2 - v_1}, \tag{6.26}$$

which highlights that the pressure $P$ only depends on the temperature $T$ along the coexistence curve. Consequently, the isothermal curves in the $(P, V)$ plane when two phases are in equilibrium ought to be *horizontal*.

Having recognized that the pressure should be constant along the coexistence curve of two phases at a given temperature, we shall soon see how to amend the curves in Figure 6.3 accordingly. Before doing this, however, let us note that one can write

$$s_2 - s_1 = \frac{\Delta H}{T}, \tag{6.27}$$

where $\Delta H$ is usually called *latent heat*. For the liquid–vapor transition $v_2 \gg v_1$, and for $v_2$, one can use the ideal gas law, so that

$$\frac{dP}{dT} \simeq \frac{P\,\Delta H}{k_B\,T^2}. \tag{6.28}$$

This equation integrates to

$$P = P_0\, e^{-\frac{\Delta H}{k_B T}} \tag{6.29}$$

if $\Delta H$ does not vary appreciably along the transition curve. This result is qualitatively interesting, although $\Delta H$ does decrease with increasing $T$, to finally vanish at the critical point where the coexistence curve terminates and the very nature of the transition changes. Anticipating some terminology, the curve of first-order transitions terminates at $T_c$, where the transition becomes of second order, and the specific

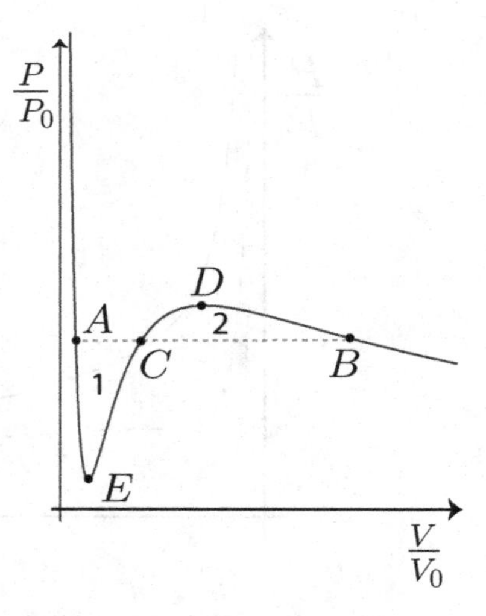


**Figure 6.4** The inner portion of the van der Waals isothermal curves in the $(P, T)$ plane $T < T_c$ does not correspond to stable configurations and must be replaced by the horizontal segment $ACB$, whose height is determined by the condition that the areas of regions 1 and 2 in the figure coincide. This "Maxwell construction" is tantamount to imposing the equality of the chemical potentials $\mu_L$ and $\mu_V$ of the two "pure" phases at the endpoints $A$ and $B$. The other branches $AE$ and $BD$ correspond to metastable phases.

volumes of the two phases coincide. The critical point for the water–vapor system occurs for $T_c \simeq 650$ K and $P_c \simeq 218$ Atm.

As we have stressed, the isothermal curves in the $(P, V)$ plane should contain horizontal segments in regions where two phases coexist. Therefore, in order to amend the original van der Waals curves, for any given $T < T_c$ one must determine the vertical positions $P_M(T)$ of these horizontal segments. This can be done by the Maxwell construction, which grants that the chemical potentials of the two "pure" phases at $A$ (liquid) and $B$ (vapor) be equal, and demands that the two areas in Figure 6.4 coincide. At a given temperature $T$, the variation of $\mu$ is indeed

$$\Delta\mu = \int_{P_1}^{P_2} v\, dP, \tag{6.30}$$

where the integral must be computed along the (original) van der Waals curve: the negative contributions to the area from the $AE$ and $DB$ portions of the curve in Figure 6.3 are thus to compensate the positive one from $ED$. Note that Eq. (6.30) can be cast in the equivalent form

$$\Delta\mu = \int_A^B d(vP) - \int_{v_L}^{v_V} P\, dv, \tag{6.31}$$

and therefore the Maxwell construction translates into the convenient conditions

$$P_M(T)(v_V - v_L) = \int_{v_L}^{v_V} P(T, v)\, dv, \tag{6.32}$$

$$P_M(T) = P(T, v_V) = P(T, v_L), \tag{6.33}$$

which determine the coexistence pressure $P_M(T)$ and the two specific volumes $v_V$ and $v_L$ corresponding to the points B and A in Figure 6.4.

Let us now discuss how the Maxwell prescription affects the free energy. To this end, it is convenient to refer to the Helmholtz free energy per particle,

$$a(T, v) = \frac{A(T, V, N)}{N}, \tag{6.34}$$

whose isothermal variations are given by

$$a(T, v_1) - a(T, v_2) = -\int_{v_1}^{v_2} P(T, v)\, dv. \tag{6.35}$$

In the coexistence region, the actual pressure is constant and we have denoted by $P_M(T)$ its value at a given temperature $T$. Therefore, $a(T, v)$ is not simply obtained from Eq. (6.14), but for $v_L < v < v_V$ must be defined according to

$$a(T, v) = a(T, v_L) - P_M(T)(v - v_L). \tag{6.36}$$

In this fashion, the continuity of $a(T, v)$ at $v = v_L$ holds manifestly, while the continuity at $v = v_V$ is guaranteed by the Maxwell construction (6.32), according to which

$$a(T, v_V) - a(T, v_L) = -P_M(T)(v_V - v_L). \tag{6.37}$$

Equation (6.36) can be cast in the more symmetrical form

$$a(T, v) = x_L\, a(T, v_L) + x_V\, a(T, v_V), \tag{6.38}$$

where $x_L$ and $x_V$ are positive numbers subject to the constraints

$$v = x_L\, v_L + x_V\, v_V, \qquad x_L + x_V = 1, \tag{6.39}$$

so that

$$x_L = \frac{v_V - v}{v_V - v_L}, \qquad x_V = \frac{v - v_L}{v_V - v_L}. \tag{6.40}$$

Note that the two ratios characterize the numbers $N_L$ and $N_V$ of particles that condense or remain in the vapor phase,

$$x_L = \frac{N_L}{N}, \qquad x_V = \frac{N_V}{N}. \tag{6.41}$$

This natural identification is consistent since, taking into account that the liquid and the vapor occupy volumes $V_L$, $V_V$, so that

$$v_L = \frac{V_L}{N_L}, \qquad v_V = \frac{V_V}{N_V}, \tag{6.42}$$

Eqs. (6.39) are simply equivalent to the conditions

$$V = V_L + V_V, \qquad N = N_L + N_V, \tag{6.43}$$

which identify the system, in the coexistence region, as an admixture of liquid and vapor, in relative amounts determined by $x_L$ and $x_V$.

Let us pause briefly to elaborate on this construction. Strictly speaking, in a simple homogeneous setting, one can only describe the two "pure" phases at the endpoints of the transition, and the van der Waals theory does this. The emergence of water droplets in vapor, or for that matter of icy islands in water, makes the system inhomogeneous and thus far more complicated. However, taking into account two important facts, namely

- that $\mu$ is a state function, which ought to be identical for the two pure phases in diffusive equilibrium,
- that the isothermal curves in the $(P, V)$ plane ought to be horizontal in the transition region,

suffices to link the relative abundances in the two-phase system to the two homogeneous limits.

Let us conclude this section with a list of thermodynamic functions for the van der Waals gas, expressed in terms of relative variables $\overline{P}$, $\overline{V}$, and $\overline{T}$:

$$\begin{aligned}
\overline{P} &= \frac{8\,\overline{T}}{3\,\overline{V}-1} - \frac{3}{\overline{V}^2}, \\
\frac{A}{N k_B T_c} &= -\,\overline{T}\log\left[n_Q(T_c)\,\frac{e\,V_c}{3N}\left(3\,\overline{V}-1\right)\overline{T}^{\frac{3}{2}}\right] - \frac{9}{8\,\overline{V}}, \\
U &= \frac{3}{2} N k_b\, T_c \left(\overline{T} - \frac{3}{4\,\overline{V}}\right), \\
\frac{S}{Nk} &= \log\left[n_Q(T_c)\,\frac{e\,V_c}{3N}\left(3\,\overline{V}-1\right)\overline{T}^{\frac{3}{2}}\right] + \frac{5}{2}, \\
\frac{\mu}{k_B T_c} &= -\,\overline{T}\log\left[n_Q(T_c)\,\frac{e\,V_c}{3N}\left(3\,\overline{V}-1\right)\overline{T}^{\frac{3}{2}}\right] - \frac{9}{4\,\overline{V}}.
\end{aligned} \tag{6.44}$$

## 6.5 The Gibbs Phase Rule

Let us conclude this chapter with a brief discussion of a general property of phase equilibria: the Gibbs phase rule. In general, this rule determines the maximum number of phases that can coexist in equilibrium. Consider a system involving $n$ components in $r$ phases. As we have just seen for the van der Waals gas, the equilibrium conditions identify the values of the chemical potential of the phases present in equilibrium for each component,

$$\begin{aligned}
\mu_1^1 &= \cdots = \mu_r^1, \\
&\vdots \\
\mu_1^n &= \cdots = \mu_r^n.
\end{aligned} \tag{6.45}$$

These are $n(r-1)$ equations for a system described via $2 + (n-1)r$ variables, i.e. $P$, $T$, and $n-1$ concentrations for each of the $r$ phases. Solutions to Eqs. (6.45) thus exist generically provided

$$n(r-1) \leq 2 + (n-1)r, \tag{6.46}$$

or if

$$r \le n+2. \tag{6.47}$$

For water, with only one component, there is indeed a triple point for $T \simeq 273$ K and low pressure $P \simeq 6\times 10^{-3}$ Atm (and there are actually other triple points at very high pressures, which involve solid phases with different symmetries).

## Bibliographical Notes

The van der Waals theory is nicely discussed, for instance, in [16, 23, 29, 54].

## Problems

**6.1** Compute $\left(\frac{\partial U}{\partial V}\right)_T$ for the van der Waals gas and verify the result in problem 1.1.

**6.2** Construct $S$, $H$, and $G$ for the van der Waals gas.

**6.3** Compute $C_V$, $C_P$, and their difference for the van der Waals gas, and verify Eq. (1.44). Show that $C_V$ is independent of the volume $V$.

**6.4** Compute explicitly the efficiency of a Carnot cycle for a van der Waals gas, and show that it is identical to the one obtained for an ideal gas, as expected from general principles.

**6.5** We would like to characterize the thermodynamic behavior of a liquid–vapor mixture slightly below the critical temperature $T_c$.

1. Show that, by letting $\overline{T} = 1 - \epsilon^2$ and

$$\overline{V}_L = 1 - a\epsilon + b\epsilon^2 - c\epsilon^3 + \cdots,$$
$$\overline{V}_V = 1 + a\epsilon + b\epsilon^2 + c\epsilon^3 + \cdots,$$

for the specific volumes $v_{L,R} = \frac{V_c}{N}\overline{V}_{L,R}$ corresponding to the points A and B in Figure 6.4 since the equations are even in $\epsilon$, the perturbative solutions of the Maxwell construction are

$$\overline{V}_L = 1 - 2\epsilon + \frac{18}{5}\epsilon^2 - \frac{147}{25}\epsilon^3 + \cdots,$$
$$\overline{V}_V = 1 + 2\epsilon + \frac{18}{5}\epsilon^2 + \frac{147}{25}\epsilon^3 + \cdots.$$

2. Use the preceding results to derive corresponding expansions, slightly below the critical temperature, for $A$, $U$, and $C_V$.
3. Show that $C_V$ experiences at $T_c$ a jump discontinuity with respect to its value above the critical temperature, which is $\frac{3}{2}Nk_B$ according to Eq. (6.16), and

$$C_V|_{T\to T_c^+} - C_V|_{T\to T_c^-} = -\frac{9}{2}Nk_B.$$

**6.6** Using the results of the preceding problem:

1. Use the Clausius–Clapeyron equation to obtain an expression for the latent heat slightly below $T_c$, and show that it vanishes at $T_c$.
2. Study the behavior of the specific heat $C_P$ across the critical temperature.
3. Study the behavior of the compressibilities $\kappa_T$ and $\kappa_S$ of Eq. (1.59) across the critical temperature.

**6.7** Assume that a gas of $N$ classical particles in $d$ dimensions is described by the Hamiltonian

$$H = \sum_{i=1}^{N} \frac{\mathbf{p}_i^2}{2m} + U(\mathbf{q}_1, \ldots, \mathbf{q}_N),$$

where the potential $U$ satisfies the scaling relation

$$U(\lambda \mathbf{q}_1, \ldots, \lambda \mathbf{q}_N) = |\lambda|^s U(\mathbf{q}_1, \ldots, \mathbf{q}_N).$$

Assume in addition that, in the thermodynamic limit, the quantity

$$a(n, T) = - \lim_{N,V \to \infty} \frac{\log Z}{N},$$

where $n = \frac{N}{V}$, is well defined.

1. Prove that the system has an equation of state of the following form:

$$P(n, T) = T^{1-\frac{d}{s}} \varphi(nT^{\frac{d}{s}}),$$

where $\varphi$ is a suitable function.

2. If the system undergoes a first-order phase transition at a certain temperature, can a critical point exist above which the transition disappears?

**6.8** Consider phase transitions in which:

1. the volume does not change;
2. the enthalpy does not change.

Derive the corresponding variants of the Clausius–Clapeyron equation, which are known as Ehrenfest equations.

**6.9** Consider a spherical droplet of liquid of radius $r$ surrounded by the corresponding vapor, and let $\gamma$, a constant, be the surface tension of the droplet, so that the Gibbs free energy for the system is

$$G = N_d \, \mu_d + N_v \, \mu_v + 4\pi\gamma \, r^2,$$

where $\mu_d$ and $\mu_v$ are the chemical potentials for the droplet and the vapor.

a. Assuming that the droplet be homogeneous, show that the equilibrium condition can be cast in the form

$$\mu_d - \mu_v + \frac{2\,\gamma}{n_d \, r} = 0,$$

where $n_d$ is the number of molecules per unit volume in the droplet.

b. Using Eq. (1.34), show that thermal and mechanical equilibrium imply

$$\frac{1}{n_d} - \frac{1}{n_v} - \frac{2\gamma}{n_d r^2}\left(\frac{\partial r}{\partial P}\right)_T = 0.$$

c. Integrate this equation, taking into account that $n_v \ll n_d$, and show that

$$P = P_\infty \, e^{\frac{2\gamma}{k_B T n_d r}},$$

which connects the typical size of the droplets to the temperature, for a given pressure $P$. The value $P_\infty$ can be regarded as the vapor pressure where the transition to the liquid phase occurs, along the lines of what we have seen elsewhere in this chapter. Or equivalently, the value of the vapor pressure at a planar interface.

**6.10** Study the screening of a charge $Q$ in a neutral plasma at a temperature $T$.

1. Consider the Poisson equation in the presence of Boltzmann distributions of charged particles, and show that it takes the form

$$\nabla^2 V = 4\pi \frac{\sum_i n_i^{(0)} q_i^2}{k_B T} V - 4\pi Q \delta(\mathbf{r}).$$

2. Show that the solution is of the form

$$V = \frac{Q}{r} e^{-\frac{r}{\rho}},$$

and identify the Debye–Hückel length $\rho$.

# 7 The Grand Canonical Ensemble

In the Chapter 6 we described how the van der Waals gas and the Maxwell construction can model the equilibrium between liquid and vapor phases. In the coexistence region, the numbers of liquid and vapor molecules undergo large fluctuations, which motivates one to address systems with variable numbers of components. This leads to the grand canonical ensemble, which will also prove very convenient in later chapters, for ideal quantum gases.

## 7.1 The Grand Canonical Equations

Our starting point is now

$$\sigma = -\sum_{I,N} p_{I,N} \log p_{I,N} - \lambda_1 \left( \sum_{I,N} p_{I,N} - 1 \right) - \lambda_2 \left( \sum_{I,N} p_{I,N} E_{I,N} - U \right) - \lambda_3 \left( \sum_{I,N} p_{I,N} N - \mathcal{N} \right), \tag{7.1}$$

where the various quantities carry two labels, $I$ and $N$, which refer here to the $I$th energy level in a system with $N$ particles. Moreover, the third Lagrange multiplier, $\lambda_3$, identifies $\mathcal{N}$ with the average number of particles present in the ensemble. As a result,

$$\frac{\partial \sigma}{\partial p_{I,N}} \equiv -\log p_{I,N} - 1 - \lambda_1 - \lambda_2 E_{I,N} - \lambda_3 N = 0, \tag{7.2}$$

so that

$$p_{I,N} = z^N e^{-(1+\lambda_1)} e^{-\lambda_2 E_{I,N}}, \tag{7.3}$$

where

$$z = e^{-\lambda_3}. \tag{7.4}$$

This extremum is again a maximum for $\sigma$, since

$$\frac{\partial^2 \sigma}{\partial p_{I_1,N_1} \partial p_{I_2,N_2}} = -\delta_{I_1 I_2} \delta_{N_1 N_2}, \tag{7.5}$$

and will be identified again with $\frac{S}{k_B}$, where $S$ is the entropy and $k_B$ is Boltzmann's constant.

In fixing the three Lagrange multipliers that enter Eq. (7.3), one is led to define the "grand partition function"

$$\mathcal{Q} = e^{1+\lambda_1} = \sum_{N=0}^{\infty} z^N \sum_{I} e^{-\lambda_2 E_{I,N}} = \sum_{N=0}^{\infty} z^N Z_N, \tag{7.6}$$

a weighted sum of partition functions

$$Z_N(\lambda_2) = \sum_{I} e^{-\lambda_2 E_{I,N}}, \tag{7.7}$$

for systems involving all possible numbers $N$ of particles. Or, if you will, a generating function for them. We shall soon find important cases where $\mathcal{Q}$ is far simpler to compute than the $Z_N$. One can now express $U$ and $\mathcal{N}$ in terms of $\mathcal{Q}$ as

$$U = -\frac{\partial \log \mathcal{Q}}{\partial \lambda_2}, \qquad \mathcal{N} = \frac{\partial \log \mathcal{Q}}{\partial \log z}, \tag{7.8}$$

and consequently

$$\frac{S}{k_B} = -\sum_{I,N} p_{I,N} \left(N \log z - \log \mathcal{Q} - \lambda_2 E_{I,N}\right), \tag{7.9}$$

or

$$\frac{S}{k_B} = -\mathcal{N} \log z + \log \mathcal{Q} + \lambda_2 U. \tag{7.10}$$

From the first and second laws of thermodynamics we know that

$$\frac{dS}{k_B} = \beta\, dU + \beta P dV - \beta \mu\, d\mathcal{N}, \tag{7.11}$$

where now the (average) total number of particles is denoted by $\mathcal{N}$ as above, and therefore

$$\frac{1}{k_B}\left(\frac{\partial S}{\partial U}\right)_{V,\mathcal{N}} = \beta, \qquad \frac{1}{k_B}\left(\frac{\partial S}{\partial \mathcal{N}}\right)_{U,V} = -\beta \mu. \tag{7.12}$$

Proceeding as in Chapter 2, one can now relate $\lambda_2$ and $z$ to thermodynamic functions. To begin with, from Eq. (7.9)

$$\begin{aligned}\frac{1}{k_B}\left(\frac{\partial S}{\partial U}\right)_{V,\mathcal{N}} = &-\mathcal{N}\frac{\partial \log z}{\partial U} + \frac{\partial \log \mathcal{Q}}{\partial \lambda_2}\frac{\partial \lambda_2}{\partial U} \\ &+ \frac{\partial \log \mathcal{Q}}{\partial \log z}\frac{\partial \log z}{\partial U} + \lambda_2 + U\frac{\partial \lambda_2}{\partial U},\end{aligned} \tag{7.13}$$

and using Eqs. (7.8) and (7.11), this expression leads to

$$\lambda_2 = \beta. \tag{7.14}$$

In a similar fashion,

$$\begin{aligned}\frac{1}{k_B}\left(\frac{\partial S}{\partial \mathcal{N}}\right)_{U,V} = &-\log z - \mathcal{N}\frac{\partial \log z}{\partial \mathcal{N}} + \frac{\partial \log \mathcal{Q}}{\partial \lambda_2}\frac{\partial \lambda_2}{\partial \mathcal{N}} \\ &+ \frac{\partial \log \mathcal{Q}}{\partial \log z}\frac{\partial \log z}{\partial \mathcal{N}} + U\frac{\partial \lambda_2}{\partial \mathcal{N}}.\end{aligned} \tag{7.15}$$

links the "fugacity" $z$ to $\beta$ and the chemical potential $\mu$, according to

$$z = e^{\beta \mu}. \tag{7.16}$$

Moreover, Eq. (7.10) can be turned into

$$\log \mathcal{Q} = \beta \left( \mu \mathcal{N} - U + TS \right) = -\beta \Omega, \tag{7.17}$$

which relates $\mathcal{Q}$ to the grand-potential, $\Omega$, discussed in Chapter 1.

Summarizing, the basic quantity of the grand canonical ensemble is the grand partition function,

$$\mathcal{Q}(z, V, T) = \sum_{N=0}^{\infty} z^N \sum_{I} e^{-\beta E_{I,N}} = \sum_{N=0}^{\infty} z^N Z_N(\beta), \tag{7.18}$$

here also expressed in terms of the partition functions $Z_N$, which determines $\Omega$ according to

$$\Omega = -k_B T \log \mathcal{Q}(z, V, T). \tag{7.19}$$

The grand potential, $\Omega$, then determines the internal energy $U$, the average number of particles $\mathcal{N}$, and the equation of state according to Eqs. (1.38), so that

$$\begin{aligned} U &= -\left( \frac{\partial \log \mathcal{Q}(z, V, T)}{\partial \beta} \right)_{V,z}, \\ \mathcal{N} &= \left( \frac{\partial \log \mathcal{Q}(z, V, T)}{\partial \log z} \right)_{T,V}, \\ P &= k_B T \left( \frac{\partial \log \mathcal{Q}(z, V, T)}{\partial V} \right)_{T,z}. \end{aligned} \tag{7.20}$$

Inverting the second of these equations will be difficult in general, but it would determine $z\left(T, \frac{V}{N}\right)$, thus yielding more conventional forms for the internal energy and the equation of state:

$$\begin{aligned} U &= - \left. \frac{\partial \log \mathcal{Q}(z, V, T)}{\partial \beta} \right|_{z=z\left(T, \frac{V}{N}\right)}, \\ P &= k_B T \left. \frac{\partial \log \mathcal{Q}(z, V, T)}{\partial V} \right|_{z=z\left(T, \frac{V}{N}\right)}. \end{aligned} \tag{7.21}$$

When $\Omega$ is extensive, one can determine the equation of state from

$$\frac{PV}{k_B T} = \log \mathcal{Q}(z, V, T)|_{z=z\left(T, \frac{V}{N}\right)}. \tag{7.22}$$

In the grand canonical ensemble, there are fluctuations in the internal energy $U$, but also in the number of particles $\mathcal{N}$. Let us now describe them: while the former are as in the canonical ensemble, the fluctuations of $\mathcal{N}$ present an interesting subtlety. To begin with, for the energy fluctuations one finds indeed that

$$(\Delta U)^2 \equiv \langle E^2 \rangle - (\langle E \rangle)^2 = -\left(\frac{\partial U}{\partial \beta}\right)_{V,z} = k_B T^2 C_V, \tag{7.23}$$

and since $C_V$ scales like $\mathcal{N}$,

$$\frac{\Delta U}{U} \sim \mathcal{N}^{-\frac{1}{2}}. \tag{7.24}$$

As in the canonical ensemble, the energy fluctuations disappear in the thermodynamic limit $\mathcal{N} \to \infty$.

On the other hand, away from phase coexistence regions, the fluctuations in $\mathcal{N}$ are determined by

$$(\Delta \mathcal{N})^2 \equiv \langle \mathcal{N}^2 \rangle - (\langle \mathcal{N} \rangle)^2 = \left(\frac{\partial \mathcal{N}}{\partial \log z}\right)_{T,V} = k_B T V \left(\frac{\partial n}{\partial \mu}\right)_{T,V}, \tag{7.25}$$

and it is instructive to take a closer look at this quantity, which one can express in terms of the specific volume as

$$(\Delta \mathcal{N})^2 \equiv -\frac{k_B T \mathcal{N}}{v}\left(\frac{\partial v}{\partial \mu}\right)_{T,V}, \tag{7.26}$$

using Eq. (1.35). Equation (1.34) and the chain rule then lead to

$$\left(\frac{\partial v}{\partial \mu}\right)_{T,V} = \left(\frac{\partial v}{\partial P}\right)_{T,V}\left(\frac{\partial P}{\partial \mu}\right)_{T,V} = \frac{1}{v\left(\frac{\partial P}{\partial v}\right)_{T,V}}, \tag{7.27}$$

and finally

$$(\Delta \mathcal{N})^2 = \frac{\mathcal{N} k_B T}{v^2}\,\frac{1}{-\left(\frac{\partial P}{\partial v}\right)_{T,V}}. \tag{7.28}$$

The sign of the last factor was ascertained in Section 1.4, and we see that insofar as this expression is not singular, it leads to

$$\frac{\Delta \mathcal{N}}{\mathcal{N}} \sim \mathcal{N}^{-\frac{1}{2}}. \tag{7.29}$$

As we have stressed, all these considerations hold away from coexistence regions where fluctuations are large. However, there are subtleties even in simple cases, as we can now see in detail.

## 7.2 Two Instructive Examples

Our first example is the ideal gas of Chapter 2, for which, as we have seen in Eq. (2.81),

$$Z_N = \frac{V^N}{N!}\left(\frac{2\pi m}{h^2 \beta}\right)^{\frac{3N}{2}}, \tag{7.30}$$

so that

$$\mathcal{Q} = e^{z\,V\left(\frac{2\,\pi\,m}{h^2\,\beta}\right)}. \tag{7.31}$$

Equations (7.20) thus take the form

$$\begin{aligned} U &= \frac{3}{2\,\beta}\,z\,V\left(\frac{2\,\pi\,m}{h^2\,\beta}\right), \\ \mathcal{N} &= z\,V\left(\frac{2\,\pi\,m}{h^2\,\beta}\right), \\ \frac{P\,V}{k_B\,T} &= z\,V\left(\frac{2\,\pi\,m}{h^2\,\beta}\right), \end{aligned} \tag{7.32}$$

and combining them one can simply recover the previous results of Eqs. (2.46), (2.47), (2.85), and (2.86):

$$\begin{aligned} U &= \frac{3}{2}\,\mathcal{N}\,k_B\,T, \\ \mu &= k_B\,T\,\log\left(\frac{n}{n_Q(T)}\right), \\ P &= \frac{k_B\,T}{v}. \end{aligned} \tag{7.33}$$

Hence

$$\left(\frac{\partial P}{\partial\,v}\right)_T = -\,\frac{k_B\,T}{v^2} \tag{7.34}$$

and

$$(\Delta\,\mathcal{N})^2 = \mathcal{N}, \tag{7.35}$$

with no surprises.

This example entails a general lesson: the equation of state in (7.33) holds for all noninteracting systems of identical particles for which

$$Z_N = \frac{Z_1^N}{N!} \tag{7.36}$$

in terms of the one-particle partition function $Z_1 = Z_1(T, V)$. In these cases the grand partition function is

$$\mathcal{Q} = \sum_{N=0}^{\infty} \frac{(z\,Z_1)^N}{N!} = \exp\,(z\,Z_1), \tag{7.37}$$

and then Eqs. (7.20) determine

$$\mathcal{N} = z\,Z_1, \tag{7.38}$$

which yields indeed the equation of state for the perfect gas via Eq. (7.22),

$$P\,V = \mathcal{N}\,k_B\,T, \tag{7.39}$$

independently of $Z_1(T, V)$, which can generically encompass single-particle Hamiltonians of the type

$$H = \frac{\mathbf{p}^2}{2m} + U(\mathbf{r}) \tag{7.40}$$

for which $\Omega$ is properly extensive. Note also that the number of constituents $N$ that enter these sums is Poisson distributed, with

$$P(N) = \frac{\mathcal{N}^N}{N!} e^{-\mathcal{N}}. \tag{7.41}$$

However, surprises emerge when considering the classical gas of distinguishable oscillators introduced in Chapter 2. In this case, as we have seen in Eq. (2.88),

$$Z_N = \frac{1}{(\hbar \omega \beta)^{3N}}, \tag{7.42}$$

and consequently

$$\log \mathcal{Q} = -\log\left(1 - \frac{z}{(\hbar \omega \beta)^3}\right). \tag{7.43}$$

For this simple system, Eqs. (7.20) yield

$$\mathcal{N} = \frac{\frac{z}{(\hbar \omega \beta)^3}}{1 - \frac{z}{(\hbar \omega \beta)^3}}, \qquad U = \frac{3}{\beta} \frac{\frac{z}{(\hbar \omega \beta)^3}}{1 - \frac{z}{(\hbar \omega \beta)^3}}, \tag{7.44}$$

and

$$P = 0. \tag{7.45}$$

The first two combine nicely to recover the equipartition law

$$U = 3 \mathcal{N} k_B T, \tag{7.46}$$

while the first can be cast in the form

$$\mathcal{N} + 1 = \frac{1}{1 - \frac{z}{(\hbar \omega \beta)^3}}, \tag{7.47}$$

so that

$$\mu = -k_B T \log\left[\frac{1 + \frac{1}{\mathcal{N}}}{(\beta \hbar \omega)^3}\right]. \tag{7.48}$$

Consequently,

$$\Omega = -k_B T \log(\mathcal{N} + 1), \tag{7.49}$$

which is sub-extensive. Proceeding along these lines, one cannot use Eq. (7.28), since the pressure vanishes identically. However, one can still resort to Eq. (7.25), which gives

$$(\Delta \mathcal{N})^2 = \mathcal{N}^2 + \mathcal{N}, \tag{7.50}$$

so that there are large fluctuations in $\mathcal{N}$ in the grand canonical ensemble for this simple system.

Alternatively, one can *define* the pressure via (7.22), obtaining

$$P = \frac{k_B T}{V} \log(\mathcal{N} + 1), \tag{7.51}$$

which also vanishes in the thermodynamic limit, and in this fashion one can recover Eq. (7.50) also from Eq. (7.28). The peculiar chemical potential in Eq. (7.48), which becomes independent of $\mathcal{N}$ in the thermodynamic limit, may be held responsible for the large fluctuations present in this system. We already encountered this behavior in Eq. (4.61), in connection with the coexistence of particle-like and wave-like aspects in the photon gas.

## Bibliographical Notes

Pedagogical introductions to the grand canonical ensemble can be found in many excellent books, including [23, 36, 49, 50, 51, 54].

# 8 Quantum Statistics

While in classical mechanics one could distinguish particles with identical attributes following their trajectories determined by the initial conditions, this is impossible in quantum mechanics. As a result, the quantum description of identical particles entails novel subtleties and brings along exotic types of behavior.

The splitting of spectral lines in a magnetic field led Pauli to conceive the electron "spin," a two-valued intrinsic angular momentum, together with what is now called the Pauli principle. The behavior of noble gases and the special numbers associated to their shells, and more generally the gross features of the periodic table, are explained if two electrons in an atom cannot possess identical orbital and spin attributes. The thermal behavior of this first type of particle, when mutual interactions are negligible, is described by the ideal Fermi gas. Identity is thus a deeper concept in the microscopic world, and we now recognize that there are two families of particles in the quantum realm. The first family comprises particles with half odd integer spin, like the electrons, which are usually called "fermions." The second family comprises particles with integer spin, which are usually called "bosons," and includes photons, whose random emissions and absorptions in a cavity underlie Planck's treatment of blackbody radiation. The thermal behavior of bosons is described by Bose–Einstein statistics, and when their number is conserved a novel phenomenon, Bose–Einstein condensation, emerges.

The modern treatment of spin rests on multi-component Schrödinger wavefunctions, which originate from relativistic fields. The relativistic treatment afforded by quantum field theory makes the correspondence between half odd integer spins and Fermi–Dirac statistics on the one hand, and between integer spins and Bose–Einstein statistics on the other, inevitable for systems of particles that transform under one-dimensional representations of the permutation group $S_N$. These are apparently the only options that are realized in nature, in the general case. This chapter is devoted to exploring systems of Fermi and Bose particles in the simplest settings, and to elucidate their strikingly different thermal behaviors.

## 8.1 Identical Particles in Quantum Mechanics

Let us begin by considering the Schrödinger equation for a pair of identical particles,

$$H(1,2)|\psi(1,2)\rangle = E|\psi(1,2)\rangle, \tag{8.1}$$

while also taking into account that, for two *identical Fermi particles*,

$$|\psi(1,2)\rangle = -|\psi(2,1)\rangle, \tag{8.2}$$

while for two *identical Bose particles*

$$|\psi(1,2)\rangle = |\psi(2,1)\rangle. \tag{8.3}$$

This is the correct prescription in general, but when mutual interactions are negligible, so that

$$H(1,2) \simeq H(1) + H(2), \tag{8.4}$$

the solutions of the single-particle problem,

$$H(1)|\psi_\alpha(1)\rangle = E_\alpha|\psi_\alpha(1)\rangle, \tag{8.5}$$

where $\alpha = 0, 1, 2, \ldots$ labels the one-particle states, also determine the corresponding solutions for two-particle systems. Thus, for a pair of Fermi particles the allowed eigenvectors corresponding to the energy eigenvalues $E_\alpha + E_\beta$ read

$$|\psi_{\alpha,\beta}(1,2)\rangle = \frac{1}{\sqrt{2}}\Big[|\psi_\alpha(1)\rangle\,|\psi_\beta(2)\rangle - |\psi_\beta(1)\rangle|\psi_\alpha(2)\rangle\Big], \tag{8.6}$$

and the independent states are all captured if $\alpha > \beta$. Note also that these state vectors vanish identically if $\alpha = \beta$, which reflects the standard form of the Pauli principle: two Fermi particles can never have all identical quantum numbers.

On the other hand, for Bose particles, again neglecting mutual interactions,

$$|\psi_{\alpha,\beta}(1,2)\rangle = \frac{1}{\sqrt{2}}\Big[|\psi_\alpha(1)\rangle\,|\psi_\beta(2)\rangle + |\psi_\beta(1)\rangle|\psi_\alpha(2)\rangle\Big], \tag{8.7}$$

and the independent states of this type correspond again to the range $\alpha > \beta$. However, in this case, there are additional state vectors,

$$|\psi_{\alpha,\alpha}(1,2)\rangle = |\psi_\alpha(1)\rangle\,|\psi_\alpha(2)\rangle, \tag{8.8}$$

since the two Bose particles can also occupy the same state, for any available value of $\alpha$.

Now we would like to compare the partition functions for a pair of Fermi or Bose particles with the distinguishable case, where the result, as we saw in previous chapters, would be simply the square of the single-particle contribution. The density matrix for two distinguishable (Boltzmann) particles,

$$\rho_{\text{Boltzmann}} = \sum_{\alpha_1,\alpha_2} e^{-\beta(E_{\alpha_1}+E_{\alpha_2})}\,|\alpha_1\rangle\,|\alpha_2\rangle\,\langle\alpha_1|\,\langle\alpha_2|, \tag{8.9}$$

where we have used an evident shorthand notation for the states, implies the result

$$Z_{\text{Boltzmann}} = \text{Tr}\left(\rho_{\text{Boltz}}\right) = \left[Z_1(\beta)\right]^2, \tag{8.10}$$

where

$$Z_1(\beta) = \sum_\alpha e^{-\beta E_\alpha}. \tag{8.11}$$

The density matrix for the Fermi case is different, and reads

$$\rho_{\text{Fermi}} = \frac{1}{2}\sum_{\alpha_1>\alpha_2} e^{-\beta(E_{\alpha_1}+E_{\alpha_2})} \times \Big(|\alpha_1\rangle\,|\alpha_2\rangle - |\alpha_2\rangle\,|\alpha_1\rangle\Big)\Big(\langle\alpha_1|\,\langle\alpha_2| - \langle\alpha_2|\,\langle\alpha_1|\Big), \tag{8.12}$$

and the corresponding partition function is thus

$$Z_{\text{Fermi}} = \sum_{\alpha_1 > \alpha_2} e^{-\beta\,(E_{\alpha_1}+E_{\alpha_2})} = \frac{1}{2}\sum_{\alpha_1 \neq \alpha_2} e^{-\beta\,(E_{\alpha_1}+E_{\alpha_2})}$$
$$= \frac{1}{2}\left[Z_1(\beta)\right]^2 - \frac{1}{2}Z_1(2\,\beta), \tag{8.13}$$

and in a similar fashion one can show that

$$Z_{\text{Bose}} = \sum_{\alpha_1 \geq \alpha_2} e^{-\beta\,(E_{\alpha_1}+E_{\alpha_2})} = \frac{1}{2}\left[Z_1(\beta)\right]^2 + \frac{1}{2}Z_1(2\,\beta), \tag{8.14}$$

since it also includes the "diagonal" contribution of Eq. (8.8).

Note that the first term in Eqs. (8.13) and (8.14) coincides with the result for a pair of Boltzmann particles in Eq. (8.10), up to the factor $\frac{1}{2}$ that reflects their identity, and dominates at high temperatures. One can thus foresee the general trend that emerged in Eq. (2.105), which holds in the nondegenerate limit. However, the intricacy of the correction terms grows rapidly with the number of particles. While general closed-form expressions exist, and we shall discuss a quantum example where $Z_N$ can be nicely computed in closed form in the next section, for these systems it is more convenient to resort to the grand canonical ensemble.

In general, the state vectors for noninteracting Bose particles involve sums over all permutations $P$ of the state labels involved,

$$\psi_B(\alpha_1, \cdots, \alpha_N) \sim \sum_P |\alpha_{P(1)}\rangle \cdots |\alpha_{P(N)}\rangle, \tag{8.15}$$

so that two or more labels can coincide, while the state vectors for noninteracting Fermi particles involve sums over permutations $P$ with signs determined by their (even or odd) nature,

$$\psi_F(m_1, \cdots, m_N) \sim \sum_P (-1)^P\, |Pm_1(1)\rangle \cdots |Pm_N(N)\rangle, \tag{8.16}$$

so that no two labels can coincide, in compliance with the Pauli principle.

All in all, the information encoded in these state vectors concerns merely the numbers of particles that occupy each of the available single-particle states. This number can be arbitrarily large for bosons, but can be only 0 or 1 for fermions. As a result, the corresponding partition functions can be recast into sums over the occupation numbers of single-particle states, according to

$$Z_{N,F} = \sum_{\{\sum n_\alpha = N,\, n_\alpha = 0,1\}} e^{-\beta \sum_\alpha n_\alpha\, \epsilon_\alpha}, \tag{8.17}$$

$$Z_{N,B} = \sum_{\{\sum n_\alpha = N\}} e^{-\beta \sum_\alpha n_\alpha\, \epsilon_\alpha}. \tag{8.18}$$

When computing the corresponding grand partition functions, the power $z^N$ distributes itself among the levels, and the sum over $N$ removes the constraint on the total occupation number, so that

$$\mathcal{Q}_F = \sum_{\{n_\alpha=0,1\}} e^{-\beta \sum_\alpha n_\alpha \epsilon_\alpha} z^{\sum_\alpha n_\alpha} = \prod_\alpha \left[1 + z e^{-\beta\epsilon_\alpha}\right], \tag{8.19}$$

$$\mathcal{Q}_B = \sum_{\{n_\alpha\}} e^{-\beta \sum_\alpha n_\alpha \epsilon_\alpha} z^{\sum_\alpha n_\alpha} = \prod_\alpha \left[\frac{1}{1 - z e^{-\beta\epsilon_\alpha}}\right]. \tag{8.20}$$

More generally, one could allow for degeneracy factors $g_\alpha$ associated to the various levels, which would turn the grand partition function into

$$\mathcal{Q} = \prod_\alpha \left[\sum_{n_\alpha} e^{\beta(\mu-\epsilon_\alpha)n_\alpha}\right]^{g_\alpha}. \tag{8.21}$$

In particular, for Fermi particles

$$\mathcal{Q}_{\rm F} = \prod_\alpha \left[1 + e^{\beta(\mu-\epsilon_\alpha)}\right]^{g_\alpha}, \tag{8.22}$$

while for Bose particles

$$\mathcal{Q}_{\rm B} = \prod_\alpha \left(1 - e^{\beta(\mu-\epsilon_\alpha)}\right)^{-g_\alpha}. \tag{8.23}$$

Note that, in view of the Taylor expansions,

$$(1+t)^{g_\alpha} = \sum_{n_\alpha=0}^{g_\alpha} \binom{g_\alpha}{n_\alpha} t^{n_\alpha}, \qquad (1-t)^{-g_\alpha} = \sum_{n_\alpha=0}^{\infty} \binom{g_\alpha + n_\alpha - 1}{n_\alpha} t^{n_\alpha}, \tag{8.24}$$

the grand partition functions (8.22) and (8.23) are the generating functions of the following Fermionic and Bosonic partition functions:

$$Z_{N,F} = \sum_{\{\sum n_\alpha = N\}} \binom{g_\alpha}{n_\alpha} e^{-\beta \sum_\alpha n_\alpha \epsilon_\alpha}, \tag{8.25}$$

$$Z_{N,B} = \sum_{\{\sum n_\alpha = N\}} \binom{g_\alpha + n_\alpha - 1}{n_\alpha} e^{-\beta \sum_\alpha n_\alpha \epsilon_\alpha}, \tag{8.26}$$

where the binomial coefficients reflect the options for placing the particles in the various energy levels, while also allowing for single or multiple occupancy (see Appendix B). From these expressions, one can simply recover the case of Boltzmann particles, and thus Eq. (7.37), since in the low-density limit

$$\left(1 \pm e^{\beta(\mu-\epsilon_\alpha)}\right)^{\pm g_\alpha} \simeq \exp\left(g_\alpha\, e^{\beta(\mu-\epsilon_\alpha)}\right), \tag{8.27}$$

and then both Eqs. (8.22) and (8.23) lead to

$$\mathcal{Q} = e^{z Z_1}. \tag{8.28}$$

Alternatively, one can recover this result from the power series in Eqs. (8.25) and (8.26), in the nondegenerate limit, proceeding as in Appendix B, which leads to the Boltzmann partition function

$$Z_N = \sum_{\{\sum n_\alpha = N\}} \frac{g_\alpha^{n_\alpha}}{n_\alpha!} e^{-\beta \sum_\alpha n_\alpha \epsilon_\alpha}. \tag{8.29}$$

## 8.2 Identical Oscillators in the Canonical Ensemble

Before moving to a detailed discussion of bosonic and fermionic systems within the grand canonical ensemble, it is instructive and amusing to consider a gas of indistinguishable one-dimensional quantum harmonic oscillators. For this special system one can indeed obtain relatively handy closed-form expressions for the partition functions $Z_N$. As we have seen, the single-particle energy levels are $\epsilon_\alpha = \hbar\omega(\alpha + \frac{1}{2})$, for $\alpha = 0, 1, 2, \ldots$, with no degeneracy, and hence the partition function for a corresponding Fermi gas reads

$$Z_N = q^{\frac{N}{2}} \sum_{0 \leq \alpha_1 < \alpha_2 < \cdots < \alpha_N} q^{\alpha_1 + \alpha_2 + \cdots + \alpha_N}, \tag{8.30}$$

where we have set $q = e^{-\beta\hbar\omega}$. The constraints in the sums can be solved explicitly by letting

$$\begin{aligned} \alpha_1 &= l_1, \\ \alpha_2 &= l_2 + l_1 + 1, \\ &\vdots \\ \alpha_N &= l_N + \cdots + l_2 + l_1 + N - 1. \end{aligned} \tag{8.31}$$

The $l_i$ are now allowed to take integer values from 0 to $\infty$, independently, and this reduces the partition function to

$$Z_N = q^{\frac{N}{2}} q^{1+2+\cdots+(N-1)} \sum_{l_1, l_2, \ldots, l_N} q^{N l_1 + (N-1) l_2 + \cdots + l_N}. \tag{8.32}$$

After recognizing the standard sum

$$1 + 2 + \cdots + N - 1 = \frac{N(N-1)}{2} \tag{8.33}$$

in the second exponent, the final result reads

$$Z_N = \frac{q^{\frac{N^2}{2}}}{\prod_{n=1}^{N}(1 - q^n)}. \tag{8.34}$$

A Bose gas of harmonic oscillators can be dealt with in a similar fashion: the sums in (8.30) range over $0 \leq \alpha_1 \leq \alpha_2 \leq \cdots \leq \alpha_N$ and can be disentangled by letting

$$\begin{aligned} \alpha_1 &= l_1, \\ \alpha_2 &= l_2 + l_1, \\ &\vdots \\ \alpha_N &= l_N + \cdots + l_2 + l_1, \end{aligned} \tag{8.35}$$

where the reader should notice the absence of fixed integers of Eq. (8.31) in the right-hand sides, so that finally

$$Z_N = \frac{q^{\frac{N}{2}}}{\prod_{n=1}^{N}(1 - q^n)}. \tag{8.36}$$

The comparison with the corresponding grand partition functions,

$$\mathcal{Q}(z) = \prod_{\alpha=0}^{\infty} (1 \pm z\, q^{\alpha+1/2})^{\pm 1}, \tag{8.37}$$

proves, as a byproduct, the nontrivial Taylor expansions

$$\prod_{\alpha=0}^{\infty} (1 + z\, q^{\alpha+1/2}) = \sum_{N=0}^{\infty} \frac{z^N q^{\frac{N^2}{2}}}{\prod_{n=1}^{N}(1-q^n)} \tag{8.38}$$

and

$$\prod_{\alpha=0}^{\infty} \frac{1}{1 - z\, q^{\alpha+1/2}} = \sum_{N=0}^{\infty} \frac{z^N q^{\frac{N}{2}}}{\prod_{n=1}^{N}(1-q^n)}. \tag{8.39}$$

A closer look at these systems unveils an interesting lesson. Since the harmonic potential effectively confines the particles to a finite region, sending $N$ to infinity should correspond to a regime of very high density, where quantum effects are not negligible. Therefore, *the thermodynamic limit and the nondegenerate limit are incompatible for indistinguishable particles subject to the harmonic potential.* Indeed, the fermionic partition function in Eq. (8.34) leads to

$$A = \frac{N^2}{2}\hbar\omega + k_B T \sum_{n=1}^{N} \log(1 - e^{-n\beta\hbar\omega}), \tag{8.40}$$

and hence

$$\begin{aligned} \mu &\simeq A(N) - A(N-1) \\ &= (N - \tfrac{1}{2})\hbar\omega + k_B T \log(1 - e^{-N\beta\hbar\omega}), \end{aligned} \tag{8.41}$$

which depends on $N$ and diverges in the thermodynamic limit. As a result, $z \to \infty$ as $N \to \infty$, which signals the regime of very high degeneracy. This correspondence should be intuitively clear, since the unbounded increase of $\mu$ makes it impossible to add more particles to the system. We shall discuss this point in more detail shortly. On the other hand, for bosonic oscillators Eq. (8.36) gives

$$A = \frac{N}{2}\hbar\omega + k_B T \sum_{n=1}^{N} \log(1 - e^{-n\beta\hbar\omega}), \tag{8.42}$$

and therefore

$$\mu \simeq \frac{\hbar\omega}{2} + k_B T \log(1 - e^{-N\beta\hbar\omega}), \tag{8.43}$$

which tends to the ground-state energy $\frac{\hbar\omega}{2}$ as $N \to \infty$, where particles indeed accumulate in a highly degenerate regime.

Nondegenerate regimes can be attained by considering the high-temperature limit *before* sending $N \to \infty$. In this case, the leading behavior of the partition functions (8.34) and (8.36) as $q \to 1$ at a fixed value of $N$ is

$$Z_N \sim \frac{1}{N!\,(\beta\hbar\omega)^N} = \frac{Z_1^N}{N!}, \tag{8.44}$$

and one recovers the expected result for classical indistinguishable oscillators. This is physically sensible, because at high temperatures large thermal fluctuations lead particles to explore higher and higher excited states of the spectrum, which spreads them on average further away from the bottom of the potential well. The situation was far simpler, in this respect, for the distinguishable oscillators that we considered in previous chapters, for which the formula $Z_N = Z_1^N$ is exact.

## 8.3 Nonrelativistic Fermi and Bose Gases

Let us now apply the grand canonical formalism to the important case of nonrelativistic particles in three dimensions. Their single-particle Hamiltonian is

$$H = \frac{\mathbf{p}^2}{2m}, \tag{8.45}$$

so that

$$\log \mathcal{Q}_{F,B} = \pm (2s+1) \sum_{\mathbf{p}} \log\left(1 \pm z\, e^{-\beta \frac{\mathbf{p}^2}{2m}}\right), \tag{8.46}$$

and we are also accounting for the presence of $2s+1$ distinct states of a given $\mathbf{p}$ if the particles have spin $s$. Moreover, we shall treat Fermi and Bose particles at the same time, with the upper sign referring to the former and the lower one referring to the latter.

In the usual large-volume limit,

$$\log \mathcal{Q}_{F,B} = \pm (2s+1) \frac{V}{h^3} \int d^3\mathbf{p} \, \log\left(1 \pm z e^{-\beta \frac{\mathbf{p}^2}{2m}}\right), \tag{8.47}$$

and consequently

$$\mathcal{N}_{F,B} = (2s+1) \frac{V}{h^3} \int d^3\mathbf{p} \, \frac{1}{\frac{1}{z} e^{\beta \frac{\mathbf{p}^2}{2m}} \pm 1} \tag{8.48}$$

and

$$U_{F,B} = (2s+1) \frac{V}{h^3} \int d^3\mathbf{p} \, \frac{\frac{\mathbf{p}^2}{2m}}{\frac{1}{z} e^{\beta \frac{\mathbf{p}^2}{2m}} \pm 1}. \tag{8.49}$$

Moreover, integrating by parts one can see that

$$\int d^3\mathbf{p} \, \log\left(1 \pm z e^{-\beta \frac{\mathbf{p}^2}{2m}}\right) = \frac{4\pi}{3} \left[p^3 \log\left(1 \pm z e^{-\beta \frac{\mathbf{p}^2}{2m}}\right)\right]_0^\infty \pm \frac{2}{3} \int d^3\mathbf{p} \, \frac{\frac{\mathbf{p}^2}{2m}}{\frac{1}{z} e^{\beta \frac{\mathbf{p}^2}{2m}} \pm 1}, \tag{8.50}$$

and the first term vanishes, so that, using Eq. (7.22), for these systems

$$PV = \frac{2}{3} U. \tag{8.51}$$

Note that the two integers reflect, respectively, the exponent in the single-particle Hamiltonian and the dimension of the space where the particles live. Consequently, for massless relativistic particles, for which $E = pc$, the factor would be $\frac{1}{3}$.

Note also that these expressions identify the average occupation numbers of the single-particle levels,

$$n_{F,B} = \frac{1}{\frac{1}{z}\, e^{\beta \frac{\mathbf{p}^2}{2m}} \pm 1}, \tag{8.52}$$

from which one can learn a few interesting facts:

1. $n_F < 1$, as expected, for all $0 < z < \infty$ for Fermi particles;
2. $n_B$ can be larger than one for Bose particles, but it remains positive only if $0 < z \leq 1$, which sets an upper limit for the corresponding fugacity;
3. the classical limit of small densities and high temperatures obtains, in both cases, as $z \to 0$, where the distinction between the two types of particles fades out;
4. quantum effects will become prominent in the opposite limit of high densities, which is approached as $z \to \infty$ for the Fermi gas and as $z \to 1$ for the Bose gas.

## 8.4 High-Temperature Limits

At high temperatures and low densities, Bose and Fermi gases exhibit small deviations from the ideal Boltzmann gas, which we can now describe in detail. To this end, it is convenient to introduce in the integrals (8.48) and (8.49) the dimensionless variable

$$x = \beta\, \frac{\mathbf{p}^2}{2m}, \tag{8.53}$$

recasting them in the form

$$n \equiv \frac{\mathcal{N}}{V} = \frac{2}{\sqrt{\pi}}\,(2s+1)\, n_Q(T) \int_0^\infty dx\, x^{\frac{1}{2}}\, \frac{1}{\frac{1}{z}\, e^x \pm 1}, \tag{8.54}$$

$$\beta\, u \equiv \frac{\beta\, U}{V} = \frac{2}{\sqrt{\pi}}\,(2s+1)\, n_Q(T) \int_0^\infty dx\, x^{\frac{3}{2}}\, \frac{1}{\frac{1}{z}\, e^x \pm 1}, \tag{8.55}$$

where $n_Q(T)$ was defined in Eq. (2.86). Our next task is to extract from Eq. (8.54) the first correction to $z$ with respect to its value for the ideal gas. This will also determine the corresponding corrections to $U$ and to the equation of state. To begin with, for small values of $z$,

$$n \simeq \frac{2}{\sqrt{\pi}}\,(2s+1)\, n_Q(T) \int_0^\infty dx\, x^{\frac{1}{2}} \left(z\, e^{-x} \mp z^2\, e^{-2x}\right), \tag{8.56}$$

and the residual integrals are simply related to Euler's $\Gamma$ function as in Eq. (2.23), so that

$$n \simeq (2s+1)\, n_Q(T) \left(z \mp \frac{z^2}{2\sqrt{2}}\right). \tag{8.57}$$

One can invert this equation perturbatively, and the end result is

$$z \simeq \frac{n}{(2s+1)\,n_Q(T)} \pm \frac{1}{2\sqrt{2}}\left(\frac{n}{(2s+1)\,n_Q(T)}\right)^2 . \tag{8.58}$$

Note that the corrections are sized by the ratio

$$\eta = \frac{n}{(2s+1)\,n_Q(T)} . \tag{8.59}$$

This is the parameter introduced in Eq. (2.110), with the proper spin degeneracy, which confirms that the limiting behavior under scrutiny is approached at high temperatures but also at low densities. Note that, for He at room temperature, $n_Q(T)$ corresponds to a distance scale of about $5 \times 10^{-9}$ cm, comparable to Bohr's radius for H. In a similar fashion, the first two terms in the expansion of Eq. (8.55) give

$$\beta u = \frac{3}{2}\,(2s+1)\,n_Q(T)\left(z \mp \frac{z^2}{4\sqrt{2}}\right), \tag{8.60}$$

so that finally, making use of Eq. (8.58),

$$\beta u = \frac{3}{2}\,n\left(1 \pm \frac{\eta}{4\sqrt{2}}\right). \tag{8.61}$$

From (8.51) one can now read the equation of state

$$P \simeq n k_B T\left(1 \pm \frac{\eta}{4\sqrt{2}}\right), \tag{8.62}$$

which reveals that the pressure increases slightly for the Fermi gas and decreases slightly for the Bose gas with respect to the Boltzmann case. This result confirms the intuition drawn from the Pauli principle. The comparison with the van der Waals equation in the form (6.18) is also interesting: the Fermi gas exhibits an effective short-range repulsion, with a temperature-dependent "excluded volume"

$$b_{\text{eff}} = \frac{1}{4\sqrt{2}\,(2s+1)n_Q(T)} , \tag{8.63}$$

while the Bose gas exhibits a corresponding short-range attraction, which reflects itself in a "negative excluded volume." Both effects are quantum mechanical, and become milder as the temperature $T$ increases.

We can now present relatively simple all-order series expansions for small values of $z$. These apply to these nonrelativistic gases, but also to other examples resting on more general single-particle Hamiltonians. In the limit of low densities or high temperatures, and expanding in powers $z$,

$$\mathcal{N} = \sum_\alpha z e^{-\beta\epsilon_\alpha}\frac{g_\alpha}{1 \pm z e^{-\beta\epsilon_\alpha}} = -\sum_{n=1}^{\infty}(\mp)^{n-1} z^n \sum_\alpha g_\alpha e^{-n\beta\epsilon_\alpha} . \tag{8.64}$$

The various terms are canonical partition functions at temperatures $\frac{T}{n}$, so that the parameter $\eta$ of Eq. (8.59) takes the form

$$\eta = \frac{\mathcal{N}}{Z_1} = \sum_{n=1}^{\infty}(\mp)^{n-1} z^n \frac{Z_1(n\beta)}{Z_1(\beta)} . \tag{8.65}$$

In a similar fashion, one can write

$$\frac{PV}{k_B T} = \pm \sum_\alpha g_\alpha \log(1 \pm z e^{-\beta\epsilon_\alpha}) = -k_B T \sum_{n=1}^{\infty} (\mp)^{n-1} \frac{z^n}{n} \sum_\alpha g_\alpha e^{-n\beta\epsilon_\alpha}, \tag{8.66}$$

thus obtaining a similar power series for the equation of state:

$$\frac{PV}{N k_B T} = \frac{1}{\eta} \sum_{n=1}^{\infty} (\mp)^{n-1} \frac{z^n}{n} \frac{Z_1(n\beta)}{Z_1(\beta)}. \tag{8.67}$$

One can then invert Eq. (8.65), order by order in powers of $\eta$. Confining the attention to the first term as above, one can conclude that

$$z = \eta \pm \eta^2 \frac{Z_1(2\beta)}{Z_1(\beta)} + \cdots, \tag{8.68}$$

and hence finally

$$\frac{PV}{\mathcal{N} k_B T} = 1 \pm \frac{\eta}{2} \frac{Z_1(2\beta)}{Z_1(\beta)} + \cdots, \tag{8.69}$$

which relates the result in Eq. (8.62) to the single-particle partition function. Summarizing, the two types of quantum gases behave in a similar fashion at high $T$, with minor deviations from the more standard case of distinguishable particles.

## 8.5 The Free Fermi Gas at Low Temperatures

At low temperatures or high densities, when the ratio $\eta$ in Eq. (8.59) is large, deep novelties emerge for the Fermi gas. In this case the corresponding fugacity $z$ ought to grow arbitrarily large, and therefore we shall begin by studying the large-$z$ behavior of the class of integrals

$$f_\lambda(z) = \int_0^\infty \frac{x^{\lambda-1}\, dx}{\frac{1}{z} e^x + 1}, \tag{8.70}$$

since the equations for the Fermi gas can be recast in the form

$$\begin{aligned} \frac{n}{(2s+1) n_Q(T)} &= \frac{2}{\sqrt{\pi}} f_{\frac{3}{2}}(z), \\ \frac{\beta u}{(2s+1) n_Q(T)} &= \frac{2}{\sqrt{\pi}} f_{\frac{5}{2}}(z). \end{aligned} \tag{8.71}$$

The important observation, in this respect, is that the factor

$$\frac{1}{\frac{1}{z} e^x + 1} \tag{8.72}$$

has a step-like behavior: it is about one for $x < \log z$ and becomes about zero for $x > \log z$, after a narrow transition region. Its derivative is therefore appreciably large only close to $x = \log z$, and this suggests we replace Eq. (8.70) by the more convenient expression

$$f_\lambda(z) = \frac{1}{4\lambda} \int_0^\infty \frac{dx\, x^\lambda}{\left[\cosh\left(\frac{x - \log z}{2}\right)\right]^2} \tag{8.73}$$

where we have integrated by parts. This is indeed a convenient starting point to arrive at simple approximate results. After shifting the integration variable according to

$$x \to x + \log z, \tag{8.74}$$

one can expand the numerator in a power series in $\log z$, while also extending the integral to the whole real axis, which only introduces exponentially small errors in $\log z$, thanks to the denominator. For the moment, we shall only consider the first two terms, which leads to

$$f_\lambda(z) \simeq \frac{(\log z)^\lambda}{4\lambda} \int_{-\infty}^{\infty} dx \, \frac{1 + \frac{\lambda(\lambda-1)}{2}\left(\frac{x}{\log z}\right)^2}{\left[\cosh\left(\frac{x}{2}\right)\right]^2}, \tag{8.75}$$

since the symmetry of the denominator implies that corrections only involve *even* powers of $\log z$. The first term is an elementary integral, since

$$\int_{-\infty}^{\infty} \frac{dx}{4\left[\cosh\left(\frac{x}{2}\right)\right]^2} = \frac{1}{2} \tanh\left(\frac{x}{2}\right)\Big|_{-\infty}^{\infty} = 1, \tag{8.76}$$

while the second can be related to some preceding results. Indeed

$$\int_{-\infty}^{\infty} \frac{dx\, x^2}{4\left[\cosh\left(\frac{x}{2}\right)\right]^2} = 2 \int_0^{\infty} \frac{dx\, x^2\, e^{-x}}{(1+e^{-x})^2} = 4 \sum_{n=1}^{\infty} \frac{(-1)^{n-1}}{n^2}, \tag{8.77}$$

and the sum can be related to $\zeta(2)$, which is discussed in Appendix D, since

$$\sum_{n=1}^{\infty} \frac{(-1)^{n-1}}{n^2} = \sum_{n=1}^{\infty} \left(\frac{1}{n^2} - \frac{2}{(2n)^2}\right) = \frac{\pi^2}{12}. \tag{8.78}$$

In conclusion,

$$f_\lambda(z) \simeq \frac{(\log z)^\lambda}{\lambda} \left[1 + \frac{\pi^2}{6} \frac{\lambda(\lambda-1)}{(\log z)^2}\right], \tag{8.79}$$

and therefore, for large values of $z$, Eqs. (8.71) become

$$\begin{aligned} \frac{n}{(2s+1)\, n_Q(T)} &\simeq \frac{4}{3\sqrt{\pi}} (\log z)^{\frac{3}{2}} \left[1 + \frac{\pi^2}{8} \frac{1}{(\log z)^2}\right], \\ \frac{\beta u}{(2s+1)\, n_Q(T)} &\simeq \frac{4}{5\sqrt{\pi}} (\log z)^{\frac{5}{2}} \left[1 + \frac{5\pi^2}{8} \frac{1}{(\log z)^2}\right], \end{aligned} \tag{8.80}$$

whose ratio, within the same approximations, reads

$$\beta u \simeq \frac{3}{5}\, n \log z \left[1 + \frac{\pi^2}{2} \frac{1}{(\log z)^2}\right]. \tag{8.81}$$

Let us now discuss the physical meaning of these expansions, starting from the leading terms, while also recalling Eqs. (4.68) and (2.86). To begin with, the dominant behavior of the first of Eqs. (8.80) for large values of $z$ gives

$$\log z = \frac{\mu}{k_B T} \simeq \frac{1}{k_B T} \frac{\hbar^2}{2m} \left[\frac{6\pi^2 n}{2s+1}\right]^{\frac{2}{3}}, \tag{8.82}$$

so that $\log z$ is proportional to $\eta^{\frac{2}{3}}$, with $\eta$ as in Eq. (8.59). Note that the chemical potential $\mu$ approaches at low temperatures a finite limit: this is usually called "Fermi energy" or "Fermi level," and we shall denote it by $E_F$ in the following. For the Fermi gas of nonrelativistic particles

$$E_F = \frac{\hbar^2}{2\,m}\left[\frac{6\,\pi^2\,n}{2s+1}\right]^{\frac{2}{3}}, \tag{8.83}$$

where the last factor is closely related to the phase-space volume that we already met in Eq. (4.68). Equation (8.81) thus translates, to lowest order, into a nonvanishing energy density that persists even as $T \to 0$:

$$u = \frac{3}{5}\,n\,E_F. \tag{8.84}$$

Moreover, in view of Eq. (8.51), this residual energy implies the presence of a corresponding residual "Fermi pressure,"

$$P \simeq \frac{2}{5}\,n\,E_F. \tag{8.85}$$

As we shall see, this "degenerate Fermi gas" has striking manifestations in nature.

One can now obtain the first corrections by inverting perturbatively the first of Eqs. (8.80), in the spirit of the expansions that led to it, which yields

$$\mu \simeq E_F\left[1 - \frac{\pi^2}{12}\,\frac{1}{(\log z)^2}\right] \simeq E_F\left[1 - \frac{\pi^2}{12}\left(\frac{k_BT}{E_F}\right)^2\right]. \tag{8.86}$$

Making use of this result in Eq. (8.81) then gives, within the same approximation, and thus for $k_BT \ll \epsilon_F$,

$$u \simeq \frac{3}{5}\,n\,E_F\left[1 + \frac{5\,\pi^2}{12}\left(\frac{k_BT}{E_F}\right)^2\right], \tag{8.87}$$

and therefore a specific heat

$$C_V \simeq \mathcal{N} k_B\,\frac{\pi^2}{2}\left(\frac{k_B\,T}{E_F}\right), \tag{8.88}$$

which grows linearly with $T$. The quantity

$$T_F \simeq \frac{E_F}{k_B} \tag{8.89}$$

is usually called the Fermi temperature.

One can understand all this in simpler terms, starting from the "degenerate Fermi gas" of Eqs. (8.83), (8.84), and (8.85). As we have seen, these results describe the low-temperature behavior up to second-order corrections, and embody a general lesson for Fermi systems with negligible mutual interactions: *at low temperatures all available states below the Fermi level are essentially filled up, while all states above it are essentially empty.*

The key ingredient to derive the leading contributions in the low-temperature limit is what one could call the "degenerate Fermi distribution,"

$$f_F(E,T) = \frac{1}{e^{\frac{E-\mu}{k_BT}} + 1} \simeq \frac{1}{e^{\frac{E-E_F}{k_BT}} + 1} \simeq \theta(E_F - E), \tag{8.90}$$

where $\theta$ is the Heaviside step function of Eq. (3.367). One can thus define a "degenerate Fermi gas" with a single-particle Hamiltonian $H$ via the simpler equations

$$\mathcal{N} = \mathrm{Tr}\,[\theta\,(E_F - H)],$$
$$U = \mathrm{Tr}\,[H\,\theta\,(E_F - H)], \tag{8.91}$$

where the traces are over single-particle states. For example, for the nonrelativistic gas the first equation becomes

$$\mathcal{N} = \frac{(2s+1)V}{h^3}\int_{|\mathbf{p}|<\sqrt{2mE_F}} d^3\mathbf{p} = \frac{(2s+1)V}{h^3}\,\frac{4\pi\,(2mE_F)^{\frac{3}{2}}}{3}, \tag{8.92}$$

which is readily inverted to recover Eq. (8.83), while

$$U = \frac{(2s+1)V}{h^3}\int_{|\mathbf{p}|<\sqrt{2mE_F}} d^3\mathbf{p}\,\frac{\mathbf{p}^2}{2\,m} = \frac{(2s+1)}{h^3}\,\frac{4\pi E_F\,(2mE_F)^{\frac{3}{2}}}{5}, \tag{8.93}$$

and from these results one can simply recover Eq. (8.84). The pressure is then determined by the general relation induced by the Hamiltonian, as we saw in Eq. (8.51) for this type of system, which leads again to Eq. (8.85).

Finally, the low-temperature behavior of the specific heat in Eq. (8.88) affords a simple explanation. Referring to Figure 8.1, at low but nonvanishing temperatures the step function is slightly deformed, so that a fraction $\mathcal{O}\left(\frac{k_BT}{E_F}\right)$ of the particles jumps to empty levels above $E_F$, with a typical energy gain $\mathcal{O}(k_BT)$. Therefore

$$\Delta U \propto k_BT\,V\frac{k_BT}{E_F}, \tag{8.94}$$

and this simple argument suffices to recover the linear growth of its derivative $C_V$.

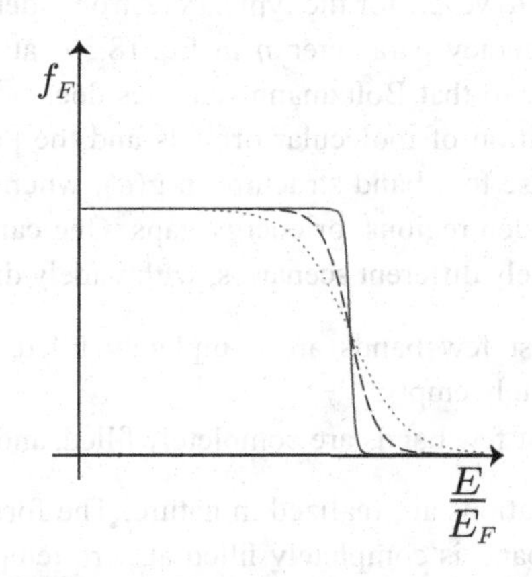


Figure 8.1 The Fermi distribution for three increasing values of $k_B T$ below $E_F$, in increasing order (solid, dashed, and dotted lines).

## 8.6 Fermi Gases in Solids

It is instructive to recast the preceding results in terms of the density of states

$$g(\epsilon) = \mathrm{Tr}\,\delta\,(H - \epsilon), \tag{8.95}$$

which will be our shorthand for $g(\epsilon; T, V)$, a function that depends generally on $V$ and $T$. This more general setup is well suited, as we shall see shortly, to model the behavior of fermions in an external potential, and in particular of conduction and valence electrons in solids. The corresponding expressions for the average number of particles and the average energy are

$$\mathcal{N} = \int_0^\infty \frac{g(\epsilon)\, d\epsilon}{e^{\beta(\epsilon-\mu)} + 1}, \qquad U = \int_0^\infty \frac{\epsilon\, g(\epsilon)\, d\epsilon}{e^{\beta(\epsilon-\mu)} + 1}, \tag{8.96}$$

where we have again set to zero the lowest available energy. Here we would like to characterize in this language the Fermi energy $\epsilon_F$. As we have seen, $\epsilon_F$ is the zero-temperature limit of the chemical potential that is reached when the Fermi–Dirac distribution attains the limiting behavior of Eq. (8.90).

In the low-temperature limit, our starting point is provided by Eqs. (8.91), which in our current language read

$$\mathcal{N} = \int_0^{\epsilon_F} g(\epsilon)\, d\epsilon, \qquad U = \int_0^{\epsilon_F} \epsilon\, g(\epsilon)\, d\epsilon. \tag{8.97}$$

In general, the first determines the Fermi energy as a function of the density $n = \frac{\mathcal{N}}{V}$ of the gas, while the second determines the total energy at zero temperature. We saw this in detail, in the previous section, for a gas of free, massive fermions, for which the Fermi level is the highest energy of the one-particle states that are filled at zero temperature. However, in systems with bands of forbidden energies the analysis of (8.96) as $T \to 0$ must be reconsidered, as we now explain.

In condensed-matter systems the outermost electrons of the constituents have widely spread wavefunctions, and yet they are well characterized as free particles, with an effective density of states $g(\epsilon)$ determined by the properties of the solid of interest. However, for the typical electronic densities, $n = 10^{22}\ \mathrm{cm}^{-3}$, of conductors, the degeneracy parameter $\eta$ in Eq. (8.59) attains large values even at room temperatures, so that Boltzmann statistics does not apply. As we saw in Chapter 3, the superposition of molecular orbitals and the periodic structure of crystals conspire to give rise to a band structure in $g(\epsilon)$, where pairs of nearby bands are separated by forbidden regions, or energy gaps. One can therefore distinguish, at $T = 0$, two qualitatively different scenarios, with widely different manifestations:

- The first few bands are completely filled, and those with higher energies are completely empty.
- The first few bands are completely filled, and the last band is only partly filled.

Both situations are realized in nature. The former corresponds to insulators, whose valence band is completely filled at zero temperature while the nearby conduction band is completely empty. The latter corresponds instead to conductors, whose conduction band is only partly filled. These two situations are portrayed in Figure 8.2.

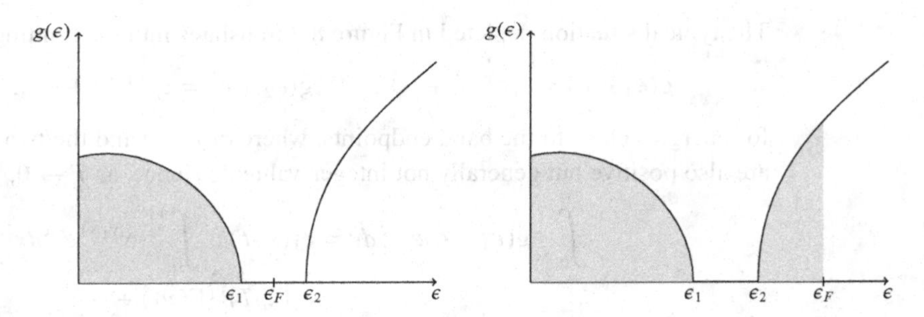


Figure 8.2 Conduction and valence bands for (left) an insulator with energy gap $\Delta = \epsilon_2 - \epsilon_1$ and (right) a conductor. The shaded areas highlight the electron distributions at zero temperature.

In a conductor the first of Eqs. (8.97), which can be recast in the form

$$\mathcal{N} = G(\epsilon_F), \qquad G(E) = \int_0^E g(\epsilon)\, d\epsilon, \tag{8.98}$$

determines $\epsilon_F$ uniquely once $\mathcal{N}$ is given, because $G(E)$ is a strictly increasing function of $E$ for $E > \epsilon_2$. The Fermi energy is then the highest energy level that the electron gas occupies at zero temperature. On the other hand, in an insulator the condition $G(\epsilon_F) = \mathcal{N}$, which must still hold at zero temperature, would appear to leave $\epsilon_F$ undetermined within the interval $[\epsilon_1, \epsilon_2]$, where $G(E)$ is constant. We can now take a closer look at this problem.

### 8.6.1 Insulators

In order to determine $\epsilon_F$ for an insulating material, let us analyze in detail the zero-temperature limit of the condition

$$\mathcal{N} = \int_0^{\epsilon_1} g(\epsilon)\, d\epsilon = \int_0^{\epsilon_1} \frac{g(\epsilon)\, d\epsilon}{e^{\beta(\epsilon-\mu)} + 1} + \int_{\epsilon_2}^{\infty} \frac{g(\epsilon)\, d\epsilon}{e^{\beta(\epsilon-\mu)} + 1}, \tag{8.99}$$

which one can recast conveniently in the form

$$\int_0^{\epsilon_1} \frac{g(\epsilon)\, d\epsilon}{e^{\beta(\mu-\epsilon)} + 1} = \int_{\epsilon_2}^{\infty} \frac{g(\epsilon)\, d\epsilon}{e^{\beta(\epsilon-\mu)} + 1} \tag{8.100}$$

combining the expression for $\mathcal{N}$ and the first integral. As $T \to 0$, one can now safely neglect the second term in both denominators, up to an exponentially small error, since $\mu$ ought to approach a value within the energy gap, $\epsilon_1 < \epsilon_F < \epsilon_2$. As a result, one obtains the condition

$$e^{-\beta\mu} \int_0^{\epsilon_1} g(\epsilon)\, e^{\beta\epsilon} d\epsilon = e^{\beta\mu} \int_{\epsilon_2}^{+\infty} g(\epsilon)\, e^{-\beta\epsilon} d\epsilon, \tag{8.101}$$

and letting $\epsilon \to \epsilon_1 - \epsilon$ on the left-hand side and $\epsilon \to \epsilon_2 + \epsilon$ on the right-hand side, one can conclude that

$$e^{2\beta\mu} = e^{\beta(\epsilon_1+\epsilon_2)} \frac{\displaystyle\int_0^{\epsilon_1} g(\epsilon_1 - \epsilon) e^{-\beta\epsilon} d\epsilon}{\displaystyle\int_0^{\infty} g(\epsilon_2 + \epsilon) e^{-\beta\epsilon} d\epsilon}. \tag{8.102}$$

The typical situation depicted in Figure 8.2 translates into the limiting behaviors

$$g(\epsilon_1 - \epsilon) = c_1 \epsilon^{\gamma_1 - 1} + \cdots , \qquad g(\epsilon_2 + \epsilon) = c_2 \epsilon^{\gamma_2 - 1} + \cdots , \tag{8.103}$$

for energies close to the band endpoints, where $c_{1,2} > 0$ and the two exponents $\gamma_{1,2}$ are also positive but generally not integer-valued[1]. Hence, as $T \to 0$,

$$\int_0^{\epsilon_1} g(\epsilon_1 - \epsilon) e^{-\beta\epsilon} d\epsilon = c_1 (k_B T)^{\gamma_1} \int_0^{+\infty} x^{\gamma_1 - 1} e^{-x} dx$$
$$= c_1 (k_B T)^{\gamma_1} \Gamma(\gamma_1) + \cdots , \tag{8.104}$$

where we have extended the upper limit to $+\infty$, at the price of an exponentially small error, and in a similar fashion

$$\int_0^{\infty} g(\epsilon_2 + \epsilon) e^{-\beta\epsilon} d\epsilon = c_2 (k_B T)^{\gamma_2} \Gamma(\gamma_2) + \cdots . \tag{8.105}$$

Making use of these results in Eq. (8.102) leads finally to

$$\mu(T) = \frac{\epsilon_1 + \epsilon_2}{2} + \frac{k_B T}{2} \log \left[ \frac{c_1 \Gamma(\gamma_1)}{c_2 \Gamma(\gamma_2)} (k_B T)^{\gamma_1 - \gamma_2} \right] + \cdots , \tag{8.106}$$

whose limiting behavior implies that the Fermi energy falls *precisely in the middle of the energy gap*:

$$\epsilon_F = \frac{\epsilon_1 + \epsilon_2}{2}. \tag{8.107}$$

Note that for a symmetric gap, with $\gamma_1 = \gamma_2$, $c_1 = c_2$, the corrections to $\epsilon_F$ in (8.106) only appear at higher order in $T$, while for completely symmetric energy bands they vanish to all orders, as can be seen from Eq. (8.102).

Proceeding along the same lines, one can obtain corresponding expressions for the low-temperature limit of the internal energy $U$, starting from

$$U(T) - U(0) = \int_0^{\epsilon_1} \frac{\epsilon\, g(\epsilon)\, d\epsilon}{e^{\beta(\mu - \epsilon)} + 1} + \int_{\epsilon_2}^{\infty} \frac{\epsilon\, g(\epsilon)\, d\epsilon}{e^{\beta(\epsilon - \mu)} + 1}. \tag{8.108}$$

One thus finds, making use of Eq. (8.106),

$$U(T) \simeq U(0) + c\,\Delta\,(k_B T)^{\gamma}\, e^{-\beta\Delta/2}, \tag{8.109}$$

$$C_V(T) \simeq \frac{c\, k_B\, \Delta^2}{2} (k_B T)^{\gamma - 2}\, e^{-\beta\Delta/2}, \tag{8.110}$$

where

$$c = \sqrt{c_1\, c_2\, \Gamma(\gamma_1)\, \Gamma(\gamma_2)}, \qquad \gamma = \frac{\gamma_1 + \gamma_2}{2}, \tag{8.111}$$

[1] There are cases where one or both of the $\gamma_i$ lie within the range $(0, 1]$, and indeed the crucial requirement is that $g(\epsilon)$ be integrable, although not necessarily finite.

and

$$\Delta = \epsilon_2 - \epsilon_1 \tag{8.112}$$

denotes the *energy gap*.

As one might have anticipated, the behavior of the specific heat at low temperatures for an insulator complies with (4.11), and is dominated by an exponential factor that involves the energy gap between the Fermi level and the next band. The number of electrons,

$$N_{\text{eff}} = \int_{\epsilon_F}^{\infty} \frac{g(\epsilon)\, d\epsilon}{e^{\beta(\epsilon-\mu)}+1} = \int_{\epsilon_2}^{\infty} \frac{g(\epsilon)\, d\epsilon}{e^{\beta(\epsilon-\mu)}+1}, \tag{8.113}$$

that are thermally excited above the Fermi energy, and are thus effectively available for electrical conduction, exhibits a similar behavior. Indeed, as $T \to 0$,

$$\mathcal{N}_{\text{eff}} = e^{\beta(\mu-\epsilon_2)} \int_0^{\infty} g(\epsilon_2+\epsilon)\, e^{-\beta\epsilon} d\epsilon = c\,(k_B T)^{\gamma} e^{-\beta\Delta/2}, \tag{8.114}$$

where we have made use of Eq. (8.106), including the leading correction. Comparing the last results one can thus conclude that

$$U(T) - U(0) = \mathcal{N}_{\text{eff}}\, \Delta, \tag{8.115}$$

since, to this order, the energy increases by $\Delta$ every time an electron manages to overcome the gap. The number $\mathcal{N}_{\text{eff}}$ and the electrical conductivity are thus exponentially suppressed for temperatures $k_B T \ll \Delta$.

In materials for which $k_B T \approx \Delta$ at room temperature, this exponential suppression is rather mild and the number of conduction electrons approximates a power-law behavior $N_{\text{eff}} \propto T^{\gamma}$. These considerations reflect a key property of semiconductors, which behave as insulators at low temperatures and as weakly conducting materials at room temperature, in the spirit of Eq. (8.94).

## 8.6.2 Conductors

In a conducting material, the density of states is a smooth function of $\epsilon$ in a neighborhood of $\epsilon_F$, and one can expand $g(\epsilon)$ in a Taylor series obtaining

$$\mathcal{N}_{\text{eff}} = \int_{\epsilon_F}^{\infty} \frac{g(\epsilon)\, d\epsilon}{e^{\beta(\epsilon-\mu)}+1} = \sum_{n=0}^{\infty} g^{(n)}(\epsilon_F) \frac{(k_B T)^{n+1}}{n!} \int_0^{+\infty} \frac{x^n\, dx}{e^x+1}, \tag{8.116}$$

or, performing the integrals as in Appendix D,

$$\mathcal{N}_{\text{eff}} = g(\epsilon_F)\, k_B T \log 2 + \sum_{n=1}^{\infty} g^{(n)}(\epsilon_F)\, (k_B T)^{n+1} (1-2^{-n}) \zeta(n+1). \tag{8.117}$$

Whenever $g(\epsilon_F) \neq 0$, as is usually the case, the number of conduction electrons grows linearly with $T$ at low temperatures, to leading order.

Proceeding as in Section 8.5, one can obtain the so-called Sommerfeld expansions of Eqs. (8.96) to all orders in $T$. To this end, let us begin by writing, after a partial integration,

$$\mathcal{N} = \beta \int_0^{+\infty} \frac{G(\epsilon)\, d\epsilon}{4\cosh^2\left(\frac{\epsilon-\mu}{2k_B T}\right)} = \int_{-\beta\mu}^{\infty} \frac{G(\mu + x k_B T)\, dx}{4\cosh^2\left(\frac{x}{2}\right)}, \tag{8.118}$$

where $G(\epsilon)$ was defined in Eq. (8.98). At low temperatures, $\mu \simeq E_F \neq 0$ and $\beta \to +\infty$, so that one can extend the integration to the whole real axis, up to exponentially small errors, and expand $G(\epsilon)$ in a Taylor series near $\mu$:

$$\mathcal{N} = \sum_{l=0}^{\infty} G^{(2l)}(\mu) \frac{(k_B T)^{2l}}{(2l)!} \int_{-\infty}^{\infty} \frac{x^{2l}\, dx}{4\cosh^2\left(\frac{x}{2}\right)}, \tag{8.119}$$

where odd terms vanish due to the parity of the integrand. This extends what we did for nonrelativistic fermions in Section 8.5, and making use of the results in Appendix D finally yields, for the general case of conductors,

$$\mathcal{N} = 2 \sum_{l=0}^{\infty} G^{(2l)}(\mu)(k_B T)^{2l}(1 - 2^{1-2l})\zeta(2l). \tag{8.120}$$

One can simply invert the truncation of this result to the first nontrivial order,

$$\begin{aligned} \mathcal{N} &= G(\mu) + \frac{\pi^2}{6} g'(\mu)(k_B T)^2 + \cdots \\ &= G(\epsilon_F) + g(\epsilon_F)\mu + \frac{\pi^2}{6} g'(\epsilon_F)(k_B T)^2 + \cdots, \end{aligned} \tag{8.121}$$

to obtain

$$\mu = \epsilon_F - \frac{\pi^2}{6} \frac{g'(\epsilon_F)}{g(\epsilon_F)} (k_B T)^2. \tag{8.122}$$

This is the generalization of Eq. (8.86) for the lowest-order correction to the chemical potential at low temperatures, provided $g(\epsilon_F)$ does not vanish. Note that this correction can be positive if $g(\epsilon)$ decreases near $\epsilon_F$. If this is the case, as the temperature grows slightly the system probes a region of the spectrum with a slightly lower density of states, which increases the cost of adding a new fermion.

One can similarly obtain the Sommerfeld expansion for the internal energy, which reads

$$U(T) = 2 \sum_{l=0}^{\infty} H^{(2l)}(\mu)\,(k_B T)^{2l} \left(1 - 2^{1-2l}\right) \zeta(2l), \tag{8.123}$$

where

$$H(E) = \int_0^E \epsilon\, g(\epsilon)\, d\epsilon. \tag{8.124}$$

Making use of Eq. (8.122) then leads to

$$U(T) = H(\mu) + \frac{\pi^2}{6} \left(g(\mu) + \mu g'(\mu)\right) (k_B T)^2 + \cdots, \tag{8.125}$$

so that, expressing $H(\mu)$ in terms of $H(\epsilon_F)$ via Eq. (8.122) gives,

$$U \simeq U(0) + g(\epsilon_F) \frac{\pi^2}{6} (k_B T)^2, \tag{8.126}$$

$$C_V \simeq g(\epsilon_F) \frac{\pi^2}{3} k_B^2 T, \tag{8.127}$$

as we saw for free fermions in Section 8.5. In conducting materials, the contribution to the specific heat from the conduction electrons is thus generically linear in $k_B T$ at low temperatures. This is the type of behavior that we already met in the free-Fermi example of Section 8.5. As a result, the conduction electrons typically overwhelm, at small temperatures, the cubic phonon contribution, which we derived within the Debye theory in Section 4.3. The specific heat can thus reveal the conducting nature of materials!

## 8.7 The Free Bose Gas at Low Temperatures

Let us now discuss the behavior of the Bose gas at low temperatures or high densities, where the ratio in Eq. (8.59) is large. We begin with a couple of significant examples, before moving again to more general considerations. Our starting point is Eq. (8.54), which we rewrite for convenience, for the nonrelativistic Bose gas in three dimensions, as

$$n \equiv \frac{\mathcal{N}}{V} = \frac{2}{\sqrt{\pi}} (2s+1)\, n_Q(T) \int_0^\infty \frac{x^{\frac{1}{2}}}{\frac{1}{z}\, e^x - 1}\, dx. \tag{8.128}$$

The reader should appreciate that, although in Chapter 4 we relied on a formalism tailored to the wave nature of photons (or phonons), the resulting construction has already led to Eqs. (8.47), (8.48), and (8.49) for Bose particles, albeit with $z = 1$ or, equivalently, with a vanishing chemical potential $\mu$.

The role of a nonvanishing chemical potential $\mu$ is precisely to grant the conservation of the number of particles, when it is demanded, as we saw already for the Fermi gas, and photons have $z = 1$ precisely because they can be emitted and absorbed at will. However, for a generic Bose gas of particles whose number is conserved, the finite available range $0 < z \leq 1$ can bring about a novel phenomenon, depending on how Eq. (8.128) behaves as $z \to 1$. One can thus envision two distinct scenarios:

1. if the integral in Eq. (8.128) were to diverge as $z \to 1$, its limiting behavior could compensate the factor $n_Q(T) \sim T^{\frac{3}{2}}$, thus maintaining the density $n$ in Eq. (8.128) constant at low temperatures;
2. if, on the other hand, as in our example, the integral converges as $z \to 1$, the factor $n_Q(T) \sim T^{\frac{3}{2}}$ would seem to drive $n$ to zero for decreasing values of $T$, thus violating the conservation of the number of particles. When this occurs, the system actually hosts an enticing phenomenon called Bose–Einstein condensation. The apparent contradiction actually originates from the approximation that turned momentum sums into integrals, which is clearly not reliable if a sizable fraction of the $\mathcal{N}$ particles occupies the single state with $\mathbf{p} = 0$, the ground state for this gas. This is precisely what happens for the nonrelativistic gas of Bose particles that we are considering: for any fixed density $n$, there is a "critical temperature" $T_c$, below which the ground state starts to acquire a macroscopic occupation number.

In general, one should thus write Eq. (8.128), more precisely, in the form

$$n = \langle n_0 \rangle + \frac{2}{\sqrt{\pi}} (2s+1)\, n_Q(T) \int_0^\infty \frac{x^{\frac{1}{2}}}{\frac{1}{z} e^x - 1} dx, \tag{8.129}$$

where $\langle n_0 \rangle$ vanishes above the critical temperature $T_c$. In this case, $T_c$ is defined implicitly in terms of $n$ by

$$n = (2s+1)\, n_Q(T_c)\, \zeta\left(\frac{3}{2}\right), \tag{8.130}$$

since Eq. (8.129) is to hold at $T_c$ with $\langle n_0 \rangle = 0$ and $z = 1$. The resulting integral is a special case of Eq. (D.15) and equals $\frac{\sqrt{\pi}}{2}\,\zeta\left(\frac{3}{2}\right)$, where $\zeta\left(\frac{3}{2}\right) \simeq 2.61$.

For $T < T_c$ the ground state hosts strikingly, in this system, a finite fraction of the total number of particles, which one can quantify as the difference

$$\langle n_0 \rangle = (2s+1)\, n_Q(T_c)\, \zeta\left(\frac{3}{2}\right) - (2s+1)\, n_Q(T)\, \zeta\left(\frac{3}{2}\right), \tag{8.131}$$

and taking into account Eqs. (2.86), (8.129) and (8.131) leads finally to

$$\langle n_0 \rangle = n \left[ 1 - \left( \frac{T}{T_c} \right)^{\frac{3}{2}} \right]. \tag{8.132}$$

There are no similar subtleties when computing the internal energy or the pressure, since particles in the ground state do not contribute to them, and therefore

$$\beta u = \frac{2}{\sqrt{\pi}} (2s+1)\, n_Q(T) \int_0^\infty dx\, x^{\frac{3}{2}} \frac{1}{\frac{1}{z} e^x - 1}. \tag{8.133}$$

Below the critical temperature, again using Eq. (D.15), this becomes

$$\beta u = \frac{3}{2} (2s+1)\, n_Q(T)\, \zeta\left(\frac{5}{2}\right), \tag{8.134}$$

where $\zeta\left(\frac{5}{2}\right) \simeq 1.34$, and therefore

$$P = (2s+1)\, n_Q(T)\, k_B\, T\, \zeta\left(\frac{5}{2}\right) \sim T^{\frac{5}{2}}. \tag{8.135}$$

Finally,

$$\frac{dP}{dT} = \frac{5}{2} k_B (2s+1)\, n_Q(T)\, \zeta\left(\frac{5}{2}\right), \tag{8.136}$$

which reveals a coexistence line of first-order transitions between the ordinary and Bose–Einstein condensed phases for $0 < T < T_c$ for this gas. The latent heat $\lambda = T\Delta s$ is determined by the Clausius–Clapeyron equation (6.26), with a specific volume $v_G$ given by

$$\frac{1}{v_G} = (2s+1)\, n_Q(T)\, \zeta\left(\frac{3}{2}\right) \tag{8.137}$$

for the gas phase, and a vanishing specific volume for the condensate, so that

$$\lambda = \frac{5}{2}\,k_B T \frac{\zeta\left(\frac{5}{2}\right)}{\zeta\left(\frac{3}{2}\right)}. \tag{8.138}$$

Note that in this system one cannot overcome the transition line in the $(P, T)$ plane, which does not terminate at a critical point: the condensed phase can appear at any temperature, provided the density is sufficiently high or, equivalently, provided the specific volume is sufficiently low. On the other hand, in the $(P, v)$ plane, with $v = \frac{1}{n}$, one can explore the coexistence region, which lies below the curve

$$P\, v^{\frac{5}{3}} = \frac{h^2 \zeta\left(\frac{5}{2}\right)}{2\pi m (2s+1)^{\frac{2}{3}} \zeta\left(\frac{3}{2}\right)^{\frac{5}{3}}} \tag{8.139}$$

obtained combining Eqs. (8.130) and (8.135) at $T = T_c$, but the pure condensed phase only emerges in the limit $v \to 0$. These two scenarios are illustrated in Figure 8.3. Note also that the condensation takes place in momentum space, so that, in ordinary conditions, the condensed phase of the Bose–Einstein gas is fully delocalized within the available volume $V$, albeit in a sub-extensive fashion.

It is also instructive to contrast these results with the behavior of a gas of nonrelativistic particles confined to a *two-dimensional* surface of area $A$, which is a simple system where Bose–Einstein condensation does not occur. Retracing the preceding steps, one could write in this case, up to the factor reflecting the spin multiplicity,

$$\begin{aligned} \frac{N}{A} &= \frac{2\pi}{h^2} \int d^2\mathbf{p}\, \frac{1}{\frac{1}{z}\, e^{\frac{\beta \mathbf{p}^2}{2m}} - 1} = \frac{m k_B T}{2\pi\hbar^2} \int_0^\infty \frac{dx}{\frac{1}{z}\, e^x - 1} \\ &= -\frac{m k_B T}{2\pi\hbar^2} \log\left(1 - z\right). \end{aligned} \tag{8.140}$$

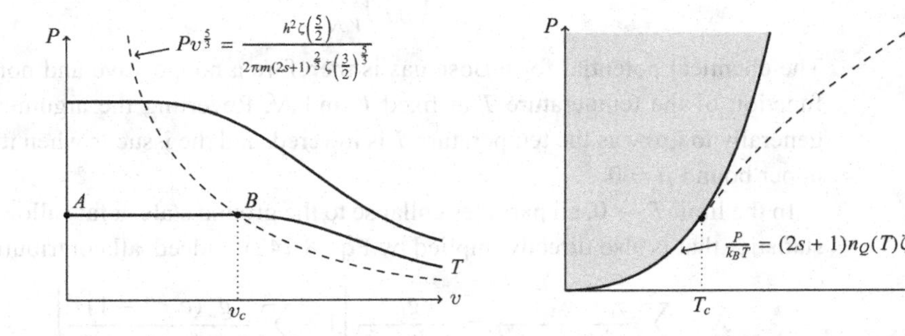


Figure 8.3 (Left) In the $(P, \mathrm{v})$ diagram for the free Bose gas the dashed line marks the appearance of the condensate while isotherms are depicted as solid black curves. At point $B$ the system is still in the pure gas phase, albeit at a critical specific volume $\mathrm{v}_c$, while at point $A$ the system is fully in the condensed phase. (Right) In the $(P, T)$ plane the pure gas phase lies below the solid line, while the system is confined to it when the condensate is present. The dashed line corresponds to a fixed total density and joins with the thick solid line at the corresponding critical temperature. The shaded region cannot be reached.

This equation can be clearly solved for all values of $T$, and inverts simply to determine the chemical potential

$$\beta\mu = \log z = \log\left(1 - e^{-\frac{2\pi\hbar^2\mathcal{N}}{mk_BTA}}\right). \tag{8.141}$$

One can therefore conclude that this system does not develop a macroscopic $\langle n_0\rangle$ for any value of $T$, and therefore does not experience Bose–Einstein condensation.

## 8.8 Bosons in an External Potential

We can now elaborate on more general Bose systems characterized by a generic density of states, while setting again to zero the ground-state energy. This modified degeneracy may be due, for instance, to the presence of an external potential, the likes of which are employed to trap atoms, thus triggering Bose–Einstein condensation in the laboratory. Let us also refer initially to discrete energy levels, so that

$$n_\alpha = \frac{g_\alpha}{e^{\beta(\epsilon_\alpha-\mu)} - 1}, \tag{8.142}$$

and

$$\mathcal{N} = \sum_\alpha \frac{g_\alpha}{e^{\beta(\epsilon_\alpha-\mu)} - 1}. \tag{8.143}$$

Our first observation is that the derivative of this expression with respect to $T$ at fixed $V$ and $\mathcal{N}$ yields the constraint

$$0 = \sum_\alpha \frac{g_\alpha}{\sinh^2(\beta(\epsilon_\alpha - \mu)/2)}\left[\epsilon_\alpha - \mu + T\left(\frac{\partial\mu}{\partial T}\right)_{V,\mathcal{N}}\right]. \tag{8.144}$$

Since $\epsilon_\alpha - \mu$ is always nonnegative in order that all $n_\alpha \geq 0$, one can conclude that

$$\left(\frac{\partial\mu}{\partial T}\right)_{V,\mathcal{N}} \leq 0. \tag{8.145}$$

The chemical potential for a Bose gas is therefore a nonpositive and nonincreasing function of the temperature $T$ at fixed $V$ and $\mathcal{N}$. Reverting the argument, $\mu$ tends generally to grow as the temperature $T$ is lowered, and the issue is when it reaches its upper bound $\mu = 0$.

In the limit $T \to 0$, all particles collapse to the ground state, a fate allowed by Bose statistics that is also directly implied by Eq. (8.142). Indeed, all contributions to

$$\mathcal{N} = \sum_{\alpha=0}^{\infty} \frac{g_\alpha}{e^{\beta(\epsilon_\alpha-\mu)} - 1} = \frac{g_0}{e^{-\beta\mu} - 1}\left[1 + \sum_{\alpha=1}^{\infty} \frac{g_\alpha(e^{-\beta\mu} - 1)}{g_0(e^{\beta(\epsilon_\alpha-\mu)} - 1)}\right] \tag{8.146}$$

from states with $\alpha = 1, 2, \ldots$, which have $\epsilon_\alpha > 0$, tend to zero as $T \to 0$. Therefore, in this limit the population $\langle n_0\rangle$ of the ground state, or "condensate," becomes extensive and $z = e^{\beta\mu} \to 1$, or equivalently $\mu \to 0$. However, as we have seen in our preceding examples, the interesting question is whether $\mu(T)$ remains negative for $T > 0$ and approaches zero only for $T \to 0$, or rather reaches zero for a positive

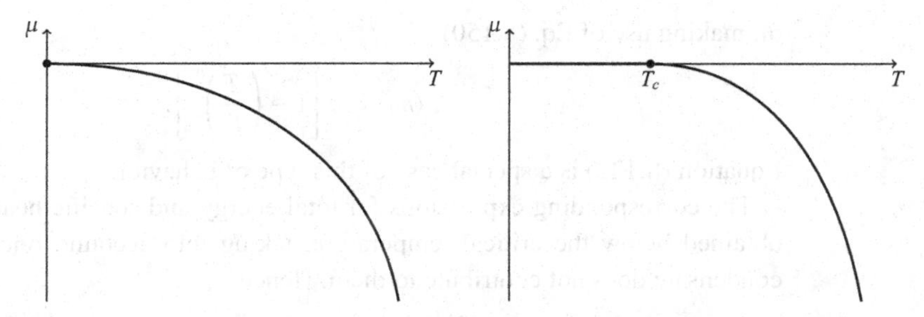


Figure 8.4 Chemical potentials for Bose systems near $T = 0$ in (left) the absence or (right) the presence of Bose–Einstein condensation.

temperature $T_c$ (see Figure 8.4). When this occurs, macroscopic numbers of particles populate the ground state for a finite temperature range $0 \leq T < T_c$.

We can now elaborate on an instructive class of examples, characterized by

$$g(\epsilon) \;=\; c\, V \epsilon^{\alpha-1}, \tag{8.147}$$

for some exponent $\alpha > 0$ and some positive constant $c$. This family of systems includes free massive bosons in three dimensions, a case that we have already analyzed in Section 8.7, for which $\alpha = \frac{3}{2}$. The corresponding expressions for the total number of bosons read

$$n = \frac{\mathcal{N}}{V} = c \int_0^\infty \frac{\epsilon^{\alpha-1}\, d\epsilon}{e^{\beta(\epsilon-\mu)} - 1} = c\,(k_B T)^\alpha \int_0^\infty \frac{x^{\alpha-1}}{e^{x-\beta\mu} - 1}\, dx, \tag{8.148}$$

and different ranges of $\alpha$ translate into different types of behavior.

### 8.8.1 Condensation at $T = T_c > 0$

If $\alpha > 1$, which is the case, as we have seen, for free massive particles in three dimensions, recalling that $\mu \leq 0$ and hence $e^{-\beta\mu} \geq 1$, one can conclude that

$$n \leq c\,(k_B T)^\alpha \int_0^\infty \frac{x^{\alpha-1}}{e^x - 1}\, dx \xrightarrow[T\to 0]{} 0, \tag{8.149}$$

since the integral is finite. As we have seen, this implies that, as $T \to 0$, the states described by $g(\epsilon)$ cannot accommodate all particles in the system, some of which therefore form a macroscopic ground–state condensate. Equation (8.147) must be modified in order to account for this phenomenon, adding to $g(\epsilon)$ a nontrivial ground–state contribution $\langle n_0 \rangle\, \delta(\epsilon)$, and the critical temperature is determined by the condition that $\mu$ vanishes with $\langle n_0 \rangle = 0$:

$$n = c\,(k_B T_c)^\alpha \int_0^\infty \frac{x^{\alpha-1}}{e^x - 1}\, dx = c\,(k_B T_c)^\alpha \Gamma(\alpha)\zeta(\alpha), \tag{8.150}$$

making use of Eq. (D.15). Above $T_c$ the condensate is not significantly populated, so that $\langle n_0 \rangle = 0$, while the chemical potential is strictly negative, and therefore Eq. (8.148) applies. Below $T_c$, instead, the contribution of the condensate becomes significant, while $\mu = 0$, so that

$$n = \langle n_0 \rangle + c\,(k_B T)^\alpha \Gamma(\alpha)\zeta(\alpha), \tag{8.151}$$

or, making use of Eq. (8.150),

$$\langle n_0 \rangle = n \left[ 1 - \left( \frac{T}{T_c} \right)^{\alpha} \right]. \tag{8.152}$$

Equation (8.132) is a special case of this type of behavior.

The corresponding expressions for total energy and specific heat are also readily obtained below the critical temperature, taking into account, once more, that the condensate does not contribute to them. Hence

$$U = c\, V (k_B T)^{\alpha+1} \int_0^\infty \frac{x^\alpha}{e^x - 1}\, dx = \mathcal{N} k_B T_c\, \alpha\, \frac{\zeta(\alpha+1)}{\zeta(\alpha)} \left( \frac{T}{T_c} \right)^{\alpha+1}, \tag{8.153}$$

where we have made use of Eq. (8.150), and

$$C_V = \mathcal{N} k_B\, \alpha(\alpha+1) \frac{\zeta(\alpha+1)}{\zeta(\alpha)} \left( \frac{T}{T_c} \right)^{\alpha}. \tag{8.154}$$

In a similar fashion,

$$P = -c\, (k_B T)^{\alpha+1} \int_0^\infty x^{\alpha-1} \log(1 - e^{-x})\, dx = \frac{U}{\alpha V}. \tag{8.155}$$

and the latent heat associated to the condensation is determined by the Clausius–Clapeyron equation

$$\frac{dP}{dT} = \frac{\lambda}{T} (n - \langle n_0 \rangle), \tag{8.156}$$

so that

$$\lambda = k_B T (\alpha + 1) \frac{\zeta(\alpha+1)}{\zeta(\alpha)}. \tag{8.157}$$

Bose–Einstein condensation is therefore a first-order phase transition for all these systems in the range $0 < T < T_c$.

It is interesting to elaborate on the behavior of the chemical potential $\mu$ around the critical temperature. To this end, let us write

$$T = T_c (1 + \theta), \tag{8.158}$$

where $\theta > 0$ is a small parameter. Comparing Eqs. (8.148) and (8.150), which give the density $n$ for $T \geq T_c$, leads to

$$(1+\theta)^{-\alpha} \int_0^\infty \frac{x^{\alpha-1}}{e^x - 1}\, dx = \int_0^\infty \frac{x^{\alpha-1}}{e^{x+\delta} - 1}\, dx, \tag{8.159}$$

where we have set

$$\delta = -\beta\mu \tag{8.160}$$

for convenience. Expanding the left-hand side to first order in $\theta$, one can recast the preceding condition in the form

$$(e^\delta - 1) \int_0^\infty \frac{x^{\alpha-1} e^x}{(e^x - 1)(e^{x+\delta} - 1)}\, dx = \alpha\, \theta \int_0^\infty \frac{x^{\alpha-1}}{e^x - 1}\, dx. \tag{8.161}$$

The integral on the right-hand side is computed in Eq. (D.15) and equals $\Gamma(\alpha)\, \zeta(\alpha)$, but in order to estimate the dominant behavior of the left-hand side one must distinguish two ranges for $\alpha$.

- If $\alpha > 2$, as is the case for free particles for $d > 4$, the integral on the left-hand side converges as $\mu \to 0$, and hence one can safely expand the factor to first order in $\delta$ while also setting $\delta = 0$ in the integrand, obtaining

$$\delta \int_0^\infty \frac{x^{\alpha-1} e^x}{(e^x - 1)^2} dx \simeq \alpha\, \theta \int_0^\infty \frac{x^{\alpha-1}}{e^x - 1} dx. \tag{8.162}$$

As a result, in this case the first correction to the chemical potential $\mu$ is linear in the parameter $\theta = \frac{T}{T_c} - 1$, and

$$\mu \simeq -\frac{\alpha\, \zeta(\alpha)\, k_B\, (T - T_c)}{\zeta(\alpha - 1)}. \tag{8.163}$$

- On the other hand, in the complementary range $1 < \alpha \leq 2$ Bose condensation still occurs below a nonvanishing critical temperature $T_c$, but the integral on the left-hand side of Eq. (8.161) contains divergent contributions from small values of $x$ in the limit $\mu \to 0$. One can still estimate its leading-order behavior by introducing an upper cutoff $\Delta \ll 1$ on the integration range in order to highlight the singular region near $x = 0$. A Taylor expansion of the left-hand side of Eq. (8.161) in both $\delta$ and $x$ then yields

$$\delta \int_0^\Delta \frac{x^{\alpha-2}}{x + \delta} dx = \delta^{\alpha-1} \int_0^{\Delta/\delta} \frac{y^{\alpha-2}}{1 + y} dy, \tag{8.164}$$

after letting $x = y\,\delta$. For $1 < \alpha < 2$ one can further approximate this quantity, noting that $\frac{\Delta}{\delta} \to \infty$ as $\delta \to 0$ for a fixed value of $\Delta$, which finally leads to

$$\delta^{\alpha-1} \int_0^\infty \frac{y^{\alpha-2}}{1 + y} dy = \delta^{\alpha-1} \frac{\pi}{\sin[\pi(\alpha - 1)]}, \tag{8.165}$$

since one can relate the integral to Euler's $\Gamma$ function, and the result follows using Eq. (D.27). On the other hand, for $\alpha = 2$,

$$\delta \int_0^{\Delta/\delta} \frac{dy}{1 + y} \simeq -\,\delta \log \delta, \tag{8.166}$$

up to higher-order corrections.

Substituting in (8.161), these results highlight a power-law dependence of $\delta$ on $\theta$, so that

$$\mu \simeq -\left\{[\alpha \sin[\pi(\alpha - 1)]\, \frac{\Gamma(\alpha)\zeta(\alpha)}{\pi}\right\}^{\frac{1}{\alpha-1}} k_B T_c \left(\frac{T}{T_c} - 1\right)^{\frac{1}{\alpha-1}}, \tag{8.167}$$

for $1 < \alpha < 2$, and a poly-logarithmic dependence with dominant behavior

$$\mu \simeq \frac{\pi^2}{3} \frac{k_B(T - T_c)}{\log(T/T_c - 1)} \tag{8.168}$$

for $\alpha = 2$.

Let us focus again, for definiteness, on the physically interesting case of free particles in $d = 3$, for which $\alpha = \frac{3}{2}$. We can now show that, amusingly, in this case the specific heat $C_V$ has a discontinuous first derivative at $T_c$, as displayed in Figure 8.5. For temperatures slightly below $T_c$, Eq. (8.154) gives

$$\frac{\partial C_V}{\partial T} = \frac{45}{8} \frac{\mathcal{N} k_B}{T_c} \frac{\zeta\left(\frac{5}{2}\right)}{\zeta\left(\frac{3}{2}\right)} \left(\frac{T}{T_c}\right)^{\frac{1}{2}}. \tag{8.169}$$

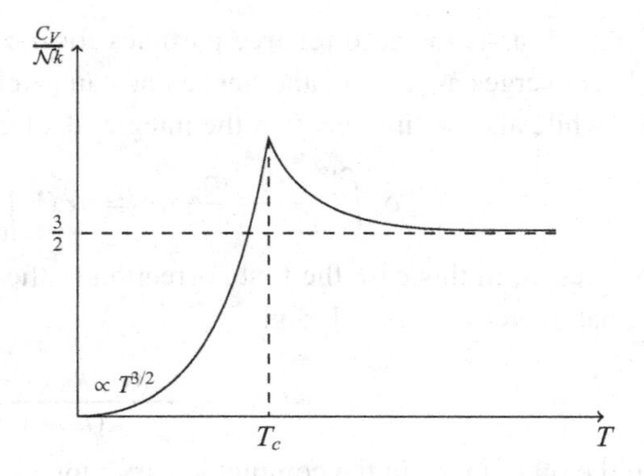


Figure 8.5 The specific heat as a function of $T$ for a free Bose gas in three dimensions.

On the other hand, for temperatures slightly above $T_c$ the total energy of the gas for small $\delta = -\beta\mu$ becomes

$$\begin{aligned} U &= c\,V(k_BT)^{\frac{5}{2}} \int_0^\infty \frac{x^{3/2}}{e^{x+\delta}-1}\,dx \\ &= c\,V(k_BT)^{\frac{5}{2}} \left[\int_0^\infty \frac{x^{3/2}}{e^x-1}\,dx \;-\; \delta \int_0^\infty \frac{x^{3/2}e^x}{(e^x-1)^2}\,dx + \cdots\right], \end{aligned} \tag{8.170}$$

where

$$c = \frac{2}{\sqrt{\pi}}\left(\frac{2\pi m}{h^2}\right)^{\frac{3}{2}} . \tag{8.171}$$

Integrating the second term by parts, one can recast the result in the form

$$U \;=\; \frac{3}{2}\,\mathcal{N}k_B\,T_c\,\frac{\zeta\left(\frac{5}{2}\right)}{\zeta\left(\frac{3}{2}\right)}\left(\frac{T}{T_c}\right)^{\frac{5}{2}} \;-\; \frac{3}{2}\,\mathcal{N}k_BT_c\left(\frac{T}{T_c}\right)^{\frac{5}{2}}\delta. \tag{8.172}$$

The first term coincides with the expression for $U(T)$ in Eq. (8.153), which we obtained working below the critical temperature, while $\delta$ for temperatures slightly above $T_c$ is given in Eq. (8.167), and in this case

$$\delta \;=\; \frac{9}{16\pi}\left[\zeta\left(\frac{3}{2}\right)\right]^2\left(\frac{T}{T_c}-1\right)^2 . \tag{8.173}$$

The specific heat $C_V$ is therefore continuous at $T_c$, but its first derivative exhibits the discontinuity

$$\lim_{T\to T_c^+}\left(\frac{\partial C_V}{\partial T}\right) - \lim_{T\to T_c^-}\left(\frac{\partial C_V}{\partial T}\right) \;=\; -\,\frac{27}{16\pi}\left[\zeta\left(\frac{3}{2}\right)\right]^2\frac{\mathcal{N}k_B}{T_c}. \tag{8.174}$$

As we saw in Eq. (8.136), Bose–Einstein condensation entails a line of first-order transitions, which reflects the emergence of a more orderly macroscopic condensate for $T \le T_c$. However, at $T_c$, the transition does not introduce singularities in the internal energy nor in the specific heat. Still, as we have just seen, even in this simple context the derivative of $C_V$ does exhibit a mildly irregular behavior when the phenomenon sets in.

### 8.8.2 Condensation at $T = 0$

If $0 < \alpha \leq 1$, condensation does not occur for $T_c > 0$. The value $\alpha = 1$, which obtains for free massive bosons confined to a plane, was already considered in Eq. (8.141) and led to an analytic expression for the chemical potential $\mu$.

A similar phenomenon occurs for $0 < \alpha < 1$. As $T \to 0$, the factor $(k_B T)^\alpha$ in

$$n = c\,(k_B T)^\alpha \int_0^\infty \frac{x^{\alpha-1}}{e^{x-\beta\mu} - 1}\,dx \tag{8.175}$$

tends to zero while the integral diverges, and the two effects can conspire to keep $n$ fixed without the presence of a macroscopic condensate. In order to estimate the behavior of $\mu$ in this limit, one can resort to the same technique employed in the previous subsection. Introducing again $\delta = -\beta\mu$, as in Eq. (8.160), and choosing a cutoff $\Delta \ll 1$, one can expand the integrand to first order in $x$ and $\delta$, obtaining

$$c\,(k_B T)^\alpha \int_0^\Delta \frac{x^{\alpha-1}}{x+\delta}\,dx \;=\; c\,(k_B T)^\alpha\,\delta^{\alpha-1} \int_0^{\Delta/\delta} \frac{y^{\alpha-1} dy}{y+1}. \tag{8.176}$$

Letting $\frac{\Delta}{\delta} \to \infty$ finally leads to

$$\mu \simeq -\left[\frac{c\pi k_B T}{n\,\sin(\pi(1-\alpha))}\right]^{\frac{1}{1-\alpha}}. \tag{8.177}$$

For example, for particles confined to a line $\alpha = \frac{1}{2}$, and therefore the chemical potential decreases quadratically as the temperature $T$ increases,

$$\mu \simeq -2m\left[(2s+1)\,\frac{\pi k_B T}{hn}\right]^2. \tag{8.178}$$

In all these cases, no Bose–Einstein condensate forms for $T > 0$.

## 8.9 Atomic and Molecular Spectra

Bohr's early success with atomic spectra revealed the universal nature of Planck's quantum of action $h$. Later applications of quantum mechanics to molecules not only confirmed this, but also provided deep indirect clues on the nature of nuclei. Having discussed the key properties of quantum statistical physics, it is time to take a look at these systems.

Let us consider a gas of identical constituents, with negligible mutual interactions but possessing an internal structure. An atom, ion, or molecule is a system of this kind, well described by a Hamiltonian of the form

$$H = \frac{\mathbf{P}^2}{2M} + H_{\text{int}}, \tag{8.179}$$

with $M$ and $\mathbf{P}$ its total mass and linear momentum. Our main focus in the following will be the Hamiltonian $H_{\text{int}}$, which accounts for the dynamics of the internal degrees of freedom. In the classical regime, where

$$\frac{n_Q(T)}{n} = \frac{V(2\pi M k_B T)^{3/2}}{N h^3} \gg 1, \tag{8.180}$$

one can deduce the thermodynamic properties of a gas from the single-constituent partition function

$$Z_1 = \frac{V}{h^3}\int e^{-\beta\frac{\mathbf{P}^2}{2M}}d^3\mathbf{P}\,\mathrm{tr}\left[e^{-\beta H_{\rm int}}\right] = \frac{V(2\pi mk_BT)^{3/2}}{Nh^3}Z_{\rm int}, \tag{8.181}$$

where

$$Z_{\rm int} = \mathrm{tr}\left[e^{-\beta H_{\rm int}}\right]. \tag{8.182}$$

While the simple integration over the classical phase space is legitimate, in the thermodynamic limit, for the first term in $H$, in computing $Z_{\rm int}$ one must take into account the discrete features introduced by quantum mechanics. Denoting as usual by $E_\alpha$ and $g_\alpha$ the internal energy levels and the associated degeneracies, one can formally write

$$Z_{\rm int} = \sum_{\alpha=0}^{\infty} g_\alpha e^{-\beta E_\alpha}, \tag{8.183}$$

and the internal Hamiltonian must account for a number of effects that we now turn to illustrate.

In atoms, $Z_{\rm int}$ reduces to $Z_{\rm el}$, the contribution of the electron cloud bound to the nucleus. As we discussed in Section 2.4.2, the correct way of treating the electronic degrees of freedom at room temperature is to truncate the partition function to its first few terms. Actually the first term

$$Z_{\rm el} \simeq g_0\, e^{-\beta E_0}, \tag{8.184}$$

which is the ground-state contribution, already conveys some relevant insights. The corresponding degeneracy $g_0$ typically reflects the total spin $s$ of the electron cloud, which is determined by the outermost electrons, and contributes a factor $(2s + 1)$. For example, in the hydrogen atom or in vapors of alkali atoms, whose outermost orbitals host a single electron, $g_0 = 2$.

The nuclear spin $I$ is another source of degeneracy. More precisely, it gives rise to the hyperfine splitting of electron levels, whose magnitude becomes comparable with $k_BT$ only at temperatures of about 270 K, where however most gases usually undergo phase transitions. At room temperatures, therefore, this difference is too small to be detected, and nuclear spins contribute simply degeneracy factors $(2I + 1)$ to atomic ground states.

Nonvanishing contributions from electron spin $s$ and orbital angular momentum $\ell$ in the electronic ground state conspire to produce fine structure splittings, as discussed in Section 3.6 for the hydrogen atom. As a result, the ground-state contribution in Eq. (8.184) should be replaced, in general, by a sum

$$Z_{\rm el} = \sum_j (2j+1)e^{-\beta E_j}, \tag{8.185}$$

where $j$ runs over the eigenvalues of the total angular momentum for the given spin $s$ and orbital angular momentum $\ell$. At high temperatures, where $k_BT$ is large compared to the fine-structure splittings, $e^{-\beta E_j} \simeq e^{-\beta E_0}$, and Eq. (8.185) reduces to Eq. (8.184) with a degeneracy factor $(2s + 1)(2\ell + 1)$, equal to the number of independent spin-orbit assignments. However, in some cases, $k_B T$ can be small compared to the fine-structure splittings, and then the only relevant contribution originates from the lowest of the $E_j$'s.

In molecules, electron clouds are bound together with nuclei by electrostatic forces. Since the electrons are far lighter, one can describe their dynamics by treating the positions of nuclei as fixed parameters, in the so-called Born–Oppenheimer approximation. In this fashion, one can separate the electronic contribution $H_{\text{el}}$ from the internal Hamiltonian $H_{\text{int}}$. This identifies a corresponding factor $Z_{\text{el}}$ in $Z_{\text{int}}$, which reflects again the most relevant energy eigenstates of the electron cloud, as discussed for atoms in Eq. (8.184). Spin degeneracies and spin-orbit interactions are also relevant for the electronic ground states of molecules. In particular, the electron cloud of a diatomic molecule is symmetric under rotations about the molecule's axis, say the $z$ axis, so that the $z$-component $L_z$ of the angular momentum commutes with the Hamiltonian, and is also symmetric under a reflection with respect to any plane containing the $z$ axis. Since $L_z$ is reverted by this operation, all energy eigenstates with a nonvanishing angular momentum are doubly degenerate, and this situation can occasionally present itself for the ground state.

A proper account of the internal dynamics of a molecule, however, must also take into account the motion of its nuclei with respect to the center of mass. As a first approximation, one can distinguish further the contributions $H_{\text{rot}}$ and $H_{\text{vib}}$, associated to rigid rotations of the molecule as a whole and to harmonic vibrations of its atoms or ions about their equilibrium positions. The intuitive picture behind this approximation is the following. The electrostatic interactions among nuclei and electron clouds determine the molecular structure, which undergoes rapid oscillations about equilibrium positions with typical angular frequencies $\omega = 10^{13}$–$10^{14}$ Hz, so that $\frac{\hbar\omega}{k_B} = 10^2$–$10^3$ K. Rotations are typically far slower, by about three orders of magnitude: vibrational effects thus average to zero during a rotation, and the rotational spectra are essentially those of rigid bodies, with characteristic temperatures of order of a few K.

For instance, the electrostatic interactions of diatomic molecules afford an effective description in terms of the Lennard-Jones potential of Eq. (6.6) (see Figure 6.1), and small harmonic vibrations about the equilibrium position, which we denote here by $r_0$,[2] are characterized by

$$\mu\,\omega^2 = \left(\frac{\partial^2 U_{\text{eff}}(r)}{\partial r^2}\right)_{r_0}, \tag{8.186}$$

with $\mu = \frac{m_1 m_2}{m_1+m_2}$ for two atoms of masses $m_1$ and $m_2$. The ratio $\frac{\hbar\omega}{k_B}$ is about $10^3$ K for molecular hydrogen $H_2$, nitrogen $N_2$ and oxygen $O_2$. Therefore, thermal fluctuations can only induce small oscillations near the bottom of the Lennard-Jones potential well, and the harmonic approximation is well justified. On the other hand, rotational energies are far smaller than $k_B T$ at room temperatures. The only exceptions are molecular hydrogen $H_2$ or deuterium $D_2$, where they correspond to temperature scales of about 350 K. This is still one order of magnitude below the corresponding

[2] In this section, $r_0$ identifies the equilibrium position, while in Chapter 6 we used the same symbol to denote the core size of the molecules. For instance, in the Lennard-Jones potential of Eq. (6.6), the equilibrium position would be actually $2^{\frac{1}{6}}\,r_0$.

vibrational scales, but in these systems rotational energies are comparable to $k_B T$ at room temperatures, which brings along interesting effects, as we shall see.

### 8.9.1 Vibrational Spectra

Let us denote collectively by $z^a$, with $a = 1, \ldots, 3n$, the Cartesian coordinates for the $n$ atoms of the molecule. After rescaling them by the atomic masses, the kinetic energy takes the form

$$T = \sum_{a=1}^{3n} \frac{1}{2} (\dot{z}^a)^2 . \tag{8.187}$$

In the harmonic approximation, which we argued is well justified, expanding the interaction energy with respect to the equilibrium positions $z_0^a$ yields

$$U = U_0 + \frac{1}{2} \sum_{a,b=1}^{3n} q^a B_{ab}\, q^b + \cdots , \tag{8.188}$$

with $q^a = z^a - z_0^a$ and

$$B_{ab} = \left( \frac{\partial^2 U}{\partial z^a \partial z^b} \right)_{z_0} . \tag{8.189}$$

By an orthogonal transformation of the coordinates $q^a$, one can diagonalize the symmetric quadratic form in Eq. (8.188), arriving at the normal modes $Q^a$, with

$$T = \frac{1}{2} \sum_{a=1}^{3n} (\dot{Q}^a)^2 , \qquad U = U_0 + \frac{1}{2} \sum_{a=1}^{3n} \omega_a^2 (Q^a)^2 . \tag{8.190}$$

These steps identify decoupled harmonic oscillators with normal frequencies $\omega_a$, but some of these vanish due to the overall translational and rotational invariance of the interactions. For any molecule, three zero-modes exist, which reflect the invariance under translations. Moreover, for nonlinear molecules three more frequencies vanish due to rotational symmetry, which leaves only $3n - 6$ nonvanishing ones. On the other hand, for linear molecules one of the three rotations has trivial effects, and there are $3n - 5$ nonvanishing normal frequencies. The degrees of freedom associated to translational zero-modes describe the free evolution of the center of mass, while the rotational zero-modes reflect the rigid rotations of the molecule, which we shall address in the next section.

The normal modes corresponding to $\omega_a \neq 0$ result in independent harmonic oscillations described by Hamiltonians of the type

$$H_{\text{vib}} = \sum_a \left[ \frac{P_a^2}{2} + \frac{\omega_a^2}{2} (Q^a)^2 \right] . \tag{8.191}$$

The corresponding quantized spectra

$$\sum_a \hbar \omega_a \left( \alpha_a + \tfrac{1}{2} \right) , \tag{8.192}$$

with $\alpha_a = 0, 1, \ldots$ (degeneracies may also occur, due to special symmetries of the molecular structure), thus contribute a factor

$$Z_{\text{vib}} = \prod_a \frac{e^{-\beta\hbar\omega_a/2}}{1 - e^{\beta\hbar\omega_a}}, \tag{8.193}$$

to the single-molecule partition function, where the product is restricted to nonvanishing $\omega_a$. It is important to stress again that the vibrational frequencies have typical values around $10^{13}$–$10^{14}$ Hz, so that at room temperature $\beta\hbar\omega \sim 1-10$, which cannot be regarded as a small quantity. The harmonic approximation is thus justified, while taking the continuum limit of the preceding expression is not correct. On the other hand, the rapid convergence of the geometric series justifies the extension of the sum to the whole harmonic spectrum.

At higher temperatures $\hbar\omega$ becomes smaller than $k_B T$, so that both the anharmonic features of the potential energy and the centrifugal force due to rotations become relevant. In order to take a look at these effects, let us focus on a diatomic molecule comprising two atoms with masses $m_1, m_2$. Proceeding as in Section 3.6, this can be modeled by the Hamiltonian

$$H = \frac{\mathbf{P}^2}{2M} + \frac{\mathbf{p}^2}{2\mu} + U(|\mathbf{x}|), \tag{8.194}$$

where $M$ and $\mu$ are the total and reduced mass. Expanding the potential about the equilibrium position and retaining the first two corrections to the harmonic case, one can write

$$U = \frac{\mu\omega^2 r_0^2}{2}(\xi^2 - \alpha_3\,\xi^3 + \alpha_4\xi^4), \tag{8.195}$$

with

$$\xi = \frac{|\mathbf{x}|}{r_0} - 1. \tag{8.196}$$

When $k_B T$ is large with respect to $\hbar\omega$ it suffices to integrate over phase space, so that

$$Z_1 = \frac{V}{h^6}\int d^3\mathbf{P}\, d^3\mathbf{p}\, d\mathbf{x}\, e^{-\frac{1}{k_B T}H} = \frac{V(2\pi k_B T\sqrt{M\mu}\,)^3}{h^6}\int_0^\infty 4\pi r^2\, dr\, e^{-\frac{1}{k_B T}U(r)}. \tag{8.197}$$

Let us now consider in detail the last integral,

$$\int_0^\infty 4\pi r^2\, dr\, e^{-\frac{1}{k_B T}U(r)} = 4\pi r_0^3\int_{-1/\theta}^\infty \theta\,(1+\theta\tau)^2\, d\tau\, e^{-\tau^2+\alpha_3\theta\tau^3-\alpha_4\theta^2\tau^4}, \tag{8.198}$$

where we have let $r = r_0\,(1+\theta\tau)$ and we have defined

$$\theta = \sqrt{\frac{2k_B T}{I\omega^2}}, \qquad I = \mu r_0^2. \tag{8.199}$$

Although we are focusing on the regime $T > \frac{\hbar\omega}{k_B}$, the temperature should still be sufficiently small so that $\theta \ll 1$ and thermal fluctuations remain smaller than the typical depth of the harmonic potential well. One can then replace the lower integration limit with $-\infty$, up to exponentially small errors, while also expanding the exponential as

$$e^{\alpha_3\theta\tau^3-\alpha_4\theta^2\tau^4} = 1+\alpha_3\theta\tau^3-\alpha_4\theta^2\tau^4+\frac{1}{2}\alpha_3^2\theta^2\tau^6, \tag{8.200}$$

up to corrections of order $\theta^3$. Recalling that, for even $n$,

$$\int_{-\infty}^{\infty}\tau^n e^{-\tau^2}d\tau = \Gamma\left(\frac{n+1}{2}\right), \tag{8.201}$$

while this integral vanishes for odd $n$, one finds

$$\int_0^{\infty} 4\pi r^2\, dr\, e^{-\beta U(r)} = 4\pi r_0^3\,\theta\left[1+\tfrac{1}{2}\theta^2\left(1+3\alpha_3-\tfrac{3}{2}\alpha_4+\tfrac{15}{8}\alpha_3^2\right)\right], \tag{8.202}$$

up to corrections of order $\theta^4$. Finally, collecting these results, the partition function for the molecule reads

$$Z_1 = \left[V\left(\frac{2\pi M}{\hbar^2\beta}\right)^{3/2}\right]\cdot\left[\frac{2I}{\hbar^2\beta}\right]\cdot\left[\frac{1}{\hbar\omega\beta}\right]\cdot\left[1+\frac{1+3\alpha_3-\frac{3}{2}\alpha_4+\frac{15}{8}\alpha_3^2}{I\omega^2\beta}\right]. \tag{8.203}$$

The first factor is the usual translational term, while the second is the purely rotational contribution for a diatomic rotor in the semiclassical approximation, as we shall see shortly. The third factor is the high-temperature limit of a harmonic term as in Eq. (8.193). Finally, the fourth factor collects higher-order roto-vibrational effects due to the anharmonic potential and the centrifugal force.

### 8.9.2 Rotational Spectra

The rotational dynamics of a molecule is well modeled by a rigid body with a fixed center of mass. It is therefore sufficient to specify a rotation, in the form of a time-dependent orthogonal transformation $R(t)$ relating the initial positions $\mathbf{y}_i$ of the points of the rigid body to their evolutions $\mathbf{x}_i(t)$,

$$\mathbf{x}_i(t) = R(t)\mathbf{y}_i. \tag{8.204}$$

The Euler angles provide a convenient parametrization of $R(t)$ via the product of three matrices, as

$$R = R_z(\phi)\,R_y(\theta)\,R_z(\psi), \tag{8.205}$$

where $R_z$ and $R_y$ denote rotations about the $z$ and $y$ axes, so that

$$R_z(\phi) = \begin{pmatrix}\cos\phi & -\sin\phi & 0\\ \sin\phi & \cos\phi & 0\\ 0 & 0 & 1\end{pmatrix} \qquad R_y(\theta) = \begin{pmatrix}\cos\theta & 0 & \sin\theta\\ 0 & 1 & 0\\ -\sin\theta & 0 & \cos\theta\end{pmatrix} \tag{8.206}$$

and $0\leq\phi<2\,\pi$ and $0\leq\psi<2\pi$, while $0\leq\theta<\pi$. Note that $R(t)$ is an orthogonal matrix,

$$R^T R = 1, \tag{8.207}$$

and consequently

$$\Omega = R^T\dot{R} \tag{8.208}$$

is an antisymmetric matrix, which can be parametrized as

$$\Omega = \begin{pmatrix} 0 & -\omega_3 & \omega_2 \\ \omega_3 & 0 & -\omega_1 \\ -\omega_2 & \omega_1 & 0 \end{pmatrix}. \tag{8.209}$$

Therefore, the kinetic energy of rigid body rotations takes the form

$$T = \frac{1}{2}\sum_{i=1}^{n} m_i \dot{\mathbf{x}}_i^2 = \frac{1}{2}\sum_{i=1}^{n} m_i \mathbf{y}_i^T \Omega^T \Omega \mathbf{y}_i. \tag{8.210}$$

In terms of the three independent entries of $\Omega$, which can be identified with the three components of the axial vector angular velocity $\boldsymbol{\omega} = (\omega_1, \omega_2, \omega_3)^T$, one finds

$$\Omega\, \mathbf{y}_i = ! \times \mathbf{y}_i, \tag{8.211}$$

where, in terms of the Euler angles of Eq. (8.205),

$$\begin{aligned} \omega_1 &= -\dot{\phi}\sin\theta\cos\psi + \dot{\theta}\sin\psi, \\ \omega_2 &= \dot{\phi}\sin\theta\sin\psi + \dot{\theta}\cos\psi, \\ \omega_3 &= \dot{\phi}\cos\theta + \dot{\psi}. \end{aligned} \tag{8.212}$$

Making use of Eq. (8.211) and of the identity

$$(\boldsymbol{\omega} \times \mathbf{y}_i) \cdot (\boldsymbol{\omega} \times \mathbf{y}_i) = ! \cdot [\mathbf{y}_i \times (! \times \mathbf{y}_i)] = \boldsymbol{\omega}^2 \mathbf{y}_i^2 - (\boldsymbol{\omega} \cdot \mathbf{y}_i)^2, \tag{8.213}$$

the kinetic energy of Eq. (8.210) takes the form

$$T = \frac{1}{2}\, \boldsymbol{\omega}^T I\, \boldsymbol{\omega}, \tag{8.214}$$

where $I$ is the inertia tensor of the molecule, a real symmetric matrix with entries

$$I^{ij} = \sum_{\ell=1}^{n} m_\ell \left(\mathbf{y}_\ell^2\, \delta^{ij} - y_\ell^i y_\ell^j\right). \tag{8.215}$$

When referred to its "principal" axes, $I$ becomes diagonal and

$$T = \frac{I_1\omega_1^2}{2} + \frac{I_2\omega_2^2}{2} + \frac{I_3\omega_3^2}{2}, \tag{8.216}$$

where $I_1$, $I_2$, and $I_3$ are usually called principal moments of inertia. Making use of Eqs. (8.212), one is thus led to the Lagrangian

$$\begin{aligned} \mathcal{L} = \frac{I_1}{2}\left(-\dot{\phi}\sin\theta\cos\psi + \dot{\theta}\sin\psi\right)^2 + \frac{I_2}{2}\left(\dot{\phi}\sin\theta\sin\psi + \dot{\theta}\cos\psi\right)^2 \\ + \frac{I_3}{2}\left(\dot{\phi}\cos\theta + \dot{\psi}\right)^2, \end{aligned} \tag{8.217}$$

whose equations of motion take the elegant Euler form

$$\begin{aligned} I_1\, \dot{\omega}_1 &= (I_2 - I_3)\, \omega_2\, \omega_3, \\ I_2\, \dot{\omega}_2 &= (I_3 - I_1)\, \omega_3\, \omega_1, \\ I_3\, \dot{\omega}_3 &= (I_1 - I_2)\, \omega_1\, \omega_2, \end{aligned} \tag{8.218}$$

when expressed in terms of the $\omega_i$. Equation (8.217) defines the conjugate momenta

$$\begin{aligned} p_\phi &= -I_1\omega_1 \sin\theta\cos\psi + I_2\omega_2 \sin\theta \sin\psi + I_3\omega_3\cos\theta, \\ p_\theta &= I_1\omega_1\sin\psi + I_2\omega_2\cos\psi, \\ p_\psi &= I_3\omega_3, \end{aligned} \tag{8.219}$$

and thus the Hamiltonian

$$\begin{aligned} H_{\rm rot} = & \frac{1}{2I_1(\sin\theta)^2}(p_\theta\sin\theta\sin\psi + (p_\psi\cos\theta - p_\phi)\cos\psi)^2 \\ & + \frac{1}{2I_2(\sin\theta)^2}(p_\theta\sin\theta\cos\psi + (p_\phi - p_\psi\cos\theta)\sin\psi)^2 + \frac{p_\psi^2}{2I_3}, \end{aligned} \tag{8.220}$$

which takes the simple form

$$H_{\rm rot} = \frac{L_1^2}{2I_1} + \frac{L_2^2}{2I_2} + \frac{L_3^2}{2I_3} \tag{8.221}$$

when expressed in terms of

$$L_i = I_i\,\omega_i \qquad (i = 1,2,3). \tag{8.222}$$

For each principal axis, the typical energy spacing between two consecutive energy levels is $\hbar^2/2I_i$, and therefore the continuum approximation is legitimate, at room temperatures, for the typical moments of inertia of *nonlinear* polyatomic molecules, which translate into spacings of about $10^{-3}$ eV, or into temperature of about 10 K. Let us also assume, to begin with, that the molecule is not invariant under any discrete rotation, which is the case, for instance, when all its atoms are distinguishable. Taking into account that the phase-space measure is independent of the specific choice of canonical coordinates and momenta, the rotational contribution to the partition function is

$$Z_{\rm rot} = \frac{1}{h^3}\int dp_\theta\, dp_\phi\, dp_\psi\, d\theta\, d\phi\, d\psi\, e^{-\beta H}, \tag{8.223}$$

but from Eqs. (8.219) and (8.222) one can show that

$$dp_\theta\, dp_\phi\, dp_\psi = \sin\theta\, dL_1\, dL_2\, dL_3 \tag{8.224}$$

and therefore the Hamiltonian in Eq. (8.221) leads simply to

$$Z_{\rm rot} = \frac{(2k_BT)^{3/2}}{\hbar^3}\sqrt{\pi I_1 I_2 I_3}, \tag{8.225}$$

which rests on the determinant of the inertia tensor $I$. Referring to Eq. (8.181), this result translates into an additional constant contribution of $\frac{3k_B}{2}$ to the specific heat per molecule, reflecting the equipartition of its rotational modes. As a result, for a gas composed of $N$ (nonlinear) polyatomic molecules the specific heat at room temperature is

$$C_V = 3Nk_B. \tag{8.226}$$

If instead $\frac{\hbar^2}{2I_i} \gg k_BT$ for some $I_i$, the corresponding degree of freedom is effectively frozen in the ground state. This situation always presents itself for linear (diatomic or polyatomic) molecules, for which,

$$I_1 = I_2 = I, \qquad I_3 \simeq 0, \tag{8.227}$$

when it is aligned along the $z$ axis. The preceding setup simplifies considerably in this case. In order to specify the configuration of the molecule, it is only necessary to assign the rotation $R = R_z(\phi)R_y(\theta)$, letting $\psi = 0$ and simply aligning $(0, 0, 1)$ to $(\sin\theta\cos\phi, \sin\theta\sin\phi, \cos\theta)$. The resulting configuration space is that of a particle constrained to the surface of a sphere. Accordingly, Eqs. (8.212) reduce to

$$\omega_1 = -\dot{\phi}\sin\theta, \qquad \omega_2 = \dot{\theta}, \qquad \omega_3 = \dot{\phi}\cos\theta, \tag{8.228}$$

and then

$$\mathcal{L} = \frac{I}{2}\left(\dot{\theta}^2 + (\sin\theta)^2\dot{\phi}^2\right) \tag{8.229}$$

while

$$H_{\text{rot}} = \frac{1}{2I}\left(p_\theta^2 + \frac{p_\phi^2}{(\sin\theta)^2}\right), \tag{8.230}$$

or

$$H = \frac{1}{2I}(L_1^2 + L_2^2). \tag{8.231}$$

For instance, for a diatomic molecule of length $r_0$ with atoms of mass $m_1$, $m_2$, one has $I = \mu r_0^2$, with $\mu = \frac{m_1 m_2}{m_1+m_2}$. Integrating over phase space, which is justified if the molecule is sufficiently heavy and long, so that $\frac{2Ik_BT}{\hbar^2} \gg 1$, and assuming that no symmetry is present,

$$Z_{\text{rot}} \simeq \frac{1}{h^2}\int dp_\phi\, dp_\theta\, d\phi\, d\theta\, e^{-\beta H} = \frac{2Ik_BT}{\hbar^2}, \tag{8.232}$$

where the Gaussian momentum integrals generate a factor $\sin\theta$ needed to build the surface element on the sphere. The condition $\frac{2Ik_BT}{\hbar^2} \gg 1$ is always satisfied at room temperature except for diatomic hydrogen and deuterium, two noteworthy cases that we shall discuss at the end of this section. With these exceptions, Eqs. (8.181) and (8.232) translate into an additional constant contribution equal to $k_B$ for the specific heat per diatomic molecule, reflecting the equipartition of its rotational modes. As a result, for a gas of $N$ diatomic molecules the specific heat at room temperatures is

$$C_V = \frac{5}{2}Nk_B. \tag{8.233}$$

Up to this point, we have referred to molecules that are not symmetric under any discrete rotation. If this is instead the case, the integration over phase space must be restricted accordingly, since configurations differing by such transformations should be identified. For instance, a linear molecule may be mapped to itself by a reflection about its midpoint (i.e. a rotation of $\pi$ about an axis perpendicular to the molecule's axis and passing through its midpoint), a common feat for diatomic molecules with two identical atoms, or for more complicated linear molecules such as acetylene, $C_2H_2$. In this case, clearly $0 \leq \theta < \frac{\pi}{2}$, hence (8.232) must be rescaled by a factor $\frac{1}{2}$ and replaced by

$$Z_{\text{rot}} = \frac{Ik_BT}{\hbar^2}. \tag{8.234}$$

For nonlinear polyatomic molecules there are more options, and one must divide (8.225), in general, by a suitable symmetry factor $\sigma$. The water molecule $H_2O$ is a first notable example: it has the shape of an isosceles triangle and is thus symmetric under $\pi$ rotations about its axis, so that in this case $\sigma = \frac{2\pi}{\pi} = 2$. The ammonia molecule $NH_3$ has the shape of an isosceles triangular pyramid, and is symmetric under rotations of $2\pi/3$ about its axis, so that $\sigma = \frac{2\pi}{2\pi/3} = 3$. More complicated examples are the methane molecule $CH_4$, which has the shape of a tetrahedron, whose symmetry group leads to $\sigma = 12$, and the benzene molecule $C_6H_6$, which has the shape of a regular hexagon, whose symmetry group leads again to $\sigma = 12$.

Let us conclude this section with a discussion of linear molecules for which the approximation $\frac{2Ik_BT}{\hbar^2} \gg 1$ fails at room temperature. As we have remarked, in practice this reduces the options to molecular hydrogen $H_2$ and deuterium $D_2$, and these cases do exhibit peculiar behaviors. Although nuclear spin interaction energies are small compared to $k_BT$, the presence of identical components can still induce an effective dependence of the molecular spectrum on nuclear spins.

Before addressing this point directly, let us briefly discuss the quantization strategy for a rigid symmetric diatomic molecule ($I_1 = I_2$) with a fixed center of mass. A quick way to solve this problem rests on the "particle" Lagrangian in Eq. (8.229), which translates into the Hamiltonian

$$H_{\text{rot}} = \frac{\mathbf{L}^2}{2I}, \tag{8.235}$$

with $I = \mu r_0^2$. Since the squared angular momentum operator $\mathbf{L}^2$ has eigenvalues

$$\hbar^2\, \ell(\ell+1), \tag{8.236}$$

with $\ell = 0, 1, 2, \ldots$ and degeneracy $g_\ell = 2\ell + 1$, one is led to the spectrum

$$E_\ell = \frac{\hbar^2}{2I}\, \ell(\ell+1), \tag{8.237}$$

and therefore

$$Z_{\text{rot}} = \sum_{\ell=0}^{\infty} (2\ell+1) e^{-\frac{\hbar^2}{2Ik_BT}\ell(\ell+1)}. \tag{8.238}$$

In the high-temperature regime Eq. (8.238) yields

$$Z_{\text{rot}} = \int_0^\infty 2\ell e^{-\frac{\hbar^2}{2Ik_BT}\ell^2}\, dK = \frac{2Ik_BT}{\hbar^2}, \tag{8.239}$$

in agreement with (8.232), while at very low temperatures, where rotational degrees of freedom are effectively frozen, $Z_{\text{rot}}$ reduces to unity. Let us now consider a diatomic molecule with two identical nuclei, keeping in mind the examples of $H_2$ and $D_2$. In both cases there are two electrons in the ground state, which is denoted by ${}^1\Sigma_g^+$ in standard notation.[3] The interactions of nuclear spins can be safely neglected, but the allowed states must be symmetric or antisymmetric under the interchange of the two identical nuclei, as dictated by the spin-statistics restrictions discussed at length in this chapter.

[3] This state has total spin equal to zero ($s = 0$, superscript $2s + 1 = 1$), vanishing total orbital angular momentum (Greek letter $\Sigma$). It is invariant under reflections about the molecule axis (superscript +) and is symmetric under parity (subscript g, which stands for "gerade," German for even).

For $H_2$, each nucleus is a proton, a spin-$\frac{1}{2}$ Fermi particle. There are four possible spin states for the two protons,

$$|++\rangle, \qquad |+-\rangle, \qquad |-+\rangle, \qquad |--\rangle, \tag{8.240}$$

where $\pm$ indicates spin projections equal to $\pm\frac{\hbar}{2}$. The antisymmetric singlet state

$$|0,0\rangle = \frac{1}{\sqrt{2}}(|+-\rangle - |-+\rangle), \tag{8.241}$$

with total spin equal 0, and the three symmetric triplet states

$$|1,-1\rangle = |--\rangle \qquad |1,0\rangle = \frac{1}{\sqrt{2}}(|+-\rangle + |-+\rangle) \qquad |1,+1\rangle = |++\rangle \tag{8.242}$$

with total spin equal to one, provide a basis with definite parity under the interchange of the two nuclei. For the singlet, the fermionic nature of the two nuclei requires an orbital wave function that is symmetric under parity, and thus rotational states (8.237) with *even* values of $\ell$, while for the triplet it requires *odd* values of $\ell$.

The singlet form of the $H_2$ molecule is called *para*-hydrogen and has access to the lowest-energy rotational state, with vanishing orbital angular momentum for the two protons and a spherically symmetric orbital wavefunction. On the other hand, the triplet spin states give rise to the so-called *ortho*-hydrogen, for which spatially symmetric states are prohibited by the exclusion principle. The prefixes ortho and para are due to the different statistical weight of the two: at high temperatures, the former occurs more often, 3 out of 4 times, while the latter occurs 1 out of 4 times, as we shall see shortly.

A similar discussion applies to deuterium $D_2$, whose nuclei are however proton-neutron bound states with total spin 1, and thus effectively Bose particles. In this case, the decomposition gives five symmetric states of total spin 2, three antisymmetric states of total spin 1 and a symmetric singlet state of total spin 0. The six states with total spin 0 and 2 combine with parity-even orbital wave functions, with *even* $\ell$. This gives the ortho-deuterium form, with a statistical weight of 6/9 at high temperatures. The remaining 3 options, with total spin 1, combine with parity-odd orbital wavefunctions, with *odd* $\ell$, and are termed para-deuterium.

Taking into account the spin-statistic constraints, the rotational partition functions of these systems take the form:

$$Z_{\text{rot}} = g_g Z_g + g_u Z_u, \tag{8.243}$$

where $g_g$ is the number of states with *even* ("gerade" in German) relative angular momentum and

$$Z_g = \sum_{\ell=0,2,4,\ldots} (2\ell+1) e^{-\frac{\hbar^2}{2Ik_BT}\ell(\ell+1)}, \tag{8.244}$$

while $g_u$ is the number of states with *odd* ("ungerade") relative angular momentum and

$$Z_u = \sum_{\ell=1,3,5,\ldots} (2\ell+1) e^{-\frac{\hbar^2}{2Ik_BT}\ell(\ell+1)}. \tag{8.245}$$

At very low temperatures,

$$Z_g \sim 1, \qquad Z_u \sim 0, \tag{8.246}$$

so that $Z_{\text{rot}} \sim g_g$, while at high temperatures

$$Z_g \sim Z_u \sim \frac{1}{2}\int_0^{\infty} 2Ke^{-\frac{\hbar^2}{2Ik_BT}K^2}\,dK = \frac{Ik_BT}{\hbar^2}, \tag{8.247}$$

and $Z_{\text{rot}} = (g_g + g_u)\frac{Ik_BT}{\hbar^2}$ which is (8.234), with the proper overall symmetry factor $\frac{1}{2}$, multiplied by the total nuclear spin degeneracy.

If a catalyst, e.g. charcoal, is present, so that the mixture of ortho and para forms can be regarded as a gas in thermal equilibrium (otherwise, spin exchanges would be extremely rare), for this gas one can construct the partition function

$$Z = \frac{(g_gZ_g)^{N_g}}{N_g!}\frac{(g_uZ_u)^{N_u}}{N_u!}, \qquad N_g + N_u = N, \tag{8.248}$$

and thus diffusive equilibrium requires

$$\frac{N_g}{N_u} = \frac{g_g\,Z_g}{g_u\,Z_u}. \tag{8.249}$$

The system will therefore fall largely into the even form (para-hydrogen and ortho-deuterium) at low temperatures. On the other hand, as the temperature increases, more and more molecules will start reverting from the even to the odd form, as the energy disadvantage of the latter becomes negligible. In the high-temperature regime, all available states will be effectively degenerate, and the ratio of even to odd molecules will be governed only by the relative multiplicity of the spin states $\frac{g_g}{g_u}$. Therefore,

$$\frac{N_{\text{para-}H_2}}{N_{\text{ortho-}H_2}} = \frac{1}{3}, \qquad \frac{N_{\text{ortho-}D_2}}{N_{\text{para-}D_2}} = 2, \tag{8.250}$$

as anticipated, and the relative intensity of absorption and emission lines reflects the spin-statistics relations for the gases.

Before concluding, let us note that one can simply count, in general, the spin states corresponding to a given spin $S$ of two identical nuclei. The total spin space can be decomposed as

$$S \otimes S = 0 \oplus 1 \oplus 2 \oplus \cdots \oplus (2S), \tag{8.251}$$

so that the total number of states $g = g_g + g_u$ is simply

$$g = \sum_{s=0}^{2S}(2s+1) = (2S+1)^2. \tag{8.252}$$

If $S$ is an integer, $2S$ is even and the number $g_g$ of states with even total spin is

$$g_g = \sum_{s=0}^{S}(2(2s)+1) = (S+1)(2S+1), \tag{8.253}$$

whereas $g_u = S(2S+1)$. Conversely, if $2S$ is an odd number,

$$g_u = \sum_{s=0}^{\frac{2S-1}{2}}(2(2s+1)+1) = (S+1)(2S+1), \tag{8.254}$$

while $g_g = S(2S+1)$. Therefore, for integer $S$, and thus for bosonic nuclei,

$$Z_{\rm rot} = (2S+1)\left[(S+1)Z_g + SZ_u\right], \qquad \frac{g_u}{g_g} = \frac{S}{S+1}, \tag{8.255}$$

while for half-odd integer $S$, and thus for fermionic nuclei,

$$Z_{\rm rot} = (2S+1)\left[SZ_g + (S+1)Z_u\right], \qquad \frac{g_u}{g_g} = \frac{S+1}{S}. \tag{8.256}$$

## 8.10 Some Applications

We can now describe a few important applications of Fermi–Dirac statistics. We begin with the phenomenon of adsorption and then turn to the Thomas–Fermi theory, before concluding with a discussion of white dwarfs, neutron stars and the corresponding Chandrasekhar limits.

### 8.10.1 Adsorption

A surface that is able to form bound states with gas particles, trapping them, provides an interesting case study. The resulting phenomenon is called adsorption.

A simple model of this process can be obtained by associating to the adsorbing surface $N_s$ available sites, which can be occupied by gas molecules. For simplicity, let us assume that each site can only accommodate one particle at a time (the so-called single-layer adsorption), and that this bound state is characterized by a negative energy $\epsilon = -|\epsilon|$. The partition function for a system of $N_g$ gas particles and $N_a$ adsorbed particles is therefore

$$Z = Z_a \frac{Z_g^{N_g}}{N_g!}, \tag{8.257}$$

where

$$Z_g = V\left(\frac{2\pi m k_B T}{h^2}\right)^{\frac{3}{2}} \tag{8.258}$$

is the single-particle partition function for the unbound gas, while

$$Z_a = \binom{N_s}{N_a} e^{\beta N_a |\epsilon|} \tag{8.259}$$

is the contribution of $N_a$ adsorbed particles, which occupy $N_a$ of the $N_s$ available binding sites. Minimizing the Helmholtz free energy for constant

$$N = N_g + N_a \tag{8.260}$$

while making use of the Stirling approximation of Eq. (2.32) yields the condition

$$\left(\frac{1}{x} - 1\right) e^{\beta|\epsilon|} = \frac{Z_g}{N_g} \tag{8.261}$$

for the occupation fraction

$$x = \frac{N_a}{N_s}. \tag{8.262}$$

Note that this is equivalent to imposing that the chemical potentials of gas and adsorbed molecules coincide, the standard condition for diffusive equilibrium. Using the equation of state

$$PV = N_g k_B T \tag{8.263}$$

for the gas component, this result can be recast in the form

$$x = \frac{P}{P + P_0}, \tag{8.264}$$

in terms of the effective pressure

$$P_0 = \frac{(2\pi m)^{\frac{3}{2}}}{h^3} (k_B T)^{\frac{5}{2}} e^{-\beta|\epsilon|}. \tag{8.265}$$

Equation (8.264) is usually called Langmuir adsorption isothermal equation. It indicates that, if the gas is subject to high pressures, almost all adsorbent sites will be occupied, while for rarefied gases the surface will bind almost no particles. On the other hand, for fixed pressure, adsorption becomes more likely if the binding energy $|\epsilon|$ becomes larger.

There is an instructive alternative to this derivation. As we have just seen, in single-layer adsorption the one-particle states on the adsorbing surface have energy $-|\epsilon|$ and degeneracy $N_s$, and can be occupied at most once. Therefore, the corresponding "fermionic" grand partition function reads

$$\mathcal{Q} = (1 + e^{\beta(\mu - \epsilon)})^{N_s}, \tag{8.266}$$

and the average number of adsorbed particles is therefore

$$x = \frac{N_a}{N_s} = \frac{k_B T}{N_s} \frac{\partial \log \mathcal{Q}}{\partial \mu} = \frac{1}{e^{\beta(\epsilon - \mu)} + 1}. \tag{8.267}$$

Substituting in this expression the chemical potential (2.84) of the free gas one can simply recover the Langmuir equation (8.264).

### 8.10.2 Thomas–Fermi Theory

The Thomas–Fermi model provides an effective description of large atoms via a self-consistent potential. The idea is that, in a large atom, the electrons form a degenerate Fermi gas around the nucleus that fills all energy levels between a negative, spherically symmetric, effective potential energy $\mathcal{U}(r)$ and a Fermi level $\epsilon_F$.

Our starting point is thus Eq. (8.92), which we rewrite in the form

$$n(r) = \begin{cases} \frac{1}{3\pi^2} \left[ \frac{2m}{\hbar^2} (\epsilon_F - \mathcal{U}(r)) \right]^{\frac{3}{2}} & \text{within the electronic cloud} \\ 0 & \text{elsewhere} \end{cases} \tag{8.268}$$

and the next step is to determine $\mathcal{U}(r)$ self-consistently from the Poisson equation

$$\nabla^2 \mathcal{U} = -4\pi e^2 n(r). \tag{8.269}$$

The "bare" Coulomb potential energy of an electron close to the nucleus,

$$\mathcal{U}(r) \sim -\frac{Ze^2}{r}, \tag{8.270}$$

ought to be screened, for increasing values of $r$, by the inner electrons distributed according to Eq. (8.268). One is thus led, within the electronic cloud, to

$$\nabla^2 \mathcal{U} \equiv \frac{1}{r}\frac{d^2(r\mathcal{U})}{dr^2} = -\frac{4\pi e^2}{3\pi^2}\left[\frac{2m}{\hbar^2}\left(\epsilon_F - \mathcal{U}(r)\right)\right]^{\frac{3}{2}}. \tag{8.271}$$

The next step is the convenient redefinition

$$\epsilon_F - \mathcal{U}(r) = \frac{Ze^2}{r}\,\Phi(r), \tag{8.272}$$

where $\Phi \geq 0$, which translates the singular limiting behavior of Eq. (8.270) into the finite initial condition

$$\Phi(0) = 1, \tag{8.273}$$

and leads to

$$\frac{d^2\Phi}{dr^2} = \frac{4}{3\pi}\sqrt{\frac{Z}{r}}\left(\frac{2}{a_0}\Phi\right)^{\frac{3}{2}}, \tag{8.274}$$

where $a_0$ is the Bohr radius for hydrogen,

$$a_0 = \frac{\hbar^2}{m e^2}, \tag{8.275}$$

which already appeared in Eq. (3.171). The last step is the introduction of a dimensionless radial variable $\rho$, via

$$r = \rho\,\frac{a_0}{2}\left(\frac{9\pi^2}{16Z}\right)^{\frac{1}{3}}, \tag{8.276}$$

which indicates that atomic sizes scale inversely to a power of $Z$. This step casts the Thomas–Fermi equation in its standard form

$$\frac{d^2\Phi}{d\rho^2} = \frac{1}{\sqrt{\rho}}\,\Phi^{\frac{3}{2}}, \tag{8.277}$$

which holds within the electronic cloud. Note, however, that Eqs. (8.268) and (8.276) imply that

$$n(r) = \frac{4Z^2}{9\pi^3}\left(\frac{2}{a_0}\right)^3\left(\frac{\Phi}{\rho}\right)^{\frac{3}{2}}, \tag{8.278}$$

and therefore, in view of Eq. (8.276), the actual electron distribution scales with $Z$ according to

$$n(r)\,r^2\,dr \sim Z^2\left(Z^{-\frac{1}{3}}\right)^3 \sim Z. \tag{8.279}$$

All in all, atomic sizes scale proportionally to $Z^{-\frac{1}{3}}$, due to the combined effects of nuclear charge and Pauli exclusion principle, but the fraction of the $Z$ electrons of a neutral atom that lie within a certain distance from the nucleus is independent of $Z$.

One can compute the total number of electrons within a radius $r_0$ as

$$\mathcal{N}(r_0) = 4\pi \int_0^{r_0} r^2\, dr\, n(r). \tag{8.280}$$

Making use of the preceding substitutions gives

$$\mathcal{N}(\rho_0) = Z \int_0^{\rho_0} d\rho\, \sqrt{\rho}\, \Phi^{\frac{3}{2}}, \tag{8.281}$$

and taking into account the Thomas–Fermi equation (8.277) and the initial condition (8.273) finally leads to

$$\mathcal{N}(\rho_0) = Z\Big(\rho_0\, \Phi'(\rho_0) - \Phi(\rho_0) + 1\Big), \tag{8.282}$$

after integrating by parts.

For a neutral atom, the first two terms on the right-hand side must cancel each other at the edge of the electron cloud. This can be attained with a diffuse electron cloud if both $\Phi(\rho_0)$ and $\Phi'(\rho_0)\, \rho_0$ tend to zero as $\rho_0 \to \infty$, which leads to the standard Thomas–Fermi function. Equation (8.272) then implies that $\mathcal{U}(r)$ tends to zero faster than $\frac{1}{r}$ as $r \to \infty$, which is the expected behavior for a neutral atom (See Figure 8.7), provided $\epsilon_F = 0$. The corresponding $\Phi$ and $\mathcal{U}$ are illustrated in Figure 8.6 (dashed curve) and Figure 8.7, respectively. Note, incidentally, that

$$\Phi(\rho) = \frac{144}{\rho^3} \tag{8.283}$$

is an exact solution of the Thomas–Fermi equation (8.277) with this type of behavior as $\rho \to \infty$, which however does not satisfy the boundary condition (8.273).

The Thomas–Fermi equation, however, admits two other types of solutions. The first starts with a larger value of $\Phi'(0)$ and corresponds to the dash-dotted curve in Figure 8.6, while the third starts with a smaller value of $\Phi'(0)$ and corresponds to the solid curve in the figure. They both afford physical interpretations, up to some amendments that we can now describe.

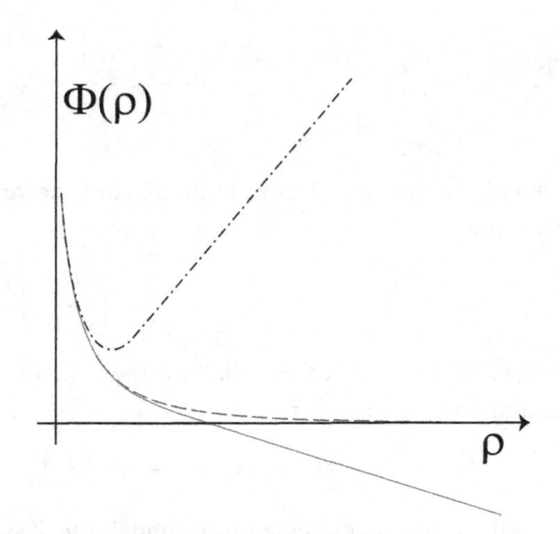


Figure 8.6 The three types of solutions of the Thomas–Fermi equation. They can be associated to (dashed line) a neutral atom, (dash-dotted line) a compressed neutral atom and, (solid line) a positive ion.

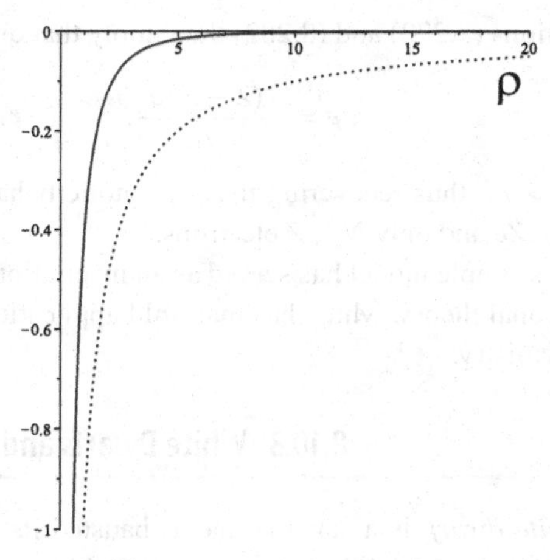


Figure 8.7 The Thomas–Fermi potential energy $\frac{a_0 \mathcal{U}}{Ze^2}$ for (solid line) a neutral atom and (dotted line) the unscreened contribution $-\frac{1}{\rho}$. In this case, as we have seen, $\epsilon_F = 0$.

The solutions with a larger value of $\Phi'(0)$ describe atoms whose electrons are compressed to a finite region. This is the case because the two terms in Eq. (8.282) compensate each other at a finite value of $\rho_0$,

$$\Phi'(\rho_0) = \frac{\Phi(\rho_0)}{\rho_0}. \tag{8.284}$$

Beyond $\rho_0$ there are no more electrons, so that $\Phi$ obeys

$$\frac{d^2\Phi}{d\rho^2} = 0 \qquad (\rho > \rho_0), \tag{8.285}$$

and the consistent choice is

$$\Phi(\rho) = \Phi(\rho_0)\,\frac{\rho}{\rho_0} \qquad (\rho > \rho_0). \tag{8.286}$$

Equation (8.272) implies that choosing

$$\epsilon_F = Ze^2\,\frac{\Phi(\rho_0)}{r_0} > 0 \tag{8.287}$$

leads in this fashion to $\mathcal{U}(r) = 0$ for $r > r_0$. This system can be regarded as a neutral atom whose electron cloud is compressed to a finite size.

Finally, positive ions can be modeled by the third type of solutions, which start from the origin with values of $\Phi'(0)$ below the special one corresponding to diffused neutral atoms, and therefore

$$\Phi(\rho_0) = 0, \qquad \Phi'(\rho_0) < 0, \tag{8.288}$$

for a finite $\rho_0$, are natural conditions. Beyond $\rho_0$, Eq. (8.285) holds and

$$\Phi(\rho) = \Phi'(\rho_0)\,\rho_0\left(\frac{\rho}{\rho_0} - 1\right). \tag{8.289}$$

Equations (8.272) and (8.282) then imply that one can choose the Fermi level so that

$$\epsilon_F = -\frac{(Z-N)e^2}{r_0}, \qquad \mathcal{U}(r) = -\frac{(Z-N)e^2}{r}, \tag{8.290}$$

for $r > r_0$, thus recovering the asymptotic behavior for a positive ion with nuclear charge $Ze$ and only $N < Z$ electrons.

This simple model has served as an inspiration and a starting point for the density-functional theory, which has manifold applications in condensed matter physics and in chemistry.

### 8.10.3 White Dwarfs and Neutron Stars

A *white dwarf* is a star that has exhausted its fuel, having transformed all H into He. One can model this system as a combination of $\frac{\mathcal{N}}{2}$ He atoms, which provide the gravitational pull, and $\mathcal{N}$ electrons, which form a highly degenerate Fermi gas. In fact, the typical density of a white dwarf is $\rho \simeq 10^{10}$ kg/m$^3$, which corresponds to about $10^7$ times the density of the sun, and its typical core temperature is comparable to the temperature of the sun's interior, about $T_\odot \simeq 10^7$ K. One can obtain a crude estimate of the resulting degeneracy substituting these values in the parameter $\eta$, which gives

$$\eta = \frac{\rho}{4m_N}\left[\frac{1}{n_Q(T_\odot)}\right]_{m=m_e} \simeq 10^5, \tag{8.291}$$

where $m_N \simeq 1.7 \times 10^{-27}$ kg is the nucleon mass and $m_e \simeq 9.1 \times 10^{-31}$ kg is the electron mass. These white dwarfs can remain stable insofar as the Fermi pressure of degenerate electrons can balance the gravitational attraction, and Chandrasekhar showed that this sets an upper bound on their masses, which is comparable to the solar mass. Neutron stars are similar, but neutrons, rather than electrons, are responsible for the Fermi pressure. This section is devoted to a brief discussion of these important classes of compact objects.

Let us begin our analysis by establishing some links between the microscopic and macroscopic parameters of a white dwarf star, which can be modeled as a spherical object of mass $M$ composed of fully ionized Helium atoms, so that

$$M \simeq 2\mathcal{N} m_N, \tag{8.292}$$

and its volume

$$V = \frac{4}{3}\pi R^3. \tag{8.293}$$

The condition on the total number of electrons reads

$$\mathcal{N} = \frac{2V}{h^3}\int_{|\mathbf{p}|\leq p_F} d^3\mathbf{p} = \frac{V}{3\pi^2\hbar^3}\, p_F^3, \tag{8.294}$$

so that

$$p_F = \hbar\left(3\pi^2\frac{\mathcal{N}}{V}\right)^{\frac{1}{3}}. \tag{8.295}$$

The internal energy for the degenerate electron gas, in this relativistic context, is then

$$U = \frac{2V}{h^3} \int_{|\mathbf{p}| \leq p_F} d^3\mathbf{p} \sqrt{(c\,\mathbf{p})^2 + m_e^2 c^4}. \tag{8.296}$$

It is now convenient to change variable to $x = \frac{|\mathbf{p}|}{m_e c}$, so that

$$U = \frac{V m_e^4 c^5}{\pi^2 \hbar^3} \int_0^{x_F} dx\, x^2 \sqrt{1 + x^2}, \tag{8.297}$$

where

$$x_F = \frac{\hbar}{m_e c} \left(3\pi^2 \frac{N}{V}\right)^{\frac{1}{3}}. \tag{8.298}$$

Note that, for the systems of interest, with radii comparable to the Earth radius, $x_F$ is slightly above 1.

Integrating by parts, one can show that

$$U = \frac{V m_e^4 c^5}{\pi^2 \hbar^3} \left[\frac{x_F^3}{4} \sqrt{1 + x_F^2} + g(x_F)\right], \tag{8.299}$$

where

$$g(x) = \frac{x_F}{8} \sqrt{1 + x_F^2} - \frac{1}{8} \log\left(x_F + \sqrt{1 + x_F^2}\right), \tag{8.300}$$

and consequently

$$P = -\frac{\partial U}{\partial V} = \frac{m_e^4 c^5}{\pi^2 \hbar^3} \left[\frac{x_F^3}{12} \sqrt{1 + x_F^2} - g(x_F)\right], \tag{8.301}$$

since one is effectively working at zero temperature. In a white dwarf, this pressure is to balance the gravitational pull that, for dimensional reasons, is of the form

$$P = \gamma \frac{G_N M^2}{4\pi R^4} \simeq \gamma\, \sigma \frac{m_e^4 c^5}{\pi^2 \hbar^3} x_F^4 \left(\frac{M}{M_\odot}\right)^{\frac{2}{3}}, \tag{8.302}$$

where $M_\odot \simeq 2 \times 10^{30}$ kg is the solar mass, $\gamma$ is a factor of order one, and

$$\sigma = \frac{2}{9} \left(\frac{8}{9\pi}\right)^{\frac{1}{3}} \frac{G_N m_N^{\frac{4}{3}} M_\odot^{\frac{2}{3}}}{\hbar c} \simeq 0.1. \tag{8.303}$$

The more convenient expression for $P$ is obtained by eliminating $R$ in terms of density and mass, making use of Eqs. (8.292) and (8.293). One is thus led to demand that

$$\frac{x_F^3}{12} \sqrt{1 + x_F^2} - g(x_F) \simeq 0.1\, \gamma\, x_F^4 \left(\frac{M}{M_\odot}\right)^{\frac{2}{3}}. \tag{8.304}$$

Note that Eq. (8.304) is well approximated, beyond $x_F \simeq 1$, by

$$\frac{1}{12}\left(x_F^4 - x_F^2\right) \simeq 0.1\, \gamma\, x_F^4 \left(\frac{M}{M_\odot}\right)^{\frac{2}{3}}, \tag{8.305}$$

which has no real solutions if

$$\frac{M}{M_\odot} > \frac{0.76}{\gamma^{\frac{3}{2}}}, \tag{8.306}$$

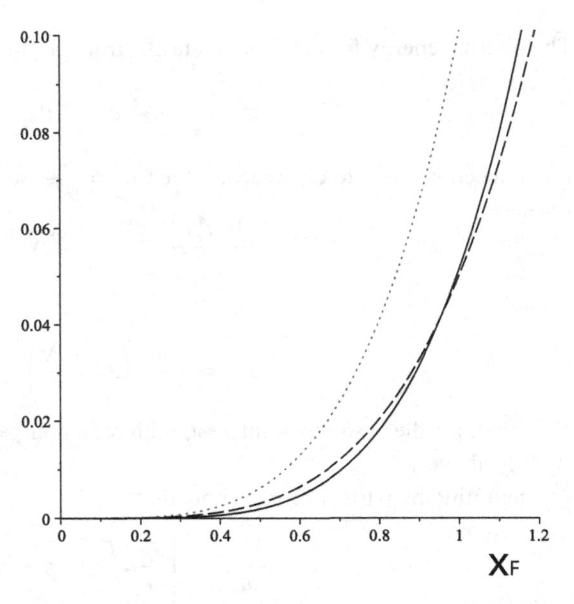


**Figure 8.8** The curve on (solid line) the left-hand side of Eq. (8.304) compared with (dashed line) $0.05 * x_F^4$ and (dotted line) $0.1 * x_F^4$. The intersection with the latter has moved sizably closer to the origin from the region $x_F \simeq 1$.

corresponding to a value of the order of the solar mass for $\gamma$ of order one. Alternatively, as illustrated in Figure 8.8, for $\gamma$ of order one the condition $x_F \simeq 1$ translates again into an estimate for $M$ of the order of the solar mass. More refined considerations would yield Chandrasekhar's result of about 1.4 solar masses.

One can also adapt these rough considerations to the case of *neutron stars*, which rest on a highly degenerate Fermi gas of neutrons, where however general relativity would play a more prominent role, starting from

$$P = \frac{m_N^4 c^5}{\pi^2 \hbar^3} \left[ \frac{x_F^3}{12} \sqrt{1 + x_F^2} - g(x_F) \right], \tag{8.307}$$

where now

$$x_F = \frac{\hbar}{m_N c} \left( 3\pi^2 \frac{\mathcal{N}}{V} \right)^{\frac{1}{3}}, \qquad M \simeq \mathcal{N} m_N. \tag{8.308}$$

Consequently, the equilibrium condition becomes, in this case,

$$\frac{x_F^3}{12} \sqrt{1 + x_F^2} - g(x_F) \simeq 0.1\, \gamma\, x_F^4 \left( \frac{M}{4 M_\odot} \right)^{\frac{2}{3}}, \tag{8.309}$$

and the predicted bound is about four times as large, with corresponding radii that are about 1800 times smaller. However, all these estimates are only qualitatively correct, since for instance a black hole of this mass would have a horizon radius of about 12 km, about twice the estimated radius for a neutron star (while it should be smaller, of course!)

## Bibliographical Notes

Pedagogical introductions to quantum gases can be found in many excellent books, including [23, 36, 49, 50, 51, 54]. A clear discussion of electrons in bands can be found in [31]. White dwarfs are discussed in detail in [10].

## Problems

**8.1** Evaluate the density of states $g(\epsilon)$ for the following systems in $d$ dimensions:

1. a particle with dispersion relation $\epsilon = c\,|\mathbf{p}|^{\alpha}$ confined to a volume $V$
2. a harmonic oscillator with mass $m$ and frequency $\omega$;
3. a nonrelativistic particle with mass $m$ subject to a constant gravitational potential, confined in a cylinder with section $A$ and axis parallel to the gravitational force.

Find, for each of these systems, the temperature $T_c$ at which Bose condensation occurs for a corresponding gas of noninteracting particles at a given density $n = \frac{N}{V}$.

**8.2** Here we would like to reconsider Eqs. (8.13) and (8.14) and generalize them.

1. Verify that Eqs. (8.13) and (8.14) hold, expressing the partition functions $Z_2$ for two fermions or bosons in terms of the single-particle partition function $Z_1$.
2. Generalize Eqs. (8.13) and (8.14) to cases where the energy levels $E_\alpha$ are $g_\alpha$-degenerate.
3. Find the corresponding equations that express $Z_3$ and $Z_4$ in terms of $Z_1$.
4. Prove that, for any $N$,

$$Z_N(\beta) = (\mp)^N \sum_{\{\ell_n\}} \prod_n \frac{\left[\mp Z_1(n\beta)\right]^{\ell_n}}{n^{\ell_n}\,\ell_n!},$$

where the sum is restricted to $\ell_n \geq 0$ such that $\sum_n n\ell_n = N$.
*Hint: it is convenient to start from* $\log \mathcal{Q}$, *written as in Eq. (8.66).*

**8.3** A number $N$ of noninteracting spin-$\frac{1}{2}$ particles is allowed to hop, with a negligible kinetic energy, between the $N_0$ sites of a lattice. In each site, a particle has two available energy levels: the ground state with zero energy and an excited state with energy $E > 0$. Show that at low temperatures, letting $x = \frac{N}{2N_0}$, the chemical potential $\mu$ behaves as follows:

$$\mu = \begin{cases} -\infty & \text{if } x = 0 \\ k_B T \log\left(\frac{x}{1-x}\right) & \text{if } 0 < x < 1,\ \ \beta E \gg 1 \\ \frac{E}{2} & \text{if } x = 1 \\ E + k_B T \log\left(\frac{x-1}{2-x}\right) & \text{if } 1 < x < 2,\ \ \beta E \gg 1 \\ +\infty & \text{if } x = 2. \end{cases}$$

**8.4** Consider a vertical cylinder containing a gas of $N$ spin–$\frac{1}{2}$ particles, which is subject to a constant gravitational field $g$.

1. Find the Fermi energy $\epsilon_F$, the total energy at zero temperature, and evaluate the first correction to the chemical potential for low temperatures.
2. Prove that there is a maximum height $h_{\max}$ that the gas particles can reach at $T = 0$, and that $\epsilon_F = mgh_{\max}$.

**8.5** A box with total volume $2V$ is divided into two parts by an ideal insulating barrier that is free to slide. One of the two parts, of volume $V$, is filled with a completely degenerate Fermi gas of particles with mass $m$ and spin $s$. The other part is filled with a Bose gas of $N$ particles with mass $m$ and spin zero, 19% of which are in the condensed phase. Determine that value $s$ of the spin so that the system is in equilibrium.

**8.6** Construct explicitly a basis of states with definite parity for the deuterium $D_2$ molecule starting from the single-nucleus basis vectors $|0\rangle$, $|+\rangle$, $|-\rangle$, where 0 and $\pm$ denote spin projections equal to 0 and $\pm\hbar$.
*Hint: it is convenient to start from*

$$S^2 = S_1^2 + S_2^2 + 2S_{1z}S_{2z} + S_{1+}S_{2-} + S_{2+}S_{1-}.$$

**8.7** Study the internal energy, the specific heat $C_V$, and the entropy for a gas of $N$ massless spin–$\frac{1}{2}$ fermions in three dimensions, not mutually interacting, for which

$$H = \sum_{i=1}^{N} c\,|\mathbf{p}_i|,$$

and whose number *is not conserved*. Show that the resulting expressions are $\frac{7}{8}$ of those obtained for the bosonic case in Chapter 4. This result plays a role in our current picture of early universe cosmology.

**8.8** Consider a system of two particles for which three single-particle states $|0\rangle$, $|1\rangle$ and $|2\rangle$ are available, with energies 0, $\epsilon$, $2\,\epsilon$ and such that

$$\langle i|j\rangle = \delta_{ij}.$$

1. Write the possible normalized state vectors for a pair of particles in terms of the available one-particle states and compute the corresponding partition function if the particles are *distinguishable*.
2. Write the possible normalized state vectors for a pair of particles in terms of the available one-particle states and compute the corresponding partition function if the particles are *identical fermions*.
3. Write the possible normalized state vectors for a pair of particles in terms of the available one-particle states and compute the corresponding partition function if the particles are *identical bosons*.

**8.9** Consider a degenerate two-dimensional gas of $N$ spin-$\frac{1}{2}$ particles, not mutually interacting and described by the single-particle Hamiltonian

$$H = \frac{\mathbf{p}^2}{2m} + \lambda(\mathbf{r}^2 - \eta^2)^2, \tag{8.310}$$

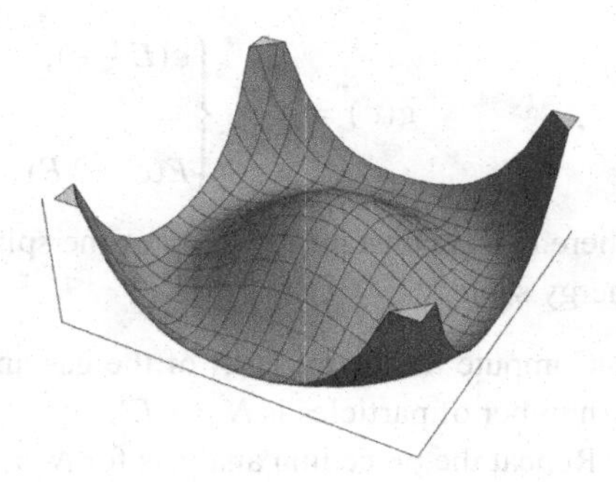

Figure 8.9 The potential in problem 8.9.

with $\lambda$ and $\eta$ two positive parameters. The potential energy is depicted in Figure 8.9. Compute the Fermi level and the internal energy, and study their behavior as a function of $N$.

**8.10** Let us consider a hypothetical gas of noninteracting particles of spin 0 in three dimensions, for which

$$H = \frac{\mathbf{p}^2}{2m},$$

and such that every level can accommodate 0, 1, or 2 particles.

1. Compute the grand partition function for this exotic gas, and derive the corresponding expressions for average number of particles, pressure, and internal energy.
2. Prove that for this system the relation

$$PV = \frac{2}{3}U$$

holds.
3. Compute the first correction to the equation of state, in the high-temperature limit, with respect to the Boltzmann gas.
4. Investigate the limiting behavior of the gas at low temperatures.

**8.11** Consider a gas of $N$ spin$-\frac{1}{2}$ particles, not mutually interacting and whose single-particle energy levels are $\epsilon_n = E_0\, n$, where $E_0$ is a given energy scale, with degeneracies $d_n = n$ (for $n = 1, 2, 3, \ldots$).

1. Discuss the behavior of the Fermi level as a function of $N$ and compute the internal energy of the gas at $T = 0$.
2. Repeat the analysis in the continuum limit.

*Hint: Recall that*

$$\sum_{k=1}^{j} k = \frac{j(j+1)}{2}, \qquad \sum_{k=1}^{j} k^2 = \frac{j(j+1)(2j+1)}{6}.$$

**8.12** The behavior of electrons in a lattice of ions is well described neglecting their mutual interactions, but assigning them a density of states with some gaps. Let us consider the simple case in which

$$g(\epsilon) = \alpha\, g_{\frac{1}{2}} \begin{cases} \epsilon(E-\epsilon), & \text{per } 0 < \epsilon < E, \\ E(\epsilon - 2E), & \text{per } \epsilon > 2E, \end{cases}$$

where $\alpha$ is a constant, $g_{\frac{1}{2}}$ denotes the spin degeneracy factor and $E$ is a given energy scale.

1. Compute the Fermi level of the gas and its internal energy at $T = 0$ if the number of particles is $N = \alpha E^3$.
2. Repeat the preceding analysis for $N = \alpha \frac{E^3}{3}$.
3. Study the behavior of $C_V$ at low temperatures in the second case.

What types of "material" do these two cases model?

**8.13** Consider a system of $N$ distinguishable molecules of mass $M$ in three dimensions, for which

$$H = \sum_{i=1}^{N} \frac{\mathbf{P}_i^2}{2M} + \sum_{i=1}^{N} \frac{\mathbf{L}_i^2}{2I},$$

where $\mathbf{L}_i$ is the *relative* orbital angular momentum of the two components of the molecule, which are spin–$\frac{1}{2}$ particles.

1. Compute the partition function of the system if the two components of each molecule are *distinguishable*.
2. Compute the partition function of the system if the two components of each molecule are *indistinguishable* fermions.
3. Study in both cases the behavior of $C_V$ in the limits of high and low $T$.

**8.14** Consider a degenerate Fermi gas of spin–$\frac{1}{2}$ particles with $m = 0$ in three dimensions.

1. Compute the internal energy and the Fermi pressure.
2. Study the static equilibrium for a star supported by this gas and determine the resulting expression for its mass.

**8.15** Consider a gas of nonrelativistic particles of mass $m$ and spin $s = 0$ in two dimensions, whose spectrum contains an additional nondegenerate level of energy $-\Delta < 0$.

1. Study the phenomenon of Bose–Einstein condensation for this system.
2. Study the limiting behavior of the specific heat at fixed area $C_A$ at low temperatures $T$.

**8.16** Consider a gas of indistinguishable particles that are not mutually interacting, and whose total number is *not* conserved. The available states for these particles are labeled by an integer $j$ (which includes the possible dependence on the spin quantum number), and let $E_j$ denote the corresponding energies.

1. If the particles are fermions, obtain the probability $p_j$ that the $j$th state be occupied at a given temperature $T$ extremizing the contribution to the free energy from the state $j$. Compute also the average occupation number $n_j$ for the state $j$.

2. If the particles are bosons, obtain the probability $p_{j,n}$ that the state $j$ be occupied by $n = 0, 1, 2, \dots$ particles at a given temperature $T$. Compute also the average occupation number $n_j$ for the state $j$.
3. Verify that in both cases the resulting average occupation numbers are special values of the standard ones for $z = 1$.

**8.17** Consider a gas of "diatomic molecules" in a volume $V$ at a temperature $T$. The Hamiltonian of a single "molecule" is

$$H = \frac{\mathbf{p}_1{}^2 + \mathbf{p}_1{}^2}{2m} + \frac{1}{2} m \omega^2 (\mathbf{r}_1 - \mathbf{r}_2)^2 ,$$

where $m$ is the mass of the molecule, and the second term accounts for the vibrations of its two constituents.

1. Study the degeneracy of the energy levels for the three-dimensional oscillator modes.
2. Write the partition function of the system and study the limiting behaviors of $C_V$ and low and high temperatures in the following three cases:

   a. The two constituents are distinguishable.
   b. The two constituents are identical fermions.
   c. The two constituents are identical bosons.

**8.18** Consider a system similar to the Debye gas, in three dimensions and with a finite number $3N$ of modes, with

$$E = \hbar v_s k,$$

but for which the Bose–Einstein statistics is replaced with the Fermi–Dirac one.

1. Study how the internal energy and the specific heat $C_V$ would depend on the temperature $T$.
2. Provide a physical interpretation for the preceding results.

**8.19** A more refined model of adsorption obtains by allowing more than one particle to condense on a given site of the surface (multi-layer adsorption): the first particle will lodge directly on the surface, with energy $\epsilon$, while each additional particle will form a bound state with the preceding ones, with an energy increase $\epsilon^*$.

1. Compute the partition function corresponding to $N_g$ gas particles, $N_a$ particles in the first layer, and $N_a^*$ particles in the higher layers.
2. Show that, in equilibrium,

$$x = \frac{N_a + N_a^*}{N_s} = \frac{\frac{P}{P_0}}{(1 - \frac{P}{P_0}^*)(1 - \frac{P}{P_0}^* + \frac{P}{P_0})},$$

   where $P_0$ is as in Eq. (8.265) and $P_0^*$ is the same expression with $\epsilon$ replaced by $\epsilon^*$. This is known as the Brunauer–Emmett–Teller adsorption isothermal equation.
3. Starting from the canonical partition function above, construct the grand partition function and obtain $x$ working in the grand canonical ensemble.

**8.20** Discuss the thermodynamics of an exotic three-dimensional system for which

$$\log Q = -\mathrm{Tr}\left[\log\left(1 - z\,e^{-\beta H} + a^2 z^2\, e^{-2\beta H}\right)\right],$$

where $s = 0$,

$$H = \frac{p^2}{2m},$$

and $a$ is a real constant.

1. Discuss how the available range for the fugacity $z$, or equivalently for the chemical potential $\mu$, depends on $a$.
2. Examine the high-temperature behavior of the system.
3. Study the possible occurrence of Bose–Einstein condensation for this system. If it occurs, write the condition that determines the critical temperature $T_c$ and obtain the temperature dependence of $\langle n_0 \rangle$ for $T < T_c$.

# 9 Magnetism in Matter, I

We can now use the tools developed so far to address, from a microscopic perspective, the behavior of matter in the presence of magnetic fields. Homogeneous magnetic fields will suffice to a large extent, since the response emerges from phenomena at atomic distance scales. For definiteness, let us consider field lines parallel to the $z$ axis of a Cartesian reference frame, so that

$$\mathbf{B} = B_z \hat{\mathbf{z}}. \tag{9.1}$$

Here, in the first of two chapters on magnetism in matter, we focus on the response to external magnetic fields, in which materials can exhibit two distinct types of behavior. Both depend crucially on quantum mechanics: the first, *diamagnetism*, is a macroscopic manifestation of the orbital motion of electrons, while the second, *paramagnetism*, is a macroscopic manifestation of the electron spin.

In general, one can characterize the behavior of materials in the presence of a uniform magnetic field $B_z$ by two quantities, the magnetization

$$\mathcal{M} = \frac{1}{\beta} \frac{\partial}{\partial B_z} \log \mathcal{Q}, \tag{9.2}$$

which identifies an average magnetic moment, and the susceptibility

$$\chi = \frac{\partial M}{\partial B_z}, \tag{9.3}$$

usually considered at a small or vanishing magnetic field. The latter quantity characterizes how variations of the external field affect the induced magnetic moment, but also, in the spirit of what we said in Section 2.3 for the specific heat, the corresponding fluctuations.

## 9.1 Orbits in a Uniform Magnetic Field

Let us begin by recalling some well-known results of classical electromagnetism on the motion of a particle of charge $-e$ in a uniform magnetic field $\mathbf{B}$, starting from

$$m \frac{d\mathbf{v}}{dt} = -\frac{e}{c} \mathbf{v} \times \mathbf{B}. \tag{9.4}$$

A first consequence of Eq. (9.4) is that, for a field aligned as in (9.1), the particle does not experience any acceleration in the $z$ direction, and

$$z = z_0 + v_z t, \tag{9.5}$$

with a constant $v_z$.

The projection on planes orthogonal to the field lines is more interesting, and

$$\frac{dv_x}{dt} = -\frac{e B_z}{m c} v_y, \qquad \frac{dv_y}{dt} = \frac{e B_z}{m c} v_x \tag{9.6}$$

combine nicely into

$$\frac{d\left(v_x + i v_y\right)}{dt} = i \omega_B \left(v_x + i v_y\right), \tag{9.7}$$

where

$$\omega_B = \frac{e B_z}{m c} \tag{9.8}$$

is often called "cyclotron frequency." The solution reads

$$v_x(t) + i v_y(t) = \left[v_x(0) + i v_y(0)\right] e^{i \omega_b t}, \tag{9.9}$$

and one more integration leads to

$$x(t) + i y(t) = [x(0) + i y(0)] + \frac{1}{i \omega_B}\left[v_x(0) + i v_y(0)\right]\left(e^{i \omega_b t} - 1\right), \tag{9.10}$$

which describes circular trajectories with center at

$$\left(x(0) - \frac{v_y(0)}{\omega_B},\, y(0) + \frac{v_x(0)}{\omega_B}\right) \tag{9.11}$$

and radius

$$R = \frac{|\mathbf{v}(0)|}{\omega_B}. \tag{9.12}$$

## 9.2 Landau Levels

Let us now proceed to quantum mechanics, considering, to begin with, the Lagrangian that describes the interactions of a particle of mass $m$ and charge $-e$ with electric and magnetic fields,

$$\mathcal{L} = \frac{m}{2} \mathbf{v} \cdot \mathbf{v} - \frac{e}{c} \mathbf{v} \cdot \mathbf{A}(t, \mathbf{r}) + e V(t, \mathbf{r}). \tag{9.13}$$

Note that $\mathcal{L}$ varies into a total derivative under the "gauge transformations"

$$\mathbf{A} \to \mathbf{A} + \nabla \Lambda, \qquad V \to V - \frac{1}{c}\frac{\partial \Lambda}{\partial t} \tag{9.14}$$

that leave invariant the electric and magnetic fields

$$\mathbf{E} = -\nabla V - \frac{1}{c}\frac{\partial \mathbf{A}}{\partial t}, \qquad \mathbf{B} = \nabla \times \mathbf{A}. \tag{9.15}$$

Moreover, $\mathcal{L}$ varies according to

$$\delta \mathcal{L} = \delta \mathbf{v} \cdot \left[m \mathbf{v} - \frac{e}{c} \mathbf{A}\right] \tag{9.16}$$

under variations of $\mathbf{v}$, so that the conjugate momenta are

$$\mathbf{p} = m\mathbf{v} - \frac{e}{c}\mathbf{A}. \tag{9.17}$$

The Hamiltonian, obtained as

$$H = \mathbf{v} \cdot \left( m\mathbf{v} - \frac{e}{c}\mathbf{A} \right) - \mathcal{L}, \tag{9.18}$$

thus takes the form

$$H = \frac{1}{2m}\left( \mathbf{p} + \frac{e}{c}\mathbf{A} \right)^2 - eV, \tag{9.19}$$

when expressed in terms of them, so that the Schrödinger equation for this system reads

$$\left[ -\frac{\hbar^2}{2m}\left( \nabla + \frac{ie}{\hbar c}\mathbf{A} \right)^2 - eV \right] \psi = i\hbar \frac{\partial \psi}{\partial t}. \tag{9.20}$$

Note, in particular, that the vector potential $\mathbf{A}$ enters via the "minimal coupling"

$$\nabla \to \nabla + \frac{ie}{\hbar c}\mathbf{A}, \tag{9.21}$$

while the gauge transformation of Eq. (9.14) leaves the Schrödinger equation invariant when it is combined with a corresponding redefinition of $\psi$ by a phase factor:

$$\psi \to e^{-\frac{ie}{\hbar c}\Lambda}\, \psi. \tag{9.22}$$

This deceivingly simple setting has proved a deep source of inspiration, and was instrumental for attaining the current understanding of the fundamental interactions.

Returning to the uniform magnetic field, among the different "gauge choices" for a vector potential $\mathbf{A}$ leading to Eq. (9.1), let us now concentrate on

$$A_y = x B_z. \tag{9.23}$$

It leads, in fact, to simple exact solutions of the time-independent Schrödinger equation for the energy levels

$$-\frac{\hbar^2}{2m}\left[ \frac{\partial^2}{\partial x^2} + \left( \frac{\partial}{\partial y} + \frac{ieB_z}{\hbar c}x \right)^2 + \frac{\partial^2}{\partial z^2} \right] \psi = E\psi, \tag{9.24}$$

which reduces to a familiar one-dimensional harmonic oscillator problem,

$$-\frac{\hbar^2}{2m}\frac{d^2\chi(x)}{dx^2} + \frac{1}{2}m\omega_B^2\left( x + \frac{p_y}{m\omega_B} \right)^2 = \left( E - \frac{p_z^2}{2m} \right)\chi(x), \tag{9.25}$$

after letting

$$\psi = \chi(x)\, e^{\frac{i}{\hbar}(yp_y + zp_z)}. \tag{9.26}$$

The energy eigenvalues are thus a discrete set of "Landau levels" for any eigenvalue of $p_z$:

$$E = \hbar\omega_B\left( n + \frac{1}{2} \right) + \frac{p_z^2}{2m}, \qquad n = 0, 1, \ldots, \tag{9.27}$$

and there is a degeneracy for any given value of $n$, which we can now determine referring, as usual, to a large cubic box of side $L$ and volume $V = L^3$. To begin with, in this case, with periodic boundary conditions,

$$\frac{p_y}{\hbar} = \frac{2\pi k_y}{L}, \qquad \frac{p_z}{\hbar} = \frac{2\pi k_z}{L}, \tag{9.28}$$

for $k_y, k_z \in \mathbb{Z}$. As a result, the position of the center of the harmonic potential in Eq. (9.25) is quantized according to

$$x_c = -\frac{h k_y}{m\,\omega_B L}, \tag{9.29}$$

and the independent values of $x_c$ are to span a distance length $L$, so that the degeneracies of all Landau levels are identical and equal to

$$\gamma = \frac{m\,\omega_B}{h} V^{\frac{2}{3}} = \frac{e B_z}{h c} V^{\frac{2}{3}} \equiv \frac{B_z}{B_0}. \tag{9.30}$$

There is another, more symmetric, choice of $\mathbf{A}$ for the uniform magnetic field of Eq. (9.1),

$$\mathbf{A}' = \frac{1}{2}(\mathbf{B} \times \mathbf{r}), \tag{9.31}$$

and since

$$\mathbf{A}' = \mathbf{A} - \frac{1}{2} B_z \nabla(x y), \tag{9.32}$$

the solutions in this gauge would be related to the previous ones according to

$$\psi' = \psi\, e^{\frac{ieB_z}{2\hbar c} x y}. \tag{9.33}$$

It is also instructive to retrace these steps following a gauge-independent algebraic route. To this end, let us return to the motion on planes orthogonal to $\mathbf{B}$, for which

$$H = \frac{1}{2m}\left(\mathbf{p} + \frac{e}{c}\mathbf{A}\right)^2 = \frac{m}{2}\mathbf{v}^2, \tag{9.34}$$

where we have defined the kinematical velocity operators

$$\mathbf{v} = \frac{1}{m}\left(\mathbf{p} + \frac{e}{c}\mathbf{A}\right), \tag{9.35}$$

which satisfy the commutation relations

$$[v_x, v_y] = -i\,\frac{\hbar\omega_B}{m}, \tag{9.36}$$

with the cyclotron frequency $\omega_B$ given in Eq. (9.8), for any vector potential such that

$$\partial_x A_y - \partial_y A_x = B_z. \tag{9.37}$$

The Hamiltonian of this system can be diagonalized explicitly defining the ladder operators

$$\Pi = \sqrt{\frac{m}{2\hbar\omega_B}}\left(v_x - i v_y\right), \qquad \Pi^\dagger = \sqrt{\frac{m}{2\hbar\omega_B}}\left(v_x + i v_y\right), \tag{9.38}$$

which satisfy indeed

$$[\Pi, \Pi^\dagger] = 1, \tag{9.39}$$

and

$$H = \hbar\omega_B \left(\Pi^\dagger \Pi + \frac{1}{2}\right). \tag{9.40}$$

The resulting spectrum thus recovers the Landau levels included in Eq. (9.27)

$$H|\alpha\rangle = \hbar\omega_B \left(\alpha + \frac{1}{2}\right) |\alpha\rangle, \qquad \alpha = 0, 1, 2, \ldots, \tag{9.41}$$

with a ground state $|0\rangle$ that is annihilated by the operator $\Pi$.

One can also discuss the degeneracy $\gamma$ of Landau levels, which reflects the two-dimensional extension of the system, along these lines, after identifying the proper translation generators in the $x$–$y$ plane. Rather than considering $p_x$ and $p_y$, which do not commute with the Hamiltonian due to the presence of **A**, let us define

$$T_x = m\,(v_x + \omega_B y)\,, \qquad T_y = m(v_y - \omega_B x). \tag{9.42}$$

These operators generate infinitesimal translations, since

$$[T_y, y] = -i\hbar, \qquad [T_y, x] = 0, \tag{9.43}$$

and similarly for $T_x$, and have the virtue of commuting with the ladder operators, and hence with the Hamiltonian:

$$[T_x, H] = 0 = [T_y, H]. \tag{9.44}$$

However, they do not commute with one another, since

$$[T_x, T_y] = i\hbar m\omega_B, \tag{9.45}$$

and therefore one can choose to simultaneously diagonalize the Hamiltonian and, say, $T_y$, with common eigenstates labelled by $\alpha$ as in (9.41) and by another quantum number $\kappa_y$:

$$T_y|\alpha, \kappa_y\rangle = \hbar\kappa_y|\alpha, \kappa_y\rangle. \tag{9.46}$$

By Eq. (9.45), the finite translation operator $e^{ilT_x}$, acting on an eigenstate of $T_y$, shifts the eigenvalue $\kappa_y$:

$$T_y e^{ilT_x}|\alpha, \kappa_y\rangle = \hbar(\kappa_y + m\omega_B l)e^{ilT_x}|\alpha, \kappa_y\rangle, \tag{9.47}$$

where we have used the Baker–Campbell–Hausdorff expansion that here reduces to

$$e^{-ilT_x} T_y e^{ilT_x} = T_y - il[T_x, T_y], \tag{9.48}$$

but does not alter the $H$ eigenvalue. Finally, we note that $e^{ilT_x}$ is a unitary operator, which preserves the normalization of $|\alpha, \kappa_y\rangle$, so that one can conclude that

$$e^{ilT_x}|\alpha, \kappa_y\rangle = |\alpha, \kappa_y + m\omega_B l\rangle. \tag{9.49}$$

Quantizing the system on a rectangular region of the plane with total area $A = L_x L_y$, whose sides $L_x$ and $L_y$ are directed along the two axes, and demanding periodicity along $y$ requires

$$|\alpha, \kappa_y\rangle \;=\; e^{iL_y T_y/\hbar}|\alpha, \kappa_y\rangle \;=\; e^{iL_y \kappa_y}|\alpha, \kappa_y\rangle, \tag{9.50}$$

so that the eigenvalues $\kappa_y$ must be quantized according to

$$\kappa_y = \frac{2\pi n_y}{L_y}, \qquad n_y \in \mathbb{Z}. \tag{9.51}$$

On the other hand, in order to enforce periodicity along $x$,

$$|\alpha, \kappa_y\rangle = e^{iL_x T_x/\hbar}|\alpha, \kappa_y\rangle = \left|\alpha, \kappa_y + \frac{m\omega_B L_x}{\hbar}\right\rangle, \tag{9.52}$$

one must identify quantum numbers $\kappa_y$ that differ by $\frac{m\omega_B L_x}{\hbar}$ or, by (9.51). Consequently, the integers $n_y$ span a range

$$\gamma = \frac{eB_z}{hc}\, A, \tag{9.53}$$

and one thus recovers Eq. (9.30).

Finally, one can repeat the analysis by resorting to the semi-classical Bohr–Sommerfeld formula,

$$\oint \mathbf{p}\cdot d\mathbf{r} \;=\; \left(n + \frac{1}{2}\right) h, \tag{9.54}$$

that we discussed in Chapter 3, which we know is exact for harmonic oscillators. In detail, working for convenience in the symmetric gauge (9.31),

$$m \oint \mathbf{v}\cdot d\mathbf{r} \;-\; \frac{e}{c}\oint \mathbf{A}'\cdot d\mathbf{r} \;=\; \left(n + \frac{1}{2}\right) h, \tag{9.55}$$

and Stokes' theorem relates the second term to the magnetic flux though the orbit via

$$\oint \mathbf{A}'\cdot d\mathbf{r} \;=\; \Phi\,(\mathbf{B})\,. \tag{9.56}$$

Moreover, integrating by parts and using Eq. (9.4), one can turn the first term in Eq. (9.55) into

$$-\,m \oint \mathbf{r}\cdot d\mathbf{v} \;=\; \frac{e}{c}\oint \mathbf{r}\cdot(d\mathbf{r}\times\mathbf{B}) \;=\; \frac{e}{c}\oint d\mathbf{r}\cdot(\mathbf{B}\times\mathbf{r}), \tag{9.57}$$

which is $\frac{2e}{c}\,\Phi\,(\mathbf{B})$. The end result,

$$\Phi\,(\mathbf{B}) \;=\; \left(n + \frac{1}{2}\right)\frac{hc}{e}, \tag{9.58}$$

is very interesting: the discreteness of the Landau levels reflects the quantization of the flux through the orbit in units of $\frac{hc}{e}$.

## 9.3 Landau Diamagnetism

Let us now turn to the equations of the grand canonical ensemble for an electron gas, starting from

$$\log \mathcal{Q} = \frac{2\, V^{\frac{1}{3}}}{h} \int dp_z \sum_{n=0}^{\infty} \gamma \, \log \left[ 1 + z\, e^{-\beta \frac{p_z^2}{2m} - \beta\hbar\omega_B \left(n+\frac{1}{2}\right)} \right], \tag{9.59}$$

where $\gamma$ is the degeneracy of Landau levels given in Eq. (9.30). In the limit of high temperatures or low densities, one can explore the phenomenon working to first order in the fugacity $z$, which gives

$$\log \mathcal{Q} \simeq \frac{V}{h^2}\, m\, \omega_B\, z \left(\frac{2\pi m}{\beta}\right)^{\frac{1}{2}} \frac{1}{\sinh\left(\frac{\beta\hbar\omega_B}{2}\right)}. \tag{9.60}$$

This expression can be recast in the convenient form

$$\log \mathcal{Q} \simeq \frac{2V}{h^3}\, z \left(\frac{2\pi m}{\beta}\right)^{\frac{3}{2}} \frac{\frac{\beta\hbar\omega_b}{2}}{\sinh\left(\frac{\beta\hbar\omega_B}{2}\right)}, \tag{9.61}$$

so that

$$\mathcal{N} \simeq \frac{2V}{h^3}\, z \left(\frac{2\pi m}{\beta}\right)^{\frac{3}{2}} \frac{\frac{\beta\hbar\omega_b}{2}}{\sinh\left(\frac{\beta\hbar\omega_B}{2}\right)}, \tag{9.62}$$

and therefore

$$\mathcal{M} \simeq -\,\mu_B\,\mathcal{N}\left[\coth\left(\frac{\beta\hbar\omega_B}{2}\right) - \frac{2}{\beta\hbar\omega_B}\right], \tag{9.63}$$

where

$$\mu_B = \frac{e\hbar}{2mc} \tag{9.64}$$

is usually called Bohr magneton. For weak magnetic fields, this result reduces to

$$\mathcal{M} \simeq -\,\frac{\mathcal{N}}{3\, k_B T}\, \mu_B^2\, B_z, \tag{9.65}$$

and consequently

$$\chi \simeq -\,\frac{\mathcal{N}}{3\, k_B T}\, \mu_B^2. \tag{9.66}$$

Note that the diamagnetic susceptibility is *negative* and decays proportionally to $\frac{1}{k_B T}$ at high temperatures, which is usually called the Curie law.

It is also instructive to consider, for any $T$, a simpler version of the problem, the case of two dimensions, where only Landau levels enter and

$$\log \mathcal{Q} = 2 \sum_{n=0}^{\infty} \gamma \log \left[1 + z\, e^{-2\beta\mu_B B_z (n+\frac{1}{2})}\right],$$
$$\mathcal{N} = 2 \sum_{n=0}^{\infty} \gamma \frac{1}{\frac{1}{z}\, e^{2\beta\mu_B B_z (n+\frac{1}{2})} + 1}. \tag{9.67}$$

This will also pave the way for the subsequent discussion of the exotic low-temperature behavior of this system. The corresponding equation for $\mathcal{M}$ reads

$$\mathcal{M} = -4\,\mu_B \sum_{n=0}^{\infty} \gamma \frac{n+\frac{1}{2}}{\frac{1}{z}\, e^{2\beta\mu_B B_z (n+\frac{1}{2})} + 1},$$
$$+\, 2\, \frac{k_B T}{B_0} \sum_{n=0}^{\infty} \log\left[1 + z\, e^{-2\beta\mu_B B_z (n+\frac{1}{2})}\right], \tag{9.68}$$

where $B_0$ was defined in (9.30). At high temperatures, $z$ is small and therefore

$$\mathcal{M} \simeq -4\,\mu_B \frac{B_z}{B_0}\, z \sum_{n=0}^{\infty} \left(n + \frac{1}{2}\right) e^{-2\beta\mu_B B_z (n+\frac{1}{2})},$$
$$+\, 2\, \frac{k_B T}{B_0}\, z \sum_{n=0}^{\infty} e^{-2\beta\mu_B B_z (n+\frac{1}{2})}$$
$$= 2\, \frac{k_B T}{B_0}\, z \left(1 + \beta \frac{\partial}{\partial \beta}\right) \sum_{n=0}^{\infty} e^{-2\beta\mu_B B_z (n+\frac{1}{2})}, \tag{9.69}$$

so that one is finally led to

$$\mathcal{N} = z\, \frac{B_z}{B_0} \frac{1}{\sinh\left(\beta\mu_B B_z\right)} \simeq z\, \frac{k_B T}{\mu_B B_0},$$
$$\mathcal{M} = \frac{k_B T}{B_0}\, z \left(1 + \beta \frac{\partial}{\partial \beta}\right) \frac{1}{\sinh\left(\beta\mu_B B_z\right)} \simeq -\frac{z\,\mu_B}{3} \frac{B_z}{B_0}. \tag{9.70}$$

Hence

$$\mathcal{M} = -\,\mathcal{N} \frac{\mu_B^2}{3\, k_B T} B_z, \tag{9.71}$$

and

$$\chi = -\,\mathcal{N} \frac{\mu_B^2}{3\, k_B T}. \tag{9.72}$$

This is the same result that we found before.

At low $T$ the behavior is more subtle: only levels below $E_F$ are filled, but in general not fully, since there is a *discrete* spectrum of Landau levels. As a result, the naive equation for $\mathcal{N}$ should be turned into an inequality,

$$\mathcal{N} \leq 2 \sum_{n=0}^{\infty} \gamma\, \theta\left[E_F - 2\,\mu_B B_z \left(n + \frac{1}{2}\right)\right], \tag{9.73}$$

since the equality cannot hold in general for integer values of both $\mathcal{N}$ and the upper level $j$. Indeed, in general the first $j + 1$ levels are completely filled while the last, the one corresponding to $n = j + 1$, is only partly filled by the remaining particles.

Let us concentrate on systems whose last Landau level is partly filled. As we have seen, the Fermi level falls on it and Eq. (9.73) holds in this form for

$$2\gamma\,(j+1) < \mathcal{N} < 2\gamma\,(j+2), \tag{9.74}$$

taking into account the spin degeneracy, which translates into the inequality

$$\frac{\mathcal{N}}{2(j+2)} < \frac{B_z}{B_0} < \frac{\mathcal{N}}{2(j+1)} \tag{9.75}$$

for the magnetic field. At the same time, the magnetization (9.68) becomes

$$\begin{aligned}\mathcal{M} = -8\,\mu_B \sum_{n=0}^{\infty} \gamma \left(n+\frac{1}{2}\right)\,\theta\left[E_F - 2\mu_B B\left(n+\frac{1}{2}\right)\right] \\ + \frac{2E_F}{B_0}\sum_{n=0}^{\infty}\theta\left[E_F - 2\mu_B B\left(n+\frac{1}{2}\right)\right],\end{aligned} \tag{9.76}$$

since, in the second term of Eq. (9.68),

$$k_B T\,\log\left(1+z\,e^{-\beta E}\right) \simeq (E_F - E)\,\theta\,(E_F - E). \tag{9.77}$$

However, in realistic situations $\gamma \gg 1$ and the second term in Eq. (9.76) gives a negligible contribution. We can therefore concentrate on the first term, taking into account that, with a partially filled last Landau level,

$$\mathcal{M} = -8\mu_B\left[\gamma\sum_{n=0}^{j}\left(n+\frac{1}{2}\right) + [\mathcal{N} - \gamma(j+1)]\left(j+\frac{3}{2}\right)\right], \tag{9.78}$$

or

$$\mathcal{M} = -4\mu_B\left[\gamma\,(j+1)^2 + 2\,[\mathcal{N} - \gamma(j+1)]\left(j+\frac{3}{2}\right)\right]. \tag{9.79}$$

Surprisingly, the resulting susceptibility is *positive*,

$$\chi = \frac{4\,\mu_B}{B_0}(j+1)(j+2), \qquad \left(\frac{\mathcal{N}}{2(j+2)} < \frac{B_z}{B_0} < \frac{\mathcal{N}}{2(j+1)}\right), \tag{9.80}$$

due to the contribution of the partly filled level. Note that in this stepwise behavior the susceptibility $\chi$ is lower for larger magnetic fields, which have the effect of reducing $j$.

In the limiting case where the last Landau level is full, the discussion in Section 8.6 tells us that the Fermi level falls in the middle of the gap to the first empty Landau level, so that

$$\mathcal{N} = 2\gamma\,(j+1), \tag{9.81}$$

and the magnetization in Eq. (9.79) becomes

$$\mathcal{M} = -4\,\mu_B\gamma\,(j+1)^2, \tag{9.82}$$

while the susceptibility is still described by (9.80), since varying the magnetic field slightly to compute it leads back to the case of a partly filled level discussed above.

## 9.4 High-$T$ Paramagnetism

Paramagnetism is a manifestation of the coupling of electron spin to an external magnetic field $B_z$. It can be described starting from

$$\log \mathcal{Q} = \frac{V}{h^3} \int d^3\mathbf{p} \log \left[ 1 + z\, e^{-\beta \frac{\mathbf{p}^2}{2m} + \beta\, \mu_B\, B_z} \right] + \frac{V}{h^3} \int d^3\mathbf{p} \log \left[ 1 + z\, e^{-\beta \frac{\mathbf{p}^2}{2m} - \beta\, \mu_B\, B_z} \right], \tag{9.83}$$

an expression that rests on a result of Appendix F, the fact the gyromagnetic ratio is two, up to small corrections, for the electron spin. At high temperatures

$$\log \mathcal{Q} \simeq 2\, V z \left( \frac{2\pi m}{h^2 \beta} \right)^{\frac{3}{2}} \cosh \left( \beta\, \mu_B\, B_z \right), \tag{9.84}$$

so that

$$\mathcal{N} \simeq 2\, V z \left( \frac{2\pi m}{h^2 \beta} \right)^{\frac{3}{2}} \cosh \left( \beta\, \mu_B\, B_z \right), \tag{9.85}$$

and therefore

$$\mathcal{M} \simeq \nu_B\, \mathcal{N} \tanh \left( \beta\, \mu_B\, B_z \right). \tag{9.86}$$

For small magnetic fields

$$\mathcal{M} \simeq \mathcal{N} \frac{\mu_B^2}{k_B T}\, B_z, \tag{9.87}$$

and

$$\chi \simeq \frac{\mathcal{N}}{k_B T}\, \mu_B^2, \tag{9.88}$$

again in accordance with the Curie law. Note that the paramagnetic susceptibility is *positive* and is larger than the diamagnetic contribution by a factor of three. Hence, when both contributions are present, the paramagnetic effect dominates.

## 9.5 Low-$T$ Paramagnetism

It is interesting to explore the behavior of paramagnetism at very low temperatures. In this limit all states are filled, up to the common Fermi level, for two gases of electrons with opposite spin, with

$$\mathcal{N}_{\pm} = \frac{V}{h^3} \int d^3\mathbf{p}\, \theta \left( E_F \pm \mu_B\, B_z - \frac{\mathbf{p}^2}{2m} \right), \tag{9.89}$$

and computing the integrals gives

$$\mathcal{N}_{\pm} = \frac{4\pi V}{3h^3}(2\,m\,E_F)^{\frac{3}{2}}\left(1 \pm \frac{\mu_B\,B_z}{E_F}\right)^{\frac{3}{2}}. \tag{9.90}$$

For weak field magnetic fields $B$,

$$\mathcal{N} = \mathcal{N}_{+} + \mathcal{N}_{-} \simeq \frac{8\pi V}{3h^3}(2\,m\,E_F)^{\frac{3}{2}} \tag{9.91}$$

while

$$\mathcal{M} = (\mathcal{N}_{+} - \mathcal{N}_{-})\,\mu_B \simeq \frac{4\pi V}{h^3}(2\,m\,E_F)^{\frac{3}{2}}\frac{\mu_B^2\,B_z}{E_F}, \tag{9.92}$$

and the ratio of these last two results finally leads to

$$\mathcal{M} \simeq \frac{3}{2}\,\mathcal{N}\,\frac{\mu_B^2}{E_F}\,B_z \tag{9.93}$$

and

$$\chi \simeq \frac{3}{2}\,\mathcal{N}\,\frac{\mu_B^2}{E_F}. \tag{9.94}$$

One could thus say that the Curie law of Eq. (9.88) saturates at $k_B T = \frac{2}{3}\,E_F$ for this system.

## Bibliographical Notes

Pedagogical introductions to diamagnetism and paramagnetism can be found in many excellent books, including [23, 36, 49, 50, 51, 54]. Clear discussions of Landau levels can be found in [18, 35, 53].

## Problems

**9.1** For a gas of 2*N* distinguishable particles with mass *m*, spin $\frac{1}{2}$ and magnetic moment $\mu$ subject to a constant magnetic field *B*:

1. Compute the internal energy *U*, the total magnetization *M*, the magnetic susceptibility $\chi$ and the entropy *S*.
2. Discuss the behavior of these quantities in the high- and low-temperature limits.
3. Repeat the analysis when the 2*N* particles form *N* distinguishable bound states with mass 2*m*, with an internal structure given by the decomposition $\frac{1}{2} \otimes \frac{1}{2} = 0 \oplus 1$.

**9.2** Consider a gas of $N$ classical "molecules," each built from two spin-$\frac{1}{2}$ particles. Suppose these molecules are subject to an external magnetic field $\mathbf{B}$, so that the single-molecule Hamiltonian is

$$H = \frac{\mathbf{p}^2}{2m} + \frac{\mathbf{S}^2}{2I} + \gamma\,\mathbf{B}\cdot\mathbf{S},$$

where $\mathbf{S}$ is the total spin of each molecule.

1. Compute the partition function of the system if the constituents can be regarded as distinguishable.
2. Repeat the analysis if the constituents are identical fermions.

**9.3** Compute the partition function for a gas of $N$ three-dimensional rotors, interacting with an external magnetic field as described by the single-rotor Hamiltonian

$$H = \frac{\mathbf{L}^2}{2I} + \gamma\,\mathbf{B}\cdot\mathbf{L},$$

where $\mathbf{L}$ is the angular momentum of each rotor. Evaluate the partition function and the magnetization in the limits of high and low temperatures.

**9.4** Consider a one-dimensional chain of $N$ "spin" variables $s_i \in \{0, 1\}$, with Hamiltonian

$$H = \sum_{i=1}^{N} g_i\, s_i,$$

where $g_i = \gamma\, 2^{i-1}$. Find the energy levels and the corresponding degeneracies, and evaluate the partition function of the system.

**9.5** A two-dimensional polymer can be modeled as a chain of $N + 1$ identical segments. Let us assume for simplicity that two consecutive oriented segments can only form mutual angles of 0 or $\pm\frac{\pi}{2}$ radians, with corresponding energies

$$E(0) = 0, \qquad E\left(\pm\frac{\pi}{2}\right) = \epsilon \pm \gamma B.$$

The coupling to an external field $B$ models a "chiral" behavior of this polymer, which favors right turns over left turns for positive $B$.

1. Compute the partition function and the susceptibility.
2. Obtain explicit expressions for the average endpoint of the chain $\langle z_N \rangle$, when the initial point $z_0$ is fixed at the origin, and for $\langle |z_N|^2 \rangle$.
3. Discuss the behavior of these quantities in the thermodynamic limit $N \to \infty$ as the temperature $T$ varies, paying attention also to the case $T = 0$.

**9.6** Consider a gas of $N$ particles of mass $m$, which are free to move within a segment of length $L$, and whose Hamiltonian is

$$H = \sum_{i=1}^{N} \frac{p_i^2}{2m} + \frac{\kappa}{2} \sum_{i=1}^{N} (x_{i+1} - x_i)^2,$$

which describes a *closed* chain of harmonic oscillators ($x_i$ is the position of the $i$th particle), for which

$$x_{N+1} = x_1 .$$

Compute the classical partition function of the system in the thermodynamic limit, making use of the identity

$$\int_{-\infty}^{+\infty} e^{-a(x-u)^2 - b(u-y)^2} du = \sqrt{\frac{\pi}{a+b}} \exp\left[-\frac{ab}{a+b}(x-y)^2\right].$$

**9.7** Consider again the gas in the previous problem.

1. Study the normal modes of the system, treating collectively the equations of motion as a single finite-difference equation. Solve it by expanding $x_j(t)$ in a Fourier series:

$$x_j(t) = \sum_{l=0}^{N-1} f_\ell(t)\, e^{i2\pi j \frac{\ell}{N}} .$$

2. Determine the proper frequencies $\omega_\ell$ that the characteristic equation associates to the corresponding values of $\ell$.
3. Compute the classical partition function from the contributions of the normal modes.
4. Note that the end result is as in the previous problem after making use of the identity

$$\prod_{\ell=1}^{N-1} \left[2 \sin\left(\frac{\pi \ell}{N}\right)\right] = N,$$

which follows from

$$\frac{z^N - 1}{z - 1} = \prod_{\ell=1}^{N-1} \left(z - e^{i\frac{2\pi\ell}{N}}\right)$$

in the limit $z \to 1$.

5. Compute the quantum partition function of the system.

**9.8** Consider a one-dimensional "polymer," a system of $N + 1$ segments of unit length joined to one another at their ends, which can be parallel or anti-parallel to the $x$ axis. The interaction energy for a pair of consecutive segments is $-\epsilon < 0$ if they have the same orientation, and $+\epsilon > 0$ is they have opposite orientations.

1. Compute the partition function for this system at a temperature $T$, if the first segment be held fixed and parallel to the $x$ axis while the last is free.
2. Compute the average value

$$\langle x_{N+1} \rangle$$

for the position of the final segment $x_{N+1}$ of the polymer, if the first segment starts from the origin of the $x$ axis, and the variance

$$\sigma^2_{N+1} = \langle x^2_{N+1} \rangle - \langle x_{N+1} \rangle^2 .$$

3. Study how the preceding quantities behave in the thermodynamic limit $N \to \infty$ for $T > 0$, and for $T \to 0$, with $N$ finite.

**9.9** Consider a gas of $N$ triatomic molecules in three dimensions, such that

$$H = \frac{\mathbf{P}^2}{2M} + H_{int},$$

where the internal degrees of freedom are described by three spin-$\frac{1}{2}$ operator, so that

$$H_{int} = \lambda\,(\mathbf{S}_1 \cdot \mathbf{S}_2 + \mathbf{S}_2 \cdot \mathbf{S}_3 + \mathbf{S}_3 \cdot \mathbf{S}_1) - \mu\, B\,(S_{1z} + S_{2z} + S_{3z}).$$

1. Find the eigenvalues of $H_{int}$ and their degeneracies.
2. Compute the partition function for the system and study the behavior of $C_V$ at low and high temperatures.
3. Compute the magnetization, and study its behavior at low and high temperatures.

**9.10** Consider a system of $N$ noninteracting constituents, with

$$H = \sum_{i=1}^{N} \frac{\mathbf{P}^2}{2M} + H_{int}.$$

The Hamiltonian $H_{int}$ of each of the constituents has three eigenstates, of energies 0, $\epsilon$, and $2\,\epsilon$, and degeneracies $(1, 3, 1)$. The system is subject to a uniform magnetic field $B$, to which these three states couple with magnetic moments $(0, \mu, 2\,\mu)$.

1. Write the partition function of the system.
2. Study the behavior of the magnetization as the temperature $T$ varies.

# 10 Magnetism in Matter, II

We can now turn to ferromagnets, macroscopic systems that display the phenomenon of hysteresis: applying an external magnetic field one can induce a magnetization that persists even when the field is removed. This type of behavior is only present up to a characteristic Curie temperature $T_c$, beyond which these systems become paramagnets, so that one can anticipate the key role that electron spins play in them.

Ferromagnetism is of great importance for theoretical physics at large, since it is a concrete manifestation of "spontaneous symmetry breaking." This is a deep phenomenon with far-reaching implications in various contexts: the general message is that, in the presence of degenerate ground states, the low-energy excitations may be less symmetric than the underlying dynamics. Mutual interactions play a crucial role here, and cannot be neglected. Therefore, we begin with a characterization of their origin that is relevant in many cases. We then move on to describe transfer-matrix techniques, the mean-field approximation and the variational method, and the Landau–Ginzburg theory. We thus meet critical phenomena, whose universal features are captured by critical exponents, and conclude with a detailed discussion of their properties.

## 10.1 Effective Spin–Spin Interactions

Let us consider a solid material with atoms at its lattice points. Remarkably, electrostatic interactions involving pairs of electrons in outer shells can manifest themselves as spin–spin couplings due to the Pauli principle. Their contribution to the energy of the system, to lowest order in perturbation theory, takes indeed the form

$$\Delta E = e^2 \int d^3\mathbf{r}_1\, d^3\mathbf{r}_2 \frac{|\psi(\mathbf{r}_1, \mathbf{r}_2)|^2}{|\mathbf{r}_1 - \mathbf{r}_2|}, \tag{10.1}$$

and gives different results depending on the symmetry of the wavefunction under the interchange of the two electrons. The Pauli principle would associate the symmetric wavefunction,

$$\psi(\mathbf{r}_1, \mathbf{r}_2) = \frac{1}{\sqrt{2}}\big(\psi_a(\mathbf{r}_1)\, \psi_b(\mathbf{r}_2) + \psi_b(\mathbf{r}_1)\, \psi_a(\mathbf{r}_2)\big), \tag{10.2}$$

to the antisymmetric spin–0 (singlet) combination

$$\frac{1}{\sqrt{2}}\big(|+-\rangle - |-+\rangle\big), \tag{10.3}$$

and the antisymmetric wavefunction

$$\psi(\mathbf{r}_1, \mathbf{r}_2) = \frac{1}{\sqrt{2}}\Big(\psi_a(\mathbf{r}_1)\,\psi_b(\mathbf{r}_2) - \psi_b(\mathbf{r}_1)\,\psi_a(\mathbf{r}_2)\Big), \tag{10.4}$$

to any of the three symmetric spin–1 (triplet) combinations

$$|++\rangle, \quad |--\rangle, \quad \frac{1}{\sqrt{2}}\Big(|+-\rangle + |-+\rangle\Big). \tag{10.5}$$

Consequently, the two wavefunctions of Eqs. (10.2) and (10.4) give rise to the different corrections

$$\begin{aligned} \Delta E_{\text{(triplet)}} &= A - B, \\ \Delta E_{\text{(singlet)}} &= A + B, \end{aligned} \tag{10.6}$$

with

$$A = e^2 \int d^3\mathbf{r}_1\, d^3\mathbf{r}_2\, \frac{|\psi_a(\mathbf{r}_1)\,\psi_b(\mathbf{r}_2)|^2}{|\mathbf{r}_1 - \mathbf{r}_2|} \tag{10.7}$$

and

$$B = e^2\, Re \int d^3\mathbf{r}_1\, d^3\mathbf{r}_2\, \frac{\overline{\psi}_a(\mathbf{r}_2)\,\overline{\psi}_b(\mathbf{r}_1)\,\psi_a(\mathbf{r}_1)\,\psi_b(\mathbf{r}_2)}{|\mathbf{r}_1 - \mathbf{r}_2|}. \tag{10.8}$$

The exclusion principle thus conspires to translate these "electrostatic exchange interactions" into the spin–spin coupling

$$\Delta E = A - \frac{1}{2}B - \frac{2}{\hbar^2}\, B\, \vec{S}_1 \cdot \vec{S}_2, \tag{10.9}$$

since

$$\left(\vec{S}_1 + \vec{S}_2\right)^2 = \begin{cases} 2\,\hbar^2 & \text{(triplet)} \\ 0 & \text{(singlet)}. \end{cases} \tag{10.10}$$

These considerations do not determine the sign of $B$. Positive values of $B$ result in ferromagnetism, while negative values result in anti-ferromagnetism. Here we shall concentrate on the first case.

This effect is captured by the so-called quantum Heisenberg model, for which

$$H = -\frac{1}{2}\sum_{\langle ij\rangle} v_{ij}\, \vec{S}_i \cdot \vec{S}_j, \tag{10.11}$$

where $v_{ij}$ is a symmetric matrix of couplings, which we assume to be positive definite, $i$ and $j$ label points on a lattice and the local nature of the interactions restricts the sum to nearest-neighbor pairs $\langle ij\rangle$, consistently with the actual suppression of the overlap $B$ at larger distances.

Note that $H$ is invariant under the $SO(3)$ group of spatial rotations. However, any state of spontaneous magnetization $\mathcal{M}$ induced from hysteresis would be characterized by a magnetic moment pointing in a given direction, and would thus break the symmetry to the $SO(2)$ group of rotations around the direction of $\mathcal{M}$. Note, additionally, that reverting to generic configurations would cost an infinite energy in the infinite-volume limit.

Here, in principle, the $\vec{S}_i$ would be quantum-mechanical spins, but one can simplify matters and treat them as classical spins, in the so-called classical Heisenberg model, since the relevant phenomena typically occur at rather high temperatures. We shall return to the one-dimensional quantum Heisenberg model in Chapter 12. In the following, instead, we shall concentrate on an even simpler formulation of problem provided by the Ising model,

$$H = -\frac{1}{2}\sum_{\langle ij\rangle} v_{ij}\,\sigma_i\,\sigma_j, \tag{10.12}$$

where $\sigma_i = \pm 1$, which is tailored to the case of uniaxial ferromagnets. Note that $H$ is now invariant under the $\mathbb{Z}_2$ transformation that reverts all the $\sigma_i$, and this symmetry would be broken in a state of spontaneous magnetization induced by hysteresis.

The two-dimensional Ising model can be solved exactly, and the famous Onsager solution clearly displays a second-order phase transition with a consequent nonanalytic behavior of thermodynamic quantities, to which we shall return in Chapter 11. Moreover, the relative simplicity of this setting allows one to collect interesting information on the Curie transition to paramagnetism by simple approximate methods. Still, exact solutions of the Ising model in the thermodynamic limit are only available in one and two dimensions, and here we begin from the former, simpler case.

## 10.2 The One-Dimensional Ising Model

Let us consider the one-dimensional Ising partition function,

$$Z = \sum_{\{\sigma_i=\pm1\}} e^{\beta\, v\, \sum_i \sigma_i\,\sigma_{i+1}}, \tag{10.13}$$

resorting initially to periodic boundary conditions, as in preceding chapters, so that $\sigma_{N+1} = \sigma_1$. Since the $\sigma_i$ square to one, one can write

$$e^{\beta\, v\, \sigma_i\,\sigma_{i+1}} = \cosh\beta v\left[1 + \sigma_i\,\sigma_{i+1}\tanh\beta v\right], \tag{10.14}$$

and therefore

$$Z = \left(\cosh\beta v\right)^N \sum_{\{\sigma_i=\pm1\}} \prod\left[1 + \sigma_i\,\sigma_{i+1}\tanh\beta v\right]. \tag{10.15}$$

The sum over spins gives

$$Z = \left(2\cosh\beta v\right)^N\left[1 + (\tanh\beta v)^N\right], \tag{10.16}$$

since all terms not involving even powers of them average to zero. However, the expression within square brackets tends to one in the thermodynamic limit $N \to \infty$ for any finite value of $\beta$, so that the limiting form of the partition function is

$$Z = \left(2\cosh\beta v\right)^N. \tag{10.17}$$

Note that there are no signs of peculiar behavior: the system is paramagnetic for all positive values of $T$. This is seen more clearly if one adds a uniform magnetic field $B$. The issue now is to compute

$$Z = \sum_{\{\sigma_i = \pm 1\}} e^{\beta\, v\, \sum_i \sigma_i\, \sigma_{i+1} \,+\, \beta\, \mu\, B\, \sum_i \sigma_i} \tag{10.18}$$

in the thermodynamic limit $N \to \infty$, which is a bit more involved but can be done conveniently by introducing the $2 \times 2$ "transfer matrix" $T$ for this system. Its elements read

$$\langle \sigma_i | T | \sigma_j \rangle = e^{\beta\, v\, \sigma_i\, \sigma_j \,+\, \frac{1}{2}\, \beta\, \mu\, B\, (\sigma_i + \sigma_j)}, \tag{10.19}$$

where we have recast the magnetic coupling into an equivalent symmetric form that leads to a Hermitian transfer matrix

$$T = \begin{pmatrix} e^{\beta(v+\mu B)} & e^{-\beta v} \\ e^{-\beta v} & e^{\beta(v-\mu B)} \end{pmatrix}, \tag{10.20}$$

with the identifications

$$\sigma_i = 1 \to \begin{pmatrix} 1 \\ 0 \end{pmatrix}, \qquad \sigma_i = -1 \to \begin{pmatrix} 0 \\ 1 \end{pmatrix}. \tag{10.21}$$

In this fashion, the sums determining the partition function are recast into the form of a matrix product that builds, in the thermodynamic limit and with periodic boundary conditions, the trace of a huge power of $T$,

$$Z = \mathrm{Tr}\left(T^N\right). \tag{10.22}$$

The reader should have noticed that the transfer matrix $T$ is a discrete counterpart of the Euclidean evolution operator $e^{-\frac{1}{\hbar} Ht}$ of Chapter 3. The partition function of interest is simply related to its eigenvalues $\lambda_+$, $\lambda_-$ according to

$$Z = (\lambda_+)^N + (\lambda_-)^N, \tag{10.23}$$

and the issue, in general, is to identify the largest eigenvalue $\lambda_+$ of $T$, since its contribution dominates the partition function in the thermodynamic limit.

In this case the eigenvalues of $T$ are

$$\lambda_\pm = e^{\beta v} \left[ \cosh \beta\mu B \pm \sqrt{\sinh^2 \beta\mu B + e^{-4\beta v}} \right], \tag{10.24}$$

and therefore in the thermodynamic limit

$$Z = e^{N\beta v} \left[ \cosh \beta\mu B + \sqrt{\sinh^2 \beta\mu B + e^{-4\beta v}} \right]^N, \tag{10.25}$$

so that the Helmholtz free energy is

$$\frac{\beta A}{N} = -\beta v - \log \left[ \cosh \beta\mu B + \sqrt{\sinh^2 \beta\mu B + e^{-4\beta v}} \right]. \tag{10.26}$$

One can now compute the magnetization,

$$\mathcal{M} = \frac{1}{\beta} \frac{\partial \log Z}{\partial B} = N\mu \frac{\sinh \mu\beta B}{\sqrt{\sinh^2 \beta\mu B + e^{-4\beta v}}}, \tag{10.27}$$

and for small magnetic fields,

$$\chi = \frac{N \mu^2}{k_B T} e^{2\beta v}, \tag{10.28}$$

which is positive and approaches the standard Curie law of paramagnetism at high temperatures.

The correlation function is another interesting quantity that one can simply compute for this system. Confining the attention to the case $B = 0$, it is defined as

$$\langle \sigma_i \sigma_j \rangle = \frac{\sum_{\{\sigma_k = \pm 1\}} \sigma_i \sigma_j e^{\beta v \sum_k \sigma_k \sigma_{k+1}}}{\sum_{\{\sigma_k = \pm 1\}} e^{\beta v \sum_k \sigma_k \sigma_{k+1}}}, \tag{10.29}$$

and proceeding as above one can reduce this expression to

$$\langle \sigma_i \sigma_j \rangle = \frac{\sum_{\{\sigma_k = \pm 1\}} \sigma_i \sigma_j \prod_k \left[1 + \sigma_k \sigma_{k+1} \tanh \beta v\right]}{\sum_{\{\sigma_k = \pm 1\}} \prod_k \left[1 + \sigma_k \sigma_{k+1} \tanh \beta v\right]}. \tag{10.30}$$

In the thermodynamic limit

$$\langle \sigma_i \sigma_j \rangle = \left(\tanh \beta v\right)^{|i-j|} = e^{-\frac{|i-j|}{\lambda_c}}, \tag{10.31}$$

so that the correlation function decreases exponentially with the minimal distance $|i - j|$ between the two spins in the periodic lattice. The correlation length,

$$\lambda_c = \frac{1}{\log \coth \beta v}, \tag{10.32}$$

is finite for any finite $T > 0$ and grows indefinitely as $T \to 0$. One does not speak of a phase transition proper, however, since the phenomenon takes place at the end of the allowed range for $T$.

The transfer matrix formalism is also a convenient tool to compute correlation functions for nonvanishing $B$. Every $\sigma_k$ in the thermal average translates into the insertion of the Pauli matrix $\sigma_z$ in the matrix product, so that for instance

$$\begin{aligned} \langle \sigma_k \rangle &= \frac{1}{Z} \mathrm{tr}\left(\sigma_z T^N\right), \\ \langle \sigma_{k+l} \sigma_k \rangle &= \frac{1}{Z} \mathrm{tr}\left(\sigma_z T^l \sigma_z T^{N-l}\right). \end{aligned} \tag{10.33}$$

One can thus begin by diagonalizing $T$ with the orthogonal transformation

$$O = \begin{pmatrix} \cos\theta & -\sin\theta \\ \sin\theta & \cos\theta \end{pmatrix}, \tag{10.34}$$

where

$$\cos 2\theta = \frac{\sinh \beta \mu B}{\sqrt{\sinh^2 \beta \mu B + e^{-4\beta v}}}, \qquad \sin 2\theta = \frac{-e^{-2\beta v}}{\sqrt{\sinh^2 \beta \mu B + e^{-4\beta v}}}. \tag{10.35}$$

Therefore,

$$\langle \sigma_k \rangle = \frac{1}{\lambda_+^N + \lambda_-^N} \mathrm{tr}\left(O \sigma_z O^T \begin{pmatrix} \lambda_+^N & 0 \\ 0 & \lambda_-^N \end{pmatrix}\right), \tag{10.36}$$

and in the thermodynamic limit

$$\langle \sigma_k \rangle = \frac{\sinh \mu\beta B}{\sqrt{\sinh^2 \beta\mu B + e^{-4\beta v}}}. \tag{10.37}$$

In a similar fashion,

$$\langle \sigma_k \sigma_{k+l} \rangle = \cos^2 2\theta + \sin^2 2\theta \left(\frac{\lambda_-}{\lambda_+}\right)^l. \tag{10.38}$$

## 10.3 The Role of Boundary Conditions

So far we have worked with periodic boundary conditions. However, the one-dimensional Ising chain is simple enough to allow a closer look at the important role played by other choices of boundary conditions, to which we now turn.

Our starting point is a variant of Eq. (10.15) where we do not sum, for the time being, over the allowed values for the first and last spins $\sigma_1$ and $\sigma_N$,

$$Z = \sum_{\{\sigma_i=\pm 1, i\neq 1,N\}} \prod_{i=1}^{N-1} \left(\cosh \beta v + \sigma_i \sigma_{i+1} \sinh \beta v\right). \tag{10.39}$$

Taking into account that terms linear in at least one $\sigma_i$, with $i = 2, 3, \ldots, N-1$, average to zero, the only nonvanishing contribution to $Z$ is

$$Z = 2^{N-2} (\cosh \beta v)^{N-1} \left[1 + \sigma_1 \sigma_N (\tanh \beta v)^{N-1}\right]. \tag{10.40}$$

We now turn to the (spontaneous) magnetization of the system, i.e. the average total magnetic moment

$$M = \sigma_1 + \sum_{k=2}^{N-1} \langle \sigma_k \rangle + \sigma_N, \tag{10.41}$$

where (for $k = 2, \ldots, N-1$)

$$\langle \sigma_k \rangle = \frac{1}{Z} \sum_{\{\sigma_i=\pm 1\}} \sigma_k \prod_{i=1}^{N-1} (\cosh \beta v + \sigma_i \sigma_{i+1} \sinh \beta v). \tag{10.42}$$

Again, only two terms survive after summing over configurations, so that

$$\langle \sigma_k \rangle = \frac{\sigma_1 y^{k-1} + \sigma_N y^{N-k}}{1 + \sigma_1 \sigma_N y^{N-1}}, \tag{10.43}$$

where we have let, for brevity,

$$y = \tanh \beta v. \tag{10.44}$$

Note that, in the thermodynamic limit $N \to \infty$, one can read the correlation length (10.32) from the one-point function with $\sigma_1 = 1$. It is worth keeping in mind this fact, particularly in view of subsequent sections, and we shall see shortly that this correspondence also follows from the cluster property.

Equation (10.41) now determines the total magnetization,

$$M = (\sigma_1 + \sigma_N)\left[1 + \frac{y - y^{N-1}}{(1-y)(1+\sigma_1\sigma_N y^{N-1})}\right], \tag{10.45}$$

and one can extract interesting information from this result. Note that, if the boundary spins are antiparallel, $\sigma_N = -\sigma_1$, the spontaneous magnetization vanishes exactly for any temperature $T$ and for any number $N$ of sites.

Letting $\sigma_1 = \sigma_N = \sigma_B$, we can now discuss the thermodynamic limit $N \to \infty$ of Eq. (10.45), with the following results:

- for any nonvanishing temperature $y < 1$, and therefore as $N \to \infty$

$$M \to \frac{2\sigma_B}{1-y}, \tag{10.46}$$

so that the local magnetization $m = \frac{M}{N}$ tends to zero in the limit. Thermal fluctuations destroy the spontaneous ordering of the system for any temperature, no matter how small;

- as the temperature tends to zero, $y \to 1^-$, so that the magnetization tends to

$$M = N\sigma_B. \tag{10.47}$$

Hence, even in the thermodynamic limit, one obtains the nonvanishing result

$$m = \frac{M}{N} = \sigma_B. \tag{10.48}$$

This was to be expected: in the absence of thermal fluctuations, all spins align themselves to the ones that have been fixed on the boundary.

In conclusion

$$m(T) = \begin{cases} \sigma_B & \text{for } T = 0, \\ 0 & \text{for } T > 0. \end{cases} \tag{10.49}$$

Note, however, that had we chosen *periodic* boundary conditions as in the last section, also summing over $\sigma_N = \sigma_1$, we would have found instead a vanishing spontaneous magnetization, even at zero temperature! This result is evident from the preceding expression, Eq. (10.45), but one can also recover it from an instructive symmetry argument. Denoting by $P$ the parity transformation that sends $\sigma_i \mapsto -\sigma_i$, one finds

$$\langle P\sigma_i \rangle = -\langle \sigma_i \rangle. \tag{10.50}$$

However, for periodic boundary conditions, the Hamiltonian and the configuration space are both symmetric under parity, and these facts imply that

$$\langle P\sigma_i \rangle = \langle \sigma_i \rangle. \tag{10.51}$$

Therefore, by comparison, $\langle \sigma_i \rangle = 0$.

The reason for this apparent contradiction is that the Hamiltonian has two translationally invariant ground states, with all spins pointing up or down. Periodic boundary conditions do not distinguish them, so that one is effectively considering a mixed phase, while open boundary conditions $\sigma_1 = \sigma_N = \sigma_B$ can select one of the two ground states, thus leading to a pure phase.

*Pure phases*, in contrast with *mixed phases*, can be characterized by the lack of long-distance correlations, i.e. from the condition that for all $k$,

$$\lim_{l\to+\infty} (\langle\sigma_{k+l}\sigma_k\rangle - \langle\sigma_{k+l}\rangle\langle\sigma_k\rangle) = 0 \tag{10.52}$$

in the thermodynamic limit. This "cluster decomposition" property plays a crucial role in rigorous discussions of symmetry breaking.

Here we shall content ourselves with verifying that Eq. (10.52) holds for open boundary conditions, even at zero temperature, while it fails for periodic boundary conditions.

To begin with, for both open and periodic boundary conditions, the two-point function is

$$\langle\sigma_{k+l}\sigma_k\rangle = \frac{y^l + y^{N-l-1}}{1 + y^{N-1}}, \tag{10.53}$$

up to a shift $N - 1 \to N$ in the latter case, which is immaterial for large values of $N$. In the thermodynamic limit, this expression exhibits for any nonzero temperature the nonvanishing correlation length of Eq. (10.32), since (here implicitly $l < \frac{N-1}{2}$)

$$\langle\sigma_{k+l}\sigma_k\rangle \approx y^l = e^{-\frac{l}{\lambda_c}}, \tag{10.54}$$

while $\lambda_c \to \infty$ in the zero-temperature limit, and consequently

$$\langle\sigma_{k+l}\sigma_k\rangle \to 1, \tag{10.55}$$

independently of $N$. One-point functions are instead given by Eq. (10.43), which becomes

$$\langle\sigma_k\rangle = \sigma_B \frac{y^{k-1} + y^{N-k}}{1 + y^{N-1}}, \tag{10.56}$$

for open boundary conditions, while they vanish for periodic ones. The left-hand side of Eq. (10.52) tends to zero exponentially fast for any finite temperature, but at zero temperature

$$\langle\sigma_{k+l}\sigma_k\rangle - \langle\sigma_{k+l}\rangle\langle\sigma_k\rangle = 1 - 1 = 0 \tag{10.57}$$

for open boundary conditions, and

$$\langle\sigma_{k+l}\sigma_k\rangle - \langle\sigma_{k+l}\rangle\langle\sigma_k\rangle = 1 - 0 = 1 \tag{10.58}$$

for periodic boundary conditions. The latter result violates the cluster property, and thus signals the presence of a mixed phase.

As we have seen, the degeneracy between the two translationally invariant ground states can be resolved by introducing an external magnetic field $B$, as in Section 10.2, thus inducing the magnetization in Eq. (10.27). This expression shows that, consistently with the preceding analysis, $m$ vanishes when $B$ tends to zero for any nonvanishing temperature. However, for fixed nonzero $B$, the zero-temperature limit yields

$$\lim_{T\to 0} m = \frac{B}{|B|} = \begin{cases} +1 & \text{for } B > 0, \\ -1 & \text{for } B < 0, \end{cases} \tag{10.59}$$

so that this "hysteresis" can leave behind a magnetized system at zero temperature. Using the results in Eqs. (10.37) and (10.38), one can verify that the cluster property holds at all temperatures in the presence of a nonvanishing magnetic field.

## 10.4 The Continuum Limit

There is a reason behind the lack of spontaneous magnetization of the one-dimensional Ising chain: the interactions do not suffice to counteract disorder for any positive value of the temperature $T$. There is another way of understanding the origin of the problem: we have seen that the ground state of a symmetric double well in quantum mechanics is also symmetric under the parity operation $x \to -x$. We can now show that the partition function for a one-dimensional Ising chain has the same form, in the thermodynamic limit and in the continuum approximation, as the Euclidean quantum kernel for a one-dimensional double well. In this setting the Euclidean time variable becomes, in the limit, a continuous coordinate along the Ising chain.

For definiteness, let us consider the Ising model on a $D$-dimensional cubic lattice, whose partition function

$$Z = \sum_{\{\sigma_i = \pm 1\}} \prod_i e^{\frac{1}{2}\beta \sum_{\langle ij \rangle} v_{ij}\, \sigma_i\, \sigma_j + \beta\, \mu\, B \sum_i \sigma_i} \tag{10.60}$$

can be cast in the form

$$Z = \int \prod_i \frac{2\, d\sigma_i}{\sqrt{\pi\, \rho}}\, e^{\frac{1}{2}\beta \sum_{\langle ij \rangle} v_{ij}\, \sigma_i\, \sigma_j + \beta\, \mu\, B \sum_i \sigma_i - \sum_i \frac{(\sigma_i^2 - 1)^2}{\rho}}\,, \tag{10.61}$$

via the replacement

$$\delta(\sigma_i - 1) + \delta(\sigma_i + 1) \;=\; 2\delta(\sigma_i^2 - 1) \;\to\; \frac{2}{\sqrt{\pi \rho}}\, e^{-\frac{(\sigma_i^2-1)^2}{\rho}}\,, \tag{10.62}$$

which becomes an identity as $\rho \to 0$. Let us note that, if all nonvanishing elements of $v_{ij}$ are equal to $v$, the terms entering the Ising Hamiltonian are of the form

$$\sigma(P) \sum_{\ell=1}^{D} \left[\sigma(P + \mathbf{e}_\ell) + \sigma(P - \mathbf{e}_\ell)\right], \tag{10.63}$$

where the $\mathbf{e}_\ell$ are orthonormal vectors parallel to the axes of a corresponding Cartesian system. Therefore

$$(\mathbf{e}_\ell)^i \;=\; a\, \delta^i_\ell, \tag{10.64}$$

where the coordinates of any lattice point $P$ are multiples of the lattice spacing $a$, and in the continuum limit the dominant behavior of the preceding term is

$$2D\, \sigma(P)^2 + a^2\, \sigma(P)\, \nabla^2\, \sigma(P). \tag{10.65}$$

The coefficient of $\sigma(P)^2$ is ill defined, since it is affected by redefinitions of the measure, and we started from a lattice model with $\sigma^2 \simeq 1$. Consequently, after an integration by parts we shall write the contribution of second term in Eq. (10.65) in the form

$$\frac{1}{a^{D-2}} \int d^D x \left[ -\frac{1}{2} \beta\, v \left( \frac{\partial\, \sigma}{\partial\, x^i} \right)^2 \right], \tag{10.66}$$

while allowing an arbitrary coefficient for the $\sigma^2$ contribution.

One can perform similar steps in the remaining terms, and insofar as the lattice spacing is small with respect to the correlation length,

$$Z \simeq \int [\mathcal{D}\sigma(\mathbf{x})]\, e^{-\beta \int \frac{d^D x}{a^D} \left[ \frac{1}{2} a^2\, v \left( \frac{\partial\, \sigma(\mathbf{x})}{\partial x^i} \right)^2 + c_1(\beta)\, \sigma^2(\mathbf{x}) + c_2\, \sigma^4(\mathbf{x}) + \sigma(\mathbf{x}) J(\mathbf{x}) \right]}. \tag{10.67}$$

The Euclidean action can be recast in canonical form by making the additional redefinitions

$$\sqrt{v}\, a^{1-\frac{D}{2}}\, \sigma(\mathbf{x}) \;\rightarrow\; \Phi(\mathbf{x}), \tag{10.68}$$

and rescaling the positive $c_2$, one is left with

$$Z \simeq \int [\mathcal{D}\Phi(\mathbf{x})]\, e^{-\beta \int d^D x \left[ \frac{1}{2} \nabla\Phi \cdot \nabla\Phi \,+\, \frac{1}{2} m^2(\beta)\, \Phi^2(\mathbf{x}) \,+\, \frac{1}{4}\, \Phi^4(\mathbf{x}) + j(\mathbf{x})\Phi(\mathbf{x}) \right]}. \tag{10.69}$$

Finally, one ends up with the "Landau–Ginzburg" potential

$$V \;=\; \frac{1}{2}\, m^2(\beta)\, \Phi^2 \;+\; \frac{1}{4}\, \Phi^4, \tag{10.70}$$

which describes a single or double well depending on the sign of $m^2(\beta)$ (this is illustrated in Figure 10.1). Note how the original $\mathbb{Z}_2$ symmetry is inherited by the continuous model: it flips the sign of $\Phi$ simultaneously at each point, and is clearly a symmetry of the Hamiltonian (or, if you will, of the Euclidean action) in Eq. (10.69).

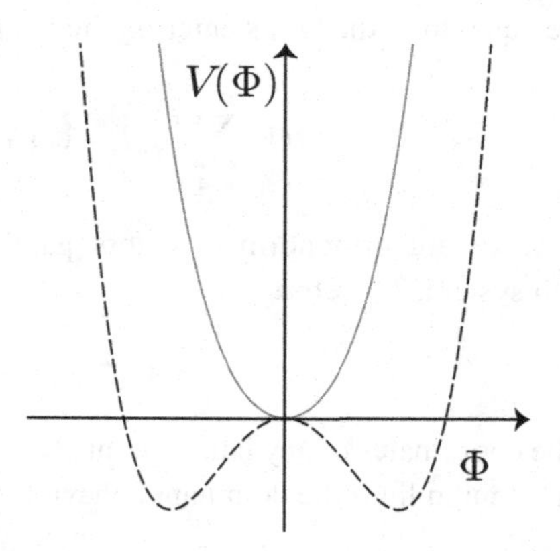


Figure 10.1 The Landau–Ginzburg potential $V$ in Eq. (10.70) for (solid line) positive and (dashed line) negative $m^2(\beta)$.

A special case of this argument links the partition function for a one-dimensional Ising chain to the Euclidean kernel of a quantum particle in a symmetric potential well. In the thermodynamic limit, the chain becomes infinitely long and this translates into the limit of large Euclidean time for the analog quantum system, which selects the ground-state contribution. The ground state of this system, as we have seen in Chapter 3, is symmetric regardless of the sign of $m^2(\beta)$, due to instanton effects. Correspondingly, $\Phi$ is bound to have a vanishing average in the thermodynamic limit at any temperature $T > 0$, and one recovers the result found in the preceding sections: the one-dimensional Ising model cannot have a spontaneous magnetization at any nonzero temperature.

## 10.5 The Infinite-Range Ising Model

In the preceding sections we have argued that the nearest-neighbor interactions of the one-dimensional Ising model do not suffice to induce an ordered phase for any positive value of $T$. In this section we study an exactly solvable model where the order–disorder transition is always present, for any value of the dimension $D$. In this "infinite-range" model, each spin interacts with all others, so that

$$Z = \sum_{\{\sigma_i=\pm 1\}} e^{\frac{\beta v}{2N} \sum_{i,j} \sigma_i \sigma_j + \beta \mu B \sum_i \sigma_i}, \tag{10.71}$$

where the rescaling in the first term is meant to compensate the growth with $N$ of the number of interactions, and moreover all pairs are included, so that

$$\sum_{i,j} \sigma_i \sigma_j = \left(\sum_i \sigma_i\right)^2 . \tag{10.72}$$

The partition function $Z$ can be simplified by resorting to the standard identity of Eq. (2.21), which we display again for the reader's convenience,

$$e^{\frac{\beta^2}{4\alpha}} = \sqrt{\frac{\alpha}{\pi}} \int_{-\infty}^{\infty} dx\, e^{-\alpha x^2 + \beta x} \tag{10.73}$$

and can thus be cast in the form

$$e^{\frac{\beta v}{2N} \left(\sum_i \sigma_i\right)^2} = \sqrt{\frac{N\beta}{2\pi v}} \int_{-\infty}^{\infty} dx\, e^{-\frac{N\beta}{2v} x^2 + \beta x \sum_i \sigma_i}, \tag{10.74}$$

which reduces the corresponding Hamiltonian to a paramagnetic-like one. This type of step, used in several contexts, is often called a Hubbard–Stratonovich transformation. Therefore

$$Z = \sqrt{\frac{N\beta}{2\pi v}} \int_{-\infty}^{\infty} dx\, e^{-\frac{N\beta}{2v} x^2} \sum_{\{\sigma_i=\pm 1\}} e^{\beta(x+\mu B) \sum_i \sigma_i}, \tag{10.75}$$

and, performing the spin sums, one is led to

$$Z = \sqrt{\frac{N\beta}{2\pi v}} \int_{-\infty}^{\infty} dx\, e^{-\frac{N\beta}{2v} x^2} \left[2 \cosh \beta\left(x + \mu B\right)\right]^N . \tag{10.76}$$

Finally, the partition function can be cast in the form

$$Z = \sqrt{\frac{N\beta}{2\pi v}} \int_{-\infty}^{\infty} dx\, e^{-Nf(x)}, \tag{10.77}$$

an integral that can be evaluated exactly by the saddle-point method in the thermodynamic limit, where

$$f(x) = \frac{\beta}{2v}x^2 - \log\left[2\cosh\beta\left(x + \mu B\right)\right]. \tag{10.78}$$

Letting

$$\xi = \beta\left(x + \mu B\right), \tag{10.79}$$

the saddle-point condition yields the "master equation"

$$\tanh\xi = \frac{\xi - \beta\mu B}{\beta v}, \tag{10.80}$$

which we shall soon recover, in more general contexts, in the mean-field approximation. As a result

$$Z \sim \sqrt{\frac{\beta}{v f''(\xi^*)}}\, e^{-Nf(x^*)}, \tag{10.81}$$

with $\xi^*$ a stable solution of the saddle-point equation.

We are confronted here with a first nontrivial example of spontaneous magnetization: starting with a nonvanishing $B$, which one can take to be positive for definiteness, the equation admits a solution corresponding to a positive value of $\xi$. Referring to Figure 10.2, if $B$ is then slowly removed and

$$T < T_c = \frac{v}{k_B}, \tag{10.82}$$

the solution approaches a positive value $\xi^\star > 0$. On the other hand, for $T > T_c$, the straight line becomes steep enough to let the intersection point $\xi^\star$ move to the origin

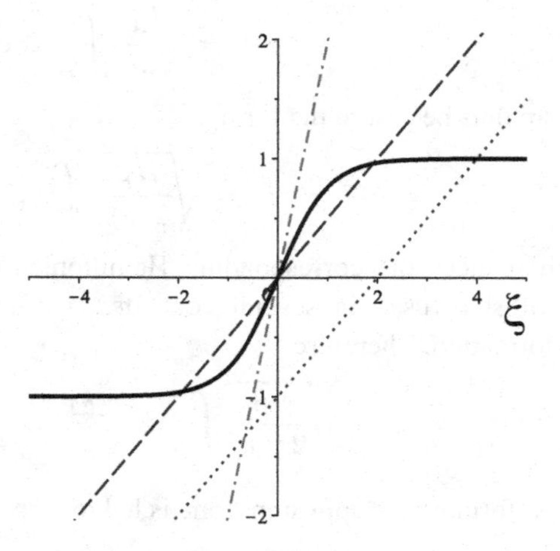


Figure 10.2 The three types of solutions of Eq. (10.80): (dotted line) $T < T_c$ with $B > 0$; (dashed line) $T < T_c$ with $B = 0$; (dash-dotted line) $T > T_c$ with $B = 0$.

as $B \to 0$. In other words, hysteresis can occur for sufficiently low temperatures, $T < T_c$. As we saw in Eq. (10.81), in the thermodynamic limit $\log Z$ is essentially $-f(x)$ computed on the solution of the master equation, therefore

$$\log Z \sim -N\frac{\beta}{2v}(x^*)^2 + N\log\left[2\cosh\beta\left(x^* + \mu B\right)\right], \tag{10.83}$$

and

$$\mathcal{M} = \frac{1}{\beta}\frac{\partial \log Z}{\partial B} = N\mu\tanh(\xi^* + \beta\mu B), \tag{10.84}$$

so that a nonvanishing value of $\xi^*$ translates into a nonvanishing magnetization, even at vanishing magnetic field. For this reason $\xi^*$ is often called the "order parameter" for the magnetization.

## 10.6 Mean-Field and Variational Method

The mean-field approximation was originally introduced for a Hamiltonian of the form

$$H = -\frac{1}{2}\sum_{i,j} v_{ij}\,\sigma_i\,\sigma_j - \beta\mu\sum_i B_i\,\sigma_i, \tag{10.85}$$

where $v_{ij} = f(|\mathbf{i} - \mathbf{j}|)$ is the short-range interaction that we met by introducing the fluctuation

$$\xi_i = \sigma_i - \langle\sigma_i\rangle \tag{10.86}$$

with respect to the mean value $\langle\sigma_i\rangle$, and retaining in the Hamiltonian only terms of first order in $\xi_i$. This leads to

$$\sigma_i\sigma_j \simeq \langle\sigma_i\rangle\xi_j + \langle\sigma_j\rangle\xi_i, \tag{10.87}$$

up to a constant, and thus to an effective paramagnetic Hamiltonian

$$H^{\text{mf}} \simeq -\sum_i \mu B_i^{\text{mf}}\sigma_i, \tag{10.88}$$

where $B_i^{\text{mf}}$ is the mean field or Curie–Weiss molecular field

$$\mu B_i^{\text{mf}} = \mu B_i + \sum_j v_{ij}\langle\sigma_j\rangle. \tag{10.89}$$

The corresponding partition function can be simply calculated, and reads

$$Z^{\text{mf}} = 2^N\prod_{i=1}^{N}\cosh\left(\mu\beta B_i^{\text{mf}}\right). \tag{10.90}$$

The Curie–Weiss field, however, must satisfy the consistency condition that $\langle\sigma_i\rangle$ be the thermal average of the spins, which leads to the master equation

$$\langle\sigma_i\rangle = \tanh\left(\mu\,\beta B_i^{\text{mf}}\right), \tag{10.91}$$

a result that is akin to Eq. (10.84).

This approximation lies within a far-reaching setup, which can be motivated as follows. In quantum mechanics, there is a very powerful technique to estimate the ground-state energy of a system: the variational method. Indeed, for any state $|\psi\rangle$ one can compute the corresponding average

$$E(\psi) = \frac{\langle\psi|H|\psi\rangle}{\langle\psi|\psi\rangle}, \tag{10.92}$$

and, expanding $|\psi\rangle$ in terms of the (unknown) eigenstates $|\phi_\alpha\rangle$ corresponding to the exact (unknown) energies $E_\alpha$ according to

$$|\psi\rangle = \sum_\alpha c_\alpha \, |\phi_\alpha\rangle, \tag{10.93}$$

one can conclude that

$$E(\psi) = \frac{\sum_\alpha |c_\alpha|^2 E_\alpha}{\sum_\alpha |c_\alpha|^2} \geq E_0. \tag{10.94}$$

This result has a striking implication: within a class of test functions, the best approximation for $E_0$ is the one yielding the lowest value for the estimate $E(\psi)$. Moreover, a test function with a first-order error,

$$|\psi\rangle = |\phi_0\rangle + \epsilon \, |\phi_1\rangle, \tag{10.95}$$

where $|\epsilon| \ll 1$ models a contamination of the true ground state by other contributions, yields

$$E(\psi) = \frac{E_0 + \epsilon^2 E_1}{1 + \epsilon^2}, \tag{10.96}$$

with an error that is reduced to second order. In this fashion, for example, one can obtain an estimate for the ground-state energy for Helium, via simple hydrogen-like test functions, with an error of about 1%.

In statistical mechanics there is a similar technique, by which one can recover, in particular, the mean-field method that we have illustrated for the Ising model. Let us therefore consider a system with a given configuration space and a "simple" Hamiltonian $H_0$, for which one can readily compute

$$Z_0 = \sum_{\text{Config}} e^{-\beta H_0}, \tag{10.97}$$

where the sum is over the possible configurations of the system while our problem concerns

$$Z = \sum_{\text{Config}} e^{-\beta H}, \tag{10.98}$$

where the sum is over the same configurations but involves a more complicated Hamiltonian $H$.

One can clearly write

$$Z = \sum_{\text{Config}} e^{-\beta H_0} \, e^{-\beta(H-H_0)}, \tag{10.99}$$

and this leads to

$$Z = Z_0 \, \langle \, e^{-\beta(H-H_0)} \, \rangle_0, \tag{10.100}$$

where the average $\langle O \rangle_0 = \sum_{\text{Config}} O\, e^{-\beta H_0}/Z_0$ is performed with the simpler Hamiltonian $H_0$. One can now find a useful approximation since $e^{-x}$ is a convex function, which therefore lies below the chord joining any two points $P \equiv (a, e^{-a})$ and $Q \equiv \left(b, e^{-b}\right)$. Hence

$$\langle\, e^{-\beta(H-H_0)}\, \rangle_0 \;\geq\; e^{-\langle\beta(H-H_0)\rangle_0}, \tag{10.101}$$

and therefore

$$Z \;\geq\; Z_0\, e^{-\langle\beta(H-H_0)\rangle_0}, \tag{10.102}$$

so that

$$A \;\leq\; A_0 \;+\; \langle\, H \;-\; H_0\, \rangle_0. \tag{10.103}$$

We shall now refer to the right-hand side of this expression as the variational estimate for the free energy, for which we shall continue to use the symbol $A$.

## 10.7 Mean-Field Analysis of the Ising Model

The variational method is a systematic procedure that can yield deep insights into complicated dynamical systems. Here we confine our attention to simple forms of $H_0$ that lead to handy analytic expressions for the Ising system, while also recovering the exact result for the infinite-range model, but this method can prove invaluable in many cases.

Our starting point is again the $D$-dimensional Ising Hamiltonian

$$H \;=\; -\,\frac{1}{2}\sum_{\langle ij\rangle} v_{ij}\,\sigma_i\,\sigma_j \;-\; \mu\, B \sum_i \sigma_i, \tag{10.104}$$

where $\langle ij\rangle$ identifies nearest-neighbor pairs. As a test Hamiltonian we choose

$$H_0 \;=\; -y \sum_i \sigma_i, \tag{10.105}$$

which yields

$$Z_0 \;=\; \left[2\,\cosh\beta y\right]^N. \tag{10.106}$$

Hence

$$\langle H_0\rangle_0 \;=\; -\frac{\partial}{\partial\beta}\,\log Z_0 \;=\; -\,N y\,\tanh\beta y, \tag{10.107}$$

while

$$\langle\sigma_i\rangle_0 \;=\; \frac{1}{N\beta}\,\frac{\partial}{\partial y}\,\log Z_0 \;=\; \tanh\beta y, \tag{10.108}$$

and therefore

$$\langle H\rangle_0 = -\frac{1}{2}\sum_{\langle ij\rangle} v_{ij}\,\langle\sigma_i\rangle_0\,\langle\sigma_j\rangle_0 - \mu B\sum_i\langle\sigma_i\rangle_0$$
$$= -\frac{1}{2}\tanh^2\beta y\sum_{\langle ij\rangle} v_{ij} - \mu B N\tanh\beta y. \tag{10.109}$$

Letting, for brevity,

$$\sum_{\langle ij\rangle} v_{ij} = v N, \tag{10.110}$$

which defines $v$ consistently with Section 10.5, one is finally led to

$$\langle H\rangle_0 = -\frac{vN}{2}\tanh^2\beta y - \mu B N\tanh\beta y. \tag{10.111}$$

The resulting variational estimate $\mathcal{A}$ for the free energy is thus

$$\frac{\beta\mathcal{A}}{N} = -\log\left[2\cosh\beta y\right] - \frac{\beta v}{2}\tanh^2\beta y$$
$$+\beta\left(y - \mu B\right)\tanh\beta y, \tag{10.112}$$

or, letting

$$\xi = \beta y, \tag{10.113}$$

$$\frac{\beta\mathcal{A}}{N} = -\log\left[2\cosh\xi\right] - \frac{\beta v}{2}\tanh^2\xi$$
$$+\left(\xi - \mu B\beta\right)\tanh\xi. \tag{10.114}$$

The best variational estimate within this class of test functions is obtained by setting to zero the derivative of this expression with respect to $\xi$, which leads to the "master equation"

$$\xi - \mu B\beta = \beta v\tanh\xi, \tag{10.115}$$

a result that we had already obtained, exactly, in the infinite-range model, in Eq. (10.80), and via the Curie–Weiss molecular field in Eq. (10.91). The corresponding critical temperature is given in Eq. (10.82).

The variational estimate for the magnetization is

$$\mathcal{M} = N\mu\tanh\xi, \tag{10.116}$$

an expression where one should use the solution of the master equation. For this reason $\xi$ is often called "order parameter" for the magnetization.

We can now derive explicit expressions for the various quantities of interest in the vicinity of the critical temperature of Eq. (10.82). Expanding Eq. (10.114) for small

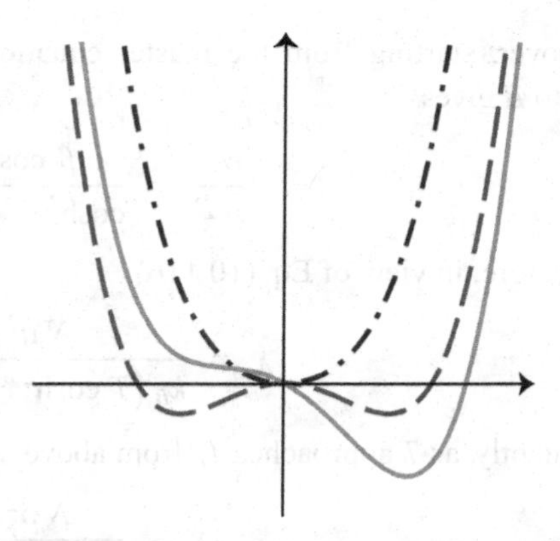

Figure 10.3 Three Landau–Ginzburg-like potential shapes that arise from the variational free energy of Eq. (10.118): (dash-dotted line) $B = 0, T > T_c$; (dashed line) $B = 0, T < T_c$; (solid line) $B \neq 0, T < T_c$.

values of $\xi$ gives

$$\frac{\beta \mathcal{A}}{N} \simeq -\log 2 + \frac{1}{2}\left(1 - \beta v\right)\xi^2 + \left(\frac{\beta v}{3} - \frac{1}{4}\right)\xi^4 - \mu B \beta\, \xi, \tag{10.117}$$

and close to the critical temperature one can set $\beta v \simeq 1$ in the last terms, which gives

$$\frac{\beta \mathcal{A}}{N} \simeq -\log 2 + \frac{1}{2}\left(1 - \beta v\right)\xi^2 + \frac{1}{12}\,\xi^4 - \frac{\mu B}{v}\,\xi. \tag{10.118}$$

Aside from the magnetic field contribution that we did not include there, these Landau–Ginzburg potentials, sketched in Figure 10.3, are along the lines of what we discussed in Section 10.4.

Note that the quadratic term changes sign at $T = T_c$, so that for $B = 0$ there are two minima at opposite values of $\xi$ below $T_c$ and a single one, at $\xi = 0$, above $T_c$. The preceding expression implies indeed the corresponding form of the master equation for $B = 0$,

$$\xi\left(\frac{\xi^2}{3} + (1 - \beta v)\right) = 0, \tag{10.119}$$

so that the only solution is $\xi = 0$ for $T > T_c$. On the other hand, there are three solutions for $T < T_c$: $\xi = 0$, which is a local maximum, and

$$\xi = \pm\sqrt{3\left(\beta v - 1\right)} \simeq \pm\sqrt{3\left(1 - \frac{T}{T_c}\right)}, \tag{10.120}$$

which are identical minima in the absence of a magnetic field. When a magnetic field is turned on, one of the minima becomes lower than the other and is thus preferred. When the field is removed, the hysteresis cycle can thus generate a residual magnetization, with a nonanalytic dependence on the temperature:

$$\mathcal{M} \simeq N\mu\sqrt{3\left(1 - \frac{T}{T_c}\right)} \qquad (T < T_c). \tag{10.121}$$

Moreover, starting from the master equation (10.115) and differentiating with respect to $B$ gives

$$\frac{\partial \xi}{\partial B} = \frac{\mu \beta \cosh^2 \xi}{\cosh^2 \xi - \beta v}, \tag{10.122}$$

and therefore, in view of Eq. (10.116),

$$\chi = \frac{N \mu^2}{k_B \left(T \cosh^2 \xi - T_c\right)}. \tag{10.123}$$

Consequently, as $T$ approaches $T_c$ from above,

$$\chi \sim \frac{N \mu^2}{k_B (T - T_c)}, \tag{10.124}$$

while as as $T$ approaches $T_c$ from below, making use of Eq. (10.120),

$$\chi \sim \frac{N \mu^2}{2 k_B (T_c - T)}. \tag{10.125}$$

There is another interesting type of behavior in the presence of a magnetic field at $T = T_c$. Starting from Eq. (10.115) and expanding the hyperbolic function gives

$$\xi \simeq \left(\frac{3 \mu B}{v}\right)^{\frac{1}{3}}, \tag{10.126}$$

and therefore

$$\mathcal{M} \simeq N \mu \left(\frac{3 \mu B}{v}\right)^{\frac{1}{3}}. \tag{10.127}$$

One can also exhibit, along the same lines, the variational or mean-field result for the specific heat. The starting point is in this case the variational expression for $\beta A$ in Eq. (10.114), which implies that, if the master equation holds,

$$U = \langle H \rangle_0. \tag{10.128}$$

All in all, for vanishing $B$ and close to the critical temperature $T_c$, one can thus conclude that

$$C_V \simeq -\frac{v N}{2} \frac{\partial \xi^2}{\partial T}. \tag{10.129}$$

For $T > T_c$, as we have seen, $\xi = 0$, and therefore $C_V = 0$, while for $T < T_c$, in the ordered phase one can use Eq. (10.120), and then

$$C_V \simeq \frac{3 v N}{2 T_c} = \frac{3}{2} N k_B, \tag{10.130}$$

so that the mean-field specific heat has a jump discontinuity at $T_c$.

Our final task now is to estimate the correlation length $\lambda$ in the mean-field approximation, following Eq. (10.43), via the large-distance behavior of the one-point function $\langle \sigma(\mathbf{x}) \rangle$ obtained with a fixed value of $\langle \sigma(0) \rangle$. To this end, one can generalize the variational ansatz, introducing a site-dependent variational parameter $\xi(i)$ to account for the effects of an inhomogeneous magnetic field $B(i)$. A large magnetic field at the origin can then fix the corresponding spin. For the small values of $\xi(i)$ that will be induced well away from the origin, one can linearize the master equation around $\xi = 0$ working, for simplicity, for $T > T_c$, which leads to a finite-difference equation for $\xi(i)$:

$$\xi(i) - \mu \beta B(i) = \frac{\beta v}{2D} \sum_{\langle ij \rangle} \xi(j). \tag{10.131}$$

In the continuum limit, proceeding as in Section 10.4, one is thus led to the differential equation

$$\left(\nabla^2 - m^2(\beta)\right) \xi(\mathbf{x}) = -\frac{2\mu D}{va^2} B(\mathbf{x}), \tag{10.132}$$

with

$$m^2(\beta) = \left(1 - \beta v\right) \frac{2D}{a^2} \sim \left(\frac{T}{T_c} - 1\right). \tag{10.133}$$

This linear equation can be solved by using a Fourier transform, obtaining at large distances

$$\xi(\mathbf{x}) = \frac{2\mu D}{va^2} \int d^D\mathbf{y}\, G(|\mathbf{x} - \mathbf{y}|)\, B(\mathbf{y}), \tag{10.134}$$

where

$$G(r) = \int \frac{d^D\mathbf{k}}{(2\pi)^D} \frac{e^{i\mathbf{k}\cdot\mathbf{x}}}{k^2 + m^2}. \tag{10.135}$$

By construction, $\xi(\mathbf{x})$ is the order parameter for the magnetization but, as we saw in Section 10.3, thanks to the cluster property one can extract from it the correlation length. However, $G(r)$ also encodes information on the susceptibility. Considering a weak uniform magnetic field $B$ and taking the derivative with respect to its strength gives

$$\chi \sim \int G(r)\, d^D\mathbf{r}, \tag{10.136}$$

which links $\chi$ to the zero-momentum limit of the Fourier transform of $G$.

Here we have confined our attention to the disordered phase, where the average of $\xi$ vanishes in the absence of a magnetic field, and $m^2 > 0$. However, for $T < T_c$ the behavior would be qualitatively similar, after defining $\xi$ as a deviation with respect to a minimum of the free energy. The identity

$$\frac{1}{k^2 + m^2} = \int_0^\infty ds\, e^{-s\left(k^2 + m^2\right)} \tag{10.137}$$

reduces the **k**-integral to $D$ copies of the standard Gaussian term of Eq. (2.21), and therefore

$$\xi(\mathbf{x}) \sim \left(\frac{m}{|\mathbf{x}|}\right)^{\frac{D}{2}-1} \int ds\, e^{-m|\mathbf{x}|\left(s + \frac{1}{4s}\right)}. \tag{10.138}$$

This integral can now be estimated using the saddle-point method, and

$$\xi \sim \left(\frac{m}{|\mathbf{x}|}\right)^{\frac{D-1}{2}} e^{-m|\mathbf{x}|}, \tag{10.139}$$

where we are only keeping track of the dependence on $x$, so that the correlation length behaves as

$$\lambda_c \sim \frac{1}{m} \sim \left(\frac{T}{T_c} - 1\right)^{-\frac{1}{2}} \tag{10.140}$$

as $T \to T_c^+$. On the other hand, at the critical temperature $m = 0$, and $\xi$ decreases with increasing $\mathbf{x}$ as a simple power. Rotational invariance and a simple dimensional argument show indeed that Eq. (10.134) implies

$$\xi \sim \frac{1}{|\mathbf{x}|^{D-2}}, \tag{10.141}$$

a result that translates into an infinite value of $\lambda$. We shall return to the significance of this fact in Chapter 13. Finally, for $T < T_c$ one ought to expand around one of the minima in Figure 10.3, arriving at a result qualitatively similar to Eq. (10.139), with

$$\lambda_c \sim \left|\frac{T}{T_c} - 1\right|^{-\frac{1}{2}}. \tag{10.142}$$

## 10.8 Critical Exponents and Scaling Relations

In general, one can parametrize various types of singular behavior close to the critical temperature $T_c$ in terms of

$$t = \frac{T - T_c}{T_c}, \tag{10.143}$$

which leads to define a number of "critical exponents." For brevity, here we list the exponents without distinguishing their values in the ordered and disordered phases, although these can be different.

1. **Specific heat:**

$$C_V \sim |t|^{-\alpha} \qquad (t \to 0). \tag{10.144}$$

In mean field Ising we found $\alpha = 0$, with a jump discontinuity,

2. **Spontaneous magnetization:**

$$\mathcal{M} \sim (-t)^{\beta} \qquad (t \to 0^-). \tag{10.145}$$

In mean field Ising we found $\beta = \frac{1}{2}$.

3. **Susceptibility:**

$$\chi \sim t^{-\gamma} \qquad (t \to 0). \tag{10.146}$$

In mean field Ising we found $\gamma = 1$.

4. **Magnetization at $T_c$:**

$$\mathcal{M} \sim |B|^{\frac{1}{\delta}} \qquad (B \to 0, t = 0). \tag{10.147}$$

In mean field Ising we found $\delta = 3$.

5. **Correlation function for $t \neq 0$:**

$$\lambda \sim |t|^{-\nu} \qquad (t \to 0), \tag{10.148}$$

where $\lambda$ is the correlation length, defined via

$$G(r) \sim \frac{e^{-\frac{r}{\lambda}}}{r^{\frac{D-1}{2}}} \qquad (r \to \infty). \tag{10.149}$$

In mean field Ising we found $\nu = \frac{1}{2}$.

6. **Correlation function at $T_c$:**

$$G(r) \sim \frac{1}{r^{D-2+\eta}} \qquad (r \to \infty, t = 0). \tag{10.150}$$

In mean field Ising we found $\eta = 0$.

All in all, the simple mean-field analysis has displayed various instances of non–analytic behavior close to the critical temperature. The results are only qualitatively correct, and ought to become more and more accurate for higher values of $D$, since this would increase the number of nearest neighbors, thus contrasting disorder more and more effectively, as exemplified by the infinite-range model. In the next Chapter, we shall illustrate Onsager's exact solution of the two-dimensional Ising model, which will display a nonanalytic behavior of this type for the specific heat, albeit with a logarithmic singularity rather than a jump discontinuity.

The critical exponents, however, are not independent, since there are links among them that originate from scaling symmetry. The two exponents $\nu$ and $\eta$, which characterize the critical behavior of the correlation function, actually determine the scaling behavior of all thermodynamic quantities.

A convenient starting point to discuss scaling relations is the natural assumption that an expression of the form

$$\mathcal{M} = |t|^{\beta} f\left(\frac{|B|}{|t|^{\beta\delta}}\right) \tag{10.151}$$

link $\mathcal{M}$, $t$ and $B$ close to the critical temperature, which corresponds to $t = 0$ according to Eq. (10.143). This type of expression indeed embodies the exponents in Eqs. (10.145) and (10.147), provided

$$f(x) \sim \begin{cases} 1 + \mathcal{O}(x) & \text{for } x \to 0, \\ x^{\frac{1}{\delta}} & \text{for } x \to \infty. \end{cases} \tag{10.152}$$

Moreover, a derivative of this expression yields a limiting form for the susceptibility,

$$\chi = |t|^{\beta(1-\delta)} f'\left(\frac{|B|}{|t|^{\beta\delta}}\right), \tag{10.153}$$

and the preceding assumptions on $f$ translate into a first link among the critical exponents:

$$\gamma = \beta(\delta - 1), \tag{10.154}$$

in view of Eq. (10.146). The form of the argument of $f$ now implies that, at vanishing $B$, the free energy must exhibit a definite power-like behavior in t,

$$A = - \int \mathcal{M} dB \sim |t|^{\beta(1+\delta)}, \tag{10.155}$$

which follows completing the argument in the integration, and therefore the singular limiting behavior of the specific heat is

$$C_V \sim - T_c \frac{\partial^2 A}{\partial t^2} \sim |t|^{\beta(1+\delta)-2}, \tag{10.156}$$

since the factor $T \simeq T_c$ that enters the definition of $C_V$ is not relevant for it. This expression leads to another link among the critical exponents,

$$\alpha = 2 - \beta(1+\delta). \tag{10.157}$$

The next step involves the correlation function. As we have seen, this quantity is also a measure of fluctuations, and can be parametrized, for dimensional reasons and under the spell of Eqs. (10.149) and (10.150), as

$$G(|x|) = \frac{g\left(\frac{|x|}{\lambda}\right)}{|x|^{D-2+\eta}} \longleftrightarrow \widetilde{G}(q) = \frac{h(q\lambda)}{q^{2-\eta}}, \tag{10.158}$$

with $g(0)$ and $h(\infty)$ finite nonzero constants, and where $\widetilde{G}$ denotes the Fourier transform of $G$. The overall powers are tailored to the critical case and to the preceding mean-field analysis, where however $\eta$ was 0. Just as Eq. (10.135), slightly away from the critical temperature $\widetilde{G}(q)$, should have a finite limit as $q \to 0$, which is possible if

$$h(q\lambda) \sim (q\lambda)^{2-\eta}, \tag{10.159}$$

and consequently

$$\widetilde{G}(q) \sim \lambda^{2-\eta} \sim |t|^{-\nu(2-\eta)} \tag{10.160}$$

as $|t| \to 0$. As we have stressed below Eq. (10.134), however, this limiting behavior must also reflect the critical exponent $\gamma$ of the susceptibility, and therefore one is led to conclude that

$$\gamma = \nu(2-\eta). \tag{10.161}$$

The last relation reflects the dependence of the free energy on a microscopic scale of the system, the lattice spacing. If the system is $D$ dimensional we expect that, rescaling the lattice spacing by a factor $b$,

$$A \sim b^{-D} \sim t^{D\nu} \tag{10.162}$$

since this reduces the number of sites within a given volume $V$ by a factor $b^{-D}$. The dependence on $t$ follows since the correlation length also scales with $b$ and is linked to $t$ by Eq. (10.148). On the other hand, Eq. (10.144) translates into

$$A \sim t^{2-\alpha}, \tag{10.163}$$

and by comparison the proper scaling is granted if

$$D\nu = 2 - \alpha. \tag{10.164}$$

Summarizing, the different scaling relations that we have collected are:

$$\begin{aligned} \gamma &= \beta(\delta - 1), \\ \alpha &= 2 - \beta(1+\delta), \\ \gamma &= \nu(2-\eta), \\ D\nu &= 2 - \alpha. \end{aligned} \qquad (10.165)$$

These relations are satisfied by the exact solution of the two-dimensional Ising model, some aspects of which we shall discuss in Chapter 11, and by its approximate solution in three dimensions, which we shall touch upon in Chapter 13. All relations, apart from the last one, are also satisfied by the mean-field values for the Ising model, while the last holds for them only in $D = 4$ where, as we shall see in Chapter 13, the large-distance behavior is captured by a Gaussian model and the mean-field results become exact.

Notice that all critical exponents are determined by $\nu$ and $\eta$. Indeed, Eqs. (10.165) can be combined and presented in the form

$$\begin{aligned} \alpha &= 2 - D\nu\,, \\ \beta &= \frac{\nu}{2}\,(D - 2 + \eta)\,, \\ \gamma &= \nu\,(2 - \eta)\,, \\ \delta &= \frac{D + 2 - \eta}{D - 2 + \eta}\,. \end{aligned} \qquad (10.166)$$

## 10.9 Landau–Ginzburg Theory

The variational method has provided a nice picture of the Ising model in mean field but, as we have stressed, has manifold applications above and beyond this simple case. In fact, all the results that we have deduced from it not involving the correlation length can be derived very simply from the Landau–Ginzburg free energy

$$A(\Phi) = \frac{t}{2}\,\Phi^2 + \frac{1}{4}\,\Phi^4 - B\,\Phi, \qquad (10.167)$$

where the positive coefficient of the $\Phi^4$ term grants the overall stability of the system. The different conventions used here do not affect the critical exponents, and this expression can be regarded as emerging from a saddle-point of the continuum model of Section 10.4 for homogeneous configurations. As we have seen, the Landau–Ginzburg field $\Phi$ determines the magnetization, and positive values of

$$t = \frac{T - T_c}{T_c}, \qquad (10.168)$$

characterize the high-temperature phase, while negative values accompany the breaking of the $\mathbb{Z}_2$ symmetry and thus characterize the low-temperature phase. The minima of $A(\Phi)$ are determined by

$$A'(\Phi) \equiv t\Phi + \Phi^3 - B = 0, \qquad (10.169)$$

to be combined with the local stability condition

$$A''(\Phi) \equiv t + 3\,\Phi^2 \geq 0. \tag{10.170}$$

In the absence of an external field $B$, for $t > 0$ there is only one extremum at the origin, which is a minimum since $A''(0) = t$. On the other hand, if $t < 0$ there are two additional ones at $\pm\Phi_+$, which are determined by the condition

$$\Phi_+^2 = |t|, \tag{10.171}$$

and in this case the extremum at the origin becomes a maximum, while these new extrema are minima since $A''(\Phi_+) = 2\,|t|$ (see the dash-dotted and dashed curves in Figure 10.3).

From Eq. (10.171), taking into account Eq. (10.168), one can simply recover the critical index $\beta = \frac{1}{2}$ for the spontaneous magnetization, since for $T < T_c$,

$$\mathcal{M} \sim \Phi \sim \pm\sqrt{|t|}. \tag{10.172}$$

In a similar fashion, in the presence of a magnetic field $B$, for $T = T_c$ Eq. (10.169) gives

$$\Phi \sim |B|^{\frac{1}{3}}, \tag{10.173}$$

so that $\delta = 3$. One can also study the susceptibility, starting from Eq. (10.169), which leads to

$$\chi \sim \left.\frac{1}{t + 3\,\Phi^2}\right|_{A'(\Phi)=0}, \tag{10.174}$$

and from this one can conclude that

$$\chi \sim \frac{1}{|t|}. \tag{10.175}$$

This limiting behavior involves two different overall coefficients, one for for $t > 0$ and one for $t < 0$, but the conclusion is that $\gamma = 1$ in both cases, as in Section 10.7. Finally, the specific heat is determined by the internal energy,

$$U = A + TS = -\frac{1}{2}\,\Phi^2 + \frac{1}{4}\,\Phi^4, \tag{10.176}$$

with

$$C = \frac{1}{2}\left(\Phi^2 - 1\right)\left.\frac{\partial\,\Phi^2}{\partial\,T}\right|_{A'(\Phi)=0}, \tag{10.177}$$

to be computed at the relevant extremum. Hence $C = 0$ for $T > T_c$, since in this case $\Phi^2$ remains fixed at the origin as $T$ varies. On the other hand, $C = \frac{1}{T_c}$ for $T < T_c$, so that there is a jump discontinuity, and the critical exponent $\alpha$ vanishes.

Let us now complicate matters a bit, considering the wider family of Landau–Ginzburg potentials

$$A(\Phi) = \frac{t}{2}\,\Phi^2 + \frac{b}{4}\,\Phi^4 + \frac{1}{6}\,\Phi^6. \tag{10.178}$$

These depend on two parameters, and the positive coefficient of the highest power again grants global stability. Now there is a richer spectrum of options, since the extrema are determined by the condition

$$A'(\Phi) \equiv t\,\Phi + b\,\Phi^3 + \Phi^5 = 0, \tag{10.179}$$

while

$$A''(\Phi) = t + 3b\Phi^2 + 5\Phi^4. \tag{10.180}$$

There is always an extremum at $\Phi = 0$, which is again a local minimum if $t > 0$ and a local maximum if $t < 0$. However, there are possibly other extrema, which are determined by the condition

$$t + b\Phi^2 + \Phi^4 = 0, \tag{10.181}$$

provided one or both formal solutions,

$$\Phi_\pm^2 = \frac{-b \pm \sqrt{b^2 - 4t}}{2}, \tag{10.182}$$

are real and positive. Note also that, when this is the case, combining Eqs. (10.180), (10.181), and (10.182), one can show that

$$A''(\Phi_\pm) = \pm 2\Phi_\pm^2 \sqrt{b^2 - 4t}, \tag{10.183}$$

so that the extrema at $\pm\Phi_+$, when they are real, correspond to local minima, while those at $\pm\Phi_-$, when they are real, correspond to local maxima.

Let us now discuss in detail the behavior of this system, with reference to Figure 10.4.

- In the quadrant $(t > 0, b > 0)$ there is a single real extremum at the origin.
- In the half plane $(t < 0)$ there is an extremum at $\Phi = 0$, which is a local maximum, and two more at $\Phi = \pm\Phi_+$, which are the minima of $A(\Phi)$. The dashed half-line

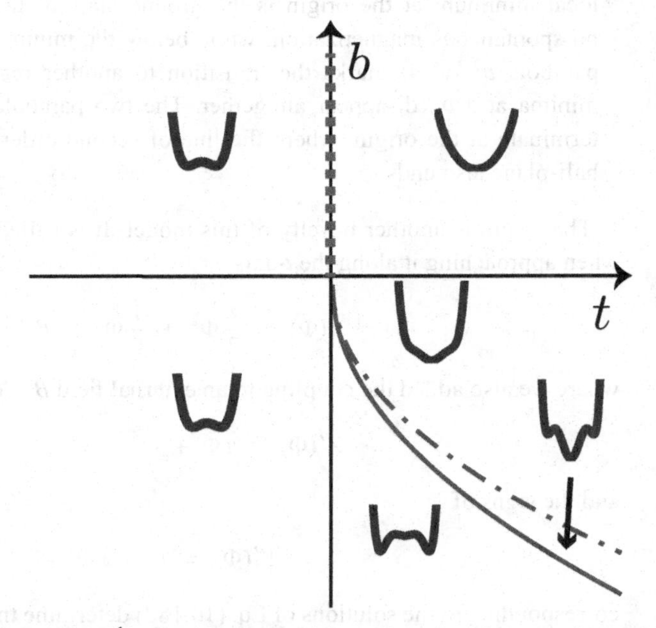


Figure 10.4 The phase diagram of the $\Phi^6$ Landau–Ginzburg model comprises a line of second-order transitions (the dashed vertical half-line $t = 0, b > 0$), a line of first-order transitions (the solid line $3b^2 = 16t, b < 0$), and above the dash-dotted line $b^2 = 4t, b < 0$ there is no spontaneous magnetization. The arrow is meant to stress that the special form of the free energy with a lower minimum at the origin and two higher ones on the sides applies to the region between the two parabolic branches. The origin is a tricritical point. The cartoons indicate the different shapes of the Landau–Ginzburg potential in the various regions.

$t = 0, b > 0$ corresponds to second-order phase transitions, akin to those described by the quartic potential.

- The quadrant $t > 0$, $b < 0$ contains some interesting novelties. In the region $b^2 > 4t$, aside from the local minimum at $\Phi = 0$, there are positive solutions $\Phi_+^2$ and $\Phi_-^2$ of eq (10.182). As we have seen in eq. (10.183), $\Phi_+^2$ corresponds to a pair of local minima at $\pm\,\Phi_+$, while $\Phi_-^2$ corresponds to a pair of local maxima at $\pm\,\Phi_-$. The issue is whether the local minima lie below or above the local minimum that is present at the origin for $t > 0$, where $A(\Phi)$ vanishes. Combining the definition of $A$ with Eq. (10.179) one can recast it, at the non–vanishing extrema, in the form

$$A(\Phi_\pm) \;=\; \frac{1}{12}\,\Phi_\pm^4\left(|b| \,\mp\, 2\sqrt{b^2 - 4\,t}\right), \tag{10.184}$$

and therefore the minima at $\Phi_+$ and $-\Phi_+$ lie below (above, respectively) the one at the origin if

$$|b| \;-\; 2\sqrt{b^2 - 4\,t} \;<\; 0 \qquad (\,>\,0\,). \tag{10.185}$$

The two regions are separated by a line of first-order transitions where the left-hand side of Eq. (10.185) vanishes, the branch of the parabola

$$t \;=\; \frac{3}{16}\,b^2 \tag{10.186}$$

that lies in the half-plane $b < 0$ (solid curve in the Figure 10.4). The reason is that the location of the minima is discontinuous across this curve. Above the line the local minimum at the origin is the ground state of the system, which displays no spontaneous magnetization, while below the minima lie at $\pm\,\Phi_+$. Finally, the parabola $b^2 = 4t$ marks the transition to another region where the two local minima at $\pm\Phi_+$ disappear altogether. The two parabolas in the fourth quadrant terminate at the origin, where the line of second-order transitions in the upper-half-plane also ends.

The origin is another novelty of this model. It is called a "tricritical" point, and when approaching it along the $t$-axis

$$A(\Phi) \;=\; \frac{t}{2}\,\Phi^2 \;+\; \frac{1}{6}\,\Phi^6 \;-\; B\,\Phi, \tag{10.187}$$

where we also added the coupling to an external field $B$. Consequently the condition

$$A'(\Phi) \;\equiv\; t\,\Phi \;+\; \Phi^5 \;-\; B \;=\; 0 \tag{10.188}$$

and the signs of

$$A''(\Phi) \;=\; t \;+\; 5\,\Phi^4 \tag{10.189}$$

corresponding to the solutions of Eq. (10.188) determine the relevant extrema for the system.

It is interesting to compute the critical exponents for the tricritical point, starting from Eqs. (10.187) and (10.188). To begin with, for $t = 0$,

$$\mathcal{M} \;=\; \Phi \;=\; |B|^{\frac{1}{5}} \tag{10.190}$$

so that $\delta = 5$. Moreover, solving Eq. (10.188) for $B = 0$ yields, for $t < 0$, a spontaneous magnetization

$$\mathcal{M} = \Phi = \pm |t|^{\frac{1}{4}}, \tag{10.191}$$

so that $\beta = \frac{1}{4}$. From Eq. (10.188) one can also obtain

$$\chi = \frac{\partial \Phi}{\partial B} = \frac{1}{t + 5\Phi^4} \sim \frac{1}{|t|}, \tag{10.192}$$

so that $\gamma = 1$. Finally $A(\Phi)$ determines the internal energy,

$$U = \frac{1}{2}\Phi^2 + \frac{1}{6}\Phi^6, \tag{10.193}$$

for $B = 0$, and consequently as $t \to 0^-$,

$$C \sim |t|^{-\frac{1}{2}} \tag{10.194}$$

Therefore $\alpha = \frac{1}{2}$ when approaching the tricritical point for $t < 0$.

## 10.10 A Toy Model of a Phase Transition

As we have seen, phase transitions bring along singularities in thermodynamic functions. At first sight, the occurrence of such singularities may seem paradoxical, since the partition function,

$$Z(T, V, N) = \sum_I e^{-\beta E_I}, \tag{10.195}$$

seems to be a rapidly converging series of positive terms for $\beta > 0$ and hence the free energy

$$F = -k_B T \log Z \tag{10.196}$$

looks like an analytic function for positive values of $\beta$. This naive argument, however, runs into trouble in the thermodynamic limit: as $N$ and $V$ increase, singularities that may be present in $F$ for complex values of $\beta$ can accumulate close to the positive real axis, giving rise to irregular points in the free energy per particle

$$f(T, n) = -k_B T \lim_{N,V \to \infty} \frac{\log Z(T, V, N)}{N}, \tag{10.197}$$

where $n = \frac{N}{V}$.

This phenomenon can be nicely illustrated by referring to a simple example, a system whose finite-volume grand-partition function is

$$\mathcal{Q} = (1 + z)^V \frac{1 - z^V}{1 - z}, \tag{10.198}$$

where for simplicity $V$ is a positive integer representing the volume in suitable units. The zeros of this function, and therefore the singularities of the grand potential $\Omega = -k_B T \log \mathcal{Q}$ lie at $z = e^{i2\pi k/V}$, where $k = 1, 2, \ldots, V - 1$, and at $z = -1$. No singularities are thus present on the physical portion $z > 0$ of the real axis, but as

$V \to \infty$ these singularities do clump near $z = 1$. The pressure and total number of particles per unit volume are, in this limit,

$$\frac{P}{k_B T} = \begin{cases} \log(1+z) & \text{for } z < 1, \\ \log[z(1+z)] & \text{for } z > 1, \end{cases} \tag{10.199}$$

and

$$n \equiv \frac{1}{v} = \begin{cases} \dfrac{z}{1+z} & \text{for } z < 1, \\ \dfrac{1+2z}{1+z} & \text{for } z > 1. \end{cases} \tag{10.200}$$

The particle density, displayed in Figure 10.5, is thus discontinuous across $z = 1$, with

$$\lim_{z \to 1^-} n = \frac{1}{2}, \qquad \lim_{z \to 1^+} n = \frac{3}{2}. \tag{10.201}$$

These results display interesting similarities to the behavior close to a vapor–liquid transition.

The equation of state can be obtained inverting the relation between $n$ and $z$ and substituting into the expression for $P$. In terms of the specific volume $v = \frac{1}{n}$, one thus finds

$$\frac{P}{k_B T} = \begin{cases} \log\left[\dfrac{v(1-v)}{(2v-1)^2}\right] & \text{for } \dfrac{1}{2} < v < \dfrac{2}{3} \\ \log 2 & \text{for } \dfrac{2}{3} < v < 2 \\ \log\left(\dfrac{v}{v-1}\right) & \text{for } v > 2. \end{cases} \tag{10.202}$$

The isothermal curves in the $P$-$v$ plane, displayed in Figure 10.5, agree qualitatively with those of a first-order phase transition, with a coexistence of the two phases in the region $\frac{2}{3} < v < 2$. We already encountered this type of behavior when we discussed the van der Waals gas in Chapter 6.

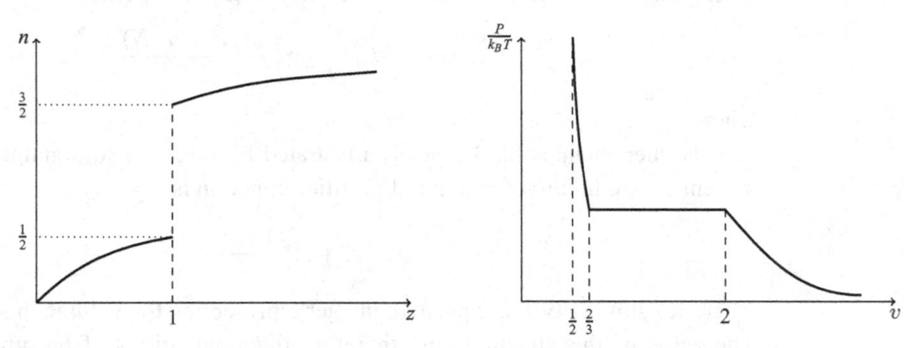


**Figure 10.5** (Left) The dependence of the particle density on the fugacity $z$, and (right) an isothermal curve for the toy model.

## Bibliographical Notes

Detailed discussions of the one-dimensional Ising model and of the Landau–Ginzburg model can be found in many excellent books, including [6, 25, 36, 43, 45, 53]. The role of boundary conditions and pure phases in connection with symmetry breaking is discussed in detail in [56]. The variational method in statistical physics is discussed in [17], with emphasis on the quantum case. The critical exponents for the Ising system and the corresponding scaling relations can be found, for instance, in [36, 39, 54].

## Problems

**10.1** Consider the thin tube displayed in Figure 10.6, whose two end sections, of area $S$, rest on a horizontal plane. A partition of mass $M$ and negligible volume, free to move along it, divides the tube into two portions. Each portion hosts a perfect gas of $N$ nonrelativistic particles of mass $m$, maintained at a temperature $T$. Taking into account the effects of gravity, which acts vertically on the partition:

1. Discuss the equilibrium configurations of the system and their stability as the temperature $T$ of the gas varies;
2. Derive the free energy of the system in equilibrium;
3. Discuss the behavior of the specific heat as $T$ varies.

**10.2** Write the transfer matrices for the following one-dimensional systems:

1. An Ising model with second-neighbor interactions

$$H = -J\sum_{i=1}^{N}\sigma_i\sigma_{i+1} - K\sum_{i=1}^{N}\sigma_i\sigma_{i+2} - h\sum_{i=1}^{N}\sigma_i.$$

2. A one-dimensional Ising chain where each site can be vacant or, alternatively, can host a particle of spin $\pm 1$, for which

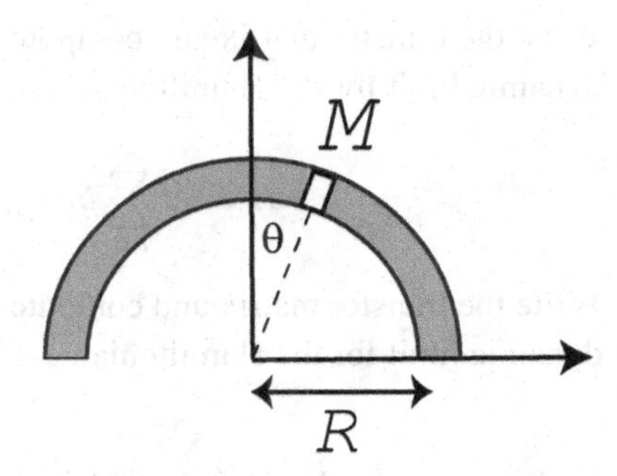


Figure 10.6 A tube with gas on both sides and a moving partition.

$$H = -J\sum_{i=1}^{N} \sigma_i\sigma_{i+1} - B\sum_{i=1}^{N} \sigma_i,$$

where $\sigma_i \in \{0, \pm 1\}$.

3. Two coupled Ising chains

$$H = -J\sum_{i=1}^{N} \sigma_i\sigma_{i+1} - K\sum_{i=1}^{N} \tau_i\tau_{i+1} - h\sum_{i=1}^{N} \sigma_i\tau_i\,,$$

where $\sigma_i \in \{-1, 1\}$ and $\tau_i \in \{-1, +1\}$.

**10.3** The Potts model is a generalization of the Ising model defined by the Hamiltonian

$$H = -J\sum_{i=1}^{N} \delta_{n_i,n_{i+1}},$$

in which the variables $n_i$ can take integer values from 1 to $p$.

1. Write the transfer matrix for this system, and compute the partition function in the thermodynamic limit, with periodic boundary conditions.
2. Compute the average $\langle n_i \rangle$ for this system.
3. Compute the two-point function $\langle n_i\, n_{i+k} \rangle$.

**10.4** Repeat the analysis of the Potts system of Problem 10.3, with

$$H = -J\sum_{i=1}^{N} \delta_{n_i,n_{i+1}},$$

but now with a different choice of boundary conditions.

1. Compute the partition function in the thermodynamic limit, assuming that $n_1$ and $n_N$ are fixed to one.
2. Compute the average $\langle n_i \rangle$ for the system with these new boundary conditions.

**10.5** Extend the treatment of the Potts model to cases in which an external field is also present.

1. Write the transfer matrix and compute the partition function in the thermodynamic limit for the Hamiltonian

$$H = -J\sum_{i=1}^{N} \delta_{n_i,n_{i+1}} + B\sum_{i=1}^{N} \delta_{\sigma_i,1}.$$

2. Write the transfer matrix and compute the partition function in the thermodynamic limit for the Hamiltonian

$$H = -J\sum_{i=1}^{N} \delta_{n_i,n_{i+1}} + \Lambda\sum_{i=1}^{N} \delta_{\sigma_i,1}\,\delta_{\sigma_{i+1},1}.$$

**10.6** Consider a linear chain of $N$ segments in a plane. Each segment can lie along three possible directions with respect to the preceding one, "L," "S," and "R," in such a way that the first and last segments lie in the same direction. The Hamiltonian of the system is

$$H = \sum_i H_{(i,i+1)},$$

and the interaction energies among the segments are as follows:
a. $H_{(i,i+1)} = 0$ if the two consecutive segments are in the same direction (case $S$);
b. $H_{(i,i+1)} = \epsilon > 0$ if the move is $L$;
c. $H_{(i,i+1)} = +\infty$ in all other cases.

1. Write the transfer matrix $T$ of the system.
2. Compute the partition function in the thermodynamic limit and study the behavior of the internal energy as a function of the temperature.

**10.7** Consider again a chain as in Problem 10.6, where now
a. $H_{(i,i+1)} = 0$ if the two consecutive segments are in the same direction (case $S$);
b. $H_{(i,i+1)} = \epsilon > 0$ if the move is $L$ or $R$.
c. $H_{(i,i+1)} = +\infty$ in all other cases.

1. Write the transfer matrix $T$ of the system.
2. Compute the partition function in the thermodynamic limit and study the behavior of the internal energy as a function of the temperature.

**10.8** Consider a linear chain of $N$ segments in the plane with four possible direction, labeled by $s_j = 1, i, -i, -1$, and such that the first and last elements are in the same direction. The Hamiltonian of the system is

$$H = \sum_i H_{(j,j+1)},$$

and the interaction energies between successive links are

a. $H_{(j,j+1)} = 0$ if $s_j s_{j+1} = 1$;
b. $H_{(j,j+1)} = \epsilon > 0$ if $s_j s_{j+1} = \pm i$;
c. $H_{(j,j+1)} = +\infty$ if $s_j s_{j+1} = -1$.

1. Write the transfer matrix $T$ of the system.
2. Compute the partition function, $Tr(T^N)$, in the thermodynamic limit and study the behavior of the internal energy and the specific heat $C_V$ as functions of the temperature $T$.

**10.9** Consider a one-dimensional Ising chain where each site can be vacant or, alternatively, can host a particle of spin $\pm 1$. The Hamiltonian of the system, which is supplemented by the periodic boundary condition $\sigma_{N+1} \equiv \sigma_1$, is thus

$$H = -J \sum_{i=1}^{N} \sigma_i \sigma_{i+1},$$

where $\sigma_i = 0, \pm 1$.

1. Write the transfer matrix $T$ of the system, and compute the partition function in the thermodynamic limit. Hint: One of the eigenvalues of $T$ is

$$\lambda = e^{\beta J} - e^{-\beta J}.$$

2. Compute the free energy using a variational approach for the case of a hypercubic lattice in $D$ dimensions with $N$ sites, writing $H$ in the form

$$H = -\frac{J}{2} \sum_{\langle ij \rangle} \sigma_i \sigma_j,$$

where the interaction is among nearest neighbors, and using the reference Hamiltonian

$$H_0 = -x \sum_{i=1}^{N} \sigma_i.$$

3. Discuss the corresponding master equation for $x$ and determine the corresponding critical temperature.

**10.10** Consider a $d$-dimensional hypercubic lattice, on whose sites there are "spins" $\sigma_i$ that can take the values $0, \pm 1, \pm 2$. The Hamiltonian of the system is

$$H = -\frac{J}{2} \sum_{\langle ij \rangle} \sigma_i \sigma_j,$$

with nearest-neighbor interactions. Compute the free energy of the system in the mean-field approximation, using the test Hamiltonian

$$H_0 = -x \sum_i \sigma_i,$$

and discuss the corresponding "master equation" for $x$.

**10.11** Consider a one-dimensional spin chain for which $\sigma_i = 1, \ldots, p$ and

$$H = -J \sum_i \delta_{\sigma_i, \sigma_{i+1}} - \mu B \sum_i \delta_{\sigma_i, 1}.$$

Write down the variational free energy and the master equation using the test Hamiltonian

$$H_0 = -x \sum_i \delta_{\sigma_i, 1}.$$

**10.12** Consider again a one-dimensional spin chain for which $\sigma_i = 1, \ldots, p$, and

$$H = -J \sum_i \delta_{\sigma_i, \sigma_{i+1}} - \Lambda \sum_i \delta_{\sigma_i, 1} \, \delta_{\sigma_{i+1}, 1}.$$

Write out the variational free energy and the master equation for this system, using the test Hamiltonian

$$H_0 = -x \sum_i \delta_{\sigma_i,1}.$$

**10.13** Consider a one-dimensional with quartic interactions, with $\sigma_i = \pm 1$ and

$$H = -J \sum_{i=1}^{N} \sigma_i \sigma_{i+1} \sigma_{i+2} \sigma_{i+3}.$$

1. Write the variational free energy using the test Hamiltonian

$$H_0 = -x \sum_{i=1}^{N} \sigma_i.$$

2. Write the corresponding master equation and discuss the nature of its solutions as the temperature $T$ varies.

**10.14** Consider a one-dimensional Ising model for which

$$H = -\epsilon \sum_i \sigma_i \sigma_{i+1} \qquad (\sigma_i = \pm 1),$$

but whose spins are subject to the constraint

$$\sum_{i=1}^{N} \sigma_i = 0.$$

1. Recalling that for an integer-valued variable $x$

$$\delta_{x,0} = \int_{-\pi}^{\pi} \frac{d\theta}{2\pi} e^{i\theta x},$$

show that the partition function of the system can be cast in the form

$$Z_N = \int_{-\pi}^{\pi} \frac{d\theta}{2\pi} \left( \lambda_1(\theta)^N + \lambda_2(\theta)^N \right).$$

where

$$\lambda_{1,2}(\theta) = e^{\beta\epsilon} \left( \cos\theta \pm \sqrt{e^{-4\beta\epsilon} - \sin^2\theta} \right).$$

2. Show that the partition function $Z_N$ vanishes for odd $N$, while for even $N$

$$Z_N = \frac{1}{\pi} e^{N\beta\epsilon} \int_{-\pi}^{\pi} e^{Nf(\theta)} \, d\theta,$$

where

$$f(\theta) = \log \left( \cos\theta + \sqrt{e^{-4\beta\epsilon} - \sin^2\theta} \right).$$

3. Show that the corresponding "master equation" $f'(\theta) = 0$ has two saddle points at $\theta = 0, \pi$. Adding the two corresponding contributions, obtain $Z_N$ in the thermodynamic limit.

**10.15** Consider a one-dimensional *open* chain with

$$H = -J \sum_{i=1}^{N} \cos(\theta_i - \theta_{i+1}).$$

1. Compute the partition of the system, expressing it in terms of modified Bessel functions (see [1] and [62]).
2. Study the behavior of $C_V$ as a function of the temperature $T$.

**10.16** Consider a one-dimensional spin chain with

$$H = -\sum_i \frac{v\,\sigma_i\,\sigma_{1+1}}{1 - w\,\sigma_i\,\sigma_{i+1}},$$

where $\sigma_i = \pm 1$ and $v$ and $w$ ($< 1$) are two positive constants.

1. Write the transfer matrix $T$ and deduce from it the Helmholtz free energy.
2. Repeat the analysis expanding the denominator in a power series and connecting $H$ to a conventional Ising Hamiltonian.

**10.17** We want to return to an issue that we touched upon at the beginning of this chapter. So far we have assumed that spin–spin interactions tend to align nearest neighbors, which results in ferromagnetism. However, some materials exhibit the opposite phenomenon, called antiferromagnetism, and here we would like to discuss some features of it in the simplest possible terms. Consider, to this end, a three-spin chain with Hamiltonian

$$H = v\,(\sigma_1\,\sigma_2 + \sigma_2\,\sigma_3 + \sigma_3\,\sigma_1),$$

where the sign of the mutual coupling is opposite to one considered so far, and a similar four-spin chain, with

$$H = v\,(\sigma_1\,\sigma_2 + \sigma_2\,\sigma_3 + \sigma_3\,\sigma_4 + \sigma_4\,\sigma_1),$$

with $\sigma_i = \pm 1$.

1. What are the available energies in the two cases, and what are their multiplicities? The presence of multiple ground state is usually termed "frustration."
2. What limiting values do you expect for the entropy at low temperature for a chain of $2N$ spins and for one of $2N + 1$ spins?
3. Using the transfer matrix, compute the susceptibility for a one-dimensional chain with

$$H = v \sum_i \sigma_i\,\sigma_{i+1}.$$

What peculiar feature does it exhibit as a function of $T$?

**10.18** Study a one-dimensional antiferromagnetic chain of $2N$ sites with the variational method, using the test Hamiltonian

$$H_0 = -x \sum_i \sigma_{2i} - y \sum_i \sigma_{2i+1}.$$

1. Show that this leads to the master equations

$$2\,\beta\,v\,\tanh\xi = -\,\eta,$$
$$2\,\beta\,v\,\tanh\eta = -\,\xi,$$

where $\xi = \beta x$ and $\eta = \beta y$.

2. Compare the two values $\frac{A(\xi,\xi)}{N}$ and $\frac{A(\xi,-\xi)}{N}$, which reduce the preceding equations to

$$\beta v \tanh x = \mp x.$$

Which choice yields a better estimate for the Helmholtz free energy? What types of configurations emerge in this fashion?

**10.19** Consider a one-dimensional Ising chain whose coupling $v$ is a random variable, distributed according to

$$d\Pi = \frac{1}{\Delta^{\gamma}\,\Gamma(\gamma)}\, e^{-\frac{v}{\Delta}}\, v^{\gamma-1}\, dv.$$

Compute the expression for the resulting averaged specific heat, and show that

$$\overline{C}_V(x,\gamma) = k_B\,\gamma(\gamma+1)\sum_{n=1}^{\infty}\frac{(-1)^{n-1}}{n}\frac{\left(\frac{x}{n}\right)^{\gamma}}{\left[1+\frac{x}{n}\right]^{\gamma+2}},$$

where

$$x = \frac{k_B\,T}{2\,\Delta}.$$

Discuss qualitatively how the behavior of the averaged specific heat changes when $\gamma$ increases.

**10.20** Consider the classical Heisenberg model on a cubic lattice, for which

$$H = -J\sum_{\langle ij\rangle}\mathbf{S}_i\cdot\mathbf{S}_i - \mu B\sum_i S_{z\,i},$$

where the $\mathbf{S}_i$ are unit vectors in three dimensions, so that

$$\mathbf{S}_i = \sin\theta_i\,\cos\phi_i\,\hat{\mathbf{x}} + \sin\theta_i\,\sin\phi_i\,\hat{\mathbf{y}} + \cos\theta_i\,\hat{\mathbf{z}},$$

and the test Hamiltonian

$$H_0 = -x\sum_i \cos\theta_i.$$

1. Construct the variational estimate for the Helmholtz free energy, write the master equation, and determine the corresponding estimate for the critical temperature.
2. Determine the estimates for the critical exponents $\alpha$, $\beta$, $\gamma$, and $\delta$.

# PART II

# 11 The 2D Ising Model

The two-dimensional Ising model is a remarkable system that, despite its apparent simplicity, can illustrate in detail the deep and far-reaching idea of weak–strong coupling duality. This model also affords the famous Onsager solution, which displays clearly the nonanalytic behavior of the exact specific heat at the phase transition.

## 11.1 Closed Polygons in Two Dimensions

Our starting point is the partition function

$$Z = \sum_{\{\sigma_i=\pm 1\}} e^{H \sum_{\langle ij \rangle} \sigma_i \sigma_j}, \tag{11.1}$$

where

$$H = \frac{\beta v}{2}, \tag{11.2}$$

and, as before, each of the $\sigma$'s takes the values $\pm 1$. We shall now associate them to the $N$ sites of a two-dimensional square lattice, working again with periodic boundary conditions. As before, one can use the identity

$$e^{H \sigma_i \sigma_j} = \cosh H \left[ 1 + \sigma_i \sigma_j \tanh H \right], \tag{11.3}$$

and since there are $2N$ links in a two-dimensional lattice with $N$ sites,

$$Z = [\cosh H]^{2N} \sum_{\{\sigma_i=\pm 1\}} \prod_{\langle ij \rangle} \left[ 1 + \sigma_i \sigma_j \tanh H \right], \tag{11.4}$$

but in two dimensions nontrivial contributions from the terms within square brackets proportional to finite powers of $\tanh H$ can emerge from *closed polygons of finite size* (see Figure 11.1), so that

$$\sum_{\{\sigma_i=\pm 1\}} \prod_{\langle ij \rangle} \left[ 1 + \sigma_i \sigma_j \tanh H \right] = 2^N \sum_L g(L) \left[ \tanh H \right]^L, \tag{11.5}$$

where $g(L)$ is the number of closed polygons (connected or disconnected) of length $L$ that can be drawn on the lattice, with $g(0) = 1$. We shall return to their nature, but for the time being one can conclude that

$$Z = \left[ 2 \cosh^2 H \right]^N \sum_L g(L) \left[ \tanh H \right]^L. \tag{11.6}$$

Note that this is a *high-temperature* (or *weak-coupling*) expansion, whose first few terms yield a good approximation for $Z$ when $H$ is small.

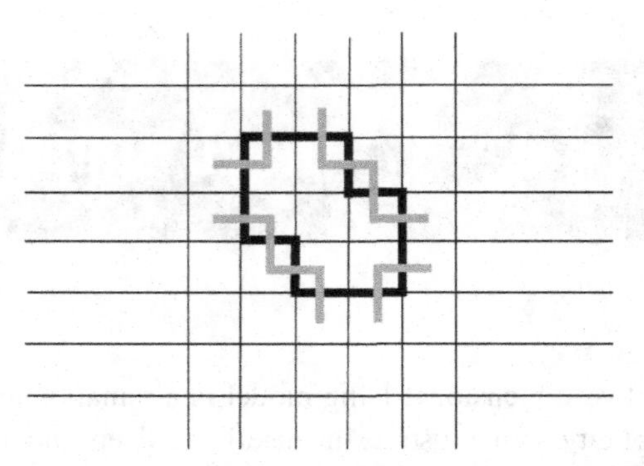

Figure 11.1 A closed path in the two-dimensional Ising model (black) and the corresponding "dual" links (gray).

## 11.2 Kramers–Wannier Duality

This section is devoted to illustrating a deep correspondence between strong and weak coupling descriptions, which we shall discuss in two cases that exhibit different types of behavior. In the first case both descriptions involve the Ising model on a two-dimensional square lattice, while in the second they relate two-dimensional Ising models on triangular and hexagonal lattices.

### 11.2.1 The Square Lattice

One can consider another Ising model, whose spins lie at the centers of the original plaquettes, in the so-called dual lattice, and the corresponding partition function reads

$$Z' = \sum_{\{\mu_i = \pm 1\}} e^{h \sum_{\langle ij \rangle} \mu_i \mu_j} . \tag{11.7}$$

As can be seen from Figure 11.1, closed contours drawn on the original lattice separate the dual lattice into two regions, depending on whether the new spins $\mu_i$ lie inside or outside of them. This suggests we organize the dual partition function with reference to this distinction, associating the value $\mu_i = +1$ to the dual spins inside the contours and $-1$ to those outside them. Hence, by construction,

$$Z' = \sum_{L} g(L)\, e^{-hL}\, e^{h\,(2N-L)}, \tag{11.8}$$

where the two factors take into account the $L$ negative contributions from nearby flipped spins and the $2N - L$ remaining positive ones. Note that this is a *low-temperature* (or *strong-coupling*) expansion, whose first few terms provide a good approximation for $Z'$ when $h$ is large.

Since the two lattices are of the same type, one can identify the nontrivial portions of the two partition functions of Eqs. (11.6) and (11.8) if the corresponding couplings are related according to

$$\tanh H = e^{-2h}. \tag{11.9}$$

Remarkably, this transformation, which, as one can verify, coincides with its own inverse, interchanges the roles of low and high temperatures, or, if you will, of strong and weak coupling: small $H$ translates into large $h$ and vice versa. Comparing the two complete expressions for $Z$ and $Z'$ and making use of Eq. (11.9), one can now conclude that

$$Z = \left[\sinh 2H\right]^N Z', \tag{11.10}$$

where the prefactor is analytic for any finite values of $H$ or $h$.

Moreover, if only one singularity existed, which should be the case since one is modeling the Curie transition, it ought to correspond to the fixed point of Eq. (11.9), in order that $Z$ and $Z'$ be simultaneously singular. One is thus led to the condition

$$\tanh H = e^{-2H} \tag{11.11}$$

for the critical temperature, and hence to the quadratic equation

$$e^{-4H} + 2\,e^{-2H} - 1 = 0, \tag{11.12}$$

whose real solution is

$$H_c = \frac{\beta_c v}{2} = \frac{1}{2}\log\left(1 + \sqrt{2}\right) \simeq 0.44, \tag{11.13}$$

or

$$\beta_c v = \log(1 + \sqrt{2}) \simeq 0.88. \tag{11.14}$$

In order to compare with the mean-field result, one should take into account that the definition of $v$ in Eq. (10.110) differs from the present one by a factor $2D$, i.e. by a factor of 4 in two dimensions, so that in the current notation

$$\beta_{\text{m.f.}}\, v = 0.25, \tag{11.15}$$

and the mean-field estimate for $T_c$ is thus higher by a factor 3.5, consistent with the fact that the mean-field method underestimates fluctuations.

Note that Eq. (11.9) can be recast in the equivalent symmetrical form

$$\sinh 2H \sinh 2h = 1, \tag{11.16}$$

which is reminiscent of what we saw for electric–magnetic duality in Problem 3.14.

### 11.2.2 Triangular and Hexagonal Lattices

Triangular and hexagonal lattices provide other interesting examples, since a weak–strong coupling duality connects them pairwise. To begin with, there are $L = 3N$ links in a triangular lattice $G$ with $N$ sites, since each site has 6 nearest neighbors

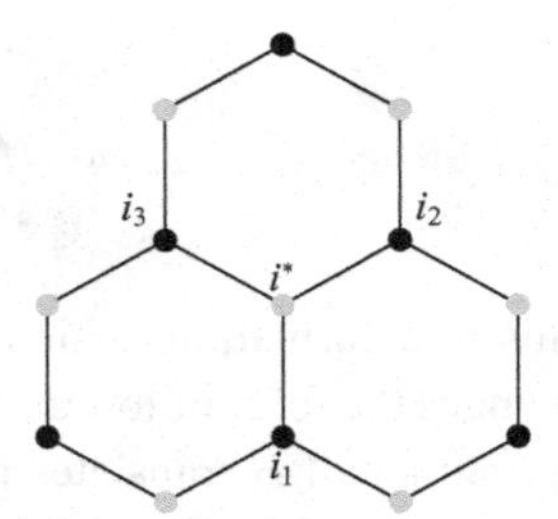


Figure 11.2 Decomposition of a hexagonal lattice into two triangular lattices (indicated by the black and gray points).

and each link connects two sites. The corresponding high-temperature expansion is therefore

$$Z_G(H_{(3)}, N) = (2 \cosh^3 H_{(3)})^N \sum_L g_3(L) (\tanh H_{(3)})^L, \tag{11.17}$$

where $g_3(L)$ is the number of polygons with perimeter $L$ that can be drawn on the triangular lattice.

Now consider a hexagonal lattice $\tilde{G}$ with $2N$ sites, which can be decomposed into a pair of triangular sub-lattices $G_1$ and $G_2$, corresponding to the black and gray points in Figure 11.2, in such a way that each site of $G_1$ has three sites of $G_2$ as nearest neighbors in $\tilde{G}$ and vice versa. In terms of these two sub-lattices, the partition function for the Ising model on the hexagonal lattice $\tilde{G}$ reads

$$Z_{\tilde{G}}(H_{(6)}, 2N) = \sum_C \sum_{C^*} \exp\Big(H_{(6)} \sum_{\langle i,i^* \rangle} \sigma_i \sigma_{i^*}\Big), \tag{11.18}$$

where $C$ and $i$ label the spin configurations and the sites on $G_1$, while $C^*$ and $i^*$ label those on $G_2$. The peculiar form of this expression makes it possible to relate the hexagonal and triangular partition functions via the identity

$$e^{a(\sigma_1+\sigma_2+\sigma_3)} + e^{-a(\sigma_1+\sigma_2+\sigma_3)} = F(a)\, e^{f(a)\,(\sigma_1\sigma_2+\sigma_2\sigma_3+\sigma_3\sigma_1)}, \tag{11.19}$$

where

$$\begin{aligned} f(a) &= \frac{1}{4} \log\left[1 + 4 \sinh^2 a\right], \\ F(a) &= 2 \cosh a \left[1 + 4 \sinh^2 a\right]^{\frac{1}{4}}. \end{aligned} \tag{11.20}$$

This result is a rearrangement of the product of the three simpler identities,

$$e^{a\sigma_i} = \cosh a + \sigma_i \sinh a, \qquad (i = 1, 2, 3), \tag{11.21}$$

and one can prove it by noting that

$$x = 1 + \sigma_1\sigma_2 + \sigma_2\sigma_3 + \sigma_3\sigma_1 \tag{11.22}$$

satisfies

$$x^2 = 4x. \tag{11.23}$$

Making use of Eq. (11.19) one can write

$$\begin{aligned} Z_{\tilde{G}}(H_{(6)}, 2N) &= \sum_{C}\sum_{C^*}\prod_{i^*} e^{H_{(6)}\,\sigma_{i^*}(\sigma_{i_1}+\sigma_{i_2}+\sigma_{i_3})} \\ &= \sum_{C}\prod_{i^*}\left[e^{H_{(6)}(\sigma_{i_1}+\sigma_{i_2}+\sigma_{i_3})} + e^{-H_{(6)}(\sigma_{i_1}+\sigma_{i_2}+\sigma_{i_3})}\right] \\ &= F(H_{(6)})^N \sum_{C} \exp\left[f(H_{(6)}) \sum_{\langle i,j\rangle} \sigma_i\sigma_j\right], \end{aligned} \tag{11.24}$$

where in the last line $i$ and $j$ refer to points in the $G_1$ triangular lattice, and any reference to the $G_2$ lattice has disappeared.

The final expression in Eq. (11.24) is the partition function for the Ising model defined on a triangular lattice $G$ with effective coupling $f(H_{(6)})$ and $N$ sites. In conclusion, we have reduced the partition function for the Ising model on an hexagonal lattice to the partition function on the triangular lattice with a modified coupling:

$$Z_{\tilde{G}}(H_{(6)}, 2N) = F(H_{(6)})^N Z_G(f(H_{(6)}), N). \tag{11.25}$$

This is a crucial step, since the low-temperature expansion for a hexagonal lattice $\tilde{G}$ with $2N$ sites is also a sum over polygons on the dual triangular lattice $G$ that enters Eq. (11.17),

$$Z_{\tilde{G}}(h_{(6)}, 2N) = e^{3Nh_{(6)}} \sum_{L} g_3(L)\, e^{-2L\,h_{(6)}}, \tag{11.26}$$

since in this case the number of links is $3N$.

Comparing Eqs. (11.25) and (11.26), one can conclude that the high-temperature expansion for the Ising model defined on the hexagonal lattice is linked to the low-temperature expansion on the dual triangular lattice via

$$\tanh f(H_{(6)}) \;=\; e^{-2\,h_{(6)}}, \tag{11.27}$$

or equivalently

$$\sinh\left[2 f(H_{(6)})\right]\, \sinh\left[2\,h_{(6)}\right] \;=\; 1. \tag{11.28}$$

The counterpart of the square-lattice self-duality condition,

$$\tanh f(H_{(6)}) \;=\; \frac{e^{2f(H_{(6)})} - 1}{e^{2f(H_{(6)})} + 1} \;=\; e^{-2\,H_{(6)}}, \tag{11.29}$$

now implies the simpler relation

$$\frac{e^{4f(H_{(6)})} + 1}{e^{4f(H_{(6)})} - 1} \;=\; \cosh 2\,H_{(6)}, \tag{11.30}$$

and finally, using Eq. (11.20),

$$\cosh 2H_{(6)}\,(\cosh 2H_{(6)} \;-\; 2) \;=\; 0. \tag{11.31}$$

The only available option, the vanishing of the second factor, translates into

$$e^{4H_{(6)}} - 4e^{2H_{(6)}} + 1 = 0. \tag{11.32}$$

This determines the critical temperature of the hexagonal lattice via

$$\beta_c^{(6)}\, v_6 = 2\,H_{(6)} = \log(2+\sqrt{3}) \simeq 1.32. \tag{11.33}$$

The critical temperature of the triangular lattice is instead determined by

$$\beta_c^{(3)}\, v_3 = 2 f(H_{(6)}) = \log(\sqrt{3}) \simeq 0.55. \tag{11.34}$$

Note that the relative ordering of the critical temperatures in Eqs. (11.33), (11.14), and (11.34) conforms to the intuitive idea that higher numbers of nearest neighbors favor the persistence of the ordered phase.

## 11.3 The Onsager Solution

Let us now turn to explain how one can compute exactly, in the thermodynamic limit,

$$\widetilde{Z} = \sum_L g(L)\, u^L \tag{11.35}$$

on a square lattice, where, in order to lighten the notation, we let

$$u = \tanh H \tag{11.36}$$

and

$$Z = \left[\frac{2}{1-u^2}\right]^N \widetilde{Z}. \tag{11.37}$$

$Z$ is the partition function (11.37) of the two-dimensional Ising model, while $\widetilde{Z}$ is a sum over closed polygons, and a first convenient step is the exponential map,

$$\widetilde{Z} = e^{\widetilde{\mathcal{A}}}. \tag{11.38}$$

This is a standard step in more advanced treatments of field theory: while $\widetilde{Z}$ is a sum over closed polygons, $\widetilde{\mathcal{A}}$ involves *connected* closed polygons only. As we shall see shortly, this choice can simplify, to a large extent, our current task.

The next issue is to build, via an iterative procedure, contributions to $\widetilde{\mathcal{A}}$ that are allowed in $\widetilde{Z}$ by the original Ising Hamiltonian in Eq. (11.1). These interactions involve at most quartic vertices at any given lattice point $P$ of coordinates $(i,j)$, since there are at most four contributions in the product of Eq. (11.4) that link the spin $\sigma_{i,j}$ to its four nearest neighbors. However, in this system one can actually trade quartic vertices for pairs of quadratic ones, which allows one to reinterpret any connected polygon in terms of closed paths. This process, however, is fraught with ambiguities: the example in Figure 11.3 shows clearly that different paths can be associated to a given vertex. Luckily, one can count paths by a recursive algorithm, and therefore it is crucial to resolve the ambiguity. One can show that this can be done, for the two-dimensional Ising model, with the following prescription:

1. a factor $\alpha = e^{\frac{i\pi}{4}}$ for every left turn;
2. a factor $\frac{1}{\alpha} = e^{-\frac{i\pi}{4}}$ for every right turn;

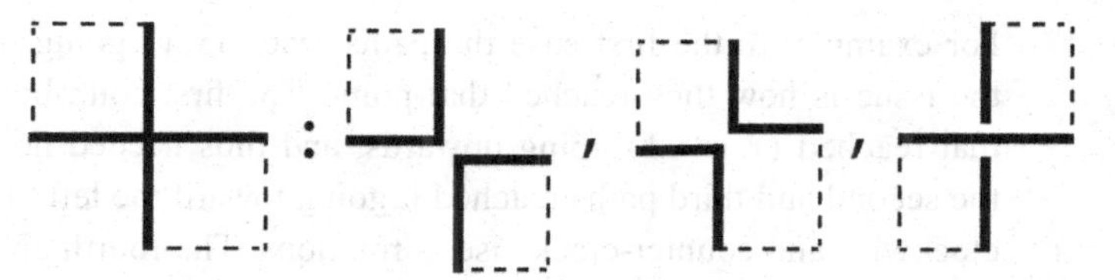

Figure 11.3 A quartic vertex affords three different interpretations in terms of paths. One should count only one path, not three, for any such contribution.

3. no backtracking;
4. a "minus" sign for every closed loop.

For example, the overall factors for the three paths in Figure 11.3 are

1. $(-1)^2_{\text{tangent}} \times (-1)^2_{\text{loop}} = 1,$
2. $(-1)_{\text{tangent}} \times (-1)_{\text{loop}} = 1,$
3. $(-1)^0_{\text{tangent}} \times (-1)_{\text{loop}} = -1,$

and the sum of the three contributions counts, as it should, as a single one, as pertains to the single polygon that led to the three distinct resolutions. These prescriptions assign to each path an overall sign $(-1)^n$, where $n$ is the number of self-intersections, and the exponential map in Eq. (11.38) then builds diagrams that are allowed by the Ising Hamiltonian, with two-point and four-point vertices and no multiple links. In the above example, the first two paths do not contain a self-intersection, and the last contains one.

With this proviso, one can recast Eq. (11.38) in terms of paths, and actually in terms of a more convenient set of objects, $W_r$, the set of paths built from a given starting point in $r$ steps, as

$$\widetilde{\mathcal{A}} = -N \sum_r \frac{u^r}{2r} \sum_{\mu=1}^{4} W_r(0,0;\mu). \tag{11.39}$$

The "minus" sign reflects the rule for closed contours, and the factor $\frac{1}{2r}$ takes into account a manifest overcounting: the choice of the starting point is arbitrary and there are two directions to cover a given closed path in $r$ steps. Moreover, there are in general four distinct components $W_r(x,y;\mu)$, which count the number of ways in which the path crosses the point $(x,y)$: going up ($\mu = 1$), down ($\mu = 2$), left ($\mu = 3$), or right ($\mu = 4$). In the thermodynamic limit, the choice of the origin for the paths is clearly immaterial, and is thus chosen to be $(0,0)$.

We can now define recursion relations for the four $W_r$. Recalling that $\alpha = e^{\frac{i\pi}{4}}$,

$$W_{r+1}(x,y;1) = W_r(x,y-1;1) + \frac{1}{\alpha}\,W_r(x,y-1;3) + \alpha\,W_r(x,y-1;4),$$

$$W_{r+1}(x,y;2) = W_r(x,y+1;2) + \alpha\,W_r(x,y+1;3) + \frac{1}{\alpha}\,W_r(x,y+1;4),$$

$$W_{r+1}(x,y;3) = \alpha\,W_r(x+1,y;1) + \frac{1}{\alpha}\,W_r(x+1,y;2) + W_r(x+1,y;3),$$

$$W_{r+1}(x,y;4) = \frac{1}{\alpha}\,W_r(x-1,y;1) + \alpha\,W_r(x-1,y;2) + W_r(x-1,y;4).$$

For example, in the first case the paths reach $(x, y)$ going up from $(x, y-1)$, and the issue is how they reached that point. The first contribution arises from a path that reached $(x, y-1)$ going upwards, and thus needed no further rotation, while the second and third paths reached it going toward the left or right, and thus needed clockwise and counter-clockwise $\frac{\pi}{2}$ rotations. The fourth option is missing because backtracking is not allowed.

One can recast this system in a more convenient matrix form, introducing the Fourier decomposition for a generic function $f$,

$$f(x,y) = \frac{1}{L^2} \sum_{(p,q)=0}^{L-1} e^{\frac{2\pi i}{L}(xp+yq)} \widetilde{f}(p,q), \tag{11.40}$$

whose inverse is

$$\widetilde{f}(p,q) = \sum_{(x,y)=0}^{L-1} e^{-\frac{2\pi i}{L}(xp+yq)} f(x,y), \tag{11.41}$$

since

$$\sum_{k=0}^{L-1} e^{\frac{2\pi i}{L} k(x-y)} = L\,\delta_{x,y}. \tag{11.42}$$

Letting

$$\sigma = e^{\frac{2\pi i}{L}}, \tag{11.43}$$

and

$$\widetilde{\mathbf{W}}_r = \begin{pmatrix} \widetilde{W}_r(p,q;1) \\ \widetilde{W}_r(p,q;2) \\ \widetilde{W}_r(p,q;3) \\ \widetilde{W}_r(p,q;4) \end{pmatrix}, \tag{11.44}$$

one is thus led to the recursion relation

$$\widetilde{\mathbf{W}}_{r+1} = \mathbf{\Lambda}(p,q)\,\widetilde{\mathbf{W}}_r, \tag{11.45}$$

where the matrix $\Lambda(p,q)$ is

$$\Lambda(p,q) = \begin{pmatrix} \frac{1}{\sigma^q} & 0 & \frac{1}{\alpha\,\sigma^q} & \frac{\alpha}{\sigma^q} \\ 0 & \sigma^q & \alpha\,\sigma^q & \frac{\sigma^q}{\alpha} \\ \alpha\sigma^p & \frac{\sigma^p}{\alpha} & \sigma^p & 0 \\ \frac{1}{\alpha\,\sigma^p} & \frac{\alpha}{\sigma^p} & 0 & \frac{1}{\sigma^p} \end{pmatrix}. \tag{11.46}$$

Therefore

$$\widetilde{\mathbf{W}}_r = \mathbf{\Lambda}^r(p,q)\,\widetilde{\mathbf{W}}_0, \tag{11.47}$$

and the four choices for $\widetilde{\mathbf{W}}_0$ correspond to the four possible ways of crossing the origin,

$$\begin{pmatrix} 1 \\ 0 \\ 0 \\ 0 \end{pmatrix}, \begin{pmatrix} 0 \\ 1 \\ 0 \\ 0 \end{pmatrix}, \begin{pmatrix} 0 \\ 0 \\ 1 \\ 0 \end{pmatrix}, \begin{pmatrix} 0 \\ 0 \\ 0 \\ 1 \end{pmatrix}, \tag{11.48}$$

so that, in conclusion ($N = L^2$),

$$\widetilde{\mathcal{A}} = -N \sum_r \frac{u^r}{2r} \sum_{p,q} \mathrm{Tr}\, \mathbf{\Lambda}^r(p,q). \tag{11.49}$$

The series can easily be summed, using a familiar relation for determinants

$$\begin{aligned} \widetilde{\mathcal{A}} &= \frac{N}{2} \sum_{p,q} \mathrm{Tr} \log \left[1 - u\, \mathbf{\Lambda}(p,q)\right] \\ &= \frac{N}{2} \sum_{p,q} \log \det \left[1 - u\, \mathbf{\Lambda}(p,q)\right], \end{aligned} \tag{11.50}$$

and one can verify that the determinant of interest is

$$\left(1 + u^2\right)^2 - 2u(1 - u^2) \left[\cos\left(\frac{2\pi p}{L}\right) + \cos\left(\frac{2\pi q}{L}\right)\right]. \tag{11.51}$$

In the thermodynamic limit the double series turns into a double integral. Rescaling the variables $p$ and $q$ so that their range becomes $[0, 2\pi]$, and combining this result with Eq. (11.37) yields the Helmholtz free energy

$$\begin{aligned} \frac{\beta A}{N} &= \log\left(\frac{1 - u^2}{2}\right) \\ &- \frac{1}{2} \int_0^{2\pi} \int_0^{2\pi} \frac{dp\, dq}{(2\pi)^2} \log\left[\left(1 + u^2\right)^2 - 2u(1 - u^2)(\cos p + \cos q)\right]. \end{aligned} \tag{11.52}$$

Note that this expression can be recast in the more convenient form

$$\frac{\beta A}{N} = -\frac{1}{2} \int_0^{2\pi} \int_0^{2\pi} \frac{dp\, dq}{(2\pi)^2} \log\left[4y\left(y + \frac{1}{y} - \cos p - \cos q\right)\right], \tag{11.53}$$

where

$$y = \sinh 2H = \sinh \beta v, \tag{11.54}$$

so that the invariance under Kramers–Wannier duality of the nontrivial portion of $A$ becomes manifest. From this expression one can now obtain the internal energy density

$$u = -\frac{v\sqrt{1 + y^2}}{2y}\left[1 + \int_0^{2\pi} \int_0^{2\pi} \frac{dp\, dq}{(2\pi)^2} \frac{y - \frac{1}{y}}{y + \frac{1}{y} - \cos p - \cos q}\right], \tag{11.55}$$

whose singular behavior arises from the origin, where the denominator vanishes at $T_c$. One can therefore make the singularity manifest by expanding the trigonometric functions, while also deforming the integration region to a circular one, with an upper bound that will play no role. These simplifying steps lead to

$$u_{\text{sing}} = -\frac{v\sqrt{2}}{4\pi} \int_0^a d\xi\, \frac{y - \frac{1}{y}}{y + \frac{1}{y} - 2 + \xi}, \tag{11.56}$$

and the singular part can finally be extracted from

$$u_{\text{sing}} = \frac{v\sqrt{2}}{2\pi} (y - 1) \log (y - 1)^2, \tag{11.57}$$

where we have dropped some contributions that are regular at $y = 1$. Close to the critical temperature

$$y \simeq 1 + 2\sqrt{2}H_c\left(1 - \frac{T}{T_c}\right), \tag{11.58}$$

and finally $C_V$ is logarithmically divergent at the critical temperature,

$$C_V \simeq -\frac{4\,k_B\,H_c^2}{\pi}\log\left(1 - \frac{T}{T_c}\right)^2 \simeq -\frac{k_B}{4}\log\left(1 - \frac{T}{T_c}\right)^2, \tag{11.59}$$

where $H_c$ is defined in Eq. (11.13). Note that the corresponding critical exponent $\alpha$ vanishes, as in the mean-field treatment of Chapter 10, but the specific heat diverges logarithmically at $T_c$.

## Bibliographical Notes

Detailed discussions of Kramers–Wannier duality can be found in many excellent books, including [6, 25, 36, 43, 45, 53]. Discussions of the Onsager solution along the lines of the one presented here can be found in [17, 36], while a discussion based on the transfer matrix can be found in [23].

# 12 The Heisenberg Spin Chain

This chapter is meant as an initial look into some widely used techniques for the exact solution of many-body systems. Only for a few of them can one determine the exact spectrum and thermodynamic functions in the large-volume limit, without resorting to approximations at one stage or another, and the cases that afford exact solutions are thus key beacons for further progress.

Noninteracting many-body systems are clearly the simplest instances of exactly solvable ones. Although they constitute a somewhat degenerate case, we briefly reconsider them in the next section to prepare the ground for the ensuing discussion. We then illustrate some aspects and techniques arising in the study of interacting systems, focusing on a physically relevant and technically nontrivial example. This is the one-dimensional quantum Heisenberg model, which describes isotropic nearest–neighbor interactions of quantum spins lying on a one-dimensional lattice. In this sense, it is a quantum counterpart of the classical Ising spin chain discussed in Chapter 10. We first illustrate the solution of the corresponding spectral problem by means of the coordinate Bethe ansatz, and then briefly address its thermal behavior relying on the thermodynamic Bethe ansatz.

## 12.1 Noninteracting Systems

Let us begin our analysis of integrable systems by referring to their simplest possible incarnation: noninteracting identical particles on a one-dimensional lattice. This discussion will allow us to reconsider several facts touched upon in previous chapters, while also preparing the ground for the ensuing analysis of the Heisenberg spin chain.

### 12.1.1 Single-particle States

Let us consider a one-dimensional lattice with $L$ sites, with unit lattice spacing, for simplicity. Let us also denote by $|n\rangle$ a state in which the particle is localized at the $n$th site, with $n = 1, 2, \ldots, L$, so that for any constant $a$

$$e^{iaX}|n\rangle = |n\rangle e^{ian}, \tag{12.1}$$

where $X$ is the position operator. On the other hand, in view of the canonical commutation relations, the momentum operator $P$ obeys the identity

$$e^{iaX}e^{i\frac{bP}{\hbar}} = e^{i\frac{bP}{\hbar}}e^{iaX}e^{-i\,ab}, \tag{12.2}$$

for any constant $b$, and therefore

$$e^{iaX}e^{i\frac{bP}{\hbar}}|n\rangle = e^{i\frac{bP}{\hbar}}|n\rangle e^{ia(n-b)}. \tag{12.3}$$

Since $e^{i\frac{bP}{\hbar}}$ is unitary and the eigenspaces of $e^{iaX}$ are one-dimensional, this shows that it shifts the position of the particle by $-b$ sites,

$$e^{i\frac{bP}{\hbar}}|n\rangle = |n-b\rangle, \tag{12.4}$$

which leads us to consider integer values for $b$.

Let us now consider closed chains, so that the states satisfy the periodic boundary conditions

$$|n+L\rangle \equiv |n\rangle. \tag{12.5}$$

The operator that translates the particle by $L$ sites, or integer multiples thereof, should act as the identity,

$$e^{i\frac{LP}{\hbar}} = 1, \tag{12.6}$$

so that the states defined by

$$|\psi_k\rangle = \sum_{n=1}^{L} e^{ink}|n\rangle \tag{12.7}$$

satisfy

$$e^{imP/\hbar}|\psi_k\rangle = |\psi_k\rangle e^{ikm}. \tag{12.8}$$

They are thus momentum eigenstates with eigenvalue $\hbar k$, and the periodicity condition (12.6) requires that

$$e^{ikL} = 1, \tag{12.9}$$

so that the allowed values of $k$ are

$$k_\ell = \frac{2\pi\ell}{L}, \qquad \ell = 0, 1, \ldots, L-1. \tag{12.10}$$

As we shall see shortly, these integers $\ell$ can be regarded as the simplest instance of Bethe quantum numbers.

### 12.1.2 Free Fermions

Let us consider a gas of noninteracting identical "spinless fermions" that can move on the chain, with single-particle energies $E(k_\ell)$ depending on the momentum $k_\ell$. The total energy is determined by how many Fermions $n(k_\ell)$ occupy each of the available single-particle states $k_\ell$ and reads

$$E = \sum_{\ell=1}^{L} E(k_\ell)n(k_\ell). \tag{12.11}$$

Of course, each $n(k_\ell)$ can be either 0, if the state is empty, or 1, if it is filled, in view of the Pauli exclusion principle. As is often done in condensed matter physics, we find it convenient here to say that a state not hosting a fermion is occupied by a *hole*, so that the occupation numbers for holes, $\bar{n}(k_\ell)$, satisfy the conditions

$$n(k_\ell) + \bar{n}(k_\ell) = 1. \tag{12.12}$$

In the large-$L$ limit, as the chain becomes very long, the available single-particle states become dense in the momentum interval $0 \le k_\ell \le 2\pi$, and the system can be treated in the continuum limit. The density of available single-particle states is then determined by the derivative of Eq. (12.10) so that

$$\rho_s(k_\ell) = \frac{1}{L}\frac{d\ell}{dk_\ell} = \frac{1}{2\pi}. \tag{12.13}$$

Correspondingly, the densities of particles, $\rho(k)$, of particles and holes, $\bar{\rho}(k)$, are

$$\rho(k_\ell) = n(k_\ell)\,\rho_s(k_\ell), \qquad \bar{\rho}(k_\ell) = \bar{n}(k_\ell)\,\rho_s(k_\ell), \tag{12.14}$$

and satisfy

$$\rho(k) + \bar{\rho}(k) = \rho_s(k). \tag{12.15}$$

Let us now turn to the thermodynamics of this system. The number of ways of placing $dN = \rho(k)L\,dk$ particles and $d\bar{N} = \bar{\rho}(k)L\,dk$ holes in the $dN_s = \rho_s(k)L\,dk$ available states in a small region near $k$ in momentum space is formally determined by the binomial coefficient (see Appendix B)

$$\binom{dN_s}{dN} = \frac{[\rho_s(k)L\,dk]!}{[\rho(k)L\,dk]!\,[\bar{\rho}(k)L\,dk]!}. \tag{12.16}$$

As a result, the contribution to the total entropy from this small region can be extracted from the Stirling approximation, and reads

$$dS = k_B[\rho_s(k)\log\rho_s(k) - \rho(k)\log\rho(k) - \bar{\rho}(k)\log\bar{\rho}(k)]L\,dk, \tag{12.17}$$

since for $L \gg 1$ the number of such states is very large. The internal energy receives a contribution

$$dU = E(k)\,dN = \rho(k)E(k)L\,dk, \tag{12.18}$$

while the grand-potential is given by

$$d\Omega = dU - T\,dS - \mu\,dN. \tag{12.19}$$

Demanding that the total grand-potential,

$$\Omega = \int_0^{2\pi} \frac{d\Omega}{dk}\,dk, \tag{12.20}$$

be stationary, one finds the condition

$$\int_0^{2\pi} \Big[\beta(E(k)-\mu)\delta\rho(k) - \delta\rho_s(k)\log\rho_s(k) + \delta\rho(k)\log\rho(k) + \delta\bar{\rho}(k)\log\bar{\rho}(k) \\ - \delta\left(\rho_s(k) - \rho(k) - \bar{\rho}(k)\right)\Big]dk = 0. \tag{12.21}$$

Taking into account that $\rho_s(k)$ is in fact a constant and that the constraint (12.15) implies

$$\delta\rho(k) + \delta\bar{\rho}(k) = 0, \tag{12.22}$$

Eq. (12.21) then leads to

$$\beta\left(E(k) - \mu\right) + \log\frac{\rho(k)}{\bar{\rho}(k)} = 0. \tag{12.23}$$

Solving this equation for $\bar{\rho}(k)$ and substituting in the constraint (12.15), one recovers once more the familiar Fermi–Dirac distribution for the equilibrium particle density

$$\rho(k) = \frac{1}{2\pi}\frac{1}{e^{\beta(E(k)-\mu)} + 1}. \tag{12.24}$$

Before concluding this preliminary discussion, let us mention that Bose particles can be treated in a similar way, taking into account, however, that the occupation numbers may be arbitrarily large $n(k_\ell) = 0, 1, 2, \ldots$. Therefore the concept of a hole does not apply and the expression (12.16) for the number of available microstates should be replaced by

$$\frac{[(\rho_s(k) + \rho(k) - 1)L\,dk]!}{[\rho(k)L\,dk]!\,[(\rho_s(k) - 1)L\,dk]!}. \tag{12.25}$$

Retracing the familiar steps (see Appendix B), these considerations lead to the Bose–Einstein distribution

$$\rho(k) = \frac{1}{2\pi}\frac{1}{e^{\beta(E(k)-\mu)} - 1}. \tag{12.26}$$

## 12.2 The Spectrum of the Heisenberg Model

We can now turn to the interacting system of interest, a chain whose $L$ sites are occupied by quantum spin variables $\mathbf{S}_n = (S_n^x, S_n^y, S_n^z)$, where the subscript $n$ specifies the lattice site. These variables satisfy the commutation relations

$$[S_n^i, S_m^j] = i\hbar\,\epsilon^{ijk}\delta_{nm}S_n^k. \tag{12.27}$$

We shall focus on the spin-$\frac{1}{2}$ case, so that

$$\mathbf{S}_n^2 = \frac{3\hbar^2}{4}\begin{pmatrix}1 & 0\\ 0 & 1\end{pmatrix}_n \tag{12.28}$$

and the spin components can be represented in terms of the Pauli matrices

$$S_n^x = \frac{\hbar}{2}\begin{pmatrix}0 & 1\\ 1 & 0\end{pmatrix}_n \qquad S_n^y = \frac{\hbar}{2}\begin{pmatrix}0 & -i\\ i & 0\end{pmatrix}_n \qquad S_n^z = \frac{\hbar}{2}\begin{pmatrix}1 & 0\\ 0 & -1\end{pmatrix}_n, \tag{12.29}$$

which act on two-dimensional spaces associated to the corresponding sites.

Let us now consider the Heisenberg Hamiltonian

$$H = -J\sum_{n=1}^{L}\mathbf{S}_n\cdot\mathbf{S}_{n+1} + \frac{\hbar^2 JL}{4}, \tag{12.30}$$

where the constant term is added for convenience, with periodic boundary conditions

$$\mathbf{S}_{n+L} = \mathbf{S}_n. \tag{12.31}$$

This is usually referred to as the Heisenberg *XXX* chain in the integrability literature, where the three identical labels are meant to stress that the *J*-couplings of the three spin components are identical.

We can now present a detailed analysis of the energy eigenvalues of this Hamiltonian and of the corresponding eigenstates, following a strategy known as the coordinate Bethe ansatz. To this end, it is convenient to introduce the combinations

$$S_n^{\pm} = S_n^x \pm i S_n^y, \tag{12.32}$$

which satisfy

$$[S_n^z, S_n^{\pm}] = \pm\hbar S_n^{\pm}, \qquad [S_n^+, S_n^-] = 2\hbar S_n^z \tag{12.33}$$

and have the matrix representation

$$S_n^+ = \hbar \begin{pmatrix} 0 & 1 \\ 0 & 0 \end{pmatrix}_n \qquad S_n^- = \hbar \begin{pmatrix} 0 & 0 \\ 1 & 0 \end{pmatrix}_n. \tag{12.34}$$

Here $S_n^+$ is a spin-raising operator for the *n*th site while $S_n^-$ is the corresponding spin-lowering operator. Note that these operators are nilpotent,

$$(S_n^+)^2 = 0, \qquad (S_n^-)^2 = 0, \tag{12.35}$$

while, for example,

$$[H, S_n^-] = -\hbar J(S_{n-1}^- S_n^z + S_n^z S_{n+1}^-) + \hbar J(S_{n-1}^z S_n^- + S_n^- S_{n+1}^z). \tag{12.36}$$

The Hamiltonian commutes with the total spin

$$\mathbf{S} = \sum_{n=1}^{L} \mathbf{S}_n, \tag{12.37}$$

which is the generator of spin rotations, and also with the translation operator defined by

$$e^{i\frac{mP}{\hbar}}\, \mathbf{S}_n\, e^{-i\frac{mP}{\hbar}} = \mathbf{S}_{n-m}. \tag{12.38}$$

To reiterate

$$[H, \mathbf{S}] = 0, \qquad e^{imP/\hbar} H e^{-imP/\hbar} = H, \tag{12.39}$$

and in view of these rotational and translational symmetries we begin by considering the translation-invariant state $|\Omega\rangle$ in which all spins are aligned, say, along the positive *z* direction, so that

$$S_n^z |\Omega\rangle = \frac{\hbar}{2} |\Omega\rangle, \qquad n = 1, 2, \ldots, L. \tag{12.40}$$

This is a state of zero energy,

$$H|\Omega\rangle = 0, \tag{12.41}$$

in view of the additive constant introduced in Eq. (12.30). Let us also note that, from the point of view of the $su(2)$ symmetry defined by the total spin (12.37), this state satisfies

$$S^+|\Omega\rangle = \sum_{n=1}^{L} S_n^+|\Omega\rangle = 0, \tag{12.42}$$

and is therefore called a *highest-weight state*. It belongs to a spin-$\frac{L}{2}$ representation, since

$$\mathbf{S}^2|\Omega\rangle = |\Omega\rangle\,\hbar^2\frac{L}{2}\left(\frac{L}{2}+1\right), \tag{12.43}$$

as can be seen for instance using $\mathbf{S}^2 = S^-S^+ + S^z(S^z+1)$. The other members can be obtained by repeatedly applying the total spin-lowering operator $S^-$.

Since the operator

$$\frac{\hbar L}{2} - S^z \tag{12.44}$$

defines another conserved quantity, one can diagonalize the Hamiltonian by considering separately the subspaces labeled by its allowed eigenvalues $\hbar r$. The value of $r$ is the number of spins that are flipped to the down position when starting from the state $|\Omega\rangle$, which has $r = 0$. These eigenspaces are reached by acting repeatedly on $|\Omega\rangle$ with the spin-lowering operators $S_n^-$. Finally, letting

$$|n_1,\ldots,n_r\rangle = S_{n_1}^- S_{n_2}^- \cdots S_{n_r}^-|\Omega\rangle, \tag{12.45}$$

one finds that

$$\left(\frac{\hbar L}{2} - S^z\right)|n_1,\ldots,n_r\rangle = \hbar r\,|n_1,\ldots,n_r\rangle. \tag{12.46}$$

Note that at most one spin flip can occur at a given site, because otherwise the corresponding states would vanish, due to Eq. (12.35).

### 12.2.1 Quasi-Particle States

Let us begin from the $r = 1$ subspace. A generic $|\psi\rangle$ in this subspace is a linear combination of the states

$$|n\rangle = S_n^-|\Omega\rangle, \tag{12.47}$$

obtained by performing on $|\Omega\rangle$ a single spin flip,

$$|\psi\rangle = \sum_{n=1}^{L} a_n|n\rangle. \tag{12.48}$$

Compatibility with the periodic boundary conditions,

$$|n+L\rangle = |n\rangle, \tag{12.49}$$

implies that one can define the $a_n$ beyond the range $n = 0, 1, \ldots, L-1$ via the identifications

$$a_{n+L} = a_n, \qquad a_{n-L} = a_n. \tag{12.50}$$

Before addressing the eigenvalue equation, let us note that the rotational symmetry of $H$ implies that this subspace contains a zero-energy eigenstate, which is obtained by applying to $|\Omega\rangle$ the total spin-lowering operator,

$$|\psi_0\rangle = S^- |\Omega\rangle = \sum_{n=1}^{L} |n\rangle, \tag{12.51}$$

up to a suitable normalization. This state is clearly invariant under translations, and therefore has zero momentum. In view of the above relation, $|\psi_0\rangle$ is called a *descendant*, and satisfies

$$S^+ |\psi_0\rangle = |\Omega\rangle \hbar^2 L \neq 0. \tag{12.52}$$

In other words, it is not a highest-weight state for the total $su(2)$ symmetry. However, there are other energy eigenstates with $r = 1$ aside from this descendant. To find them, let us resort to the commutation relation (12.36) to recast the eigenvalue equation

$$H|\psi\rangle = E|\psi\rangle \tag{12.53}$$

in the form of the recursion relations

$$\hbar^2 J(2a_n - a_{n-1} - a_{n+1}) = 2Ea_n. \tag{12.54}$$

This is actually a linear difference equation with constant coefficients, whose left-hand side involves a discretized second derivative. An equation of this type can be solved by letting

$$a_n = e^{ink}, \tag{12.55}$$

where $a_n$ is subject to the boundary conditions (12.50).

This leads to

$$k = \frac{2\pi\ell}{L}, \qquad E(k) = \hbar^2 J(1 - \cos k), \tag{12.56}$$

for $\ell = 0, 1, \ldots, L - 1$. The result is the full set of $r = 1$ eigenstates

$$|\psi_k\rangle = \sum_{n=1}^{L} e^{ink} |n\rangle, \tag{12.57}$$

which includes the descendant for $k = 0$. We know that Eq. (12.55) was able to capture the full set of solutions, because the dimension of the $r = 1$ space is clearly $L$ and we have found $L$ independent eigenstates. These states are called "magnons," and identify quasi-particles, in analogy with the discussion of one-particle states presented in the previous section.

Taking into account the invariance of $|\Omega\rangle$ under translations

$$e^{imP/\hbar} |n\rangle = |n - m\rangle, \tag{12.58}$$

and hence

$$e^{imP/\hbar} |\psi_k\rangle = |\psi_k\rangle \, e^{imk}. \tag{12.59}$$

Magnon states thus represent spin waves that travel along the chain with a well-defined momentum $\hbar k$. Note also that for $k \neq 0$ the states $|\psi_k\rangle$ are highest-weight states, since

$$S^+|\psi_k\rangle = |\Omega\rangle\hbar^2 \sum_{n=1}^{L} e^{ink} = |\Omega\rangle\hbar^2 e^{ik}\frac{1-e^{iLk}}{1-e^{ik}} = 0. \tag{12.60}$$

Therefore $|\psi_0\rangle$ is the only descendant belonging to the $r = 1$ sector, and furthermore each of the $L-1$ states with $k \neq 0$ gives rise to a spin-$\left(\frac{L}{2}-1\right)$ representation for the total spin, via repeated applications of the total spin-lowering operator.

### 12.2.2 The Two-Body Spectrum

Consider now the $r = 2$ subspace, spanned by the vectors

$$|n_1, n_2\rangle = S^-_{n_1} S^-_{n_2} |\Omega\rangle. \tag{12.61}$$

Note that $|n, n\rangle = 0$, and the order of the labels in each state is irrelevant, since $|n_1, n_2\rangle = |n_2, n_1\rangle$. Therefore, the most general state belonging to the $r = 2$ sector takes the form

$$|\psi\rangle = \sum_{1 \leq n_1 < n_2 \leq L} a_{n_1,n_2} |n_1, n_2\rangle, \tag{12.62}$$

where the ordering on the labels in $a_{n_1,n_2}$ restricts the sum to the $\frac{L(L-1)}{2}$ states that, in view of the preceding considerations, are actually independent. In order to comply with this convention and with the periodic boundary conditions

$$|n_1 + L, n_2\rangle = |n_1, n_2\rangle, \qquad |n_1, n_2 + L\rangle = |n_1, n_2\rangle, \tag{12.63}$$

we extend the definition of $a_{n_1,n_2}$ letting, for $1 \leq n_1 < n_2 \leq L$,

$$\begin{aligned} a_{n_1,n_2+L} &= a_{n_1,n_2}, \qquad & a_{n_1-L,n_2} &= a_{n_1,n_2}, \\ a_{n_2,n_1+L} &= a_{n_1,n_2}, \qquad & a_{n_2-L,n_1} &= a_{n_1,n_2}. \end{aligned} \tag{12.64}$$

As before, a first set of eigenstates is already known by symmetry considerations: they are the descendants obtained acting with the spin-lowering operator on magnon states

$$S^-|\psi_k\rangle = 2 \sum_{1 \leq n_1 < n_2 \leq L} e^{in_1 k} |n_1, n_2\rangle, \tag{12.65}$$

where $k = \frac{2\pi\ell}{L}$ for $\ell = 0, 1, \ldots, L-1$. These solutions represent the superposition of a magnon with zero momentum and a magnon with momentum $k$. The rationale behind their existence is that, when a moving spin wave hits one at rest, their roles are simply interchanged without any energy expense, so that the overall configuration is stationary.

More general eigenstates can be identified by returning to the eigenvalue equation (12.53), which now gives rise to two distinct sets of recursion relations for the coefficients $a_{n_1,n_2}$:

$$\begin{aligned} &\hbar^2 J(2a_{n_1,n_2} - a_{n_1+1,n_2} - a_{n_1-1,n_2}) \\ &\quad + \hbar^2 J(2a_{n_1,n_2} - a_{n_1,n_2+1} - a_{n_1,n_2-1}) = 2E a_{n_1,n_2}, \end{aligned} \tag{12.66}$$

if $n_2 > n_1 + 1$, together with

$$\hbar^2 J(2a_{n,n+1} - a_{n,n+2} - a_{n-1,n+1}) = 2E a_{n,n+1}. \tag{12.67}$$

The first relation (12.66) can be solved by a factorized ansatz of the type $a_{n_1,n_2} = e^{ik_1 n_1} e^{ik_2 n_2}$, which gives

$$E = \hbar^2 J(1 - \cos k_1) + \hbar^2 J(1 - \cos k_2) = E(k_1) + E(k_2). \tag{12.68}$$

However, in general the second relation (12.67) does not admit solutions of this form. We therefore consider a linear combination of solutions of the above type: the *Bethe ansatz* coefficients (or wavefunctions),

$$a_{n_1,n_2} = A_{12} e^{ik_1 n_1 + ik_2 n_2} + A_{21} e^{ik_2 n_1 + ik_1 n_2}, \tag{12.69}$$

which clearly still solve (12.66) with the eigenvalue (12.68). Additionally, enforcing Eq. (12.67) now demands that the coefficients $A_{12}$ and $A_{21}$ satisfy

$$A_{12}\left(1 - 2e^{ik_2} + e^{i(k_1+k_2)}\right) + A_{21}\left(1 - 2e^{ik_1} + e^{i(k_1+k_2)}\right) = 0, \tag{12.70}$$

while periodic boundary conditions, in the form (12.64), require in addition that

$$A_{21}\, e^{ik_1 L} = A_{12}, \qquad A_{12}\, e^{ik_2 L} = A_{21}. \tag{12.71}$$

Equations (12.70) and (12.71) are the *Bethe Ansatz Equations* (BAE) for the $r = 2$ sector. In particular, they are satisfied by the descendants (12.65) with $k_1 = \frac{2\pi\ell_1}{L}$ for $\ell_1 = 0, 1, \ldots, L-1$ and $k_2 = 0$, and in this case $A_{12} = A_{21}$.

These equations are usually written in terms of the parameter $\theta_{12}$ defined by

$$e^{i\theta_{12}} = \frac{A_{12}}{A_{21}}, \qquad \mathrm{Re}\,(\theta_{12}) \in (-\pi, \pi], \tag{12.72}$$

so that they take the form

$$e^{i\theta_{12}} = -\frac{1 - 2e^{ik_1} + e^{i(k_1+k_2)}}{1 - 2e^{ik_2} + e^{i(k_1+k_2)}}, \tag{12.73}$$

provided the denominator is nonzero ($k_1, k_2 \neq 0$), and

$$e^{ik_1 L} = e^{i\theta_{12}}, \qquad e^{ik_2 L} = e^{i\theta_{21}}, \tag{12.74}$$

where, away from $\pi$, $\theta_{21} = -\theta_{12}$. Nontrivial values of $\theta_{12}$ reflect the presence of interactions, while the spectrum remains additive, as in Eq. (12.68).

Equations (12.74) determine the allowed $k_1$ and $k_2$, and can be recast in a form reminiscent of the expressions for allowed one-particle momenta

$$k_1 = \frac{2\pi\ell_1}{L} + \frac{\theta_{12}}{L}, \qquad k_2 = \frac{2\pi\ell_2}{L} + \frac{\theta_{21}}{L}, \tag{12.75}$$

with $\ell_1, \ell_2 = 0, 1, \ldots, L-1$. For this reason, $k_1$ and $k_2$ are termed *quasi-momenta* and the integers $\ell_{1,2}$ represent the allowed Bethe quantum numbers. One can also rewrite the Bethe ansatz (12.69) as

$$a_{n_1,n_2} = e^{ik_1n_1+ik_2n_2+\frac{i\theta_{12}}{2}} + e^{ik_2n_1+ik_1n_2+\frac{i\theta_{21}}{2}}, \tag{12.76}$$

up to a normalization constant.

This form of the coefficients $a_{n_1,n_2}$ and the expression (12.68) for the total energy suggest to interpret the Bethe ansatz wavefunction as representing the scattering of two magnons with energies $E(k_1)$ and $E(k_2)$ given by the single-particle dispersion relation (12.56), which takes into account a phase shift $\theta_{12}$ due to their interaction. Note, however, that in general the allowed quasi-momenta $k_1$ and $k_2$ are no longer given by a simple expression of the type (12.56), but must be determined solving the Bethe ansatz equations. They may even be complex numbers, as we shall see shortly. The total momentum, however, given by $\hbar k$ with

$$k = k_1 + k_2 = \frac{2\pi(\ell_1 + \ell_2)}{L}, \tag{12.77}$$

has the same form as for a free particle, owing to translational invariance, and the same can be said for descendant states, as we have seen.

An important feature of the Bethe ansatz equations is that, if $k_1 = k_2 \neq 0$, then by (12.70) $A_{12} = -A_{21}$ and $a_{n_1,n_2} = 0$, as follows from Eq. (12.69), so that *the two magnons cannot have identical nonzero quasi-momenta*. This can be regarded as an analog of the Pauli exclusion principle for quasi-particles. However, we have also seen that the descendant state with $k_1 = k_2 = 0$ and $A_{12} = A_{21}$ does actually provide an eigenvector, as it solves (12.70) and (12.71), and is therefore an "exception" to this rule. The physical picture that emerges is that two quasi-particles cannot move around the chain with the same quasi-momentum, while they are allowed to be simultaneously at rest.

The BAE (12.73) and (12.74) can be used to determine the two-body spectrum of the Hamiltonian in terms of allowed quasi-momenta and phase shifts. While some special solutions of these equations can be found analytically, for instance in the case of descendants, exhibiting the complete set of solutions typically requires the use of numerical methods. From a more abstract standpoint, one may wonder whether the Bethe ansatz captures all $L(L-1)/2$ independent solutions of the secular equation: we have given above the explicit expression for the $L$ descendants, but we have not shown that the remaining solutions do provide the missing $L(L-3)/2$ eigenstates (after all, the Bethe ansatz is just an ansatz!). This issue goes under the name of completeness problem, and has been fully clarified in the Heisenberg model: a theorem guarantees that the Bethe ansatz does determine the full set of solutions. We shall return briefly to this point in Section 12.2.3.

Fortunately, the analysis of the solutions simplifies in the thermodynamic limit $L \to \infty$, as we now illustrate. If $k_1$ and $k_2$ are real, one can neglect the terms $\pm\theta_{12}/L$ on the right-hand sides of Eqs. (12.75) when $L$ becomes very large. Therefore, the two quasi-momenta essentially become free momenta and the corresponding energy, which one may recast in the form

$$E = 2\hbar^2 J\left(1 - \cos\frac{k}{2}\,\cos\frac{\Delta k}{2}\right), \tag{12.78}$$

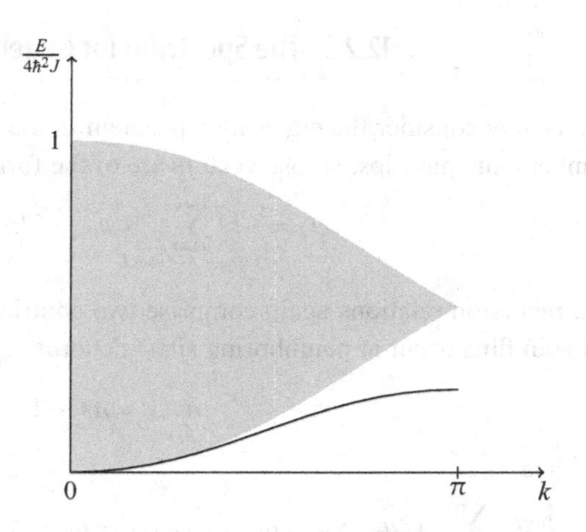


Figure 12.1 A schematic representation of the two-body spectrum for large $L$. The shaded area represents the continuous band of allowed free quasi-particle states, while the solid line corresponds to a bound state of two magnons.

with $k = k_1 + k_2$ and $\Delta k = k_1 - k_2$, describes a continuous band delimited by the curves $2\hbar^2 J(1 \pm \cos \frac{k}{2})$, as depicted in Figure 12.1.

If one instead allows complex $k_1$ or $k_2$, clearly one can no longer consistently neglect $\theta_{12}/L$, even as $L \to \infty$, so that the phase shift $\theta_{12}$ must develop a singularity in this limit. In order to highlight this phenomenon, let us recast expression (12.73) for the phase shift in the form

$$-e^{i\theta_{12}} = \frac{\cos \frac{k}{2} - e^{\frac{i\Delta k}{2}}}{\cos \frac{k}{2} - e^{-\frac{i\Delta k}{2}}}. \tag{12.79}$$

Singularities in the phase shift occur if either the numerator or the denominator of expression (12.79) vanish, so that

$$e^{\pm \frac{i\Delta k}{2}} = \cos \frac{k}{2}. \tag{12.80}$$

In both cases,

$$\cos \frac{\Delta k}{2} = \frac{1}{2}\left(\cos \frac{k}{2} + \frac{1}{\cos \frac{k}{2}}\right), \tag{12.81}$$

and consequently the dispersion relation characterizing these configurations is

$$E = \frac{\hbar^2 J}{2}(1 - \cos k), \tag{12.82}$$

as can be seen by direct substitution in Eq. (12.78).

We interpret these types of solutions as bound states of two magnons, to begin with, because their energy is always lower than the energy of superposed free states (12.78) (also see Figure 12.1). Moreover, one can show that their wavefunctions $a_{n_1,n_2}$ exhibit an exponential decay along the chain due to the occurrence of complex quasi-momenta.

### 12.2.3 The Spectrum for Generic Values of $r$

Let us now consider the eigenvalue problem in the sector characterized by a generic number $r$ of spin flips, whose vectors are of the form

$$|\psi\rangle = \sum_{1\leq n_1<\cdots<n_r\leq L} a_{n_1,\ldots,n_r}|n_1,\ldots,n_r\rangle. \tag{12.83}$$

The recursion relations again comprise two contributions, depending on whether or not spin flips occur at neighboring sites: denoting by $i^*$ the values of $i$ for which

$$n_{i^*+1} = n_{i^*} + 1, \tag{12.84}$$

we have

$$\begin{aligned} &\hbar^2 J \sum_{i\neq i^*,i^*+1} (2a_{n_1,\ldots,n_r} - a_{n_1,\ldots,n_i+1,\ldots,n_r} - a_{n_1,\ldots,n_i-1,\ldots,n_r}) \\ &\quad + \hbar^2 J \sum_{i^*} \left(2a_{n_1,\ldots,n_r} - a_{n_1,\ldots,n_i^*-1,n_{i^*+1},\ldots,n_r} - a_{n_1,\ldots,n_{i^*},n_{i^*+1}+1,\ldots,n_r}\right) \\ &= 2Ea_{n_1,\ldots,n_r}. \end{aligned} \tag{12.85}$$

The Bethe ansatz for this sector can be taken as follows:

$$a_{n_1,\ldots,n_r} = \sum_{\pi\in\mathcal{S}_r} \Big(\prod_{i<j} A_{\pi_i\pi_j}\Big) \exp\Big(i\sum_{i=1}^{r} n_i k_{\pi_i}\Big), \tag{12.86}$$

where $\pi$ runs over all permutations of $r$ labels, namely the elements of the symmetric group $\mathcal{S}_r$. For instance, for $r = 3$,

$$\begin{aligned} a_{n_1,n_2,n_3} &= A_{12}A_{13}A_{23}\, e^{in_1k_1+in_2k_2+in_3k_3} \\ &+ A_{23}A_{21}A_{31}\, e^{in_1k_2+in_2k_3+in_3k_1} \\ &+ A_{31}A_{32}A_{12}\, e^{in_1k_3+in_2k_1+in_3k_2} \\ &+ A_{21}A_{23}A_{13}\, e^{in_1k_2+in_2k_1+in_3k_3} \\ &+ A_{13}A_{12}A_{32}\, e^{in_1k_1+in_2k_3+in_3k_2} \\ &+ A_{32}A_{31}A_{21}\, e^{in_1k_3+in_2k_2+in_3k_1}. \end{aligned} \tag{12.87}$$

The Bethe wavefunction solves the eigenvalue problem with energy

$$E = \sum_{i=1}^{r} E(k_i), \tag{12.88}$$

provided the quasi-momenta solve the Bethe ansatz equations

$$\begin{aligned} e^{ik_iL}\prod_{j\neq i} A_{ji} &= \prod_{j\neq i} A_{ij}, \\ A_{ij}(1 - 2e^{ik_j} + e^{i(k_i+k_j)}) + A_{ji}(1 - 2e^{ik_i} &+ e^{i(k_i+k_j)}) = 0. \end{aligned} \tag{12.89}$$

As for the two-body case, one can recast the Bethe ansatz in terms of the phase shifts

$$e^{i\theta_{ij}} = \frac{A_{ij}}{A_{ji}}, \tag{12.90}$$

obtaining

$$a_{n_1,\ldots,n_r} = \sum_{\pi \in S_r} \exp\left(i \sum_{i=1}^{r} n_i k_{\pi_i} + \frac{i}{2} \sum_{i<j} \theta_{\pi_i \pi_j}\right) \tag{12.91}$$

for the Bethe ansatz wavefunctions and

$$e^{ik_i L} = \exp\left(i \sum_{j\neq i} \theta_{ij}\right), \qquad e^{i\theta_{ij}} = -\frac{1 - 2e^{ik_i} + e^{i(k_i+k_j)}}{1 - 2e^{ik_j} + e^{i(k_i+k_j)}}, \tag{12.92}$$

for the Bethe ansatz equations. Nontrivial values of $\theta_{ij}$ again reflect the presence of interactions, while the spectrum remains additive, as displayed in Eq. (12.88).

If $k_i = k_j \neq 0$ for $i \neq j$, one can prove again that the Bethe ansatz wavefunction vanishes, in accordance with a "Pauli exclusion principle" of sorts. Indeed, assuming without loss of generality that $k_1 = k_2 = q \neq 0$,

$$\begin{aligned} a_{n_1,\ldots,n_r} = {} & A_{12} \sum_{\ell<m} e^{iq(n_\ell+n_m)} \sum_{\substack{\pi_\ell=1 \\ \pi_m=2}} \left( \prod_{\substack{i<j \\ (i,j)\neq(\ell,m)}} A_{\pi_i\pi_j} \right) \exp\left( i \sum_{i\neq\ell,m}^{r} n_i k_{\pi_i} \right) \\ & + A_{21} \sum_{\ell>m} e^{iq(n_\ell+n_m)} \sum_{\substack{\pi_\ell=1 \\ \pi_m=2}} \left( \prod_{\substack{i<j \\ (i,j)\neq(m,\ell)}} A_{\pi_i\pi_j} \right) \exp\left( i \sum_{i\neq\ell,m}^{r} n_i k_{\pi_i} \right), \end{aligned} \tag{12.93}$$

but $A_{12} = -A_{21}$ by the second BAE in Eq. (12.89), and therefore this expression vanishes.

However, an argument of this type does not exclude possible nontrivial solutions in which $k_i = k_j = 0$ for some $i \neq j$, which clearly violate this "Pauli" criterion. Moreover, one should bear in mind that the existence of such solutions is actually somewhat hidden by the form (12.92) of the ansatz, due to the spurious singularity appearing in the phase shift for such configurations, so that one must rely on (12.89) in order to investigate their properties. Let us thus consider, as an example, a case in which $k_1 = k_2 = 0$ while the other quasi-momenta are nonzero. By the second BAE in Eq. (12.89), this implies $A_{1j} = A_{j1}$ and $A_{2j} = A_{j2}$ for all $j \neq 1, 2$, while $A_{12} = A_{21}$ by the first BAE in Eq. (12.89). Therefore, amplitudes $A_{ij}$ involving 1 or 2 effectively factor out of all remaining BAE and can be reabsorbed in the overall normalization constant of the Bethe ansatz wavefunction, thereby reducing the problem to the study of the BAE equations in the $r - 2$ sector. Nontrivial solutions with vanishing quasi-momenta thus do exist, and are often termed singular in the literature.

A detailed analysis of the solution space for fixed $r$ and $L$ lies beyond the scope of this introduction. However, we can mention a result that allows one to address the completeness of the Bethe ansatz. It has been proved that the number of independent nonsingular solutions of the Bethe ansatz equations in the $r$-sector is precisely $\binom{L}{r} - \binom{L}{r-1}$, and the corresponding eigenstates are highest-weight states of spin-$\left(\frac{L}{2} - r\right)$ representations. We then see that, in each $r$-sector, there are additionally $\binom{L}{r-1}$ independent descendant states obtained by the action of the total spin lowering operator on sectors with lower values of $r$. We have seen examples of this mechanism in the $r = 1$ subspace, which included one descendant and $L - 1$ highest-weight

**Table 12.1.** A schematic representation of the number of independent solutions of the Bethe ansatz equation for the first few $r$-sectors. The arrow represents the action of the spin-lowering operator, giving rise to descendant states.

| $r$ | Scheme of procedure | | | | Number of solutions |
|---|---|---|---|---|---|
| 0 | $\binom{L}{0}$ | | | | $\binom{L}{0}$ |
| | ↓ | | | | |
| 1 | $\binom{L}{0}$ | $\binom{L}{1}-\binom{L}{0}$ | | | $\binom{L}{1}$ |
| | ↓ | ↓ | | | |
| 2 | $\binom{L}{0}$ | $\binom{L}{1}-\binom{L}{0}$ | $\binom{L}{2}-\binom{L}{1}$ | | $\binom{L}{2}$ |
| | ↓ | ↓ | ↓ | | |
| 3 | $\binom{L}{0}$ | $\binom{L}{1}-\binom{L}{0}$ | $\binom{L}{2}-\binom{L}{1}$ | $\binom{L}{3}-\binom{L}{2}$ | $\binom{L}{3}$ |
| ⋮ | ⋮ | ⋮ | ⋮ | ⋱ | ⋮ |

states, and in the $r = 2$ subspace, which included $(L - 1) + 1$ descendants and $\frac{L(L-3)}{2}$ new (highest-weight) states. The general picture is illustrated in Table 12.1, and the final conclusion of this line of reasoning is that the Bethe ansatz provides $\sum_{r=0}^{L}\binom{L}{r} = 2^L$ independent energy eigenstates, i.e. the complete set of solutions to the spectral problem.

The spectrum can be conveniently analyzed in the limit $L \to \infty$ for *fixed* $r$ and shows marked differences in the cases of real and complex quasi-momenta, as we have already seen for $r = 2$. With real quasi-momenta, the $k_i$ give rise to a continuum of free momenta, and taking the logarithm of the first Bethe equation in (12.92),

$$k_i = \frac{2\pi\ell_i}{L} + \frac{1}{L}\sum_{j\neq i}\theta_{ij}, \qquad \ell_i = 0, 1, \ldots L-1. \tag{12.94}$$

The second term on the right-hand side becomes negligible in the large-$L$ limit, while the first effectively gives rise to a quasi-continuous set of allowed values,

$$k_i = \frac{2\pi\ell_i}{L} \tag{12.95}$$

for arbitrarily large integers $\ell_i$. Equivalently, one may note that $e^{ik_iL}$ oscillates rapidly on the unit circle as $L \to \infty$, thus giving rise to a dense distribution thereon, which makes Eq. (12.92) trivially satisfied.

In the complex case, however, the Bethe equations develop singularities in the large-$L$ limit. Suppose, say, that $k_1$ has an imaginary part. Then $e^{ik_1L}$ will either vanish or tend to infinity in the large-$L$ limit. By the first equation in (12.92)

$$e^{ik_1L} = \exp\Big(i\sum_{j\neq 1}\theta_{1j}\Big), \tag{12.96}$$

and therefore some phase shift $\theta_{1j}$ must also be singular as $L \to \infty$, for instance $\theta_{12}$. Multiplying the corresponding equations,

$$e^{i(k_1+k_2)L} = \exp\left(i\sum_{j\neq 1,2}\theta_{1j} + i\sum_{j\neq 1,2}\theta_{2j}\right), \tag{12.97}$$

and there are again two possibilities. If $k_1 + k_2$ is real, this combination effectively gives rise to a bound state of two magnons behaving as a free particle, while if it is still complex one must look for some other $j$ that gives rise again to a compensation of the singularity occurring on the left-hand side as $L \to \infty$. Clearly this reasoning may terminate after any integer number $2M + 1 < r$ of steps, giving rise to a bound state of $2M + 1$ magnons, also called an $M$-complex or $M$-string (here $M$ may denote an integer or half-integer), whose total quasi-momentum $k$ is real and satisfies

$$e^{ikL} = \exp\left(i \sum_{i\in\text{string}} \sum_{j\notin\text{string}} \theta_{ij}\right). \tag{12.98}$$

The total momentum of a given $M$-complex will be approximately free, by the same reasoning we had followed for a single quasi-particle.

In order to study the properties of bound states, it is convenient to introduce the so-called rapidity variable, which is defined in terms of the quasi-momentum by

$$\lambda = \frac{1}{2}\cot\frac{k}{2}, \qquad k = -i\log\frac{\lambda + \frac{i}{2}}{\lambda - \frac{i}{2}} \tag{12.99}$$

and leads to

$$e^{i\theta_{ij}} = \frac{\lambda_i - \lambda_j + i}{\lambda_i - \lambda_j - i}, \qquad E(k_j) = \frac{2\hbar^2 J}{4\lambda_j^2 + 1} = -\frac{\hbar^2 J}{2}\frac{dk_j}{d\lambda_j}. \tag{12.100}$$

Note that this allows us to recast Eq. (12.96) in the form

$$e^{ik_1L} \equiv \left(\frac{\lambda_1 + \frac{i}{2}}{\lambda_1 - \frac{i}{2}}\right)^L = \prod_{j\neq 1} \frac{\lambda_1 - \lambda_j + i}{\lambda_1 - \lambda_j - i}. \tag{12.101}$$

This expression links $Im k_1 < 0$ to $Im\lambda_1 > 0$, which is crucial for the existence of the string solutions described above.

The main advantage of the new variables is that the singularities in the phase shift $\theta_{ij}$ occur when $\lambda_i - \lambda_j = \pm i$, so that the rapidities of the constituents of a given complex differ by integer multiples of $i$: for an $M$-complex,

$$\lambda_m = \lambda + im, \qquad m = -M, -M+1, \ldots, M-1, M. \tag{12.102}$$

This structure resembles a vertical "chain" in the complex plane, which also explains the name "string" given to these bound states. The total momentum of the complex is then easily calculated via the telescopic sum

$$k = -i\sum_{m=-M}^{M} \log\frac{\lambda + im + \frac{i}{2}}{\lambda + im - \frac{i}{2}} = -i\log\frac{\lambda + i(M + \frac{1}{2})}{\lambda - i(M + \frac{1}{2})} \tag{12.103}$$

so that $\lambda$ must be real in order for $k$ to be real and the corresponding energy is

$$E = \sum_{m=-M}^{M} E(k_m) = \frac{\hbar^2 J(M + \frac{1}{2})}{\lambda^2 + (M + \frac{1}{2})^2}, \tag{12.104}$$

where the result follows since the total energy also rests on a telescopic sum.

The dispersion relation of an $M$-complex is then obtained by eliminating $\lambda$ from the above equations, and reads

$$E = \frac{\hbar^2 J}{2M+1}(1 - \cos k), \tag{12.105}$$

where $k$ is the total momentum of the string. Furthermore, the Bethe ansatz equation for the string can be recast in the form

$$\begin{aligned} e^{ikL} &= \exp\left( i \sum_{j \notin \text{string}} \theta_{M1}(k, k_j) \right), \\ \exp\left( i\, \theta_{M1}(k, k_j) \right) &= \frac{\lambda - \lambda_j + i(M+1)}{\lambda - \lambda_j - i(M+1)} \frac{\lambda - \lambda_j + iM}{\lambda - \lambda_j - iM}, \end{aligned} \tag{12.106}$$

by explicitly evaluating Eq. (12.98) and noting that many terms cancel in the product. Rearranging the roots in terms of strings, the Bethe equations for a given $M$-string take the form

$$\begin{aligned} e^{ikL} &= \exp\left( i \sum_{M'} \theta_{MM'}(k, k') \right), \\ \exp\left( i\, \theta_{MM'}(k, k') \right) &= \prod_{\ell = |M - M'|}^{M+M'} \frac{\lambda - \lambda' + i(\ell+1)}{\lambda - \lambda' - i(\ell+1)} \frac{\lambda - \lambda' + i\ell}{\lambda - \lambda' - i\ell}, \end{aligned} \tag{12.107}$$

where the index $M'$ runs over all other strings, possibly with the same length.

## 12.3 Thermodynamic Bethe Ansatz

Having gained some control of the spectrum of the system, whose properties are fully characterized by the Bethe equations (12.92), we would like to address the equilibrium quasi-momentum populations for a spin chain kept in a thermal bath at a given temperature $T$.

Let us begin the analysis by first recasting Eq. (12.94) in a form that lends itself to the discussion of a generic number of spin flips, focusing on nonvanishing quasi-momenta,

$$k_\ell = \frac{2\pi \ell}{L} + \frac{1}{L} \sum_{\ell' \neq \ell} \theta(k_\ell, k_{\ell'})\, n(k_{\ell'}), \tag{12.108}$$

which provides the direct analog of (12.10) in the interacting setup. Here we have introduced two important novelties. First, we now label the allowed quasi-momenta by the Bethe quantum numbers, no longer by the index $i$ running over quasi-particles, and we have modified the notation for the $\theta_{ij}$ accordingly, which we can now assume to be known from the preceding analysis. Moreover, we have introduced the occupation number function $n(k_\ell)$ that gives 1 if the quasi-momentum identified by its argument is assigned to a quasi-particle and 0 otherwise. As we have seen, states with nonzero quasi-momentum cannot be occupied more than once. Of course the corresponding occupation function $\bar{n}(k_\ell)$ for "holes" is then determined by

$$n(k_\ell) + \bar{n}(k_\ell) = 1, \tag{12.109}$$

in analogy with (12.12) for free fermions.

However, Eq. (12.108) contains a very important difference with respect to the dynamics of free fermions. Not only does the presence or absence of quasi-particles in given states affect their availability, on account of the Pauli principle, but it actually determines the very existence of other states, since only configurations of quasi-momenta respecting the above BAE are allowed! Furthermore, one expects that the number of spin flips be comparable to the number of sites, or equivalently that an extensive number of quasi-particles be present, in the thermodynamic limit. Therefore, at any finite temperature, one can no longer neglect the second term appearing on the right-hand side of (12.108), despite the factor $\frac{1}{L}$.

In the thermodynamic limit, it is again convenient to introduce the total density of states $\rho_s(k)$ defined as

$$\rho_s(k_\ell) = \frac{1}{L}\frac{d\ell}{dk_\ell}, \tag{12.110}$$

together with the densities $\rho$ and $\bar{\rho}$ of particles and holes, related to $\rho_s$ by

$$\rho(k_\ell) = n(k_\ell)\,\rho_s(k_\ell), \qquad \bar{\rho}(k_\ell) = \bar{n}(k_\ell)\rho_s(k_\ell), \tag{12.111}$$

and subject to the constraint

$$\rho(k) + \bar{\rho}(k) = \rho_s(k), \tag{12.112}$$

again in complete analogy with Eqs. (12.14) and (12.15) for the free case.

The simple expression for $\rho_s(k)$ in Eq. (12.13), however, no longer holds as the density of states receives nontrivial corrections encoded in the last term of Eq. (12.108). Moreover, as we have discussed in detail, only string-like configurations have real quasi-momenta $k \in [0, 2\pi]$. However, for ease of presentation, let us first formulate the thermodynamic problem in terms of quasi-particle densities $\rho(k)$.

Starting from (12.108), the sum over $\ell'$ can be turned into an integral via

$$\frac{1}{L}\sum_{\ell'} \to \int_0^L \frac{d\ell'}{L} = \int_0^{2\pi} \rho_s(k')dk', \tag{12.113}$$

where (12.110) is used in the last step, and taking a derivative with respect to $k_\ell$ leads to the integral equation

$$\rho(k) + \bar{\rho}(k) = \rho_s(k) = \frac{1}{2\pi} - \int_0^{2\pi} \Theta(k,k')\rho(k')\frac{dk'}{2\pi}, \tag{12.114}$$

where

$$\Theta(k,k') = \frac{d}{dk}\theta(k,k'). \tag{12.115}$$

Equation (12.114) provides the proper density of states for the *interacting*, dynamical spectral problem under consideration, which replaces (12.13). Aside from this crucial difference, the basic thermodynamic analysis performed for free fermions can be repeated verbatim, building the grand potential $\Omega$ according to Eqs. (12.19) and (12.20). Varying $\Omega$ one thus obtains the condition

$$\int_0^{2\pi} \Big[\beta(E(k)-\mu)\delta\rho(k) - \delta\rho_s(k)\log\rho_s(k) + \delta\rho(k)\log\rho(k) + \delta\bar{\rho}(k)\log\bar{\rho}(k)$$
$$- \delta\left(\rho_s(k) - \rho(k) - \bar{\rho}(k)\right)\Big] = 0, \tag{12.116}$$

which is formally identical to (12.21). The second line of (12.116) vanishes as before, due to (12.112), but now this constraint implies

$$\delta\rho(k) + \delta\bar{\rho}(k) = \delta\rho_s(k) = -\int_0^{2\pi} \Theta(k,k')\delta\rho(k')\frac{dk'}{2\pi}, \tag{12.117}$$

in view of (12.114). Eliminating $\delta\bar{\rho}(k)$ and $\delta\rho_s(k)$ from (12.116) by means of (12.117), one is lead to

$$\int_0^{2\pi} dk\left[\beta(E(k)-\mu) + \log\frac{\rho(k)}{\bar{\rho}(k)}\right]\delta\rho(k)$$
$$+\frac{1}{2\pi}\int dk\int dk' \log\frac{\rho_s(k)}{\bar{\rho}(k)}\Theta(k,k')\delta\rho(k') = 0. \tag{12.118}$$

Finally, interchanging the integration variables $k$ and $k'$ in the second line, one obtains for the most likely equilibrium quasi-particle distribution the condition

$$\beta(E(k)-\mu) + \log\frac{\rho(k)}{\bar{\rho}(k)} + \int_0^{2\pi} \log\frac{\rho_s(k')}{\bar{\rho}(k')}\Theta(k',k)\frac{dk'}{2\pi} = 0. \tag{12.119}$$

We can now take into account the $M$-strings by assuming that the spectrum is organized in $M$-strings, even in the thermodynamic limit.

Regrouping the allowed states entering our expressions in terms of strings leads to

$$\rho_M(k) + \bar{\rho}_M(k) = \frac{1}{2\pi} - \int_0^{2\pi} \sum_{M'} \Theta_{MM'}(k,k')\rho_{M'}(k')\frac{dk'}{2\pi}, \tag{12.120}$$

which generalizes Eq. (12.114). The equilibrium density $\rho_M(k)$ at a given temperature is obtained varying the grand potential in the form of Eqs. (12.19) and (12.20), now expressed in terms of $\rho_M$ and $\bar{\rho}_M$, while enforcing the constraint (12.120), in analogy with our approach above. The resulting coupled equations for the most likely equilibrium distributions are

$$\beta(E_M(k)-\mu) + \log\frac{\rho_M(k)}{\bar{\rho}_M(k)} + \int_0^{2\pi} \sum_{M'} \log\left(\frac{\rho_{M'}(k')}{\bar{\rho}_{M'}(k')}+1\right)\Theta_{M'M}(k',k)\frac{dk'}{2\pi} = 0, \tag{12.121}$$

where $E_M(k)$ denotes the dispersion relation of an $M$-complex given by (12.105). Equations (12.120) and (12.121) are the *thermodynamic Bethe ansatz* equations, and determine the equilibrium populations of quasi-particles and holes in the thermodynamic limit for a chain at a given temperature $T$.

Strictly speaking, spurious configurations that are not organized in strings do exist even when $r$ and $L$ are arbitrarily large. However, it is now widely believed that strings provide the dominant contribution to the thermodynamic behavior. Furthermore, as we have already remarked, the Pauli principle for quasi-particles fails to account for the statistics of particles populating the $k = 0$ state. Although this state represents only a discrete exception in the continuum of available ones, it might still play a role in the low-temperature regime.

## Bibliographical Notes

The discussion presented in this chapter is based on [3] and [58], and also on J. Lamers, "Introduction to Quantum Integrability," POS 232 (2015) 1. The completeness proof for the Bethe ansatz is given in E. Mukhin, V. Tarasov and A. Varchenko, "Bethe Algebra of Homogeneous XXX Heisenberg Model has Simple Spectrum," Commun. Math. Phys. 288 (2009) 1.

# 13 Conformal Invariance and the Renormalization Group

This chapter deals with two closely connected topics. The first has to do with the continuous rotational symmetries that emerge from the asymptotic behavior of correlation functions and their enhancement to conformal symmetries at second-order phase transitions. The second topic is the renormalization group, a most effective tool for detecting points where systems exhibit this type of critical behavior. This crucially important idea emerged in quantum field theory as a way to estimate the asymptotic behavior of scattering processes at high energies, and then blossomed into a different format in statistical physics, as a way to compute complicated partition functions in steps. Its development, and its ties with conformal invariance, have led to a cross-fertilization between the two fields whose deep lessons are still under scrutiny today, after decades of intense research activity. We illustrate these ideas with reference to the one-dimensional Ising model and to vortices in the two-dimensional XY model, thus taking a first look at topological phase transitions. We conclude with a discussion of the $\epsilon$-expansion for a continuum version of the Ising model.

## 13.1 Conformal Invariance

In Chapter 10 we saw that, in the mean-field approximation, one can deduce the large-distance behavior of the spin–spin correlation function of the Ising model from the Fourier integral

$$\langle \Phi(\mathbf{r}_1)\, \Phi(\mathbf{r}_2) \rangle = \int \frac{d^D k}{(2\pi)^D} \frac{e^{i\mathbf{k}\cdot(\mathbf{r}_1 - \mathbf{r}_2)}}{k^2 + m^2}, \tag{13.1}$$

and the relevant datum here is that $m^2$ depends on the temperature $T$ and vanishes at the critical point proportionally to $T - T_c$ as $T \to T_c^+$.

Our notation here, as compared for instance to (10.134), is meant to stress that this is also the two-point correlation function of the Landau–Ginzburg model, within the same approximation, but quantities of this type are ubiquitous in quantum field theory and the quantum mechanics of relativistic systems, where they are known as Wick–rotated Feynman propagators. The Wick rotation turns the product $ct$ of a time coordinate and the speed of light into the last component $x^D$ of a Euclidean $D$-dimensional position vector $\mathbf{x}$, via $ct \to -ix^D$. This analytic continuation underlies the correspondence between quantum and statistical mechanics that we highlighted in Chapter 3.

Equation (13.1) makes no reference to the original lattice, whose details do not play a role in the vicinity of the critical point, when the correlation length is far larger than the lattice spacing. As a result, it inherits the continuous symmetries of the ambient $D$-dimensional Euclidean space. These comprise, to begin with, the *translations*

$$\mathbf{r} \rightarrow \mathbf{r} + \mathbf{a}, \tag{13.2}$$

with $\mathbf{a}$ any $D$-dimensional vector, which are manifest symmetries of the correlation function, since they do not affect the difference $\mathbf{r}_1 - \mathbf{r}_2$ in Eq. (13.1). They also include rotations, which transform the Cartesian coordinates of $D$-dimensional Euclidean space, and more generally the components of any $D$-dimensional vector, according to

$$\begin{pmatrix} x^1 \\ \vdots \\ x^D \end{pmatrix} \rightarrow R \begin{pmatrix} x^1 \\ \vdots \\ x^D \end{pmatrix}, \tag{13.3}$$

via orthogonal matrices $R$, i.e. such that

$$R^T R = 1. \tag{13.4}$$

These transformations clearly preserve the measure of the Fourier integral, since their Jacobian, $|\det(R)|$, is equal to one. As a result, one can directly conclude that the correlation function in Eq. (13.1) depends only on the distance $|\mathbf{r}_1 - \mathbf{r}_2|$, so that

$$\langle \Phi(\mathbf{r}_1)\, \Phi(\mathbf{r}_2) \rangle = G\left(|\mathbf{r}_1 - \mathbf{r}_2|\right), \tag{13.5}$$

which is manifestly invariant under rotations.

When the mass $m$ vanishes, however, the problem acquires a larger symmetry. For example, the integral is then also invariant, up to an overall factor, under the scale transformation

$$\mathbf{r} \rightarrow \lambda\, \mathbf{r}, \tag{13.6}$$

where $\lambda$ is a constant. One can absorb $\lambda$ in the exponent, letting

$$\mathbf{k} \rightarrow \frac{1}{\lambda}\, \mathbf{k}, \tag{13.7}$$

so that

$$G(\lambda r)\big|_{m=0} = \lambda^{2-D} G(r)\big|_{m=0}. \tag{13.8}$$

On the other hand, a nonvanishing $m$ would introduce an additional scale in the problem, which would be affected by the scale transformation.

There is more here, however. Let us now focus momentarily on the circle displayed in Figure 13.1, a section of the standard stereographic link between the Euclidean plane and the Riemann sphere. One can clearly project, as indicated, any point $P$ of the unit circle, aside from its north pole $N$, onto the horizontal line through its center, thus also linking it to a corresponding point $P'$ on the horizontal line, according to

$$|\mathbf{r}| = \cot\left(\frac{\theta}{2}\right) = \frac{\sin\theta}{1 - \cos\theta} = \frac{x}{1 - y}. \tag{13.9}$$

Here $|\mathbf{r}|$ is the distance between the origin and $P'$. Moreover, the pair $(x, y)$, which uniquely identifies the point $P$ on the circle, with $y$ its height above the horizontal line, is constrained to satisfy $x^2 + y^2 = 1$. A similar projection from the south pole $S$

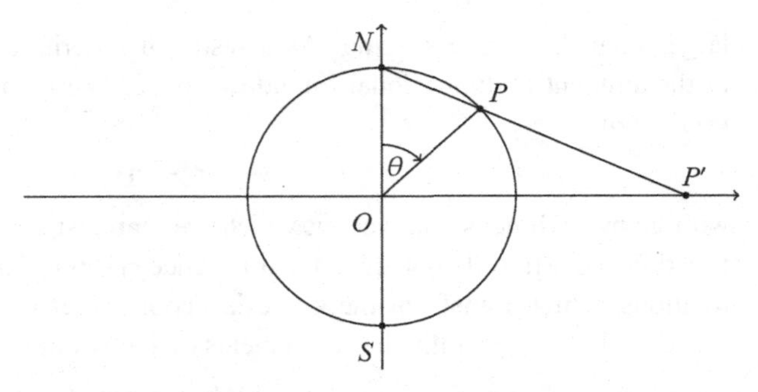


Figure 13.1 A section of the stereographic projection of a sphere on the plane from its north pole $N$. A similar projection is possible from the south pole $S$.

would result in the second chart commonly used for this surface, and the virtue of this presentation is that it extends very naturally to the $D$-dimensional case, where it becomes simply

$$\mathbf{r} = \frac{\mathbf{x}}{1 - y}, \tag{13.10}$$

with the vector $\mathbf{x}$ and the remaining coordinate $y$ constrained according to

$$\mathbf{x}^2 + y^2 = 1. \tag{13.11}$$

Homogenous coordinates are a convenient device when discussing algebraic surfaces. Letting

$$\mathbf{x} \longrightarrow \frac{\mathbf{x}}{x^-}, \qquad y \longrightarrow \frac{x^+}{x^-}, \tag{13.12}$$

so that

$$\mathbf{r} = \frac{\mathbf{x}}{x^- - x^+}, \tag{13.13}$$

the original equation of the unit sphere turns into

$$(\mathbf{x})^2 + (x^+)^2 - (x^-)^2 = 0, \tag{13.14}$$

which describes a cone. The symmetries of interest now correspond manifestly to the Lorentz group in $D + 2$ dimensions, $SO(1, D + 1)$, which is clearly the symmetry group of the quadratic form in Eq. (13.14), and is usually called Euclidean conformal group. Its $SO(D)$ subgroup of rotations among the first $D$ coordinates corresponds to the rotations of the original Euclidean space. Let us also note, in passing, that $SO(1, D + 1)$ is also the symmetry group of the family of quadratic curves

$$\mathbf{x}^2 + (x^+)^2 - (x^-)^2 = \rho^2, \tag{13.15}$$

for arbitrary values of $\rho$. Figure 13.2 shows that nonzero values of $\rho$ deform a cone into a smooth hyperboloid. This procedure is usually called a "blow up."

As we have anticipated, one can translate these results into corresponding ones in quantum field theory via an analytic continuation. The original Euclidean group

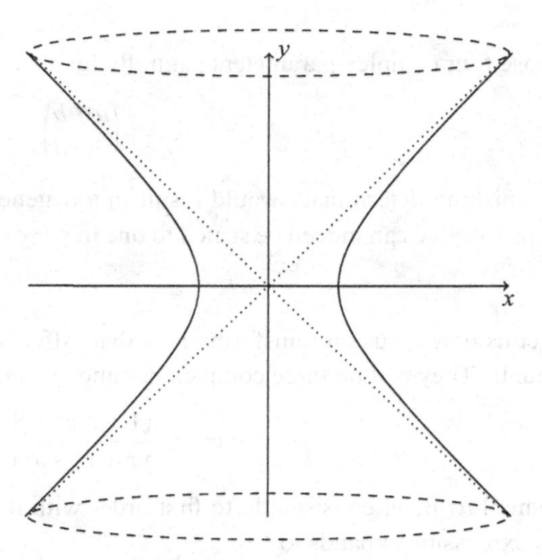


Figure 13.2 A hyperboloid with the corresponding cone and its "circles at infinity."

then becomes the Lorentz group proper, $SO(1, D-1)$, while the analytic continuation of $\mathbf{x}^2$ introduces an additional negative contribution, $-c^2t^2$, in the constraint (13.15). Consequently, the Euclidean conformal group $SO(1, D+1)$ becomes $SO(2, D)$, which is usually called conformal group in $D$ spacetime dimensions. This remains a point symmetry of the deformed hyperboloid, which now describes a $(D+1)$-dimensional manifold called anti-de Sitter space, or $AdS_{D+1}$. The reader should notice that this space possesses the curious feature of emerging from a pseudo-Euclidean space with "two times" in $D+2$ dimensions.

Conformal invariance on the original $D$-dimensional Minkowski space is thus akin to the point symmetry in $AdS_{D+1}$, and the original $D$-dimensional plane is somehow the boundary of $AdS_{D+1}$.[1] These cursory remarks are meant to arouse the reader's curiosity on a celebrated link, the $AdS_5/CFT_4$ correspondence, between theories in the bulk of $AdS_5$ and conformally invariant theories on its boundary. This correspondence emerged from string theory, a powerful extension of quantum field theory, and has found manifold applications in different contexts.

Focusing now on the two-sphere, it is convenient to work in the Euclidean plane with the complex combination

$$z = x + iy, \tag{13.16}$$

of its two Cartesian coordinates. It is well known that the (extended) complex plane affords the one-to-one continuous maps

$$w = \frac{az+b}{cz+d}, \qquad ad - bc = 1, \tag{13.17}$$

[1] The boundary comprises the two "end circles at infinity" of the hyperbola in the example of Figure (13.2). The one-dimensional case is special, however, since for $D > 1$ the boundary only has one connected component.

whose four complex parameters naturally fill an $SL(2, \mathbb{C})$ matrix

$$M = \begin{pmatrix} a & b \\ c & d \end{pmatrix}. \tag{13.18}$$

A vanishing determinant would result in a degenerate transformation, while any nonzero value can indeed be scaled to one in view of the fractional transformation.

Let us now focus on transformations that differ from the identity by infinitesimal amounts. They rest on three complex parameters, and can be cast into the form

$$w = \frac{(1+\alpha)z + \beta}{\gamma z + (1-\alpha)}, \tag{13.19}$$

for small $\alpha$, $\beta$, $\gamma$, consistently to first order with the constraint on the determinant. This expression expands to

$$w = z + 2\alpha z + \beta - \gamma z^2, \tag{13.20}$$

keeping only terms up to first order in the parameters, and one can readily recognize two translations associated to the real and imaginary parts of $\beta$ and the rotation associated to the imaginary part of $\alpha$. The new transformations that complete the so-called Euclidean group into the conformal group $SO(1,3)$ are thus a scale transformation, associated to the real part of $\alpha$, and finally two "special conformal transformations" associated to the real and imaginary parts of $\gamma$. This is what we alluded to when we stated that scale transformations would not be the end of the story. These considerations highlight the general pattern, but in the special case of two dimensions there is indeed much more. Let us therefore dwell further on this special case.

It is interesting to note that the fractional linear transformation (13.17) is a particular class of the wider family of general conformal mappings of the form

$$w = f(z), \tag{13.21}$$

where $f$ is a holomorphic function. The transformations in Eq. (13.17) are indeed *global* conformal mappings of the Riemann sphere onto itself, while general conformal mappings work only in portions of it. However, during the last decades infinitesimal transformations of the general type have also proved of great relevance for both statistical physics and quantum field theory. For example, they have led to exact results for correlations functions in the Ising model and in other two-dimensional models at their critical points.

The reader will not have failed to notice that we are confronted with something unusual here: Eq. (13.21) involves infinitely many powers of $z$, since one can write

$$f(z) = \sum_{n \in \mathbb{Z}} c_n z^n \tag{13.22}$$

within some domain of convergence, which thus encodes infinitely many different types of transformations. It is thus convenient to introduce the differential operators

$$L_m = -z^{m+1} \frac{\partial}{\partial z}, \tag{13.23}$$

which satisfy the algebra

$$[L_m, L_n] = (m-n)L_{m+n}. \tag{13.24}$$

There is actually an important family of variants of this algebra

$$[L_m, L_n] = (m-n)L_{m+n} + \frac{c}{12}\, m(m^2-1)\, \delta_{m+n,0}, \tag{13.25}$$

where $c$ is a constant, called the central charge.

This is an infinite-dimensional analog of the angular-momentum algebra, whose representations underlie many key results in quantum mechanics, and played a role in our discussion of the Keplerian problem in Chapter 3. It is called Virasoro algebra, and plays a very important role in two-dimensional statistical physics. Its representation theory has led to a detailed characterization of different universality classes of two-dimensional models at their critical points, and in particular to the exact values of the critical exponents for the two-dimensional Ising model.

One can prove that, for all values of $D > 2$, the conformal group is simply $SO(1, D+1)$, a finite-dimensional extension of the manifest symmetry of Euclidean space (or $SO(2, D)$, a finite-dimensional extension of the Lorentz group, in the Minkowski case). Infinitesimal transformations of this type can be parametrized as

$$x^\mu \to x^\mu + a^\mu + \omega^\mu{}_\nu x^\nu + \alpha x^\mu + b^\nu \left(\delta^\mu_\nu x^2 - 2 x_\nu x^\mu\right), \tag{13.26}$$

where the vector $a^\mu$ corresponds to translations, the matrix $\omega^\mu_\nu$ to rotations, $\alpha$ to dilatations, and finally the vector $b^\mu$ corresponds to special conformal transformations.

One can show that all two-point and three-point correlation functions are fully determined at a critical point, while complicated nonlinear constraints are enforced on higher-point functions. These have long been deemed too complicated to be of use, until recently they proved valuable in setting numerical constraints on the features of critical points.

## 13.2 1D Ising Model and Renormalization Group

Let us consider the 1D Ising partition function for $2N$ sites,

$$Z_{2N}(J, K) = \sum_{\{\sigma_i = \pm 1\}} e^{J \sum_i \sigma_i \sigma_{i+1} + K \sum_i \sigma_i}, \tag{13.27}$$

where $J = v\beta$ and $K = \mu\beta B$ in the notation of the preceding chapters. One can try to compute $Z$ in steps grouping together nearby sites, thus trying to relate the partition function to that of a different, "coarse grained" lattice. One may for instance perform the sum over spins at points with even labels $2i$. This step entails the evaluation of

$$\sum_{\sigma_{2i}=\pm 1} e^{\sigma_{2i}[J(\sigma_{2i-1}+\sigma_{2i+1})+K]} = 2\cosh\left[J(\sigma_{2i-1}+\sigma_{2i+1})+K\right], \tag{13.28}$$

an expression that can be recast in a form similar to the original one, letting

$$2\cosh\left[J(\sigma_{2i-1}+\sigma_{2i+1})+K\right] = e^{A\sigma_{2i-1}\sigma_{2i+1}+B(\sigma_{2i-1}+\sigma_{2i+1})+C}. \tag{13.29}$$

The coefficients $A$, $B$, and $C$ are determined considering the possible combinations of values $\sigma_{2i+1}$ and $\sigma_{2i-1}$, which leads to the conditions

$$\begin{aligned} 2\cosh K &= e^{C-A}, \\ 2\cosh(2J+K) &= e^{A+2B+C}, \\ 2\cosh(-2J+K) &= e^{A-2B+C}. \end{aligned} \tag{13.30}$$

The three parameters are thus

$$\begin{aligned} e^{A} &= \left[\cosh^2 2J\left(1-\tanh^2 K\tanh^2 2J\right)\right]^{\frac{1}{4}}, \\ \tanh 2B &= \tanh 2J\tanh K, \\ e^{C} &= 2\cosh K\left[\cosh^2 2J\left(1-\tanh^2 K\tanh^2 2J\right)\right]^{\frac{1}{4}}, \end{aligned} \tag{13.31}$$

so that, using the coarse-graining relation (13.29), one can recover the original partition function from a sum over a lattice with half as many sites, up to an overall constant,

$$Z_{2N}(J,K) = e^{NC} Z_N(A, 2B+K). \tag{13.32}$$

Hence, in the new lattice $A$ plays the same role as the coupling constant $J$ in the original one, while $2B+K$ identifies the new external field. One may interpret this transformation as a map between different theories: this step can be applied repeatedly, leading to the iterations

$$\begin{aligned} J_{n+1} &= \frac{1}{2}\log\left[\cosh 2J_n\left(1-\tanh^2 K_n\tanh^2 2J_n\right)^{\frac{1}{2}}\right], \\ K_{n+1} &= K_n + \operatorname{arctanh}\left(\tanh 2J_n\tanh K_n\right). \end{aligned} \tag{13.33}$$

The coarse graining procedure thus gives rise to a "renormalization group flow" in the space of theories.

In this simple example we have been lucky, since the coarse graining mapped an Ising model to another Ising model, but in more general situations the renormalization group flow can induce new interactions, for instance cubic or quartic couplings among spins.

A critical point should be characterized by a divergent correlation length $\lambda(J_c, K_c)$. Therefore, at criticality the system ought to be scale invariant, and hence insensitive to the coarse graining procedure: *critical points are thus fixed points of the renormalization group*. Furthermore, the critical limit should entail a singular behavior for $\frac{1}{N}\log Z$. The overall factor $e^{NC}$ clearly plays no role in this effect, since it merely adds a well-behaved constant to the free energy. Therefore we are naturally led to enforce the conditions

$$B = 0, \qquad J = A, \tag{13.34}$$

which characterize the fixed points of (13.31). The first of these translates into $K=0$, so that the magnetic field must be absent. This is actually a general feature of the renormalization group flow: external fields need to be fine tuned to zero in order to allow the coarse graining procedure to converge to fixed points.

The other condition then becomes

$$J_{n+1} = \frac{1}{2}\log\cosh 2J_n, \tag{13.35}$$

whose two sides become very large and close to one another as $J \to \infty$, up to subleading terms. This fixed point corresponds to the zero-temperature limit, and we know already that the 1D Ising model approaches an ordered state as $T \to 0$. The condition has also an exact solution, $J = 0$, which corresponds to infinite temperature, and this example suffices to illustrate the interesting difference between repulsive and attractive fixed points. Indeed, for any positive value of $J$,

$$\frac{1}{2}\log(\cosh 2J) < J, \tag{13.36}$$

and therefore the renormalization group flow moves $J$ toward smaller and smaller values, and thus *away from* the zero-temperature fixed point $J = \infty$ and *toward* the infinite-temperature fixed point $J = 0$. This reflects the fact that, at the larger scales exposed after each step, the system appears ever more within the high-temperature disordered phase. According to a suggestive terminology, $J = \infty$ is therefore called a UV fixed point and the renormalization group drives the system toward the IR fixed point $J = 0$.

Let us conclude the analysis of this rather simple example with a closer look at the correlation length. Once $K = 0$, the correlation length $\lambda$ is linked to the spin-spin coupling $J$ via Eq. (10.32), namely

$$\lambda(J) = \frac{1}{\log(\coth J)} \tag{13.37}$$

in our current notation. Then, performing a coarse graining procedure corresponds to halving the correlation length, because Eq. (13.35) can be recast as

$$\coth J_{n+1} = (\coth J_n)^2, \tag{13.38}$$

from which

$$\lambda(J_{n+1}) = \frac{1}{2}\lambda(J_n). \tag{13.39}$$

This conforms to the physical intuition that the renormalized lattice spacing should be twice the original one after summing over half the sites. Furthermore, recasting (13.35) as

$$J_{n+1} = f(J_n), \qquad f(J) = \frac{1}{2}\log\cosh 2J, \tag{13.40}$$

one may turn (13.39) into a prediction for the critical exponent $\nu$ characterizing the scaling of the correlation length

$$\lambda(J) \sim |J_c - J|^{-\nu} \tag{13.41}$$

in the vicinity of the critical point.

A Taylor expansion of $f(J)$ at $J_c$ gives

$$J_{n+1} - J_c \simeq f'(J_c)(J_n - J_c), \tag{13.42}$$

taking into account that $f(J_c) = J_c$, since the critical coupling is a fixed point, and combining this observation with Eq. (13.41) written for $J_n$ and for $J_{n+1}$ gives

$$\nu = \frac{\log\left[\frac{\lambda(J_n)}{\lambda(J_{n+1})}\right]}{\log|f'(J_c)|}. \tag{13.43}$$

In this example, the numerator is simply $\log 2$ due to Eq. (13.39), while the denominator rests on $f'(J) = \tanh 2J$, whose logarithm diverges at the critical points, so that $\nu$ vanishes in both cases. However, this line of reasoning will yield nontrivial predictions in the following examples.

## 13.3 Percolation

The one-dimensional Ising model is extremely simple, which makes it suitable for a first look at the renormalization group equations, but it does not display any fixed point at a finite nonzero temperature, so that it is not a compelling example. A slightly less naive one, which does display a physically relevant critical point, is the following "percolation" model.

Let us consider a triangular lattice whose individual sites have a probability $p$ of being occupied at a given time. One can define a new coarse-grained lattice with vertices at the centers of the shaded triangles in Figure 13.3, while also associating to the new vertices the occupation probability

$$p' = p^3 + 3p^2(1-p). \tag{13.44}$$

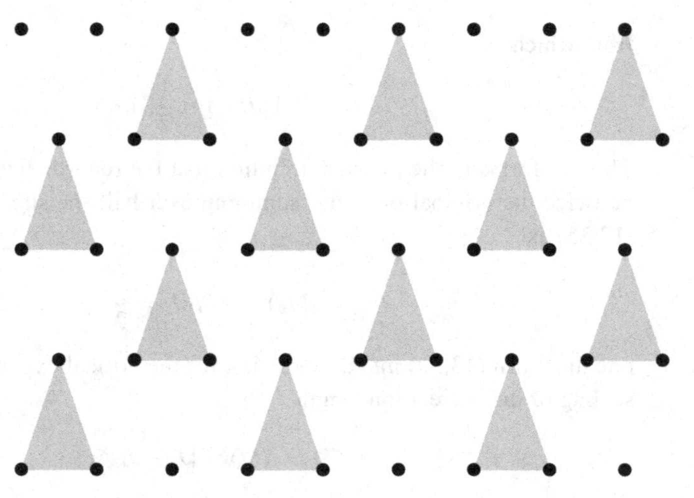

Figure 13.3 Coarse graining procedure for the percolation model on a triangular lattice.

This corresponds to building configurations for the new lattice according to the following rule: a new vertex is occupied if at least two of the three nearby sites of the original lattice were occupied, and empty otherwise.

Iterating this procedure yields the recursion relation

$$p_{n+1} = p_n^3 + 3p_n^2(1 - p_n). \tag{13.45}$$

This simple flow has three fixed points, the three real solutions $p_c = 0, \frac{1}{2}, 1$ of the equation

$$p_c = p_c^3 + 3p_c^2(1 - p_c). \tag{13.46}$$

A situation of dynamical balance obtains for $p_c = \frac{1}{2}$ (the other two correspond to the empty and the full lattice), which is our first example of a nontrivial renormalization group fixed point.

Furthermore, one may argue for geometric reasons that, since the lattice spacing increases by a factor of $\sqrt{3}$ under each coarse graining, the correlation length must satisfy

$$\lambda(p_{n+1}) = \frac{1}{\sqrt{3}}\,\lambda(p_n), \tag{13.47}$$

in analogy with Eq. (13.39) for the one-dimensional Ising model. Recasting the renormalization group equation in the form

$$p_{n+1} = f(p_n), \qquad f(p) = p^3 + 3p^2(1 - p) \tag{13.48}$$

and linearizing around the critical point leads to

$$p_{n+1} - p_c = \frac{3}{2}\,(p_n - p_c). \tag{13.49}$$

Therefore, Eq. (13.43) again provides a prediction for the critical exponent $\nu$ that governs the scaling

$$\lambda(p) \sim |p_c - p|^{-\nu}, \tag{13.50}$$

namely

$$\nu = \frac{\log 3}{2\log\frac{3}{2}} \simeq 1.35, \tag{13.51}$$

which compares nicely with the exponent $\nu = \frac{4}{3} \simeq 1.33$, which can be obtained from the exact solution of the percolation problem in two dimensions.

## 13.4 The XY Model

Our next example concerns a two-dimensional square lattice, whose sites host classical spin variables $\mathbf{s}_i$, vectors of unit length that can rotate on the lattice plane, as depicted in Figure 13.4, with the nearest-neighbor interactions

$$H = E_0 - J\sum_{\langle i,j\rangle} \mathbf{s}_i \cdot \mathbf{s}_j. \tag{13.52}$$

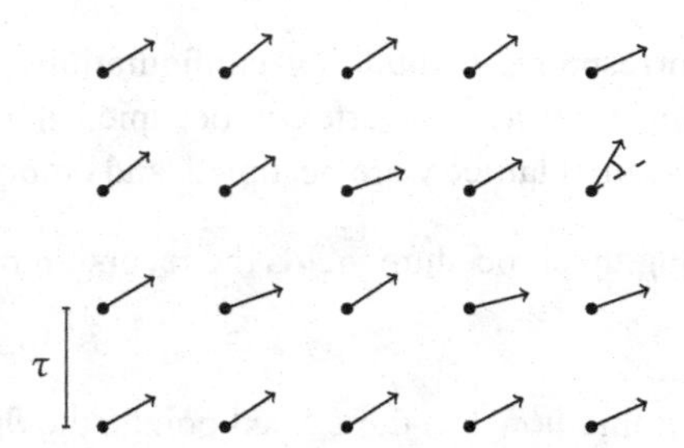


Figure 13.4 A schematic representation of the XY model.

This is the so-called XY model. Here $E_0$ is a constant introduced for later convenience, while $J > 0$ determines as usual the strength of the interactions, which favor the alignment of neighboring spins. Introducing the angle $\phi_i$ between the spin $\mathbf{s}_i$ and a reference direction on the plane, one can recast the Hamiltonian of the model (13.52) in the form

$$H = E_0 - J \sum_{\langle i,j \rangle} \cos(\phi_i - \phi_j). \tag{13.53}$$

We adopt periodic boundary conditions and observe that the lowest possible energy is attained when all spins are aligned in a given direction, so that $E = E_0 - 4NJ$ with $N$ the total number of sites.[2] In the following, we set $E_0 = 4NJ$.

### 13.4.1 Spin Waves

In the low-temperature regime, we expect the spins to experience small deviations from this preferred configuration, and we may therefore expand the Hamiltonian up to quadratic order, obtaining

$$H \simeq \frac{J}{2} \sum_{\langle i,j \rangle} (\phi_i - \phi_j)^2. \tag{13.54}$$

It can be convenient to label the sites by their lattice coordinates, i.e. by a pair of integers $n = (n_x, n_y)$, each running from 0 to $L - 1$ with $L^2 = N$. In this fashion, the approximate low-energy Hamiltonian becomes

$$H \simeq -J \sum_{n,m} \phi_n \Delta_{n,m} \phi_m, \tag{13.55}$$

where

$$-\Delta_{n,m} = \left(2\delta_{n_x,m_x} - \delta_{n_x,m_x+1} - \delta_{n_x,m_x-1}\right) \delta_{n_y,m_y} + (x \leftrightarrow y) \tag{13.56}$$

is the discretized Laplace operator on the lattice (which we already encountered in Chapter 12).

This operator becomes diagonal in Fourier space, i.e. in terms of the so-called spin waves. For any pair of integers $\ell = (\ell_x, \ell_y)$ each running from 0 to $L - 1$, one has

$$-\sum_m \Delta_{n,m} e^{\frac{i2\pi n \cdot \ell}{L}} = \lambda_\ell \, e^{\frac{i2\pi n \cdot \ell}{L}}, \qquad \lambda_\ell = 4 - 2\left(\cos \tfrac{2\pi \ell_x}{L} + \cos \tfrac{2\pi \ell_y}{L}\right), \tag{13.57}$$

[2] Each nearest-neighbor pair is counted twice.

where $n \cdot \ell = n_x \ell_x + n_y \ell_y$. Note that the zero eigenvalue, $\lambda_0 = 0$, occurs because the average of $\Delta_{n,m}$ over the whole lattice vanishes for periodic boundary conditions. Properly normalized eigenvectors, which one can also rearrange into real-valued ones, are thus given by

$$v_{\ell,n} = \frac{1}{L} e^{\frac{i2\pi n \cdot \ell}{L}}. \tag{13.58}$$

The partition function in the low-temperature limit reads

$$Z_N \simeq \int_0^{2\pi} \prod_n d\phi_n \, e^{\beta J \sum_{n,m} \phi_n \Delta_{n,m} \phi_m} \tag{13.59}$$

and letting

$$\phi_n = \sum_\ell c_\ell \, v_{\ell,n} \tag{13.60}$$

leads to

$$Z_N \simeq \int_0^{2\pi L} dc_0 \prod_{\ell \neq 0} \sqrt{\frac{\pi}{\beta J \lambda_\ell}}, \tag{13.61}$$

where we performed the Gaussian integration over all nonzero-mode coefficients, while also isolating the integration over

$$c_0 = \sum_n v_{0,n} \phi_n = \frac{1}{L} \sum_n \phi_n, \tag{13.62}$$

which ranges from 0 to $2\pi L$. All in all,

$$Z_N \simeq 2\pi L \left(\frac{\pi}{\beta J}\right)^{\frac{N-1}{2}} \frac{1}{\sqrt{\det'(-\Delta)}}, \tag{13.63}$$

where $\det'(-\Delta) = \prod_{\ell \neq 0} \lambda_\ell$ is the determinant "deprived" of the zero eigenvalue, as in Chapter 3.

Let us now discuss average observables. Since $\phi_n$ is periodic, it is enough to consider thermal averages of the type

$$\left\langle e^{i \sum_n q_n \phi_n} \right\rangle \simeq \frac{1}{Z_N} \int_0^{2\pi} \prod_n d\phi_n \, e^{\beta J \sum_{n,m} \phi_n \Delta_{n,m} \phi_m + i \sum_n q_n \phi_n} \tag{13.64}$$

with integer $q_n$. For instance, consider the two-point function

$$\langle \mathbf{s}_n \cdot \mathbf{s}_m \rangle = \langle \cos(\phi_n - \phi_m) \rangle = \Re \left\langle e^{i\phi_n - i\phi_m} \right\rangle. \tag{13.65}$$

In order to simplify Eq. (13.64), we resort again to the Fourier decomposition (13.60). Isolating the integration over $c_0$ and computing the Gaussian integrals over the remaining coefficients, as before, one finds in this case

$$\left\langle e^{i \sum_n q_n \phi_n} \right\rangle \simeq e^{-\frac{1}{4\beta J} \sum_{n,m} q_n G_{n,m} q_m} \int_0^{2\pi L} \frac{dc_0}{2\pi L} \, e^{i \frac{c_0}{L} \sum_n q_n}, \tag{13.66}$$

where

$$G_{n,m} = \frac{1}{L^2} \sum_{\ell \neq 0} \frac{1}{\lambda_\ell} e^{\frac{2\pi \ell \cdot (n-m)}{L}}. \tag{13.67}$$

Note that the zero-mode integral imposes that the correlator vanish whenever the "total charge"

$$q = \sum_n q_n \tag{13.68}$$

is nonzero, namely

$$\left\langle e^{i\sum_n q_n \phi_n} \right\rangle \simeq \delta_{q,0}\, e^{-\frac{1}{4\beta J}\sum_{n,m} q_n G_{n,m} q_m}. \tag{13.69}$$

Furthermore, $G_{n,m}$ acts as an inverse for the discretized Laplace operator in the subspace orthogonal to its kernel,

$$-\sum_m \Delta_{n,m} G_{m,k} = \delta_{n,k} - \frac{1}{L^2}, \tag{13.70}$$

consistent with the restriction to nonvanishing eigenvalues in Eq. (13.67). Let us write $G_{n,m}$, separating the contribution at coincident points, as

$$G_{n,m} = G_{n-m}, \qquad G_n = G_0 + \widetilde{G}_n, \qquad \widetilde{G}_0 = 0. \tag{13.71}$$

Using the results of Appendix G, one finds that in the thermodynamic limit

$$G_0 = \frac{1}{4\pi}\log(CN), \tag{13.72}$$

with a constant $C$, and

$$\widetilde{G}_n \simeq -\frac{1}{2\pi}\log\frac{r}{r_0}, \qquad \text{with } \frac{\tau}{r_0} = 2\sqrt{2}\, e^{\gamma_E}, \tag{13.73}$$

where $\gamma_E$ is the Euler–Mascheroni constant (D.25), whenever the length of the vector $\mathbf{r} = \tau(n_x, n_y)$ is greater than or equal to the lattice spacing $\tau$. Substituting in Eq. (13.69), the dependence on $G_0$ and on the factor $2\sqrt{2}\, e^{\gamma_E}$ drops out, thanks to the "neutrality condition" $q = 0$, so that

$$\left\langle e^{i\sum_n q_n \phi_n} \right\rangle \simeq \delta_{q,0} \prod_{n\neq m} \left(\frac{r_{nm}}{\tau}\right)^{\frac{1}{8\pi\beta J} q_n q_m} \tag{13.74}$$

with $r_{nm} = |\mathbf{r}_n - \mathbf{r}_m|$. In particular, for the spin-spin correlation function Eq. (13.65) leads to

$$\langle \mathbf{s}_n \cdot \mathbf{s}_m \rangle \simeq \left(\frac{r_{nm}}{\tau}\right)^{-\frac{1}{4\pi\beta J}}, \tag{13.75}$$

provided the two spins are at least one lattice spacing apart. Summarizing, for small fluctuations about the configuration where all spins are aligned in a given direction, spin waves give rise to a power-law behavior in correlation functions, and in particular in the two-point function. Considering only perturbations around this trivial configuration, however, one misses an important physical phenomenon, as we are about to see.

### 13.4.2 Particle-Vortex Duality

In order to illustrate this point, it is convenient to discuss the system in the continuum limit, turning finite differences into derivatives and sums into integrals, so that Eq. (13.55) takes the form

$$H = -J \int_A d^2\mathbf{r}\, \phi(\mathbf{r}) \nabla^2 \phi(\mathbf{r}). \tag{13.76}$$

Configurations that minimize the total energy can be identified by considering a small, local variation $\delta\phi(\mathbf{r})$ in the orientation of the spin variables, which induces the energy variation

$$\delta H = -2J \int_A d^2\mathbf{r}\, \delta\phi(\mathbf{r}) \nabla^2 \phi(\mathbf{r}), \tag{13.77}$$

obtained by integrating by parts. Therefore, one is naturally led to consider configurations satisfying the Laplace equation

$$\nabla^2 \phi(\mathbf{r}) = 0, \tag{13.78}$$

i.e. harmonic functions.

Equation (13.78) has strong ties with complex analysis. Given a solution $\phi(x, y)$ of the Laplace equation, where $\mathbf{r} = (x, y)$, one can always solve the Cauchy–Riemann equations

$$\partial_x \varphi = \partial_y \phi, \qquad \partial_y \varphi = -\partial_x \phi \tag{13.79}$$

for a "dual" harmonic function $\varphi(x, y)$, at least locally, thus reconstructing a holomorphic function

$$f(z) = \varphi(x, y) + i\phi(x, y) \tag{13.80}$$

of the complex variable $z = x + iy$. This follows from the Poincaré lemma applied to the differential form $\partial_y \phi\, dx - \partial_x \phi\, dy$, whose integrability condition is the Laplace equation (13.78).

Note that the field lines associated to the two potentials $\phi$ and $\varphi$ are always orthogonal to each other, since letting

$$\mathbf{u} = -\nabla\varphi, \qquad \mathbf{v} = -\nabla\phi, \tag{13.81}$$

the Cauchy–Riemann relations (13.79) imply

$$u_x = v_y, \qquad u_y = -v_x, \tag{13.82}$$

and furthermore

$$(\nabla \cdot \mathbf{u})\, \widehat{\mathbf{z}} = \nabla \times \mathbf{v}, \tag{13.83}$$

with $\widehat{\mathbf{z}}$ the unit vector orthogonal to the plane. Equations (13.82) and (13.83) have, in general, a wider domain of validity than the potentials emerging from the Cauchy–Riemann equations (13.79), since they have an intrinsic geometric meaning in terms of field lines. We are about to see an example of this fact.

These mathematical facts afford physical counterparts in the properties of electrostatics in two dimensions, if one regards $\varphi$ as the potential created by charged particles (or infinite wires, in the three-dimensional setup). Equations (13.82) and (13.83) indicate that points where the field lines of $\mathbf{u}$ converge (or diverge), which would host point *charges* in the electrostatic analogy, are *vortices* of $\mathbf{v}$ and vice

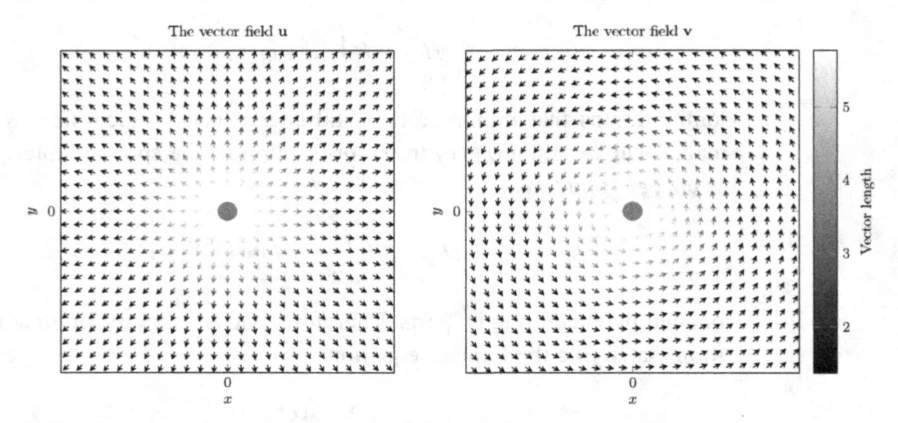


Figure 13.5 The vector fields $u$ and v in Eq. (13.88) associated to a single-vortex configuration at the origin.

versa.[3] Using the above relations, the total energy (13.76) can also be presented in the two equivalent forms

$$H = J \int_{\mathcal{A}} d^2\mathbf{r}\, |\mathbf{v}|^2 = J \int_{\mathcal{A}} d^2\mathbf{r}\, |\mathbf{u}|^2, \tag{13.84}$$

so that the distinction between particles and vortices is immaterial for $H$.

Let us consider, in particular, the Green function $G(r)$, which provides the continuum counterpart of the inverse operator $G_{n,m}$ of Eqs. (13.71), (13.72), and (13.73), up to a conventional factor of $2\pi$, so that

$$-\nabla^2 G(r) = 2\pi\delta(\mathbf{r}), \qquad G(r) = \frac{1}{2}\log(CN) + \widetilde{G}(r), \tag{13.85}$$

with

$$\widetilde{G}(r) = -\log\frac{r}{r_0}, \tag{13.86}$$

for $r > \tau$, while $\widetilde{G}(0) = 0$ by construction. One can associate $G(r)$ with the real part of

$$f(z) = -\log z, \tag{13.87}$$

up to an additive constant. The imaginary part, $-\arg(z)$, is not a smooth single-valued function for $|\mathbf{r}| > \tau$, and yet the two orthogonal vector fields

$$\mathbf{u} = \frac{\mathbf{r}}{r^2}, \qquad \mathbf{v} = \widehat{\mathbf{z}} \times \mathbf{u}, \tag{13.88}$$

are well defined and regular away from the origin, as can be seen in Figure 13.5.

### 13.4.3 A Gas of Vortices

The preceding discussion motivates one to allow, in equilibrium configurations, some singularities associated to the presence of vortices. In this case, the line integral of

[3] The reader will recognize here a two-dimensional counterpart of the electric-magnetic duality of Maxwell's theory.

the vector field $\mathbf{v}$ along closed paths does not vanish, in general, because the spin variable can undergo an integer number $q$ of revolutions around the vertical axis

$$\oint_\gamma \mathbf{v}\cdot d\boldsymbol{\ell} = 2\pi q, \qquad q = 0, \pm 1, \pm 2, \ldots . \tag{13.89}$$

One is thus allowing variations $\phi \to \phi + 2\pi q$ along the path $\gamma$, and this "winding number" translates into a nonzero flux for the dual vector $\mathbf{u}$,

$$\oint_\gamma \mathbf{u}\cdot\mathbf{n}\, d\ell = 2\pi q, \tag{13.90}$$

on account of Eq. (13.83).

Let us first consider a single vortex (or, if you will, a single particle) with unit strength (or charge) centered at the origin, that can be described via the Green function $G(r)$. Substituting (13.85) into (13.76), one can associate to this configuration the energy

$$E = \pi J \log(CN). \tag{13.91}$$

On the other hand, since the number of ways of placing the vortex is simply $N$, one can associate to it the entropy

$$S = k_B \log N. \tag{13.92}$$

The resulting free energy,

$$(\pi J - k_B T)\log N \tag{13.93}$$

up to a constant, suggests that an interesting phenomenon ought to occur at temperature

$$T_c = \frac{\pi J}{k_B}, \tag{13.94}$$

where the energy cost of isolated vortices balances the corresponding entropy gain. This simple argument indicates that above $T_c$ the system contains a macroscopic number of isolated vortices, and captures, in essence, the phenomenon of vortex deconfinement.

Let us now move on to a configuration with $2n$ of vortices centered at the positions $\mathbf{r}_i$ with $i = 1, 2, \ldots, 2n$, so that

$$\varphi(\mathbf{r}) = \sum_{i=1}^{2n} q_i G(|\mathbf{r}-\mathbf{r}_i|), \qquad \phi(\mathbf{r}) = -\sum_{i=1}^{2n} q_i \arg(\mathbf{r}-\mathbf{r}_i). \tag{13.95}$$

Using Eq. (13.76), the total energy of this configuration reads

$$H_{2n} = -J\int_A d^2\mathbf{r}\,\varphi\nabla^2\varphi = -2\pi J\sum_{i\neq j} q_i q_j \log\frac{r_{ij}}{r_0} + \pi J\sum_{i,j} q_i q_j \log(CN). \tag{13.96}$$

Let us observe that the dependence on the total number of sites drops out in the previous expression if the set of vortices is globally neutral, i.e.

$$\sum_i q_i = 0, \tag{13.97}$$

so that in this case

$$H_{2n} = -2\pi J \sum_{i\neq j} q_i q_j \log \frac{r_{ij}}{r_0}. \qquad (13.98)$$

Trading the effective short-distance cutoff $r_0$ for the lattice spacing $\tau$ via Eq. (13.73) leads to our final form for the vortex Hamiltonian,

$$H_{2n} = -2\pi J \sum_{i\neq j} q_i q_j \log \frac{r_{ij}}{\tau} + \mu \sum_{i=1}^{2n} q_i^2, \qquad (13.99)$$

where

$$2\mu = 2\pi J\,(3\log 2 + 2\gamma_E), \qquad (13.100)$$

where $\gamma_E$ is the Euler–Mascheroni constant of Eq. (D.25), is the energy of two vortices with strength $\pm 1$ placed one lattice spacing apart. All in all, this gas of vortices turns out to be equivalent to a globally neutral plasma of electrically charged particles confined to a two-dimensional world. Let us stress that the only independent parameter for this model is the coupling $J$, since $\mu$ is only a shorthand for the combination in Eq. (13.100).

In the following we shall regard $\frac{\mu}{k_B T}$ as sufficiently large to deal with a dilute gas of vortices. This crude approximation is sufficient to capture the phenomenon of interest, since, as we argued in Eq. (13.94), vortices undergo a transition when $k_B T_c \simeq \pi J$, so that $\frac{\mu}{k_B T_c} \approx 3.23$, which is large enough for our purposes.

We shall also restrict our attention to vortices of unit strength, since larger values of $|q|$ are energetically disfavored, as suggested by Eq. (13.99), thus considering $n$ vortices of strength $+1$ and $n$ "anti"-vortices of strength $-1$. The partition function of the system is then

$$Z = \sum_{n=0}^{+\infty} \frac{1}{(n!)^2} \int_{D_{2n}(\tau)} \frac{d^2\mathbf{r}_{2n}}{\tau^2} \cdots \int_{D_1(\tau)} \frac{d^2\mathbf{r}_1}{\tau^2}\, e^{-\beta H_{2n}}, \qquad (13.101)$$

where the sum over states takes into account all possible numbers $n$ of vortex-antivortex pairs and their independent permutation symmetries. The integration domains account for all possible ways of placing the vortices at mutual distances larger than the lattice spacing, i.e. for $i = 1, 2, \ldots, 2n$,

$$D_i(\tau) = \{\mathbf{r} : |\mathbf{r} - \mathbf{r}_j| > \tau, \text{ for } j = 1, 2, \ldots, i-1\}. \qquad (13.102)$$

In this fashion, the first vortex can be placed anywhere on the plane, the second one can be placed anywhere outside a small disk centered at the position of the first, and so on.

### 13.4.4 Coarse Graining and RG Equations

Introducing for convenience the dimensionless parameters

$$Q = \sqrt{2\pi\beta J}, \qquad Q_i = Q q_i, \qquad \sigma = \beta\mu, \qquad (13.103)$$

our next task is to evaluate the infinitesimal change of the sum over states in the partition function

$$Z(Q,\sigma) = \sum_n \frac{e^{-2n\sigma}}{(n!)^2(\tau^2)^{2n}} \int_{D_{2n}(\tau)} d^2\mathbf{r}_{2n} \cdots \int_{D_1(\tau)} d^2\mathbf{r}_1 \prod_{i\neq j} \left(\frac{r_{ij}}{\tau}\right)^{Q_iQ_j} \tag{13.104}$$

induced by a small change of the lattice spacing

$$\tilde{\tau} = \tau + d\tau. \tag{13.105}$$

The $\tau$-dependence would disappear if one measured lengths in units of $\tau$. This can be seen by letting $\mathbf{r}_i = \tau\,\xi_i$, which yields

$$Z(Q,\sigma) = \sum_n \frac{e^{-2n\sigma}}{(n!)^2} \int_{D_{2n}(1)} d^2\xi_{2n} \cdots \int_{D_1(1)} d^2\xi_1 \prod_{i\neq j} \xi_{ij}^{Q_iQ_j}. \tag{13.106}$$

However, it is convenient to leave $\tau$ explicitly in the partition function, in order to see how combining variations of $\tau$ and the other parameters *defines* a given theory. In this fashion, the arbitrariness in the choice of $\tau$ can be exploited to relate theories with different values of the parameters $\sigma$ and $Q$, which acquire in this fashion a dependence on $\tau$. As in previous examples, this procedure will unveil the renormalization group flow.

Let us begin by taking into account the effect of the infinitesimal change (13.105) on the integration regions. To leading order in $d\tau$, the increase in the lattice spacing from $\tau$ to $\tau + d\tau$ can be compensated for by integrating over small annular regions of width $d\tau$ around each vortex (see Figure 13.6). This can be done by rearranging the integrals in the following manner:

$$\begin{aligned}\int_{D_{2n}(\tau)} d^2\mathbf{r}_{2n} \ldots \int_{D_1(\tau)} d^2\mathbf{r}_1 &\simeq \int_{D_{2n}(\tilde{\tau})} d^2\mathbf{r}_{2n} \ldots \int_{D_1(\tilde{\tau})} d^2\mathbf{r}_1 \\ &+ \frac{1}{2}\sum_{i\neq j} \int_{D_{2n}(\tilde{\tau})} d^2\mathbf{r}_{2n} \ldots \int_{\mathcal{A}_{ij}} d^2\mathbf{r}_j \int_{\delta_i(j)} d^2\mathbf{r}_i \ldots \int_{D_1(\tilde{\tau})} d^2\mathbf{r}_1.\end{aligned} \tag{13.107}$$

The first term on the right-hand side of Eq. (13.107) is the integration with the enlarged lattice spacing while the second one accounts, to linear order in $d\tau$, for the remaining annular regions left out by the first one, as depicted in Figure 13.6. To this end, for each pair $i, j$ the original integrations are replaced by an integration over $\mathcal{A}_{ij}$, namely over the whole plane except for circles of radius $\tau$ around the other vortices $k \neq i, j$, and over the annular region $\delta_i(j)$ of radius $\tau$ and width $d\tau$ centered at $\mathbf{r}_j$. The overall factor compensates for double counting.

One can argue that the main contributions to the correction term in Eq. (13.107) originate from pairs of vortices with opposite charge, due to their attractive interactions. Let us therefore confine our attention to annular integrations for pairs $i$ and $j$ with $Q_i = -Q_j$. These terms involve

$$\int_{\delta_i(j)} d^2\mathbf{r}_i \prod_k \left(\frac{r_{ik}}{r_{jk}}\right)^{2\,Q_iQ_k} = \tau d\tau \int_0^{2\pi} d\theta \prod_k \left(1 + \frac{2\boldsymbol{\tau}\cdot\mathbf{r}_{jk}}{r_{jk}^2} + \frac{\tau^2}{r_{jk}^2}\right)^{Q_iQ_k}, \tag{13.108}$$

where $k \neq i,j$ and $\mathbf{r}_i - \mathbf{r}_j = \boldsymbol{\tau} \equiv \tau(\cos\theta, \sin\theta)$, since each contributes twice the original expression in Eq. (13.104).

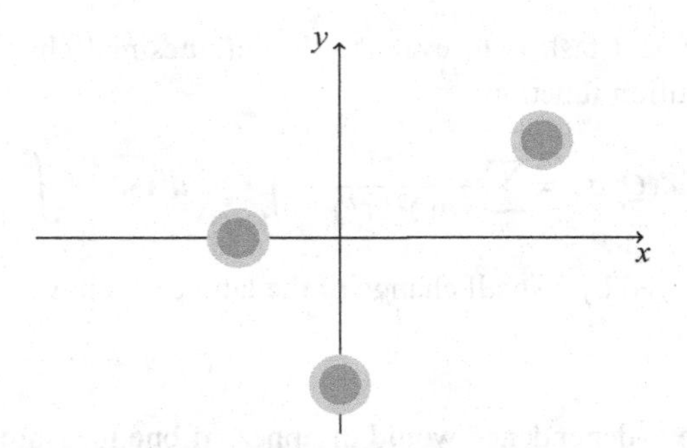


Figure 13.6 The effective size of vortices before (darker portions of discs) and after (discs including the lighter annular regions) the coarse graining procedure.

We now expand this product up to $\mathcal{O}(\frac{\tau^2}{r_{jk}^2})$, since the dilute-gas hypothesis makes it very unlikely three vortices will be close together, and therefore typically $r_{jk} \gg \tau$. Note that terms that are linear in $\tau$ will not survive the angular average, since

$$\int_0^{2\pi} \cos\theta\, d\theta = 0, \qquad \int_0^{2\pi} \cos\theta \cos(\theta - \alpha)\, d\theta = \pi \cos\alpha. \tag{13.109}$$

Consequently, the dominant terms in Eq. (13.108) are

$$\begin{aligned}
&\tau d\tau \int_0^{2\pi} d\theta \prod_k \left(1 + 2Q_iQ_k \frac{\boldsymbol{\tau}\cdot\mathbf{r}_{jk}}{r_{jk}^2} + Q_iQ_k \frac{\tau^2}{r_{jk}^2} + 2Q_iQ_k(Q_iQ_k - 1)\frac{(\boldsymbol{\tau}\cdot\mathbf{r}_{jk})^2}{r_{jk}^4}\right) \\
&\simeq \tau d\tau \int_0^{2\pi} d\theta \left(1 + \sum_k 2Q_iQ_k \frac{\boldsymbol{\tau}\cdot\mathbf{r}_{jk}}{r_{jk}^2} + Q_i^2 \sum_{k\neq l} 2Q_kQ_l \frac{\boldsymbol{\tau}\cdot\mathbf{r}_{jk}\boldsymbol{\tau}\cdot\mathbf{r}_{jl}}{r_{jk}^2 r_{jl}^2}\right. \\
&\qquad \left. + \sum_k Q_iQ_k \frac{\tau^2}{r_{jk}^2} + 2Q_i \sum_k Q_k(Q_iQ_k - 1)\frac{(\boldsymbol{\tau}\cdot\mathbf{r}_{jk})^2}{r_{jk}^4}\right) \\
&= 2\pi\tau d\tau \left(1 + Q_i^2 \tau^2 \sum_{k,l} Q_kQ_l \frac{\mathbf{r}_{jk}\cdot\mathbf{r}_{jl}}{r_{jk}^2 r_{jl}^2}\right).
\end{aligned} \tag{13.110}$$

In the first step we have expanded the product retaining the terms of the two lowest orders in $\frac{\tau}{r_{jk}}$, and in the second step we have performed the angular integration using Eq. (13.109).

One can now evaluate the remaining integral of Eq. (13.110) over $\mathcal{A}_{ij}$, noting that

$$\int_{\mathcal{A}_{ij}} d^2\mathbf{r}_j \frac{\mathbf{r}_{jk}\cdot\mathbf{r}_{jl}}{r_{jk}^2 r_{jl}^2} = \int_{\mathcal{A}_{ij}} d^2\mathbf{r}_j \nabla_j \log\frac{r_{jk}}{\tau} \cdot \nabla_j \log\frac{r_{jl}}{\tau} \tag{13.111}$$

and integrating by parts. In this way, only the boundary terms survive, since $\nabla^2 \log\frac{r}{\tau}$ vanishes away from the singularity, and one gets

$$\begin{aligned}
-\sum_{i'} \int_0^{2\pi} d\theta\, \frac{\tau^2 + \boldsymbol{\tau}\cdot\mathbf{r}_{i'k}}{|\boldsymbol{\tau} + \mathbf{r}_{i'k}|^2} \log\frac{|\boldsymbol{\tau} + \mathbf{r}_{i'l}|}{\tau} &\simeq -\sum_{i'\neq l} \int_0^{2\pi} d\theta\, \frac{\tau^2 + \boldsymbol{\tau}\cdot\mathbf{r}_{i'k}}{|\boldsymbol{\tau} + \mathbf{r}_{i'k}|^2} \log\frac{r_{i'l}}{\tau} \\
&\simeq -2\pi \sum_{i'\neq l} \delta_{i'k} \log\frac{r_{i'l}}{\tau},
\end{aligned} \tag{13.112}$$

since the diluteness condition implies that $\mathbf{r}_{i'k} \gg \tau$, after performing the angular averages as above. Note also that the annular integration has effectively removed the pair of vortices labeled by $i$ and $j$. In conclusion, the integration over $\delta_i(j)$ and $\mathcal{A}_{ij}$ leads to the remarkably simple result

$$2\pi\tau d\tau\left(A - 2\pi\tau^2 Q^2 \sum_{k\neq l} Q_k Q_l \log\frac{r_{kl}}{\tau}\right) \tag{13.113}$$

where $A$ is the area of the system and $k, l \neq i, j$.

The whole partition function resulting from the two terms in Eq. (13.107) is now

$$Z = \sum_{n=0}^{\infty} \frac{e^{-2n\sigma}}{(n!)^2(\tau^2)^{2n}} \int_{D_{2n}(\tilde{\tau})} d^2\mathbf{r}_{2n} \cdots \int_{D_1(\tilde{\tau})} d^2\mathbf{r}_1 \\ \times \left[1 + \frac{e^{-2\sigma}2\pi\tau d\tau}{(n+1)^2(\tau^2)^2} \sum_{i=1}^{n+1}\sum_{j=1}^{n+1}\left(A - 2\pi\tau^2 Q^2 \sum_{k\neq l} Q_k Q_l \log\frac{r_{kl}}{\tau}\right)\right] \prod_{k\neq l}\left(\frac{r_{kl}}{\tau}\right)^{Q_k Q_l}, \tag{13.114}$$

where we have let $n \to n+1$ in the terms where the pair $ij$ had been averaged out. Note also that, once one restricts the sum to pairs of opposite charge, the factor $\frac{1}{2}$ is not needed anymore. The sums inside the square brackets run independently from 1 to $n+1$ and give a factor $(n+1)^2$ since all dependence on $i$ and $j$ has disappeared. As a result, exponentiating the content of the square brackets, which only introduces errors to order $\mathcal{O}((d\tau)^2)$ or higher,

$$Z = e^{2\pi A e^{-2\sigma}\frac{d\tau}{\tau^3}} \sum_{n=0}^{\infty} \frac{e^{-2n\sigma}}{(n!)^2(\tau^2)^{2n}} \int_{D_{2n}(\tilde{\tau})} d^2\mathbf{r}_{2n} \cdots \int_{D_1(\tilde{\tau})} d^2\mathbf{r}_1 \\ \times \prod_{k\neq l}\left(\frac{r_{kl}}{\tau}\right)^{Q_k Q_l\left(1-e^{-2\sigma}(2\pi Q)^2\frac{d\tau}{\tau}\right)}. \tag{13.115}$$

This procedure has successfully rescaled the dependence on $\tau$ originating in the integration over vortex positions. In addition, accounting for the rescaling of the explicit powers of $\tau$ is simple, since it suffices to note that, again to leading order in $d\tau$,

$$\frac{1}{(\tau^2)^{2n}} \prod_{k\neq l} \tau^{-Q_k Q_l} = \frac{e^{-2n(Q^2-2)\frac{d\tau}{\tau}}}{(\tilde{\tau}^2)^{2n}} \prod_{k\neq l} \tilde{\tau}^{-Q_k Q_l}, \tag{13.116}$$

where we have used the charge neutrality condition (13.97), which implies that

$$\sum_{i\neq j} Q_i Q_j = -\sum_{i=1}^{2n} Q_i^2 = -2nQ^2, \tag{13.117}$$

for $2n$ (anti)vortices with $Q_i = \pm Q$.

To summarize, under the infinitesimal coarse graining of Eq. (13.105),

$$Z(Q,\sigma) = Z_0\, Z(\tilde{Q},\tilde{\sigma}), \tag{13.118}$$

with

$$\tilde{Q}^2 = Q^2\left(1 - e^{-2\sigma}(2\pi Q)^2\frac{d\tau}{\tau}\right),$$
$$\tilde{\sigma} = \sigma + (Q^2 - 2)\frac{d\tau}{\tau}, \tag{13.119}$$
$$\log Z_0 = \frac{e^{-2\sigma}}{\tau^2}\,2\pi A\,\frac{d\tau}{\tau}.$$

These equations define the renormalization group flow for the XY model. Note that although the "bare" parameters $Q^2$ and $\sigma$ were actually related to each other, the corresponding "dressed" parameters $\tilde{Q}^2$ and $\tilde{\sigma}$ evolve independently under the renormalization group flow.

We can also discuss the effect of the coarse graining on average observables for the gas of vortices. To this end, let us consider

$$\left\langle e^{i\sum_I \bar{q}_I\phi(\mathbf{R}_I)}\right\rangle = \left\langle e^{-i\sum_{i,I} q_i\bar{q}_I \arg(\mathbf{R}_I - \mathbf{r}_i)}\right\rangle \tag{13.120}$$

with suitable "sources" labeled by an uppercase index $I$, which can be obtained from the modified partition function

$$Z(Q,\sigma)[\bar{q}] = \sum_n \frac{e^{-2n\sigma}}{(n!)^2(\tau^2)^{2n}}\int_{D_{2n}(\tau)} d\mathbf{r}_{2n}\cdots\int_{D_1(\tau)} d^2\mathbf{r}_1 \prod_{i\neq j}\left(\frac{r_{ij}}{\tau}\right)^{Q_iQ_j} e^{-i\sum_{i,I} q_i\bar{q}_I \arg(\mathbf{R}_{Ii})}, \tag{13.121}$$

with $\mathbf{R}_{Ii} = \mathbf{R}_I - \mathbf{r}_i$. In this case, the domains $D_1, \ldots, D_{2n}$ also exclude small disks of radius $\tau$ around the sources. The analysis proceeds as for $Z(Q,\sigma)$, pairing vortices of opposite charge to perform the annular integrations. Vortex-source pairings could also be relevant to leading order in $d\tau$, but vanish since they would violate the neutrality condition, while multiple pairings only contribute to higher order in $d\tau$. For a vortex-antivortex pair $ij$, the counterpart of the expression in Eq. (13.108) reads in this case

$$\tau d\tau \int_0^{2\pi} d\theta \prod_k \left(1 + \frac{2\boldsymbol{\tau}\cdot\mathbf{r}_{jk}}{r_{jk}^2} + \frac{\tau^2}{r_{jk}^2}\right)^{Q_iQ_k} e^{-iq_i\sum_I \bar{q}_I\left[\arg(\mathbf{R}_{Ij} - \boldsymbol{\tau}) - \arg(\mathbf{R}_{Ij})\right]}. \tag{13.122}$$

Expanding for small $\tau$, according to the diluteness hypothesis, and performing the angular integrals leads to

$$2\pi\tau d\tau\left(1 + Q^2\tau^2\sum_{k,l} Q_kQ_l\frac{\mathbf{r}_{jk}\cdot\mathbf{r}_{jl}}{r_{jk}^2 r_{jl}^2} - \frac{\tau^2}{4}\sum_{I,J}\bar{q}_I\bar{q}_J\frac{\mathbf{R}_{Ij}\cdot\mathbf{R}_{Jj}}{R_{Ij}^2R_{Jj}^2}\right.$$
$$\left. + \frac{iQ\tau^2}{2}\sum_{k,I} Q_k\bar{q}_I\nabla_j\log\frac{r_{jk}}{\tau}\cdot\nabla_j\arg(\mathbf{R}_{Ij})\right), \tag{13.123}$$

and integrating by parts to evaluate the integral over $\mathcal{A}_{ij}$ gives

$$2\pi\tau d\tau\left(A - 2\pi\tau^2Q^2\sum_{k\neq l} Q_kQ_l\log\frac{r_{kl}}{\tau} + \frac{\pi\tau^2}{2}\sum_{I\neq J}\bar{q}_I\bar{q}_J\log\frac{R_{IJ}}{\tau}\right). \tag{13.124}$$

In particular, the imaginary cross term in the second line of Eq. (13.123) drops out because, integrating by parts, the *first* factor gives

$$\int_{A_{ij}} d^2\mathbf{r}_j \nabla_j \log\frac{r_{jk}}{\tau}\cdot\nabla_j \arg(\mathbf{R}_{Ij}) = -\sum_{\ell}\int_0^{2\pi} d\theta \, \frac{(\tau\times\mathbf{R}_{I\ell})_z}{|\mathbf{R}_{I\ell}-\tau|^2}\log\frac{|\tau+\mathbf{r}_{\ell k}|}{\tau}$$
$$-\sum_J\int_0^{2\pi} d\theta\,\frac{(\tau\times\mathbf{R}_{IJ})_z}{|\mathbf{R}_{IJ}-\tau|^2}\log\frac{|\tau+\mathbf{R}_{Jk}|}{\tau} \tag{13.125}$$

and all these terms vanish in the dilute limit. Taking the new term in Eq. (13.124) into account, one finds

$$\left\langle e^{i\sum_I \bar q_I \phi(\mathbf{R}_I)}\right\rangle_\tau = \left\langle e^{i\sum_I \bar q_I \phi(\mathbf{R}_I)}\right\rangle_{\tau+d\tau} e^{\pi^2 \frac{d\tau}{\tau} e^{-2\sigma}\sum_{I\neq J}\bar q_I \bar q_J \log\frac{R_{IJ}}{\tau}}, \tag{13.126}$$

and in particular the vortex contribution to the spin-spin correlation function satisfies

$$\left\langle \cos(\phi(\mathbf{R}_I)-\phi(\mathbf{R}_J))\right\rangle_\tau = \left\langle \cos(\phi(\mathbf{R}_I)-\phi(\mathbf{R}_J))\right\rangle_{\tau+d\tau} e^{-2\pi^2\frac{d\tau}{\tau}e^{-2\sigma}\log\frac{R_{IJ}}{\tau}}. \tag{13.127}$$

Amusingly enough, the very same scaling would be obtained by substituting the first equation of the RG system (13.119) into the spin-wave two-point function (13.75).

### 13.4.5 The BKT Transition

Equations (13.119) determine the existence of a fixed point at $Q^2 = 2$ for $\sigma \to \infty$. This fixed point signals the presence of a phase transition where the vortices, bound together in neutral pairs for small $T$ or large $J$, deconfine as $T$ increases or $J$ decreases. This peculiar critical phenomenon was first discussed by Berezinski, Kosterlitz, and Thouless (BKT), and lies at the heart of the modern understanding of topological phases of matter.

In order to discuss further the system (13.119), it is convenient to introduce the new variables

$$x = Q^2 - 2, \qquad y = 4\pi e^{-\sigma}, \tag{13.128}$$

in terms of which the renormalization group equations read

$$\tau\,\frac{dx}{d\tau} = -(x+2)^2\left(\frac{y}{2}\right)^2, \qquad \tau\,\frac{dy}{d\tau} = -xy. \tag{13.129}$$

Linearizing in the vicinity of the fixed point $x = 0$, $y = 0$, one is thus led to

$$\tau\,\frac{dx}{d\tau} = -y^2, \qquad \tau\,\frac{dy}{d\tau} = -xy. \tag{13.130}$$

Note that $d(x^2-y^2) = 0$, for this system, so that the renormalization group trajectories are the portions of the hyperbolas

$$x^2 - y^2 = x_0^2, \tag{13.131}$$

in the upper half of the $xy$ plane depicted in Figure 13.7. In particular, the critical trajectory is obtained for $x_0 = 0$ and corresponds to the asymptote in the first

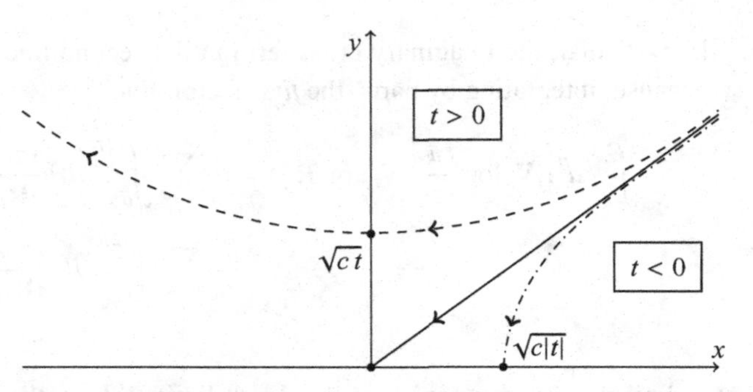


Figure 13.7 Renormalization group trajectories in a neighborhood of the critical point, which lies at the origin of the $xy$ plane. The critical trajectory is the solid line. The dashed and dash-dotted curves are two sample trajectories for $T$ above and below the critical temperature $T_c$.

quadrant. Recalling the definitions in Eqs. (13.100), (13.103), and (13.128), the critical trajectory is thus characterized by the condition

$$\frac{\pi J}{k_B T_c} - 1 = 2\pi \left(8\, e^{2\gamma_E}\right)^{-\frac{\pi J}{k_B T_c}}, \tag{13.132}$$

whose numerical solution lowers the naive critical temperature of Eq. (13.94) to

$$T_c \simeq 2.73 \times \frac{J}{k_B}. \tag{13.133}$$

Note that $T < T_c$ corresponds to the region where $x_0^2 > 0$ and $x > y$, while $T > T_c$ corresponds to the region $x_0^2 < 0$ and $|x| < y$. In fact, close to the critical temperature, where $T = T_c(1 + t)$ with $|t| \ll 1$, Eq. (13.131) yields

$$x_0^2 = -c\, t, \qquad \text{with } c \approx 2.08, \tag{13.134}$$

to leading order in $t$.

It is instructive to analyze the renormalization group curves in the vicinity of the fixed point, and to identify where significant deviations from the linear approximation are bound to occur. The maximum value attained by $\tau$ before this breakdown naturally identifies the correlation length $\lambda$.

Below the critical temperature, $t < 0$, substituting (13.131) into the first of Eqs. (13.130) one finds

$$\frac{dx}{x^2 - c|t|} = -\frac{d\tau}{\tau}, \tag{13.135}$$

and the renormalization group trajectory is therefore

$$x(\tau) = \sqrt{c|t|}\, \coth\left(\sqrt{c|t|}\, \log \frac{\tau}{\tau_\infty}\right) \tag{13.136}$$

in terms of a small scale $\tau_\infty$ such that $\tau > \tau_\infty$. This corresponds to the dash-dotted trajectory in Figure (13.7). Thus $x$ never vanishes in this region, and consequently there are no turning points in $y(\tau)$. In fact, $x$ approaches the finite value $\sqrt{c|t|}$ as $\tau \to \infty$, while $y$ tends to zero. The linearized theory in the vicinity of the fixed point is thus valid for arbitrarily large values of $\tau$ and this means that the correlation length is infinite:

$$\lambda = \infty. \tag{13.137}$$

Indeed, below the critical temperature all vortices are bound in pairs, and the two-point function is dominated by the long-range spin wave contributions of Eq. (13.75).

Above the critical temperature, $t > 0$, the counterpart of Eq. (13.135) is

$$\frac{dx}{x^2 + c\,t} = -\frac{d\tau}{\tau}, \tag{13.138}$$

from which

$$x(\tau) = \sqrt{c\,t}\,\cot\left(\sqrt{c\,t}\log\frac{\tau}{\tau_\infty}\right), \tag{13.139}$$

for $\tau_\infty < \tau < \tau_\infty e^{\frac{\pi}{\sqrt{ct}}}$. This corresponds to the dashed trajectory in fig. 13.7. This quantity vanishes at $\tau = \tau_\infty e^{\frac{\pi}{2\sqrt{ct}}}$, and blows up at

$$\tau = \tau_\infty e^{\frac{\pi}{\sqrt{ct}}}, \tag{13.140}$$

which identifies the correlation length as

$$\lambda = \tau_\infty e^{\frac{\pi}{\sqrt{ct}}} \sim e^{\frac{b}{\sqrt{t}}}, \qquad \text{with } b \simeq 2.22. \tag{13.141}$$

Note that as $t \to 0^+$ and vortices recombine into neutral pairs, the correlation length diverges as expected, but faster than any power, so that the exponent $\nu$ does not exist for this system, or formally

$$\nu = \infty. \tag{13.142}$$

One can make a different choice for the solutions (13.136) and (13.139), so that $x$ equals a given value $x_*$ at a reference scale $\tau_*$. These read

$$\frac{\tau}{\tau_*} = \exp\left[\frac{1}{\sqrt{c|t|}}\left(\coth^{-1}\frac{x}{\sqrt{c|t|}} - \coth^{-1}\frac{x_*}{\sqrt{c|t|}}\right)\right], \tag{13.143}$$

$$\frac{\tau}{\tau_*} = \exp\left[\frac{1}{\sqrt{c\,t}}\left(\cot^{-1}\frac{x}{\sqrt{c\,t}} - \cot^{-1}\frac{x_*}{\sqrt{c\,t}}\right)\right] \tag{13.144}$$

below and above $T_c$. As $t \to 0$, both Eq. (13.136) and Eq. (13.139) approach

$$x(\tau) = \frac{1}{\log\frac{\tau}{\tau_\infty}}, \tag{13.145}$$

which is exact at the critical temperature. This trajectory follows the straight (solid) line in Figure 13.7 and approaches the origin, the transition's fixed point, as $\tau \to \infty$. Similarly, at $t = 0$, (13.143) and (13.144) reduce to

$$\log\frac{\tau}{\tau_*} = \frac{1}{x} - \frac{1}{x_*}. \tag{13.146}$$

Another peculiarity of this phase transition, together with the essential singularity in $t$ exhibited by the correlation length, is that *the ordered phase is not characterized by a local microscopic order parameter*, such as for instance the local magnetization for the ferromagnetic phase. Rather, *vortex deconfinement is signaled by nonzero line integrals of $\nabla\phi$ around closed curves*, which thus represent global, topological counterparts of the usual order parameters.

### 13.4.6 Critical Exponents

Leaving aside the critical exponent $\nu$, which does not exist in a conventional sense for this system, as we have seen, let us now turn to the other microscopic exponent, $\eta$, which is defined by

$$\langle \mathbf{s}(\mathbf{r}_i) \cdot \mathbf{s}(\mathbf{r}_j) \rangle \sim r_{ij}^{-\eta} \tag{13.147}$$

as $r_{ij} \to \infty$ for $T = T_c$ in two dimensions. In order to calculate the two-point function, one must consider not only the contribution of vortex configurations, but also small fluctuations around them. The spin orientations can be thus expanded as $\phi(\mathbf{r}) = \bar{\phi}(\mathbf{r}) + \psi(\mathbf{r})$, where, as in (13.95),

$$\bar{\phi}(\mathbf{r}) = -\sum_i q_i \arg(\mathbf{r} - \mathbf{r}_i) \tag{13.148}$$

are the vortex contributions, and $\psi(\mathbf{r})$ are small spin-wave fluctuations. The corresponding dynamics can be treated as decoupled and are described by (13.55) and (13.99). The spin-spin correlation function is thus given by

$$\langle \mathbf{s}(\mathbf{r}_i) \cdot \mathbf{s}(\mathbf{r}_j) \rangle = \langle \cos(\bar{\phi}(\mathbf{r}_i) - \bar{\phi}(\mathbf{r}_j)) \rangle \langle \cos(\psi(\mathbf{r}_i) - \psi(\mathbf{r}_j)) \rangle \tag{13.149}$$

because $\langle \sin(\psi(\mathbf{r}_i) - \psi(\mathbf{r}_j)) \rangle$ vanishes for spin waves by (13.74). The spin-wave factor is given by (13.75), that is

$$\langle \cos(\psi(\mathbf{r}_i) - \psi(\mathbf{r}_j)) \rangle \simeq \left( \frac{r_{ij}}{\tau_*} \right)^{-\frac{1}{2(2+x_*)}} \tag{13.150}$$

at the scale $\tau_*$. The vortex factor is not easy to evaluate in closed form, but, since for $T \leq T_c$ the correlation length is infinite, we may start from the scaling relation (13.127) and integrate it all the way up to $\tau = r_{ij}$, obtaining

$$\langle \cos(\bar{\phi}(\mathbf{r}_i) - \bar{\phi}(\mathbf{r}_j)) \rangle = \exp\left( -\frac{1}{8} \int_{\tau_*}^{r_{ij}} \frac{d\tau}{\tau} y^2 \log \frac{r_{ij}}{\tau} \right) \tag{13.151}$$

because

$$\langle \cos(\bar{\phi}(\mathbf{r}_i) - \bar{\phi}(\mathbf{r}_j)) \rangle_{\tau = r_{ij}} = 1. \tag{13.152}$$

Using the first of Eqs. (13.130) to change integration variable,

$$\langle \cos(\bar{\phi}(\mathbf{r}_i) - \bar{\phi}(\mathbf{r}_j)) \rangle = \exp\left( \frac{1}{8} \int_{x_*}^{x(r_{ij})} dx \log \frac{r_{ij}}{\tau} \right), \tag{13.153}$$

and expressing $\tau$ in terms of $x$ at $T = T_c$ via (13.146),

$$\langle \cos(\bar{\phi}(\mathbf{r}_i) - \bar{\phi}(\mathbf{r}_j)) \rangle = \exp\left( \tfrac{x(r_{ij}) - x_*}{8} \log \frac{r_{ij}}{\tau_*} - \frac{1}{8} \log \tfrac{x(r_{ij})}{x_*} + \frac{1}{8x_*}(x(r_{ij}) - x_*) \right). \tag{13.154}$$

Since, again by Eq. (13.146), $x(r_{ij}) \simeq \left( \log \frac{r_{ij}}{\tau_*} \right)^{-1}$ for large $r_{ij}$, one finds

$$\langle \cos(\bar{\phi}(\mathbf{r}_i) - \bar{\phi}(\mathbf{r}_j)) \rangle \simeq \left( \frac{r_{ij}}{\tau_*} \right)^{-\frac{x_*}{8}} \left( x_* \log \frac{r_{ij}}{\tau_*} \right)^{\frac{1}{8}} e^{\left( 8x_* \log \frac{r_{ij}}{\tau_*} \right)^{-1}}, \tag{13.155}$$

so that, up to slowly varying logarithmic enhancement,

$$\langle\cos(\bar{\phi}(\mathbf{r}_i)-\bar{\phi}(\mathbf{r}_j))\rangle\sim\left(\frac{r_{ij}}{\tau_*}\right)^{-\frac{1}{8}x_*}. \tag{13.156}$$

The leading-order behavior of the spin-spin correlator, obtained by substituting Eqs. (13.150) and (13.156) into (13.149) at $T=T_c$, is thus

$$\langle\mathbf{s}(\mathbf{r}_i)\cdot\mathbf{s}(\mathbf{r}_j)\rangle\sim\left(\frac{r_{ij}}{\tau_*}\right)^{-\frac{1}{2(2+x_*)}}\left(\frac{r_{ij}}{\tau_*}\right)^{-\frac{1}{8}x_*}\simeq\left(\frac{r_{ij}}{\tau_*}\right)^{-\frac{1}{4}}, \tag{13.157}$$

to leading order in $x_*$, so that

$$\eta=\frac{1}{4} \tag{13.158}$$

for the XY model. Note that this is the value that we would have obtained if we only took into account the spin wave factor (13.75) at the naive critical temperature (13.94).

The susceptibility $\chi$ is linked to the spin-spin correlation function by the fluctuation-dissipation theorem that we encountered in Chapter 10 and shall discuss further in Chapter 14, according to which

$$\chi\sim\int d^2\mathbf{r}\,\langle\mathbf{s}(0)\cdot\mathbf{s}(\mathbf{r})\rangle. \tag{13.159}$$

Below $T_c$, the correlation function will decay even more slowly than $r^{-\frac{1}{4}}$, so that clearly

$$\chi=\infty. \tag{13.160}$$

On the other hand, above $T_c$, we expect a rapid falloff in the correlation function, $\langle\mathbf{s}(0)\cdot\mathbf{s}(\mathbf{r})\rangle\sim r^{-\frac{1}{4}}f(r/\lambda)$ where $f$ is rapidly decreasing and $\lambda$ is the correlation length. Therefore,

$$\chi\sim\int d^2\mathbf{r}\,r^{-\frac{1}{4}}f(r/\lambda)\sim\lambda^{\frac{7}{4}}, \tag{13.161}$$

as can be seen by rescaling the integration variable by $\lambda$.

In order to discuss the exponent $\delta$, we need to turn on an external field $h$, whose effect is to multiply $Z$ by

$$\left\langle e^{\beta h\int d^2\mathbf{r}\cos\phi(\mathbf{r})}\right\rangle=\sum_{m=0}^{\infty}\frac{(\beta h)^m}{2^m m!}\int d^2\mathbf{r}_1\cdots\int d^2\mathbf{r}_m\sum_{q_1,\ldots,q_m}\left\langle e^{\sum_j q_j\phi(\mathbf{r}_j)}\right\rangle, \tag{13.162}$$

where $q_j=\pm1$. Separating as before spin-wave and vortex factors, one obtains

$$\left\langle e^{\sum_j q_j\phi(\mathbf{r}_j)}\right\rangle=\left\langle e^{\sum_j q_j\bar{\phi}(\mathbf{r}_j)}\right\rangle\left\langle e^{\sum_j q_j\psi(\mathbf{r}_j)}\right\rangle, \tag{13.163}$$

and the analogs of Eqs. (13.150) and (13.156) in this context are, for globally neutral sets of $q_j$,

$$\left\langle e^{\sum_j q_j\psi(\mathbf{r}_j)}\right\rangle=\prod_{i\neq j}\left(\frac{r_{ij}}{\tau_*}\right)^{\frac{q_iq_j}{4(2+x_*)}},\qquad\left\langle e^{\sum_j q_j\bar{\phi}(\mathbf{r}_j)}\right\rangle\sim\prod_{i\neq j}\left(\frac{r_{ij}}{\tau_*}\right)^{\frac{q_iq_jx_*}{16}} \tag{13.164}$$

at $T = T_c$. Therefore,

$$\left\langle e^{\beta h \int d^2\mathbf{r} \cos\phi(\mathbf{r})} \right\rangle \sim \sum_{m=0}^{\infty} \frac{(\beta h)^m}{2^m m!} \int d^2\mathbf{r}_1 \cdots \int d^2\mathbf{r}_m \sum_{q_1,\ldots,q_m} \prod_{i \neq j} \left(\frac{r_{ij}}{\tau_*}\right)^{\frac{q_i q_j}{8}}, \quad (13.165)$$

where the dependence on $\beta h$ and on the size $L$ clearly enters via the combination

$$u = \beta h L^{\frac{15}{8}}, \quad (13.166)$$

as can be seen by rescaling the integration variables by $L$ and using the neutrality condition (13.97). In order to be properly extensive, the resulting free energy should therefore scale as $u^{\frac{16}{15}}$, i.e. proportionally to $h^{\frac{16}{15}}$. Therefore, taking a derivative with respect to $h$, we see that the mean magnetization at $T = T_c$ must scale as $h^{\frac{1}{15}}$, and in conclusion

$$\delta = 15. \quad (13.167)$$

In the absence of the magnetic field, the behavior of the free energy can be obtained by starting from the third of Eqs. (13.119), which one can write as

$$d \log Z = \frac{A}{8\pi} y^2 \frac{d\tau}{\tau^3}, \quad (13.168)$$

or, using the first of Eqs. (13.130),

$$d \log Z = -\frac{A}{8\pi} \frac{dx}{\tau^2}. \quad (13.169)$$

This equation can be integrated using the explicit expressions (13.143) and (13.144). Below the critical temperature, $t < 0$, we find

$$\frac{1}{A} \log Z = \int_{\tau_*}^{\infty} \frac{y^2 d\tau}{8\pi\tau^3} = \int_{\sqrt{c|t|}}^{x_*} \frac{dx}{8\pi\tau_*^2} \exp\left[-\frac{2}{\sqrt{c|t|}}\left(\coth^{-1}\frac{x}{\sqrt{c|t|}} - \coth^{-1}\frac{x_*}{\sqrt{c|t|}}\right)\right]. \quad (13.170)$$

In the limit $t \to 0^-$, we may expand the integrand of Eq. (13.170) as a power series in $|t|$, obtaining

$$\frac{1}{A} \log Z \simeq \frac{e^{\frac{2}{x_*}}}{8\pi\tau_*^2}\left(1 + \frac{2c|t|}{3x_*^3} + \cdots\right) \int_0^{x_*} e^{-\frac{2}{x}}\left(1 - \frac{2c|t|}{3x^3} + \cdots\right) dx, \quad (13.171)$$

which yields an asymptotic expansion in powers of $|t|$. By extending the integration region to 0, as in (13.171), one only introduces errors bounded by

$$\int_0^{\sqrt{c|t|}} x^{-n} e^{-\frac{2}{x}}\, dx < (c|t|)^{\frac{1}{2}-n}\, e^{-\frac{2}{\sqrt{c|t|}}}. \quad (13.172)$$

Above the critical temperature, $t > 0$, one finds instead

$$\frac{1}{A} \log Z = \frac{1}{8\pi} \int_{\tau_*}^{\bar{\lambda}} y^2 \frac{d\tau}{\tau^3} \simeq \frac{1}{8\pi} \int_{-1}^{x_*} \frac{dx}{\tau_*^2} \exp\left[-\frac{2}{\sqrt{ct}}\left(\cot^{-1}\frac{x}{\sqrt{ct}} - \cot^{-1}\frac{x_*}{\sqrt{ct}}\right)\right], \quad (13.173)$$

where $\bar{\lambda}$, such that $x(\bar{\lambda}) \simeq -1$, is a cutoff scale much smaller than the correlation length from Eq. (13.141) beyond which deviations from the linearized description become dominant. We now separate the integration interval into two regions, $-1 < x < \sqrt{c\,t}$ and $\sqrt{c\,t} < x < x_*$. The contribution due to the first region can be bounded, noting that

$$\int_{-1}^{\sqrt{c\,t}} e^{-\frac{2}{\sqrt{c\,t}} \cot^{-1} \frac{x}{\sqrt{c\,t}}}\, dx \lesssim e^{-\frac{\pi}{\sqrt{c\,t}}}. \tag{13.174}$$

The second region yields instead, proceeding as for $t < 0$, the asymptotic series

$$\frac{1}{A} \log Z \simeq \frac{e^{\frac{2}{x_*}}}{8\pi\tau_*^2} \left(1 - \frac{2c\,t}{3x_*^3} + \cdots\right) \int_0^{x_*} e^{-\frac{2}{x}} \left(1 + \frac{2c\,t}{3x^3} + \cdots\right) dx, \tag{13.175}$$

up to errors bounded by (13.172). As a result, both above and below the critical temperature, the only nonanalytic contributions to the free energy per unit area $-\frac{1}{A} k_B T \log Z$ are bounded by (13.172) and (13.174), whose derivatives are all vanishing at $t = 0$, and therefore the specific heat is regular at $T = T_c$.

Let us conclude with the following amusing observation. Although the scaling relations between critical exponents discussed in Chapter 10 cannot hold in their standard form here, simply because some of them do not exist, they actually continue to hold in a modified form. One should measure the singular parts of the specific heat $C$ and of the susceptibility $\chi$ in the vicinity of the critical point in terms of the correlation length $\lambda$, rather than in terms of $t$, so that

$$C \sim \lambda^{\tilde{\alpha}}, \qquad \chi \sim \lambda^{\tilde{\gamma}}, \tag{13.176}$$

which is equivalent, in this context, to dividing the standard exponents $\alpha$ and $\gamma$ by $\nu$ and sending $\nu \to \infty$. In this fashion, for $D = 2$,

$$\tilde{\alpha} = -D, \qquad \tilde{\gamma} = 2 - \eta, \qquad \delta = \frac{D + 2 - \eta}{D - 2 + \eta}. \tag{13.177}$$

The exponent $\beta$ is not available because there is no microscopic order parameter, as already remarked on at the end of the previous section.

## 13.5 $\epsilon$-Expansion and the $D = 3$ Ising Model

In this section we proceed further in our peek at more advanced topics, taking a closer look at the partition function of the Ising model in the continuum limit,

$$Z[j] = \int [\mathcal{D}\Phi(x)]\, e^{-\int d^D x \left[\frac{1}{2}\nabla\Phi\cdot\nabla\Phi + \frac{1}{2} m^2 \Phi^2(x) + \frac{\lambda}{4!}\Phi^4(x) - j(x)\Phi(x)\right]}. \tag{13.178}$$

Our aim will be to attain an improvement on the mean-field estimates of critical exponents, and for definiteness we shall focus on the temperature range where $m^2 > 0$, thus approaching the critical point from above. We shall also consider the model in a generic number $D$ of dimensions, a choice that will prove essential for the ensuing arguments. There are, however, a few significant differences with respect to the partition function introduced in Chapter 10: we have added a coefficient $\lambda$ in front of the quartic term, together with a different normalization, and we have also absorbed the factor $\beta$ in $\lambda$ and in the normalization of $\Phi$. The current presentation thus follows more closely the standard lines of reasoning in quantum field theory, so that energy, mass, and inverse length are measured in identical units.

Equation (13.178) can be regarded as a generating function for correlators of products of the field $\Phi(x)$ at arbitrary points,

$$\frac{1}{Z[0]}\frac{\delta^n Z[j]}{\delta j(x_1)\cdots\delta j(x_n)}\bigg|_{j=0} = \langle\Phi(x_1)\cdots\Phi(x_n)\rangle. \tag{13.179}$$

It is convenient to also consider its Gaussian counterpart,

$$Z_0[j] = \int [D\Phi(x)]\, e^{-\int d^Dx\left[\frac{1}{2}\nabla\Phi\cdot\nabla\Phi + \frac{1}{2}m^2\Phi^2(x) - j(x)\Phi(x)\right]}, \tag{13.180}$$

which can be computed directly, proceeding as in Chapter 3. To this end, it suffices to split $\Phi$ as

$$\Phi(x) = \Phi_0(x) + \delta\,\Phi(x), \tag{13.181}$$

where $\Phi_0(x)$ is a solution of the classical equation of motion

$$\left(\nabla^2 - m^2\right)\Phi_0(x) = -j(x). \tag{13.182}$$

Then

$$\Phi_0(x) = \int d^Dx'\, K(x-x')j(x'), \tag{13.183}$$

with

$$K(x-y) = \int \frac{d^Dk}{(2\pi)^D}\frac{e^{ik\cdot(x-y)}}{k^2 + m^2}. \tag{13.184}$$

If the fields decay sufficiently fast at infinity, the Euclidean action appearing in the exponent in Eq. (13.180), when evaluated on $\Phi_0(x)$, can be recast in the form

$$\begin{aligned}&\int d^Dx\left[\frac{1}{2}\nabla\Phi_0\cdot\nabla\Phi_0 + \frac{1}{2}m^2\Phi_0^2(x) - j(x)\Phi_0(x)\right]\\ &= -\frac{1}{2}\int d^Dx\, d^Dy\, j(x)\, K(x-y)\, j(y),\end{aligned} \tag{13.185}$$

after a partial integration. Therefore

$$Z_0[j] \sim e^{\frac{1}{2}\int d^Dx\, d^Dy\, j(x)\, K(x-y)\, j(y)}, \tag{13.186}$$

up to a constant overall factor resulting from the functional integral over the shifted field $\delta\,\Phi$. This normalization, however, cancels out in correlation functions, since for instance

$$\langle\Phi(x)\rangle_j = \frac{1}{Z_0[0]}\frac{\delta Z_0[j]}{\delta j(x)}, \tag{13.187}$$

and therefore we do not need to take it into account explicitly.

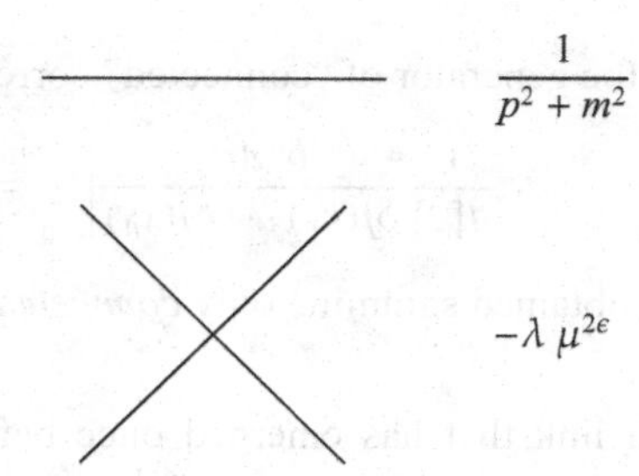


Figure 13.8 Feynman rules for the $\phi^4$ Lagrangian in Eq. (13.178).

The preceding steps also lead to a formal link between $Z[j]$ and $Z_0[j]$. Since insertions of $\Phi(x)$ in the functional integral can be traded for derivatives with respect to $j(x)$, as in (13.179), one can express the full partition function in terms of the Gaussian one, writing

$$Z[j] = e^{-\frac{\lambda}{4!}\int d^D x \frac{\delta^4}{\delta J(x)^4}}\, Z_0[j]. \tag{13.188}$$

This expression, which should be regarded as a power series in $\lambda$, is still rather formal, since the ill-defined $K(0)$ is ubiquitous in it. Later on, we shall see that the proper handling of such terms actually lies at the heart of renormalization.

Feynman's celebrated intuition turned the riddle of terms arising from the perturbative expansion of Eq. (13.188) into a suggestive set of diagrams, built from the two ingredients illustrated in Figure 13.8. The first is the "propagator,"

$$\frac{1}{p^2 + m^2}, \tag{13.189}$$

while the second is the "vertex"

$$-\lambda\, \mu^{2\epsilon}. \tag{13.190}$$

Here the theory is defined in

$$D = 4 - 2\epsilon \tag{13.191}$$

dimensions while also insisting on retaining a dimensionless $\lambda$, as in $D=4$. Therefore, $\lambda$ is accompanied by a factor $\mu^{2\epsilon}$, with $\mu$ a parameter with the dimension of mass like $m$, since the field $\Phi$ acquires a dimension $1-\epsilon$ in view of its kinetic term. Feynman's recipe for calculating $n$-point correlators in momentum space is the following: one is to sum all diagrams with $n$ external legs that can be constructed by combining the two building blocks, propagator and vertex, while also integrating over the independent momenta that are left once the external ones are fixed. Each diagram must be divided by a combinatoric factor, the dimension of the symmetry group that can turn it into itself if the external lines are held fixed.

In quantum field theory $Z[j]$ is the generating function of ordinary correlators, as in Eq. (13.179), which are obtained by adding all possible diagrams. The free energy $A[j]$, determined by $Z[j]$ according to

$$Z[j] = e^{-A[j]}, \tag{13.192}$$

is instead the generator of "connected" correlators, so that

$$\frac{1}{A[0]} \frac{\delta^n A[j]}{\delta j(x_1) \cdots \delta j(x_n)}\bigg|_{j=0} = \langle \Phi(x_1) \cdots \Phi(x_n) \rangle_c, \tag{13.193}$$

which are obtained summing only *connected* Feynman diagrams.

This is a link that has emerged once before, in our treatment of the Onsager solution in Chapter 11. It is rooted in the structure of the exponential series $e^x$, and suffice it to say, here, that the $n$th term, $\frac{x^n}{n!}$, accounts for a collection of $n$ identical objects, properly counted in view of the factor $\frac{1}{n!}$.

It is also interesting to introduce another quantity, $\Gamma[\varphi]$, which is called effective action in quantum field theory, and is defined via the Legendre transform

$$A[j] = \Gamma[\varphi] - \int d^D x j(x) \varphi(x) \tag{13.194}$$

in a manner akin to the Gibbs potential. There are two things to note here. First, the Legendre transform rests on the two relations

$$\varphi(x) = -\frac{\delta A[j]}{\delta j(x)}, \qquad j(x) = \frac{\delta \Gamma[\varphi]}{\delta \varphi(x)}, \tag{13.195}$$

so that the preceding remarks imply that $\varphi(x)$ is an average induced by the current $j(x)$ in the complete theory but still in the spirit of Eq. (13.187). The second observation is that the very structure of Eq. (13.194), and of the exponent in Eq. (13.178), suggests that $\Gamma[\varphi]$ is a corrected form of the original action $S$. The correction is in powers of the parameter $\lambda$ that characterizes the strength of the quartic interaction.

One can then show that effective action is a power series in $\lambda \mu^{2\epsilon}$ whose individual terms,

$$\frac{1}{\Gamma[0]} \frac{\delta^n \Gamma[\varphi]}{\delta \varphi(x_1) \cdots \delta \varphi(x_n)}\bigg|_{\varphi=0} = \Gamma^{(n)}(m, \lambda, x_1, \ldots, x_n), \tag{13.196}$$

can be built from a special class of diagrams with a given number of external lines. These are called 1PI or one-particle irreducible diagrams, since they cannot be split by cutting a single line. This property could be deduced from Eqs. (13.195) by an inductive argument. Note that the Feynman rules actually build $-\Gamma$, consistent with the quartic vertex in Figure 13.8.

In the following it will prove convenient to work with the momentum-space counterparts of these expressions, which can be defined as

$$\frac{1}{\Gamma[0]} \frac{\delta^n \Gamma[\varphi]}{\delta \widetilde{\varphi}(p_1) \cdots \delta \widetilde{\varphi}(p_n)}\bigg|_{\varphi=0} = (2\pi)^D \delta\Big(\sum_i p_i\Big) \widetilde{\Gamma}^{(n)}(m, \lambda, p_1, \ldots, p_n), \tag{13.197}$$

where the $\widetilde{\varphi}(p_i)$ are Fourier transforms of the $\varphi(x_i)$. Let us note, for later convenience, that the energy dimensions of the relevant quantities are

$$[\varphi] = 1 - \epsilon, \quad [\widetilde{\varphi}] = -3 + \epsilon, \quad [\Gamma] = 0, \quad [\widetilde{\Gamma}^{(n)}] = 4 - n + \epsilon(n-2). \tag{13.198}$$

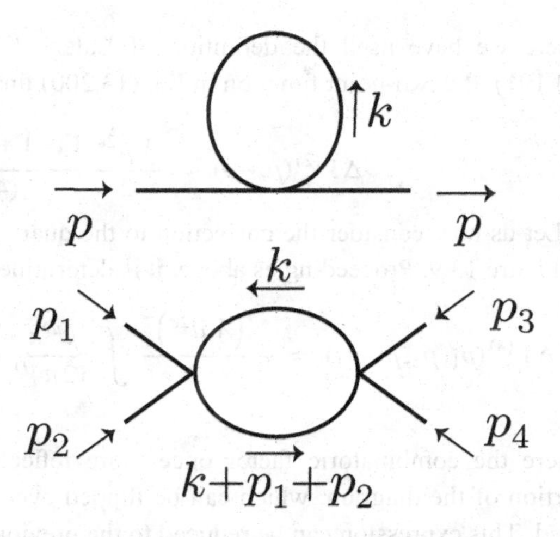


Figure 13.9 The momentum-space diagram determining (top) the one-loop contribution to the two-point coupling in the effective action, and (bottom) the corresponding contribution to the four-point coupling, where $p_1 + p_2 + p_3 + p_4 = 0$.

In this fashion the first momentum-space correction to the effective action quadratic in $\varphi$, which we shall denote $\Delta\widetilde{\Gamma}^{(2)}$, results from the top diagram in Figure 13.9, with two external lines, and reads

$$\Delta\widetilde{\Gamma}^{(2)}(p,-p) = \frac{\lambda\,\mu^{2\epsilon}}{2}\int\frac{d^Dk}{(2\pi)^D}\,\frac{1}{k^2+m^2}. \tag{13.199}$$

The combinatoric factor is in this case $\frac{1}{2}$, since a flip of the internal loop identifies a $\mathbb{Z}_2$ group that leaves the diagram unchanged. Reinstating the field $\widetilde{\varphi}(p)$, the Fourier transform of $\varphi(x)$, one can thus conclude that the order-$\lambda$ correction to the effective action includes the contribution

$$\frac{1}{2}\int\frac{d^Dp}{(2\pi)^D}\,\Delta\widetilde{\Gamma}^{(2)}(p,-p)\,\widetilde{\varphi}(p)\,\widetilde{\varphi}(-p), \tag{13.200}$$

where the additional combinatoric factor takes into account the symmetry under the interchange of the two external lines in the top portion of Figure 13.9.

Let us now compute a slight generalization of the preceding integral,

$$I(\alpha,D) = \int\frac{d^Dk}{(2\pi)^D}\,\frac{1}{[k^2+m^2]^\alpha}. \tag{13.201}$$

The first step rests on the definition of Euler's Γ function and leads to

$$I(\alpha,D) = \frac{1}{\Gamma(\alpha)}\int_0^\infty ds\,s^{\alpha-1}\,e^{-s\,m^2}\int\frac{d^Dk}{(2\pi)^D}\,e^{-sk^2}, \tag{13.202}$$

while performing the Gaussian integral leads to

$$I(\alpha,D) = \int_0^\infty\frac{ds\,s^{\alpha-1-\frac{D}{2}}\,e^{-s\,m^2}}{(4\pi)^{\frac{D}{2}}\,\Gamma(\alpha)} = \frac{\Gamma\left(\alpha-\frac{D}{2}\right)\left(m^2\right)^{\frac{D}{2}-\alpha}}{(4\pi)^{\frac{D}{2}}\,\Gamma(\alpha)}, \tag{13.203}$$

where we have used the definition of Euler's $\Gamma$ function. Hence, recalling Eq. (13.191), the two-point function in Eq. (13.200) finally becomes

$$\Delta\widetilde{\Gamma}^{(2)}(p,-p) = \frac{\lambda\,\mu^{2\epsilon}}{2}\,\frac{\Gamma(-1+\epsilon)\left(m^2\right)^{1-\epsilon}}{(4\pi)^{2-\epsilon}}. \tag{13.204}$$

Let us now consider the correction to the quartic vertex in the momentum space of Figure 13.9. Proceeding as above, it is determined by

$$\Delta\widetilde{\Gamma}^{(4)}(p_1,p_2,p_3,p_4) = -\frac{\left(\lambda\,\mu^{2\epsilon}\right)^2}{2}\int\frac{d^Dk}{(2\pi)^D}\,\frac{1}{\left[k^2+m^2\right]\left[(k+q)^2+m^2\right]}, \tag{13.205}$$

where the combinatoric factor once more reflects the symmetry of the internal portion of the diagram, which can be flipped over while keeping the external lines fixed. This expression can be reduced to the previous case resorting to the identity

$$\frac{1}{ab} = \int_0^1 dt\,\frac{1}{\left[at+b(1-t)\right]^2}, \tag{13.206}$$

and then, after translating the integration momentum $k$, becomes

$$-\frac{\left(\lambda\,\mu^{2\epsilon}\right)^2}{2}\int_0^1 dt\int\frac{d^Dk}{(2\pi)^D}\,\frac{1}{\left[k^2+m^2+t(1-t)q^2\right]^2}, \tag{13.207}$$

and using Eq. (13.203)

$$\Delta\widetilde{\Gamma}^{(4)}(p_1,p_2,p_3,p_4) = -\frac{\left(\lambda\,\mu^{2\epsilon}\right)^2\Gamma(\epsilon)}{2(4\,\pi)^{2-\epsilon}}\int_0^1 dt\left[m^2+t(1-t)q^2\right]^{-\epsilon}. \tag{13.208}$$

Consequently, the order-$\lambda$ correction to the effective action includes the contribution

$$\frac{1}{8}\prod_{i=1}^{4}\left[\int\frac{d^Dp_i}{(2\pi)^D}\,\widetilde{\varphi}(p_i)\right]\Delta\widetilde{\Gamma}^{(4)}(p_1,p_2,p_3,p_4)\,(2\pi)^D\,\delta\big(p_1+\cdots+p_4\big). \tag{13.209}$$

The factor $\frac{1}{8}=\frac{3}{4!}$ can be associated to the external lines in Figure 13.9, taking into account the interchanges of the two lines labeled by $(p_1,p_2)$, of the two lines labeled by $(p_3,p_4)$, and of the two pairs in the bottom part of the figure, which are manifest symmetries of the diagrams once it is multiplied by the external fields and is integrated over all independent momenta.

We have thus displayed the corrections, to lowest order in $\lambda$, to terms in the effective action that are, respectively, quadratic and quartic in $\varphi$. There are contributions of all orders in $\lambda$, however, to both, and in general this applies to arbitrary correlators, although only those with even numbers of legs are nonvanishing. This is a consequence of the original $\mathbb{Z}_2$ symmetry inherited from the discrete Ising model, and can be justified by noting that Eq. (13.178) is manifestly even in $j(x)$, since a change of sign can be absorbed into a redefinition of the integration variable $\Phi(x)$, and this induces a corresponding symmetry of the effective action under a sign reversal of $\varphi(x)$.

All this is seemingly nice and elegant, but the integrals diverge in general for integer values of $D$, a fact that manifests itself in the poles of the Euler $\Gamma$ functions in Eqs. (13.204) and (13.208). Which amplitudes actually diverge among those following from Eq. (13.178)? The result depends on the value of $D$, and can be determined as follows.

- In a diagram with $I$ internal lines, $V$ vertices, and $L$ closed loops, Euler's relation demands that

$$L = I - V + 1.$$

- In a diagram with $L$ closed loops in $D$ dimensions, the overall degree of divergence is

$$\Delta = DL - 2I,$$

since each internal line subtracts two powers of momentum and each closed loop is accompanied by a $D$-dimensional momentum integral. Here $\Delta \geq 0$ ($\Delta < 0$) means that the corresponding diagram has an overall divergent (convergent) behavior.

- In a diagram with $V$ vertices, $I$ internal lines and $E$ external lines,

$$4V = E + 2I,$$

since internal lines, join pairs of vertices and are thus counted twice.

Combining the preceding relations one can conclude that

$$\Delta = (D-4)\,L + 4 - E. \tag{13.210}$$

Hence, a finite number of diagrams diverge for $D < 4$, an infinite number contributing *at most* to the four-point correlator diverge for $D = 4$, and an infinite number contributing to arbitrary correlators diverge for $D > 4$. In these three cases the theory is said super-renormalizable, renormalizable, and not renormalizable.

Let us consider our model around the very rich case of four dimensions, thus focusing on the limiting behavior of Eq. (13.191) close to $\epsilon = 0$. There are then only two amplitudes that contain divergences, those with $E = 2, 4$. Each results from an infinite number of diagrams building the corresponding power series expansion in $\lambda$. We can forget here about the amplitude without external lines, since it adds a mere constant to the effective action $\Gamma$. Its puzzling effects in the presence of gravity were already highlighted by more elementary means in Chapter 5, and as we stressed there they still mark the frontier of our understanding. However, much progress has been made over the years on renormalizable models, and a first ingredient are Laurent series expansions around $\epsilon = 0$.

The pole parts in Eqs. (13.204) and (13.208) at $\epsilon = 0$ build up the divergent contribution to $\Delta\,\Gamma$,

$$-\frac{\lambda\, m^2}{64\,\pi^2\,\epsilon}\int d^Dx\,\varphi^2(x) - \frac{\lambda\,\mu^{2\epsilon}}{4!}\,\frac{3\,\lambda}{32\,\pi^2\,\epsilon}\int d^Dx\,\varphi^4(x), \tag{13.211}$$

which can be removed if the "counterterms"

$$\int d^Dx\left[\frac{m^2}{64\,\pi^2\,\epsilon}\,\Phi^2 + \frac{\lambda\,\left(\mu^2\right)^{\epsilon}}{4!}\,\frac{3\,\lambda}{32\,\pi^2\,\epsilon}\,\Phi^4\right] \tag{13.212}$$

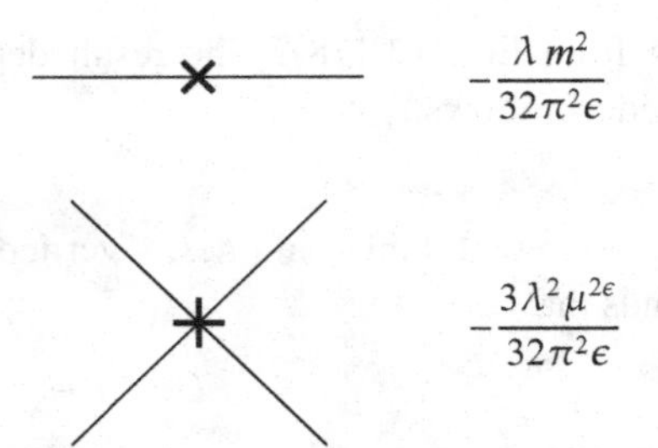


Figure 13.10 The Feynman rules for (top) the two-point counterterm, and (bottom) for the four-point one.

are added to the action in Eq. (13.178). Therefore, to this order in $\lambda$, no divergences are left if one starts from a "bare" action with mass term and quartic coupling modified according to

$$\frac{1}{2}\, m^2 \left(1 + \frac{\lambda}{32\,\pi^2\,\epsilon}\right) \Phi^2(x) + \frac{\lambda(\mu^2)^\epsilon}{4!}\left(1 + \frac{3\,\lambda}{32\,\pi^2\,\epsilon}\right) \Phi^4(x). \tag{13.213}$$

Note indeed, that the new couplings give rise to the additional vertices in Figure 13.10, which complement the Feynman rules of Figure 13.8, whose contributions are simply to be added to the two diagrams discussed above, since they are of the same order in $\lambda$.

These cursory remarks should hopefully help the reader to appreciate that iterating the procedure to all orders in $\lambda$ would result in a modified action principle of the form

$$\int d^D x \left[\frac{1}{2} Z_1\, \nabla\Phi \cdot \nabla\Phi \,+\, \frac{1}{2}\, m^2\, Z_2\, \Phi^2(x) \,+\, Z_3\, \frac{\lambda\, \left(\mu^2\right)^\epsilon}{4!}\, \Phi^4(x)\right]. \tag{13.214}$$

Here the $Z_i$ are Laurent series in $\epsilon$, which are also power series in $\lambda$ with an implicit dependence on $\mu$, as we are about to see. The essence of renormalization is that this action principle, which differs from the original one merely in the choice of coefficients, yields finite correlation functions.

Let us warn the reader that, while plausible in view of Eq. (13.210), which indicates that divergences are present in the two-point function ($\Delta = 2$) and in the four-point function ($\Delta = 0$), this is a nontrivial statement. It rests on the proof that all divergent terms are polynomials in masses and momenta, and hence local functions in position space. While this is manifest in the integral of Eq. (13.203), its higher-loop counterparts involving multiple momentum integrals do contain, in general, divergences in one sub-integration or another. These give rise to nonlocal contributions, but when they are combined with lower-order terms all nonlocal divergences disappear. Only with this proviso is Eq. (13.214) implied by dimensional analysis.

Equation (13.214) can be recast in the equivalent, albeit very useful form, of a bare action

$$\mathcal{S}_0 \;=\; \int d^D x \left[\frac{1}{2}\nabla\Phi_0 \cdot \nabla\Phi_0 \,+\, \frac{1}{2}\, m_0^2\, \Phi_0^2(x) \,+\, \frac{\lambda_0}{4!}\, \Phi_0^4(x)\right], \tag{13.215}$$

thus confining divergent contributions in $\Phi_0$, $m_0$, and $\lambda_0$, where

$$\Phi_0 = Z_1^{\frac{1}{2}}\,\Phi, \tag{13.216}$$

$$m_0 = \frac{Z_2}{Z_1}\,m, \tag{13.217}$$

$$\lambda_0 = \lambda\,\left(\mu^2\right)^\epsilon\,\frac{Z_3}{Z_1^2}. \tag{13.218}$$

One can now extract a wealth of information from all this, taking into account that the "bare" parameters *define* the theory independently of the scale $\mu$. To this end, it is convenient to introduce the functions $\beta(\lambda)$, $\gamma_\phi(\lambda)$, and $\gamma_m(\lambda)$, defined as

$$\beta = \mu\,\frac{\partial\lambda}{\partial\mu},$$
$$\gamma_\phi = \frac{1}{2\,Z_1}\,\mu\,\frac{\partial Z_1}{\partial\mu},$$
$$\gamma_m = \frac{1}{m}\,\mu\,\frac{\partial m}{\partial\mu}, \tag{13.219}$$

in order to explore how different choices of $\lambda$, $Z_1$, $m$, and $\mu$ conspire to define one and the same model.

Let us begin our analysis by taking a closer look at the $\beta$ function, starting from a generic series for Eq. (13.218):

$$\lambda_0 \;=\; \left(\mu^2\right)^\epsilon\left[\lambda \,+\, \sum_{n=1}^{\infty}\frac{a_n(\lambda)}{\epsilon^n}\right]. \tag{13.220}$$

As we have stressed, a given choice of $\lambda_0$ *defines* the coupling of the theory, so that the right-hand side of this expression links different, albeit equivalent, choices of $\lambda$ and $\mu$. Differentiating with respect to $\mu$ and taking into account that $\lambda_0$ is independent of it for a given model yields

$$\left[1 \,+\, \sum_{n=1}^{\infty}\frac{a_n'(\lambda)}{\epsilon^n}\right]\beta \;=\; -\,2\,\epsilon\,\left[\lambda \,+\, \sum_{n=1}^{\infty}\frac{a_n(\lambda)}{\epsilon^n}\right]. \tag{13.221}$$

The power of this relation lies in the fact that $\beta$, being determined by the finite quantities $\lambda$ and $\mu$, must be finite as $\epsilon \to 0$. Moreover, the preceding relation implies that it is a polynomial of first order in $\epsilon$. Solving Eq. (13.221) order by order in $\epsilon$ determines the explicit form of $\beta$,

$$\beta \;=\; -\,2\,\epsilon\,\lambda \,+\, 2\left(\lambda\,\frac{\partial}{\partial\lambda} \,-\, 1\right)a_1 \tag{13.222}$$

while the higher poles determine recursively the $a_k$ for $k \geq 2$, according to

$$\left(\lambda\,\frac{\partial}{\partial\lambda} \,-\, 1\right)a_{k+1} \;=\; \frac{\partial a_k}{\partial\lambda}\left(\lambda\,\frac{\partial}{\partial\lambda} \,-\, 1\right)a_1. \tag{13.223}$$

These additional relations place useful constraints on higher-loop diagrams. In a similar fashion, starting from a generic series for Eq. (13.217),

$$m_0^2 \;=\; m^2\left[1 \,+\, \sum_{k=1}^{\infty}\frac{b_k(\lambda)}{\epsilon^k}\right], \tag{13.224}$$

and differentiating both sides with respect to $\lambda$ gives

$$\gamma_m = \lambda \frac{\partial b_1}{\partial \lambda},$$
$$\lambda \frac{\partial b_{k+1}}{\partial \lambda} = b_k \lambda \frac{\partial b_1}{\partial \lambda} + \frac{\partial b_k}{\partial \lambda}\left(\lambda \frac{\partial}{\partial \lambda} - 1\right) a_1 . \tag{13.225}$$

Equation (13.213) shows that, for the model of interest, and up to the first order in $\lambda$ that we are considering here,

$$Z_1 = 1,$$
$$Z_2 = 1 + \frac{\lambda}{32\,\pi^2\,\epsilon},$$
$$Z_3 = 1 + \frac{3\,\lambda}{32\,\pi^2\,\epsilon}. \tag{13.226}$$

Making use of Eqs. (13.226), one can thus conclude that

$$\beta = -\,2\,\epsilon\,\lambda + \frac{3\,\lambda^2}{16\,\pi^2},$$
$$\gamma_m = \frac{\lambda}{32\,\pi^2},$$
$$\gamma_\phi = 0, \tag{13.227}$$

where the last result reflects the absence of a divergent contribution proportional to the kinetic operator in Eq. (13.204).

Let us now consider the $\beta$ function for this model in four dimensions, setting $\epsilon = 0$ in the first of Eqs. (13.227). Integrating

$$\mu \frac{\partial \lambda}{\partial \mu} = \frac{3\,\lambda^2}{16\,\pi^2} \tag{13.228}$$

then gives

$$\lambda(\mu) = \frac{\lambda(\mu_0)}{1 - \frac{3\,\lambda(\mu_0)}{16\,\pi^2}\log\left(\frac{\mu}{\mu_0}\right)}. \tag{13.229}$$

This indicates that equivalent theories are obtained if an increase (decrease) of $\mu$ is accompanied by a corresponding increase (decrease) of $\lambda$. Strictly speaking, we know only the leading term in $\beta$, but nonetheless the result that $\lambda$ decreases for decreasing values of $\mu$ becomes reliable within a region where $\lambda$ is small, which is approached asymptotically as $\mu$ decreases.

What is this telling us about the actual dynamics? Before getting to proper formal arguments, it is interesting to get an intuitive feeling of what the preceding result implies, and here dimensional analysis comes to our rescue. The dependence on external momenta has to enter via dimensionless combinations, and close to the critical temperature, where the mass $m$ is proportional to $t$, the scale $\mu$ effectively becomes the only dimensional parameter present in the model. Hence, it is natural to adapt $\mu$ to the external momenta, and in particular to reduce it when discussing large-scale properties, which are captured by lower and lower values of these momenta. Equation (13.229) thus indicates that close to the critical temperature, small values of $\lambda$ somehow characterize the large-distance behavior of correlation functions. Consequently, the mean-field approximation, where interactions were neglected, ought to be reliable for the Ising model in four dimensions. This is nicely consistent with what we found in Chapter 10, where one of the scaling relations was indeed satisfied by the mean-field Ising results precisely in four dimensions! The value $\lambda = 0$ is, for this model, an "infrared-stable Gaussian fixed point." This fixed point is infrared-stable since $\lambda$ approaches it at higher and higher distance scales (or lower and lower momentum scales), and it is also Gaussian because at that point $\lambda = 0$ and therefore the corresponding theory is free.

Let us now take a second look at these facts in a more precise way. The basic tool to this effect is the Callan–Symanzik equation, which we now illustrate. One can compare contributions to the "bare" effective action $\Gamma_0$ and corresponding contributions to the finite, renormalized effective action $\Gamma$, and the basic lesson that can be extracted from the preceding considerations is that

$$\widetilde{\Gamma}_0^{(n)}(m_0, \lambda_0, \{p_i\}) \times \widetilde{\varphi}_0^{\,n} = \widetilde{\Gamma}^{(n)}\big(m(\mu), \lambda(\mu), \mu, \{p_i\}\big) \times \widetilde{\varphi}'^{\,n}, \tag{13.230}$$

where $\widetilde{\Gamma}_0^{(n)}$ and $\widetilde{\Gamma}^{(n)}$ denote the $n$-point couplings in the effective action, as in (13.196). The notation is admittedly sketchy, since we left implicit, for brevity, the integrations and momentum conserving $\delta$ functions displayed in Eq. (13.209), but should be sufficiently clear for our purposes. Moreover, this relation should sound plausible, since it extends to the effective action what we said about the mass $m$, the coupling $\lambda$, and the field $\Phi$ for the standard action. According to Eq. (13.214), all divergences can be absorbed into renormalizations of mass, coupling, and wave function. Then, stripping out the fields and making use of Eq. (13.218) leads to a set of natural relations among the $\Gamma^{(n)}$:

$$\widetilde{\Gamma}_0^{(n)}(m_0, \lambda_0, \{p_i\}) = \widetilde{\Gamma}^{(n)}\big(m(\mu), \lambda(\mu), \mu, \{p_i\}\big)\big(Z_\phi\big)^{-\frac{n}{2}}. \tag{13.231}$$

Differentiating with respect to $\mu$ again eliminates all reference to bare quantities, and taking into account Eqs. (13.219) leads to

$$\left[\mu\frac{\partial}{\partial\mu} + m\,\gamma_m\frac{\partial}{\partial m} + \beta\frac{\partial}{\partial\lambda} - n\,\gamma_\phi\right]\widetilde{\Gamma}^{(n)}\big(m, \lambda, \mu, \{p_i\}\big) = 0. \tag{13.232}$$

On the other hand, for dimensional reasons,

$$\begin{aligned}\left[s\frac{\partial}{\partial s} + \mu\frac{\partial}{\partial\mu} + m\frac{\partial}{\partial m}\right]&\widetilde{\Gamma}^{(n)}\big(m, \lambda, \mu, \{sp_i\}\big)\\ &= (4 - n + \epsilon(n-2))\,\widetilde{\Gamma}^{(n)}\big(m, \lambda, \mu, \{sp_i\}\big),\end{aligned} \tag{13.233}$$

since the dimension of the $n$-point function is given in Eq. (13.198).

Comparing Eqs. (13.232) and (13.233) finally leads to the Callan–Symanzik equation

$$\begin{aligned}\left[-s\frac{\partial}{\partial s} + m\,(\gamma_m - 1)\frac{\partial}{\partial m} + \beta\frac{\partial}{\partial\lambda}\right]&\Gamma^{(n)}\big(m, \lambda, \mu, \{sp_i\}\big)\\ &= (n - 4 - \epsilon(n-2) + n\,\gamma_\phi)\,\Gamma^{(n)}\big(m, \lambda, \mu, \{sp_i\}\big),\end{aligned} \tag{13.234}$$

which regulates the behavior of $\Gamma^{(n)}$ when the external momenta are rescaled.

Note that the Callan–Symanzik equation is exactly solved, as expected by the preceding considerations, if

$$\Gamma^{(n)}\left(m, \lambda, \mu, \{sp_i\}\right) = s^{4-n+\epsilon(n-2)}\, e^{-n\int_1^s \gamma_\phi(s')\frac{ds'}{s'}}\, \Gamma^{(n)}\left(m(s), \lambda(s), \mu, \{p_i\}\right), \quad (13.235)$$

provided the mass $m$ and the coupling $\lambda$ are adjusted as determined by

$$s\,\frac{dm(s)}{ds} = m\left(\gamma_m - 1\right), \qquad s\,\frac{d\lambda(s)}{ds} = \beta(s). \quad (13.236)$$

These results play a central role in quantum field theory, but our more modest goal here is to extract from this line of reasoning some information on the Ising model.

Remarkably, Eq. (13.227) has a zero for a nontrivial value of $\lambda$ and $\epsilon > 0$,

$$\frac{\lambda^\star}{16\,\pi^2} = \frac{2\,\epsilon}{3}, \quad (13.237)$$

which is usually called Wilson–Fisher fixed point.

The preceding arguments tell us that a fixed point corresponds to a choice of couplings for which the theory looks the same at larger and larger scales, precisely what one would expect for the Ising model at the critical point. Aside from the $D = 4$ case, which we have discussed above and corresponds to a Gaussian fixed point, the cases that one can try to capture in this fashion concern models in (integer) dimensions $D < 4$, since $\lambda > 0$ is needed for the overall stability. In particular, the intriguing case of three dimensions would be obtained for $\epsilon = \frac{1}{2}$, but why should one trust this twist of the original four-dimensional setting? A justification is that, as we shall see shortly, the actual coupling regulating the corrections in that case is $\frac{1}{3}$, a number small enough to hope that the lowest-order approximation be somehow reasonably accurate. Moreover, we know already that, for the continuum limit of the Ising model, $Z_\phi = 1$ at first order in $\lambda$, which is tantamount to saying that $\eta = 0$ in Eq. (10.158), since the dimension of the field is not affected by the lowest-order corrections in $\lambda$ that we are considering. Therefore, proceeding as in Chapter 10, one can cast the spin-spin correlation function of Eq. (10.158) in the form

$$G(x) = \frac{1}{|x|^{D-2}}\, f\left(\mu|x|, \frac{m_*^2 t}{\mu^2}\right), \quad (13.238)$$

where we have let $m^2 = t\,m_*^2$, thus taking into account how the square mass scales as one approaches the critical temperature $t = 0$. Since $f$ is dimensionless, one can write for it the Callan–Symanzik equation

$$\left[\mu\,\frac{\partial}{\partial\mu} + \beta_2\,\frac{\partial}{\partial g_2}\right] f(\mu|x|, g_2) = 0, \quad (13.239)$$

where

$$g_2 = \frac{m_*^2\,t}{\mu^2},$$

$$\beta_2 = \mu\,\frac{\partial}{\partial\mu}\left(\frac{m_*^2\,t}{\mu^2}\right) = \frac{m_*^2\,t}{\mu^2}\left(2\,\gamma_m - 2\right). \quad (13.240)$$

At the Wilson–Fisher fixed point, Eqs. (13.227) and (13.237) imply that

$$\gamma_m = \frac{\epsilon}{3}, \tag{13.241}$$

so that

$$\beta_2 = \frac{m_*^2\, t}{\mu^2}\left(\frac{2\epsilon}{3} - 2\right) = g_2\left(\frac{2\epsilon}{3} - 2\right), \tag{13.242}$$

and therefore Eq. (13.239) reduces to

$$\left[\mu\frac{\partial}{\partial\mu} + \left(\frac{2\epsilon}{3} - 2\right) g_2 \frac{\partial}{\partial g_2}\right] f(\mu|x|, g_2) = 0, \tag{13.243}$$

or

$$\left[\frac{\partial}{\partial \log \mu} + \frac{\partial}{\partial \log g_2^{\frac{1}{\frac{2\epsilon}{3}-2}}}\right] f(\mu|x|, g_2) = 0. \tag{13.244}$$

This is a very simple partial differential equation, which demands that $f$ depend on the difference of the two logarithms, and since $g_2$ is linear in $t$ it implies that

$$G(x) = \frac{1}{|x|^{D-2}}\, f\left(\mu|x|\, t^{\frac{1}{2-\frac{2\epsilon}{3}}}\right), \tag{13.245}$$

so that the correlation length, proportional to $\frac{1}{\mu}$, scales with $t$, with the critical exponent

$$\nu = \frac{1}{2 - \frac{2\epsilon}{3}}. \tag{13.246}$$

For $\epsilon = \frac{1}{2}$, which corresponds to the $D = 3$ Ising model, one thus obtains $\nu = 0.6$, which is already rather close to the typical experimental value $\nu \simeq 0.63$, within a 5% error. On the other hand, in two dimensions $\epsilon = 1$, and one would obtain $\nu = 0.75$, with a 25% error with respect to the correct value. In this case, however, the correct critical exponent can be obtained analytically exploiting the infinite dimensional conformal symmetry discussed in Section 13.1.

Using the scaling relations of Chapter 10, the corresponding predictions for the other $D = 3$ Ising critical exponents are

$$\begin{aligned} \alpha &\simeq 0.2 && [0.08 - 0.4], \\ \beta &\simeq 0.3 && [0.3 - 0.4], \\ \gamma &\simeq 1.2 && [1.3 - 1.4], \\ \delta &\simeq 5 && [0.03 - 0.1], \\ \nu &\simeq 0.6 && [0.5 - 0.6], \end{aligned} \tag{13.247}$$

where typical ranges of experimental values are displayed within brackets. The lowest-order $\epsilon$ expansion is thus amusingly accurate.

## Bibliographical Notes

Introductions to conformal invariance can be found, for instance, in [15, 25, 44], while the renormalization group is discussed at length in most books devoted to statistical physics in the bibliography, and in particular in [6, 7, 9, 25, 26, 39, 43, 45, 54]. Percolation is discussed, for instance, in [6, 54, 55]. The XY model is discussed in [47], but our presentation is based on J.M. Kosterlitz, "The Critical Properties of the Two-Dimensional XY Model," J. Phys. **C7** (1974) 1046. Finally, the conceptual basis of the $\epsilon$ expansion is discussed in many books, including [24, 25, 48, 60]. A comprehensive summary of the critical exponents for the Ising model, with comparisons to experimental data, can be found in A. Pelissetto and E. Vicari, "Critical phenomena and renormalization group theory," Phys. Rept. **368** (2002) 549 [arXiv:cond-mat/0012164 [cond-mat]].

# 14 The Approach of Equilibrium

In all previous chapters we have described systems in thermodynamic equilibrium within the laws of statistical physics. When they became properly appreciated at the end of the nineteenth century, these principles spurred an unprecedented progress in our understanding of matter, while the mathematical correspondence with quantum mechanics has proved, in more recent years, more and more insightful and stimulating for both areas of physics. Here we would like to describe how systems approach equilibrium, referring to a large extent to particles whose mutual interactions play a subdominant role, and yet drive them toward the thermodynamic behavior of the ideal gas.

We begin by describing the Langevin equation, which encodes the key aspects of Einstein's theory of Brownian motion, and then show, via the Fokker–Planck equation, that the system does approach the Maxwell–Boltzmann distribution from generic initial conditions. We then turn to the Boltzmann equation, which links statistical physics, transport phenomena and hydrodynamics. The chapter ends with a discussion of the fluctuation–dissipation theorem.

## 14.1 The Langevin Equation

The Langevin equation,

$$m\,\dot{\mathbf{v}} + \gamma\,\mathbf{v} = \mathbf{f}(t), \tag{14.1}$$

describes the motion of a Newtonian particle of mass $m$ in a viscous medium. The damping effect of the medium is modeled by the term proportional to $\gamma$, but this is not an ordinary equation, since the driving force $\mathbf{f}(t)$ is a Gaussian-distributed random variable, with

$$\langle f_i(t)\rangle = 0, \qquad \langle f_i(t) f_j(t')\rangle = \delta_{ij}\, g\, \delta(t-t'). \tag{14.2}$$

The force $\mathbf{f}(t)$ models the effects of collisions that the particle experiences due to the random motions in the fluid. The assumption on its probability distribution thus reflects the natural requirements that the medium be isotropic and that the presence of the test mass $m$ does not perturb it significantly.

For brevity, we confine our attention to the one-dimensional case, where the formal solution of Eq. (14.1) reads

$$v(t) = v_0\, e^{-\frac{\gamma t}{m}} + \frac{1}{m}\int_0^t d\tau\, f(\tau)\, e^{-\frac{\gamma(t-\tau)}{m}}. \tag{14.3}$$

So far, this is indeed a formal solution, since $f(t)$ is not known, but one can turn it into definite predictions for the velocity distribution induced by Eq. (14.2). The first is that

$$\langle v(t)\rangle = v_0\, e^{-\frac{\gamma t}{m}}, \tag{14.4}$$

and hence, for sufficiently large times $t > \frac{m}{\gamma}$,

$$\langle v(t)\rangle \simeq 0, \tag{14.5}$$

independently of the initial conditions.

The second prediction concerns the correlation function $\langle v(t)v(t')\rangle$, and using Eq. (14.2):

$$\langle v(t)v(t')\rangle = \left(v_0^2 - \frac{g}{2\gamma m}\right) e^{-\frac{\gamma(t+t')}{m}} + \frac{g}{2\gamma m}\, e^{-\frac{\gamma|t-t'|}{m}}. \tag{14.6}$$

For a stationary process, the correlation function should only depend on the difference $t - t'$, and this would demand a fine tuning of the initial condition to

$$v_0^2 = \frac{g}{2\gamma m}, \tag{14.7}$$

which ought to characterize equilibrium configurations, but Eq. (14.6) becomes rapidly dominated by the second term for $\gamma\, t \simeq \gamma\, t' \geq m$, and in particular

$$\langle v(t)^2\rangle = \left(v_0^2 - \frac{g}{2\gamma m}\right) e^{-\frac{2\gamma t}{m}} + \frac{g}{2\gamma m} \simeq \frac{g}{2\gamma m}. \tag{14.8}$$

Note that this limit is precisely the value (14.7) that would grant a stationary behavior to the correlation function, while any reference to the initial value $v_0$ is lost after a time interval $\mathcal{O}\left(\frac{m}{\gamma}\right)$. Moreover, the classical equipartition theorem in one dimension, $\frac{1}{2}m\langle v^2\rangle = \frac{1}{2}k_B T$, implies the Einstein relation

$$\frac{g}{2\gamma} = k_B\, T. \tag{14.9}$$

Resorting to the equipartition theorem is reasonable, since the particle is immersed in the bath provided by the viscous liquid.

One more integration of Eq. (14.3) gives

$$x(t) = x_0 + \frac{v_0\, m}{\gamma}\left(1 - e^{-\frac{\gamma t}{m}}\right) + \frac{1}{\gamma}\int_0^t d\tau\, f(\tau)\left[1 - e^{-\frac{\gamma(t-\tau)}{m}}\right], \tag{14.10}$$

and interestingly,

$$\langle (x(t)-x_0)^2\rangle = \left[\left(\frac{v_0\, m}{\gamma}\right)^2 - \frac{g\, m}{2\gamma^3}\right]\left(1 - e^{-\frac{\gamma t}{m}}\right)^2 + \frac{g}{\gamma^2}\left[t - \frac{m}{\gamma}\left(1 - e^{-\frac{\gamma t}{m}}\right)\right]. \tag{14.11}$$

Therefore, for $t \gg \frac{m}{\gamma}$,

$$\langle (x(t)-x_0)^2\rangle \simeq \frac{g\, t}{\gamma^2}, \tag{14.12}$$

independently of the initial conditions, and finally, using Einstein's relation (14.9),

$$\langle (x(t) - x_0)^2 \rangle \simeq \frac{2\,k_B\,T\,t}{\gamma}. \tag{14.13}$$

This is the typical behavior of a diffusion process, whose simplest realization is the drunken-man model. If a drunken man makes random moves in steps of size $\Delta$ in time intervals $\delta$ along a line, the probability of finding our character at $x$ at time $t + \delta$ satisfies the recursion relation

$$P(x, t + \delta) = \frac{1}{2}\Big(P(x - \Delta, t) + P(x + \Delta, t)\Big), \tag{14.14}$$

since every move entails one step, within a time interval $\delta$, to the left or to the right, with equal probabilities. Letting $\delta$ and $\Delta$ approach zero, if the ratio

$$D = \frac{\Delta^2}{2\,\delta} \tag{14.15}$$

attains a finite limiting value, these considerations lead to the diffusion equation

$$\frac{\partial P}{\partial t} = D\,\frac{\partial^2 P}{\partial x^2}, \tag{14.16}$$

which is also the imaginary-time counterpart of the free Schrödinger equation. The general solution can be built via a Fourier transform, or equivalently via the Euclidean kernel of Chapter 3, and reads

$$P(x, t) = \int_{-\infty}^{\infty} \frac{dk}{2\,\pi}\, e^{ikx - Dk^2 t}\, \widetilde{P}(k, 0). \tag{14.17}$$

Therefore, when starting from

$$P(x, 0) = \delta(x), \tag{14.18}$$

one obtains

$$P(x, t) = \frac{1}{\sqrt{4\pi D t}}\, e^{-\frac{x^2}{4Dt}}, \tag{14.19}$$

a Gaussian distribution of zero average and variance

$$\langle x^2 \rangle = 2\,D\,t. \tag{14.20}$$

Comparing with Eq. (14.13) indicates that the Langevin equation describes a diffusion process with diffusion coefficient

$$D = \frac{k_B\,T}{\gamma}. \tag{14.21}$$

The main lesson to be drawn from these considerations is that diffusion processes are far less efficient than free motion, where the counterpart of Eq. (14.20) would grow quadratically with $t$. There is also an extension of the Langevin equation with a conventional driving force induced by an external potential $V(x)$,

$$m\,\dot{v} + \gamma\,v = f(t) - \frac{dV}{dx}, \tag{14.22}$$

but now we must move on to another interesting topic.

## 14.2 The Fokker–Planck Equation

In the previous section we have seen that the predictions of the Langevin equation (14.1) quickly become independent of the initial conditions, after a time interval $\Delta t = \mathcal{O}\left(\frac{\gamma}{m}\right)$. We have also deduced two properties of the velocity distribution resulting from the random force $f$, in Eqs. (14.5) and (14.8). We can now discuss the Fokker–Planck equation that determines the time evolution of the probability distribution for the velocity v, defined as

$$P(\mathrm{v}, t) \;=\; \langle \delta(\mathrm{v} - v(t)) \rangle, \tag{14.23}$$

for the Langevin equation. Note that

$$\begin{aligned} \frac{\partial P}{\partial t} &= -\frac{\partial}{\partial \mathrm{v}} \left\langle \delta(\mathrm{v} - v(t))\, \frac{dv}{dt} \right\rangle \\ &= \frac{\gamma}{m} \frac{\partial}{\partial \mathrm{v}} (\mathrm{v} P) \;-\; \frac{1}{m} \frac{\partial}{\partial \mathrm{v}} \langle \delta(\mathrm{v} - v(t)) f(t) \rangle, \end{aligned} \tag{14.24}$$

making use of the Langevin equation (14.1). One can simplify the last term taking into account that, for each value of $t$, $f(t)$ is a Gaussian noise with variance $g$ determined by the Einstein relation (14.9). As a result, the average rests on the functional integral

$$\langle \delta(\mathrm{v} - v(t)) f(t) \rangle = \int [Df(\tau)]\, \delta(\mathrm{v} - v(t)) f(t)\, e^{-\frac{1}{4\gamma k_B T} \int f^2(\tau) d\tau}. \tag{14.25}$$

Integrating by parts and taking into account that

$$f(t)\, e^{-\frac{1}{4\gamma k_B T} \int f^2(\tau) d\tau} \;=\; -\, 2\, \gamma\, k_B\, T \frac{\delta}{\delta f(t)}\, e^{-\frac{1}{4\gamma k_B T} \int f^2(\tau) d\tau}, \tag{14.26}$$

where, as in Chapter 13, the functional derivative is a limit of partial derivatives

$$f_i\, e^{-\frac{1}{4\gamma k_B T} \Sigma_i f_i^2} \;=\; -\, 2\, \gamma\, k_B\, T \frac{\partial}{\partial f_i}\, e^{-\frac{1}{4\gamma k_B T} \Sigma_i f_i^2}, \tag{14.27}$$

associated to increasingly fine discretizations of the time interval, one finds

$$\langle \delta(\mathrm{v} - v(t)) f(t) \rangle = 2\, \gamma\, k_B\, T \left\langle \frac{\delta}{\delta f(t)}\, \delta(\mathrm{v} - v(t)) \right\rangle. \tag{14.28}$$

The next step is similar to the one in the first line of Eq. (14.24), and yields

$$\langle \delta(\mathrm{v} - v(t)) f(t) \rangle \;=\; -\, 2\, \gamma\, k_B\, T \frac{\partial}{\partial\, \mathrm{v}} \left\langle \delta(\mathrm{v} - v(t))\, \frac{\delta v(t)}{\delta f(t)} \right\rangle, \tag{14.29}$$

while one can now deduce from Eq. (14.3) that

$$\frac{\delta v(t)}{\delta f(t)} \;=\; \frac{1}{m} \int_0^t d\tau\, \delta(t - \tau)\, e^{-\frac{\gamma(t-\tau)}{m}} \;=\; \frac{1}{2\, m}, \tag{14.30}$$

since the argument of the $\delta$-function vanishes at the lower end of the integration region.

In conclusion, one thus obtains the Fokker–Planck equation, which can be cast in the form

$$\frac{\partial P(\mathrm{v},t)}{\partial t} = \frac{\gamma}{m}\frac{\partial(\mathrm{v}\,P(\mathrm{v},t))}{\partial \mathrm{v}} + \frac{\gamma\,k_B\,T}{m^2}\frac{\partial^2 P(\mathrm{v},t)}{\partial \mathrm{v}^2}. \tag{14.31}$$

In solving this equation with the initial condition

$$P(\mathrm{v},0) = \delta(\mathrm{v}-v_0), \tag{14.32}$$

it is convenient to let

$$P = e^{\frac{\gamma t}{m}}\,Y, \tag{14.33}$$

which turns it into

$$\frac{\partial Y}{\partial t} = \frac{\gamma}{m}\,\mathrm{v}\,\frac{\partial Y}{\partial \mathrm{v}} + \frac{\gamma\,k_B\,T}{m^2}\frac{\partial^2 Y}{\partial \mathrm{v}^2}, \tag{14.34}$$

and to perform the change of variables

$$\mathrm{v} = \sqrt{\frac{k_B\,T}{m}}\,\frac{\rho}{\sqrt{2\,\theta+1}}, \qquad t = \frac{m}{2\,\gamma}\log\left(2\,\theta+1\right), \tag{14.35}$$

which finally leads to the diffusion equation

$$\frac{\partial Y}{\partial \theta} = \frac{\partial^2 Y}{\partial \rho^2}. \tag{14.36}$$

Consequently the solution (14.19) translates into the probability density

$$d\Pi = Y d\rho = \frac{d\,\rho}{\sqrt{4\,\pi\,\theta}}\,e^{-\frac{(\rho-\rho_0)^2}{4\,\theta}}. \tag{14.37}$$

In terms of the original variables $(\mathrm{v}, t)$,

$$d\Pi = \sqrt{\frac{m}{2\pi\,k_B\,T\,(1-e^{-\frac{2\gamma t}{m}})}}\;d\mathrm{v}\,e^{-\frac{m}{2k_BT}\frac{\left(\mathrm{v}-v_0\,e^{-\frac{\gamma t}{m}}\right)^2}{\left(1-e^{-\frac{2\gamma t}{m}}\right)}}, \tag{14.38}$$

and all information about the initial datum is thus lost after a time interval $\mathcal{O}\left(\frac{m}{\gamma}\right)$. Independently of $v_0$, sooner or later all these solutions converge to the Maxwell–Boltzmann distribution

$$d\Pi = \sqrt{\frac{m}{2\pi\,k_B\,T}}\;d\mathrm{v}\,e^{-\frac{m\mathrm{v}^2}{2k_BT}}. \tag{14.39}$$

In concluding this section, let us mention that, introducing the probability distribution

$$f(\mathrm{x},\mathrm{v},t) = \langle\delta(\mathrm{x}-x(t))\delta(\mathrm{v}-v(t))\rangle \tag{14.40}$$

for both velocity and position, one can build a corresponding Fokker–Planck equation, which reads

$$\frac{\partial f}{\partial t} + \mathrm{v}\,\frac{\partial f}{\partial \mathrm{x}} + \frac{F}{m}\frac{\partial f}{\partial \mathrm{v}} = \frac{\gamma}{m}\frac{\partial(\mathrm{v}P(\mathrm{v},t))}{\partial \mathrm{v}} + \frac{\gamma\,k_B\,T}{m^2}\frac{\partial^2 P(\mathrm{v},t)}{\partial \mathrm{v}^2}. \tag{14.41}$$

## 14.3 The Boltzmann Equation

The Boltzmann equation associates the damping effect parametrized by $\gamma$ to chaotic $2 \to 2$ particle scatterings, which dominate the scene for densities that are not too high. In this regime, as we have seen, quantum effects are typically negligible.

The Boltzmann equation describes the time evolution of the phase-space probability distribution function $f(\mathbf{x}, \mathbf{v}, t)$ under the combined effects of particle scatterings and an external force $\mathbf{F}(\mathbf{x})$. We can discuss it directly in three dimensions, where the starting point is provided by

$$f\left(\mathbf{x} + \mathbf{v}dt, \mathbf{v} + \frac{\mathbf{F}(\mathbf{x})}{m}\, dt, t + dt\right) (dXdP)\,(t + dt)$$
$$-f\left(\mathbf{x}, \mathbf{v}, t\right) (dXdP)\,(t) = \left[\frac{\partial f}{\partial t}\right]_{\text{coll}} dt\ (dXdP)\,(t). \tag{14.42}$$

Liouville's theorem grants the invariance of the phase-space measure during the dynamical evolution, and this expression can be simplified and cast into the standard form

$$\frac{\partial f}{\partial t} + \mathbf{v} \cdot \nabla_{\mathbf{x}} f + \frac{\mathbf{F}(\mathbf{x})}{m} \cdot \nabla_{\mathbf{v}} f = \left[\frac{\partial f}{\partial t}\right]_{\text{coll}}. \tag{14.43}$$

Here we recognize the same types of contributions that were present in Eq. (14.41). The novelty is the right-hand side, which replaces the phenomenological damping term with a contribution of microscopic origin. Boltzmann made for the collision term a deeply inspired guess,

$$\begin{aligned}\left[\frac{\partial f}{\partial t}\right]_{\text{coll}} = & \int d^3\mathbf{v}_2\, d^3\mathbf{v}_3\, d^3\mathbf{v}_4\ \mathcal{B}(\mathbf{v}, \mathbf{v}_2; \mathbf{v}_3, \mathbf{v}_4) \\ & \times \left[f(\mathbf{x}, \mathbf{v}_3, t) f(\mathbf{x}, \mathbf{v}_4, t) - f(\mathbf{x}, \mathbf{v}, t) f(\mathbf{x}, \mathbf{v}_2, t)\right] \\ & \times \delta\left(\mathbf{v}^2 + \mathbf{v}_2^2 - \mathbf{v}_3^2 - \mathbf{v}_4^2\right)\, \delta(\mathbf{v} + \mathbf{v}_2 - \mathbf{v}_3 - \mathbf{v}_4),\end{aligned} \tag{14.44}$$

termed the "collision integral."

Let us pause to elaborate on this expression, which rests on a highly plausible assumption usually referred to as "molecular chaos." In principle binary collisions, which dominate in the low-density regime of interest, should involve the two-particle distribution functions $f(\mathbf{x}_i, \mathbf{v}_i, \mathbf{x}_j, \mathbf{v}_j, t)$, but here these are replaced with products of one-particle distributions. This simplification is tantamount to assuming a complete lack of correlation, and has the virtue of resulting in an independent equation for $f(\mathbf{x}, \mathbf{v}, t)$. Note also that $f(\mathbf{x}, \mathbf{v}, t)$ ought to increase when a particle emerges at position $\mathbf{x}$ and time $t$ with velocity $\mathbf{v}$ from the collision with another particle, a process described by the first term in the second row. On the other hand, it ought to decrease when the particle undergoes a collision, a process described by the second term. Both conditions are guaranteed if $\mathcal{B} \geq 0$. Finally, the integrand is subject to the familiar constraints of energy and momentum conservation.

The "kernel" $\mathcal{B}$ should possess the following seemingly natural properties:

- *invariance under particle exchanges*:
$$\mathcal{B}(\mathbf{v}, \mathbf{v}_2; \mathbf{v}_3, \mathbf{v}_4) = \mathcal{B}(\mathbf{v}_2, \mathbf{v}; \mathbf{v}_4, \mathbf{v}_3),$$
which reflects the symmetry properties of the remaining terms of the integrand in Eq. (14.44);
- *invariance under continuous rotations* ($\mathbf{v}_i \to R\mathbf{v}_i$):
$$\mathcal{B}(\mathbf{v}, \mathbf{v}_2; \mathbf{v}_3, \mathbf{v}_4) = \mathcal{B}(R\mathbf{v}, R\mathbf{v}_2; R\mathbf{v}_3, R\mathbf{v}_4);$$
- *invariance under parity transformations* ($\mathbf{v}_i \to -\mathbf{v}_i$):
$$\mathcal{B}(\mathbf{v}, \mathbf{v}_2; \mathbf{v}_3, \mathbf{v}_4) = \mathcal{B}(-\mathbf{v}, -\mathbf{v}_2; -\mathbf{v}_3, -\mathbf{v}_4);$$
- *invariance under time reversal*:
$$\mathcal{B}(\mathbf{v}, \mathbf{v}_2; \mathbf{v}_3, \mathbf{v}_4) = \mathcal{B}(-\mathbf{v}_3, -\mathbf{v}_4; -\mathbf{v}, -\mathbf{v}_2).$$

From a modern vantage point, the last two assumptions clearly hold for molecular collisions, which are dominated by electromagnetic interactions in the low-density regime.

The Boltzmann equation has become a key tool in many areas, including in detailed studies of the cosmological evolution of the universe.

## 14.4 The $H$-Theorem

Let us consider the quantity

$$H = \int d^3\mathbf{x}\, d^3\mathbf{v}\, f(\mathbf{x}, \mathbf{v}, t)\, \log f(\mathbf{x}, \mathbf{v}, t), \tag{14.45}$$

which clearly captures the total entropy, up to the overall sign and an additive constant. Taking the time derivative of this expression and using the Boltzmann equation gives

$$\begin{aligned}\frac{dH}{dt} = &- \int d^3\mathbf{x}\, d^3\mathbf{v}\, \mathbf{v} \cdot \nabla_{\mathbf{x}} (f \log f) - \frac{1}{m} \int d^3\mathbf{x}\, d^3\mathbf{v}\, \mathbf{F}(\mathbf{x}) \cdot \nabla_{\mathbf{v}} (f \log f) \\ &- \frac{1}{4} \int d^3\mathbf{x}\, d^3\mathbf{v}_1\, d^3\mathbf{v}_2\, d^3\mathbf{v}_3\, d^3\mathbf{v}_4\, \mathcal{B}(\{\mathbf{v}_i\}) \left(\frac{f_1 f_2}{f_3 f_4} - 1\right) \log\left(\frac{f_1 f_2}{f_3 f_4}\right) \\ &\times f_3 f_4\, \delta(\mathbf{v}_1 + \mathbf{v}_2 - \mathbf{v}_3 - \mathbf{v}_4)\, \delta\left(\mathbf{v}_1^2 + \mathbf{v}_2^2 - \mathbf{v}_3^2 - \mathbf{v}_4^2\right).\end{aligned} \tag{14.46}$$

We have rearranged the argument of the logarithm, taking into account the symmetries of the other terms, in order to exhibit the nonnegative function $(x-1)\log(x)$ in the second line. Note also that, for brevity, we have denoted the distribution functions $f(\mathbf{x}, \mathbf{v}_i, t)$ and $\mathcal{B}(\mathbf{v}, \mathbf{v}_2; \mathbf{v}_3, \mathbf{v}_4)$ by $f_i$ and $\mathcal{B}(\{\mathbf{v}_i\})$. Now the term involving the velocity gradient can be dropped if $f$ decays sufficiently fast with $\mathbf{v}$, while the first term can be cast in the form

$$- \int d^3\mathbf{v}\, \mathbf{v} \cdot \oint d\mathbf{S}\, f \log f, \tag{14.47}$$

and clearly describes an entropy flux through the spatial boundary. Therefore, insofar as this vanishes for the large finite volumes of interest, a condition that naturally defines isolated systems is

$$\frac{dH}{dt} \leq 0. \tag{14.48}$$

A minimum value is attained if

$$\log f_1 + \log f_2 = \log f_3 + \log f_4, \tag{14.49}$$

so that $\log f_1 + \log f_2$ *should be conserved* in all particle collisions when equilibrium is attained.

In conclusion, the combinations in Eq. (14.49) ought to depend only on the microscopic conserved quantities, the total energy $\frac{1}{2}m\mathbf{v}_1^2 + \frac{1}{2}m\mathbf{v}_2^2$, and the total momentum $m\mathbf{v}_1 + m\mathbf{v}_2$ of the particles entering the scattering process or emerging from it, so that

$$\log f_1 + \log f_2 = G\left(m\mathbf{v}_1 + m\mathbf{v}_2, \frac{1}{2}m\mathbf{v}_1^2 + \frac{1}{2}m\mathbf{v}_2^2\right), \tag{14.50}$$

or $G(1+2)$ for short. However, the left-hand side of this relation is the sum of two identical functions of $\mathbf{v}_1$ and $\mathbf{v}_2$, and therefore this condition effectively demands that $G(1+2) = G(1) + G(2)$. Hence, $G$ ought to be a *linear* function of energy and momentum, the two quantities that are generally conserved in particle collisions, and one can then recast the most general linear combination of these conserved quantities,

$$\log f = a + m\mathbf{b} \cdot \mathbf{v} + \frac{c}{2}m\mathbf{v}^2, \tag{14.51}$$

into the equivalent form

$$\log f = \alpha + \frac{m}{2k_BT}(\mathbf{v} - \mathbf{u})^2, \tag{14.52}$$

for generic values of $T$ and $\mathbf{u}$, while $\alpha$ is fixed by the condition that the integral of $f$ over $\mathbf{v}$ yield the particle density $n$.

Summarizing, the $H$-theorem leads naturally to an equilibrium Maxwell–Boltzmann distribution,

$$f^{(0)} = n\left(\frac{m}{2\,\pi\, k_B\, T}\right)^{\frac{3}{2}} e^{-\frac{m(\mathbf{v}-\mathbf{u})^2}{2k_B T}}, \tag{14.53}$$

which trivializes the collision term and solves the Boltzmann equation in the absence of an external force, as pertains to standard equilibrium. This suggests to study transport phenomena starting from a deformed "local Maxwell–Boltzmann" form

$$f^{(0)}(\mathbf{x},\mathbf{v},t) = n(\mathbf{x},t)\left(\frac{m}{2\,\pi\, k_B\, T(\mathbf{x},t)}\right)^{\frac{3}{2}} e^{-\frac{m(\mathbf{v}-\mathbf{u}(\mathbf{x},t))^2}{2k_B T(\mathbf{x},t)}}, \tag{14.54}$$

with velocity and temperature profiles $\mathbf{u}(\mathbf{x}, t)$ and $T(\mathbf{x}, t)$. This type of expression does not solve identically the Boltzmann equation for generic temperature, density, and velocity profiles, but has the virtue of eliminating the collision term. As we shall see in Section 14.6, this choice for the distribution $f^{(0)}$ turns general conservation laws into the equations of nondissipative hydrodynamics.

Note that the behavior manifested by $H$ assigns a definite direction to the arrow of time. This fact stirred intense controversy at the beginning, but reflects the very form of the Boltzmann equation, whose two sides have opposite behavior under time reversal, on account of the collision term and the underlying molecular chaos hypothesis.

## 14.5 Transport Phenomena

In the previous section we saw that, once one integrates the Boltzmann equation over $\mathbf{v}$, the symmetries of the collision term imply that any function $\rho(\mathbf{x}, \mathbf{v}_1)$ inserted in it can be replaced with the combination

$$\frac{1}{4}\left(\rho(\mathbf{x}, \mathbf{v}_1) + \rho(\mathbf{x}, \mathbf{v}_2) - \rho(\mathbf{x}, \mathbf{v}_3) - \rho(\mathbf{x}, \mathbf{v}_4)\right). \tag{14.55}$$

Therefore, the scattering term disappears whenever this step builds a conserved quantity, in the sense that

$$\rho(\mathbf{x}, \mathbf{v}_1) + \rho(\mathbf{x}, \mathbf{v}_2) = \rho(\mathbf{x}, \mathbf{v}_3) + \rho(\mathbf{x}, \mathbf{v}_4). \tag{14.56}$$

For generic choices of $\mathcal{B}$ there are three quantities of this type, 1, $\mathbf{v}$, and $\mathbf{v}^2$, and we can now derive the corresponding continuity equations.

The first continuity equation follows from

$$\int d^3\mathbf{v} \left(\frac{\partial f}{\partial t} + \mathbf{v} \cdot \nabla_{\mathbf{x}} f + \frac{\mathbf{F}(\mathbf{x})}{m} \cdot \nabla_{\mathbf{v}} f\right) = 0, \tag{14.57}$$

where the last term is a total derivative, which integrates to zero if $f$ decays sufficiently fast at infinity, as is the case for the Maxwell–Boltzmann distribution. Taking into account the natural definitions of number density and average velocity,

$$n(\mathbf{x}, t) = \int d^3\mathbf{v}\, f(\mathbf{x}, \mathbf{v}, t),$$

$$n(\mathbf{x}, t)\, \mathbf{u}(\mathbf{x}, t) = \int d^3\mathbf{v}\, \mathbf{v}\, f(\mathbf{x}, \mathbf{v}, t), \tag{14.58}$$

the remaining terms give

$$\frac{\partial n}{\partial t} + \nabla_{\mathbf{x}} \cdot (n\, \mathbf{u}) = 0. \tag{14.59}$$

This first continuity equation expresses the conservation of the number of particles.

The second continuity equation follows from

$$\int d^3\mathbf{v}\, m\, \mathrm{v}_i \left(\frac{\partial f}{\partial t} + \mathbf{v} \cdot \nabla_{\mathbf{x}} f + \frac{\mathbf{F}(\mathbf{x})}{m} \cdot \nabla_{\mathbf{v}} f\right) = 0, \tag{14.60}$$

and now the last term can be simply related to the density by a partial integration, while the second leads to a new quantity, the symmetric "pressure tensor"

$$P_{ij} = m \int d^3\mathbf{v}\, f(\mathbf{x}, \mathbf{v}, t)\, (\mathrm{v}_i - \mathrm{u}_i)\, \left(\mathrm{v}_j - \mathrm{u}_j\right), \tag{14.61}$$

noting that

$$\int d^3\mathbf{v}\, \mathrm{v}_i\, \mathrm{v}_j f(\mathbf{x}, \mathbf{v}, t) = \int d^3\mathbf{v}\, (\mathrm{v}_i - \mathrm{u}_i)\, \left(\mathrm{v}_j - \mathrm{u}_j\right) f(\mathbf{x}, \mathbf{v}, t) + n\, \mathrm{u}_i\, \mathrm{u}_j, \tag{14.62}$$

since by construction the average of $\mathrm{v}_i - \mathrm{u}_i$ vanishes. Equation (14.60) can thus be cast in the form

$$m \frac{\partial\, (n\, \mathrm{u}_i)}{\partial t} + m\, \partial_j \left(n\, \mathrm{u}_i\, \mathrm{u}_j\right) = n\, F_i - \partial_j P_{ij}, \tag{14.63}$$

where the sum over repeated indices is left implicit. Using Eq. (14.59) one can obtain the equivalent form

$$m\, n \left(\frac{\partial}{\partial t} + \mathrm{u}_j \frac{\partial}{\partial x_j}\right) \mathrm{u}_i = n\, F_i - \partial_j P_{ij} \tag{14.64}$$

or, in vector-tensor notation,

$$m\, n \left(\frac{\partial}{\partial t} + \mathbf{u} \cdot \nabla\right) \mathbf{u} = n\, \mathbf{F} - \nabla \cdot \mathbf{P}, \tag{14.65}$$

where we have dropped the subscript of $\nabla$, since after the integration over $\mathbf{v}$ it refers unambiguously to $\mathbf{x}$. This second continuity equation is the counterpart of Newton's law for fluids.

The third conservation law can be found by starting from the kinetic energy, which leads to

$$\int d^3\mathbf{v}\, \frac{m}{2}\, \mathbf{v}^2 \left(\frac{\partial f}{\partial t} + \mathbf{v} \cdot \nabla_{\mathbf{x}} f + \frac{\mathbf{F}}{m} \cdot \nabla_{\mathbf{v}} f\right) = 0, \tag{14.66}$$

and this expression involves two new quantities,

$$\begin{aligned} n(\mathbf{x}, t)\, K(\mathbf{x}, t) &= \frac{m}{2} \int d^3\mathbf{v}\, f(\mathbf{x}, \mathbf{v}, t)\, (\mathrm{v} - \mathrm{u})^2, \\ \mathbf{Q}(\mathbf{x}, t) &= \frac{m}{2} \int d^3\mathbf{v}\, f(\mathbf{x}, \mathbf{v}, t)\, (\mathrm{v} - \mathrm{u})^2\, (\mathbf{v} - \mathbf{u}). \end{aligned} \tag{14.67}$$

The first accounts for the energy fluctuations in the fluid and the second for their flow, and the conservation law reads

$$\frac{\partial}{\partial t}\left(\frac{m\, n\, u^2}{2} + n\, K\right) + \nabla \cdot \left[\left(\frac{m\, n\, u^2}{2} + n\, K\right) \mathbf{u} + \mathbf{Q} + \mathbf{P} \cdot \mathbf{u}\right] = n\, \mathbf{u} \cdot \mathbf{F}, \tag{14.68}$$

where we have again dropped the subscript of $\nabla$. Now making use of Eq. (14.59), this expression can be cast in the form

$$n \left(\frac{\partial}{\partial t} + \mathbf{u} \cdot \nabla\right) \left(\frac{m\, u^2}{2} + K\right) + \nabla \cdot (\mathbf{Q} + \mathbf{P} \cdot \mathbf{u}) = n\, \mathbf{u} \cdot \mathbf{F}, \tag{14.69}$$

and one can subtract from it the scalar product of $\mathbf{u}$ with Eq. (14.65), thus removing the contribution of the convective flow, so that finally

$$n \left(\frac{\partial}{\partial t} + \mathbf{u} \cdot \nabla\right) K + \nabla \cdot \mathbf{Q} = -\, P_{ij} \partial_i u_j. \tag{14.70}$$

This third continutiy equation expresses the energy conservation condition for the fluid.

Summarizing, we have obtained the three conservation laws

$$\frac{\partial n}{\partial t} + \nabla \cdot (n\,\mathbf{u}) = 0,$$

$$m\,n\left(\frac{\partial}{\partial t} + \mathbf{u} \cdot \nabla\right)\mathbf{u} = n\,\mathbf{F} - \nabla \cdot \mathbf{P},$$

$$n\left(\frac{\partial}{\partial t} + \mathbf{u} \cdot \nabla\right)K + \nabla \cdot \mathbf{Q} = -P_{ij}\partial_i u_j. \tag{14.71}$$

We can now discuss the leading-order approximation and its consequences for fluid mechanics.

## 14.6 Nondissipative Hydrodynamics

We can now discuss the lowest-order approximation for the three conservation laws, which is commonly obtained using the local Maxwell–Boltzmann distribution (14.54) for $f$. This lowest-order distribution function does not satisfy the Boltzmann equation, but comes a long way toward this goal, since it eliminates the collision integral, and this very feature will be instrumental for taking dissipative effects into account.

This implies directly that

$$P_{ij} = \delta_{ij}\,P, \qquad P = nk_BT,$$

$$K = \frac{3}{2}\,k_B\,T, \qquad \mathbf{Q} = 0. \tag{14.72}$$

Equations. (14.71) then become

$$\frac{\partial n}{\partial t} + \nabla \cdot (n\,\mathbf{u}) = 0, \tag{14.73}$$

$$m\,n\left(\frac{\partial}{\partial t} + \mathbf{u} \cdot \nabla\right)\mathbf{u} = n\,\mathbf{F} - \nabla P, \tag{14.74}$$

$$\left(\frac{\partial}{\partial t} + \mathbf{u} \cdot \nabla\right)T = -\frac{2}{3}\,T\,\nabla \cdot \mathbf{u}. \tag{14.75}$$

This system of equations, together with Eqs. (14.72), is the setup to describe transport phenomena in ideal dilute gases or fluids. The first is the continuity equation for the particle number, the second is Euler's equation, and the last describes heat flows.

This setup does not take dissipation into account. As a result, the process is isentropic, and

$$\frac{1}{n}\,\nabla P = \nabla h, \tag{14.76}$$

with $h$ the enthalpy per particle, and letting $\mathbf{F} = -\nabla \mathcal{U}$ makes it possible to recast Euler's equation in the form

$$\frac{\partial \mathbf{u}}{\partial t} - \mathbf{u} \times (\nabla \times \mathbf{u}) = -\frac{1}{m} \nabla \left( h + \mathcal{U} + \frac{m}{2} u^2 \right), \tag{14.77}$$

where we have made use of the identity

$$\frac{1}{2} \nabla u^2 = \mathbf{u} \times (\nabla \times \mathbf{u}) + (\mathbf{u} \cdot \nabla)\, \mathbf{u}. \tag{14.78}$$

The propagation of sound waves is a simple consequence of Eqs. (14.73)–(14.75), whose linearized form for low drift velocities and vanishing external force is

$$\frac{\partial\, \delta n}{\partial t} + n \nabla \cdot \mathbf{u} = 0, \tag{14.79}$$

$$m n \frac{\partial \mathbf{u}}{\partial t} = -\nabla \delta P, \tag{14.80}$$

$$\frac{\partial\, \delta T}{\partial t} = -\frac{2}{3} T \nabla \cdot \mathbf{u}. \tag{14.81}$$

The first and last of these combine into

$$\frac{\partial}{\partial t} \left( \frac{\delta T}{T} - \frac{2}{3} \frac{\delta n}{n} \right) = 0, \tag{14.82}$$

which translates into the isentropic relation

$$\frac{\delta P}{P} = \frac{5}{3} \frac{\delta n}{n}, \tag{14.83}$$

making use of the equation of state in (14.72).

One can thus combine Eqs. (14.79)–(14.81) into the wave equation

$$\frac{\partial^2 \delta n}{\delta t^2} = \frac{5\, k_B\, T}{3\, m} \nabla^2\, \delta\, n, \tag{14.84}$$

so that the speed of sound is

$$c_s = \sqrt{\frac{5\, k_B\, T}{3\, m}}. \tag{14.85}$$

This leads to typical estimates of a few hundred m/s in gases.

Finally, in a steady flow the time dependence disappears, and Eq. (14.77), projected along $\mathbf{u}$, which identifies the stream lines for the fluid, leads to

$$\mathbf{u} \cdot \nabla \left( h + \mathcal{U} + \frac{m}{2} u^2 \right) = 0. \tag{14.86}$$

The condition that the quantity within round brackets in Eq. (14.86) be conserved along stream lines is usually called Bernoulli's theorem, and is responsible for phenomena of great interest. It is usually presented when referring to homogeneous fluids, as the statement that

$$P + \widetilde{\mathcal{U}} + \frac{1}{2} \rho\, u^2 = \text{const} \tag{14.87}$$

along streamlines, using Eq. (14.76) and where $\widetilde{\mathcal{U}} = n\, \mathcal{U}$ denotes the potential energy density. This relation contains, as a special case, the law governing the hydrostatic pressure on earth, often called Stevin's law. It follows if $\mathbf{u} = 0$ and $\mathbf{F} = -mg\hat{\mathbf{z}}$, where the unit vector points away from the interior of the Earth, so that

$$P = P_0 - \rho g z, \tag{14.88}$$

and the pressure increases as one descends into the fluid from its surface.

## 14.7 The Emergence of Viscosity

The preceding approximation leaves out dissipative effects, consistent with the fact that the lowest-order form $f \simeq f^{(0)}$, the local Maxwell–Boltzmann distribution of Eq. (14.54), eliminates the collision term altogether. Novel phenomena emerge when its effects are included. This ought to be done by letting

$$f \simeq f^{(0)} + f^{(1)} \tag{14.89}$$

in the Boltzmann equation, and expanding the collision terms to first order in $f^{(1)}$, since one expects that $\left|f^{(1)}\right| \ll f^{(0)}$ in macroscopic transport phenomena occurring in gases (and in ordinary fluids), but doing so would still result in a complicated integro-differential equation. The contribution of the linearized collision term is thus usually accounted for via the simpler "phenomenological" equation

$$\left[\frac{\partial}{\partial t} + \mathbf{v} \cdot \nabla_{\mathbf{x}} + \frac{\mathbf{F}(\mathbf{x})}{m} \cdot \nabla_{\mathbf{v}}\right]\left(f^{(0)} + f^{(1)}\right) = -\frac{1}{\tau} f^{(1)}. \tag{14.90}$$

Here, the collision integral is replaced by a local expression involving a constant $\tau$ of the order of the collision time,

$$\tau = \mathcal{O}\left(\frac{\ell_F}{\sqrt{\langle \mathbf{v}^2 \rangle}}\right) = \mathcal{O}\left(10^{-10}\,\text{s}\right), \tag{14.91}$$

where $\sqrt{\langle \mathbf{v}^2 \rangle}$ is a typical thermal speed, $\mathcal{O}$ (800m/s), and $\ell_F \simeq 2000$Å is the typical mean-free path that we encountered in Eq. (6.7). For $\left|f^{(1)}\right| \ll f^{(0)}$, Eq. (14.90) determines $f^{(1)}$ as

$$f^{(1)} \simeq -\tau \left[\frac{\partial}{\partial t} + \mathbf{v} \cdot \nabla_{\mathbf{x}} + \frac{\mathbf{F}(\mathbf{x})}{m} \cdot \nabla_{\mathbf{v}}\right] f^{(0)}, \tag{14.92}$$

and this approximation is reasonable since, in view of the separation between the microscopic time scale (14.91) and the typical macroscopic hydrodynamic time scales, one clearly expects that

$$\left|\tau \frac{\partial f^{(0)}}{\partial t}\right|, \left|\tau\, \mathbf{v} \cdot \nabla_{\mathbf{x}} f^{(0)}\right|, \left|\tau \frac{\mathbf{F}(\mathbf{x})}{m} \cdot \nabla_{\mathbf{v}} f^{(0)}\right| \ll f^{(0)}. \tag{14.93}$$

One can now explicitly compute the correction term $f^{(1)}$ using Eq. (14.92) and the lowest-order equations (14.73)–(14.75), and the end result is

$$\begin{aligned} f^{(1)} = &-\tau f^{(0)} \left[\frac{m}{2\,k_B\,T}(\mathbf{v} - \mathbf{u})^2 - \frac{5}{2}\right](\mathbf{v} - \mathbf{u}) \cdot \frac{1}{T} \nabla T \\ &- \frac{\tau\, m\, f^{(0)}}{k_B\,T}\left[(\mathbf{v} - \mathbf{u})_i\,(\mathbf{v} - \mathbf{u})_j - \frac{1}{3}\,\delta_{ij}\,(\mathbf{v} - \mathbf{u})^2\right] \partial_i u_j. \end{aligned} \tag{14.94}$$

This correction is relevant in two cases where the lowest-order contribution vanishes. The first line of Eq. (14.94), *odd* in $\mathbf{v} - \mathbf{u}$, yields the leading contribution to $\mathbf{Q}$ in Eq. (14.67), which vanished altogether to lowest order, while the second yields the dominant contribution to the traceless part of $P_{ij}$, which also vanished to lowest order. In detail

$$\mathbf{Q} = -\frac{m\,n\,\tau}{12\,T}\left(\frac{m}{2\pi k_B T}\right)^{\frac{3}{2}} \int d^3\mathbf{v}\, e^{-\frac{m\mathbf{v}^2}{2k_B T}} \left(\mathbf{v}^2\right)^2 \left(\frac{m\mathbf{v}^2}{k_B T} - 5\right) \nabla T$$
$$= -\frac{5\,n\,\tau\,k_B^2\,T}{2\,m}\,\nabla T, \tag{14.95}$$

since the symmetry of the integrand allows one to replace the product $\mathrm{v}_i \mathrm{v}_j$ with $\frac{1}{3}\,\delta_{ij}\,\mathbf{v}^2$. Now defining the viscosity coefficient, and recalling Eq. (6.7), which links the mean free path $\ell_F$ and the core radius $r_0$,

$$\mu \;=\; n\,\tau\,k_B\,T \;=\; \mathcal{O}\left(\frac{k_B\,T}{r_0^2\,\sqrt{\langle \mathbf{v}^2\rangle}}\right) \;=\; \mathcal{O}\left(\frac{\sqrt{m}\,k_B\,T}{r_0^2}\right), \tag{14.96}$$

which one can rightfully treat as a constant in standard hydrodynamic flows. One can finally write

$$\mathbf{Q} \;=\; -\,\frac{5\,\mu}{2\,m}\,k_B\,\nabla T. \tag{14.97}$$

The dependence of $\mu$ on $T$ given in Eq. (14.96) is reasonably accurate for gases, while in liquids, where interactions are more relevant, the qualitative behavior is captured by phenomenological expressions of the type $\mu = \mathcal{O}\left(e^{\frac{B}{T}}\right)$. The second line of Eq. (14.94) modifies the pressure tensor, adding to it a symmetric traceless contribution, so that

$$P_{ij} \;=\; \delta_{ij}\,P \;-\; \mu\left[\nabla_i\,\mathrm{u}_j + \nabla_j\,\mathrm{u}_i \;-\; \frac{2}{3}\,\delta_{ij}\,\nabla\cdot\mathbf{u}\right], \tag{14.98}$$

and the ratio between the correction and the lowest-order term is tiny in gases, $\mathcal{O}(\frac{\ell_F \overline{u}}{L\sqrt{\langle \mathbf{v}^2\rangle}})$, with $\overline{u}$ and $L$ being the typical macroscopic speed and distance scales of the hydrodynamic flow.

Retaining the dominant terms, the final result is then

$$\frac{\partial n}{\partial t} + \nabla\cdot(n\,\mathbf{u}) = 0, \tag{14.99}$$

$$m\,n\left(\frac{\partial}{\partial t} + \mathbf{u}\cdot\nabla\right)\mathbf{u} = n\,\mathbf{F} \;-\; \nabla\left(P - \frac{\mu}{3}\,\nabla\cdot\mathbf{u}\right) \;+\; \mu\,\nabla^2\mathbf{u}, \tag{14.100}$$

$$\left(\frac{\partial}{\partial t} + \mathbf{u}\cdot\nabla\right)T = -\,\frac{2}{3}\,T\,\nabla\cdot\mathbf{u} \;+\; \frac{5}{3}\,\frac{\mu}{n\,m}\,\nabla^2 T. \tag{14.101}$$

The continuity equation (14.99) is unaffected with respect to the lowest-order form, the Euler equation becomes the Navier–Stokes equation (14.100), and finally the energy conservation equation becomes (14.101), here written for a perfect monatomic gas.

In particular, for a fluid at rest, the convective velocity **u** vanishes, and Eq. (14.101) reduces to the diffusion equation

$$\frac{\partial T}{\partial t} = \frac{5}{3}\frac{\mu}{n m}\nabla^2 T, \tag{14.102}$$

so that, comparing with Eq. (14.16), one can conclude that

$$D = \frac{5}{3}\frac{\mu}{n m}, \tag{14.103}$$

which links the diffusion coefficient $D$ to the viscosity coefficient $\mu$.

While we derived these results with reference to the ideal gas, their range of applicability is wider, and the Navier–Stokes equation is commonly used for the motion of liquids. Our cursory remarks suffice to relate these key properties of transport phenomena (and fluid dynamics) to the Boltzmann equation, but can hardly do justice to this very interesting and important field. We are leaving out, in particular, a discussion of turbulent flows, which originate, in ways that are only partly understood, from the competition between the dissipative effects associated to $\mu$ and other terms already present in Eqs. (14.73)–(14.75) .

## 14.8 The Fluctuation–Dissipation Theorem

The fluctuation–dissipation theorem links two apparently unrelated time scales. The first characterizes how a system, initially perturbed slightly away from equilibrium, recovers it, while the second underlies its spontaneous fluctuations about the unperturbed equilibrium. Let us take a quick look at all this in its simplest incarnation, before formulating and discussing the problem in a more general setup.

Let us assume, to begin with, that a system with a fixed number of components is prepared by subjecting it for a long time to a small external force. As a result, the system initially thermalizes as demanded by the Hamiltonian $H$ augmented by

$$\Delta H = -fA, \tag{14.104}$$

and averages are computed as

$$\bar{A} = \frac{\mathrm{Tr}\left[e^{-\beta(H+\Delta H)}\,A\right]}{\mathrm{Tr}\left[e^{-\beta(H+\Delta H)}\right]}. \tag{14.105}$$

At $t = 0$ the perturbation is switched off abruptly, and the Hamiltonian $H$ determines the ensuing evolution of $A(0)$. Therefore, macroscopically, one perceives the average value

$$\bar{A}(t) = \frac{\mathrm{Tr}\left[e^{-\beta(H+\Delta H)}\,A(t)\right]}{\mathrm{Tr}\left[e^{-\beta(H+\Delta H)}\right]}, \tag{14.106}$$

where

$$A(t) = e^{\frac{iHt}{\hbar}}\,A(0)\,e^{-\frac{iHt}{\hbar}}. \tag{14.107}$$

It is indeed reasonable to expect that the small perturbation induced by the removal of $\Delta H$ will leave the system essentially in thermal equilibrium as determined by $H+\Delta H$

for a while, but its observables will evolve with $H$ only. Expanding now to first order in $f$, using[1]

$$e^{-\beta(H-fA)} = e^{-\beta H}\left(1+f\int_0^\beta e^{sH}A\,e^{-sH}ds\right)+\cdots, \tag{14.108}$$

which gives

$$\bar{A}(t) - \langle A\rangle = \beta f\langle A(0)\,A(t)\rangle_c, \tag{14.109}$$

where the thermal average $\langle A\rangle$ and the "connected two-point function" are

$$\langle A\rangle = \frac{\mathrm{Tr}\left[e^{-\beta H}A(0)\right]}{\mathrm{Tr}\left[e^{-\beta H}\right]},$$
$$\langle A(0)\,A(t)\rangle_c = \int_0^\beta \langle A(0)\,A(t+i\hbar s)\rangle\,\frac{ds}{\beta} - \langle A\rangle^2. \tag{14.110}$$

This result is usually summarized by stating that, in a given system, perturbations due to external interventions and correlations of observables at different times decay in the same manner. Note also that the dominant behavior of this expression as $\hbar\to 0$ is the standard form of the connected correlation function.

### 14.8.1 Fluctuations

It is instructive to explore the key features of these phenomena in Fourier space. To this end, let us now consider a system in a thermal equilibrium defined by a time-independent Hamiltonian $H$. In the Heisenberg picture, the time evolution of any observable $\mathcal{O}$ is determined by

$$\mathcal{O}(t) = e^{\frac{it}{\hbar}H}\mathcal{O}\,e^{-\frac{it}{\hbar}H}, \tag{14.111}$$

while thermal averages are computed according to the standard formulas

$$\langle\mathcal{O}\rangle = \frac{1}{Z}\,\mathrm{tr}\left[e^{-\beta H}\mathcal{O}\right], \qquad Z = \mathrm{tr}\left[e^{-\beta H}\right], \tag{14.112}$$

which can be also expressed in terms of the spectrum of $H$:

$$\langle O\rangle = \sum_n \frac{e^{-\beta E_n}}{Z}\mathcal{O}_{nn}, \qquad H|n\rangle = E_n|n\rangle, \qquad \mathcal{O}_{nm} = \langle n|\mathcal{O}|m\rangle. \tag{14.113}$$

The system is indeed in equilibrium since, for any observable, the average $\langle\mathcal{O}(t)\rangle = \langle\mathcal{O}\rangle$ is time independent. On the other hand, fluctuations in time can be characterized by correlators of the type

$$S_{\mathcal{O}A}(t) = \langle\mathcal{O}(t)A\rangle, \tag{14.114}$$

for any two observables $\mathcal{O}$ and $A$, which take the form

$$S_{\mathcal{O}A}(t) = \frac{1}{Z}\sum_{n,m} e^{-\beta E_n+\frac{it}{\hbar}(E_n-E_m)}\mathcal{O}_{nm}A_{mn} \tag{14.115}$$

[1] This can be seen by noting that, up to higher orders in $f$, both sides satisfy the differential equation $\frac{d}{d\beta}L(\beta) = -(H-fA)L(\beta)$ with initial condition $L(0)=1$. A similar discussion will lead to Eq. (14.126) below. In quantum mechanics this is the first-order result obtained in the interaction picture.

in terms of the spectrum of $H$. Taking a Fourier transform

$$\tilde{S}_{\mathcal{O}A}(\omega) = \int_{-\infty}^{+\infty} e^{i\omega t} S_{\mathcal{O}A}(t)\, dt, \tag{14.116}$$

one thus obtains the spectral representation

$$\tilde{S}_{\mathcal{O}A}(\omega) = \frac{1}{Z}\sum_{n,m} e^{-\beta E_n} 2\pi\hbar\, \delta(\hbar\omega + E_n - E_m)\mathcal{O}_{nm}A_{mn}. \tag{14.117}$$

For any observable $A$, Hermiticity implies that $A_{mn} = A^*_{nm}$ and using this result it is simple to verify that, on account of Eq. (14.117), $\tilde{S}_{AA}(\omega)$ is always *positive* and

$$\tilde{S}_{AA}(-\omega) = e^{-\beta\hbar\omega}\tilde{S}_{AA}(\omega). \tag{14.118}$$

### 14.8.2 Perturbations

In the spirit of what we said at the beginning of this section, let us now introduce a time-dependent perturbation. Here, however, the resulting dynamics will be described by the Hamiltonian

$$H_{f(t)} = H + f(t)A, \qquad f(t < t_0) = 0, \tag{14.119}$$

with $f(t)$ a real function and $A$ a Hermitian operator. Our next objective is to reconsider how the thermal averages (14.112) are modified when $f(t)$ can be regarded as a small disturbance of the equilibrium, while also taking a look at their properties in Fourier space.

Operators now evolve according to

$$\mathcal{O}(t, t_0) = U(t, t_0)^\dagger \mathcal{O}\, U(t, t_0), \tag{14.120}$$

where the unitary operator $U(t, t_0)$ satisfies

$$i\hbar \frac{\partial}{\partial t} U(t, t_0) = H_{f(t)} U(t, t_0), \qquad U(t_0, t_0) = 1. \tag{14.121}$$

In order to isolate the effect of the external perturbation, it is convenient to work in the interaction picture, letting

$$U(t, t_0) = e^{-\frac{i(t-t_0)}{\hbar}H} V(t, t_0). \tag{14.122}$$

Operators then evolve according to

$$\mathcal{O}(t, t_0) = V(t, t_0)^\dagger \mathcal{O}(t - t_0)\, V(t, t_0), \tag{14.123}$$

where $\mathcal{O}(t - t_0)$ is determined by the unperturbed Hamiltonian as in (14.111), while

$$i\hbar \frac{\partial}{\partial t} V(t, t_0) = f(t)A(t - t_0)V(t, t_0), \qquad V(t_0, t_0) = 1, \tag{14.124}$$

with

$$A(t - t_0) = e^{\frac{i(t-t_0)}{\hbar}H} A\, e^{-\frac{i(t-t_0)}{\hbar}H}. \tag{14.125}$$

To linear order in $f(t)$, Eq. (14.124) is solved by

$$V(t, t_0) = 1 + \frac{1}{i\hbar}\int_{t_0}^{t} f(\tau)A(\tau - t_0)d\tau. \tag{14.126}$$

Thermal averages, which we still assume to rest on the unperturbed equilibrium populations via

$$\langle \mathcal{O}(t, t_0)\rangle = \frac{1}{Z}\operatorname{tr}\left[e^{-\beta H}\mathcal{O}(t, t_0)\right], \qquad Z = \operatorname{tr}\left[e^{-\beta H}\right], \tag{14.127}$$

are no longer time-independent. Rather, by Eq. (14.126), to leading order in $f(t)$ they evolve according to

$$\langle \mathcal{O}(t, t_0)\rangle = \langle \mathcal{O}\rangle + \int_{-\infty}^{t-t_0} f(t-\tau)\, \chi_{\mathcal{O}A}(\tau) d\tau, \tag{14.128}$$

where we introduced the so-called *linear response function* or *generalized susceptibility*

$$\chi_{\mathcal{O}A}(t) = \frac{1}{i\hbar}\,\theta(t)\,\langle[\mathcal{O}(t), A]\rangle. \tag{14.129}$$

In Eq. (14.128), we used the identity

$$\langle \mathcal{O}(t-t_0)A(\tau - t_0)\rangle \;=\; \langle \mathcal{O}(t-\tau)A\rangle, \tag{14.130}$$

redefined the integration variable $\tau \to t - \tau$, and extended the range, taking into account the Heaviside function $\theta(\tau)$. Note that $\chi_{\mathcal{O}A}(t)$ is a *real* function of $t$, since $\mathcal{O}$ and $A$ are Hermitian.

We can now derive some useful properties of $\chi_{\mathcal{O}A}(t)$, starting from its spectral decomposition

$$\chi_{\mathcal{O}A}(t) = \frac{\theta(t)}{i\hbar Z}\sum_{n,m}(e^{-\beta E_n} - e^{-\beta E_m})e^{\frac{it}{\hbar}(E_n - E_m)}\mathcal{O}_{nm}A_{mn}. \tag{14.131}$$

For the corresponding Fourier transform

$$\tilde{\chi}_{\mathcal{O}A}(\omega) = \int_{-\infty}^{+\infty} e^{i\omega t}\chi_{\mathcal{O}A}(t)\, dt, \tag{14.132}$$

using

$$\int_{-\infty}^{+\infty}\theta(t)e^{i(\omega-\omega')t}dt = \int_0^{\infty} e^{i(\omega-\omega'+i\epsilon)t}dt = \frac{i}{\omega-\omega'+i\epsilon}, \tag{14.133}$$

where $+i\epsilon$ indicates a small positive imaginary part, one obtains

$$\tilde{\chi}_{\mathcal{O}A}(\omega) = \frac{1}{Z}\sum_{n,m}(e^{-\beta E_n} - e^{-\beta E_m})\frac{\mathcal{O}_{nm}A_{mn}}{\hbar\omega + E_n - E_m + i\epsilon}. \tag{14.134}$$

In particular, making use of the standard identity

$$\frac{1}{\omega-\omega'+i\epsilon} = \mathrm{PV}\frac{1}{\omega-\omega'} - i\pi\delta(\omega-\omega'), \tag{14.135}$$

where PV indicates the principal value, one obtains

$$\operatorname{Im}\tilde{\chi}_{AA}(\omega) = -\frac{\pi}{Z}\sum_{n,m}(e^{-\beta E_n} - e^{-\beta E_m})|A_{nm}|^2\delta(\hbar\omega + E_n - E_m). \tag{14.136}$$

Note that this quantity is *negative for positive frequency*, because it is localized over energies satisfying $E_m = E_n + \hbar\omega > E_m$, for which $e^{-\beta E_m} < e^{-\beta E_n}$. Moreover, it is linked to the Fourier transform of the time-correlation function (14.117) according to

$$\operatorname{Im}\tilde{\chi}_{AA}(\omega) = -\frac{1}{2\hbar}\left[\tilde{S}_{AA}(\omega) - \tilde{S}_{AA}(-\omega)\right] = -\frac{1-e^{-\beta\hbar\omega}}{2\hbar}\tilde{S}_{AA}(\omega), \tag{14.137}$$

where we used (14.118) in the last step, or equivalently,

$$\operatorname{Im}\tilde{\chi}_{AA}(\omega) = -\frac{1}{2\hbar}\tanh\frac{\beta\hbar\omega}{2}\left[\tilde{S}_{AA}(\omega) + \tilde{S}_{AA}(-\omega)\right]. \tag{14.138}$$

Equation (14.138) links two different types of processes, which depend on a priori independent time scales: the left-hand side characterizes how the system responds to an external perturbation, while the right-hand side measures self-correlations of random time fluctuations in the unperturbed system. Now taking the Fourier transform of the function $f(t)$, for which

$$f(t) = \int_{-\infty}^{+\infty} e^{-i\omega t}\tilde{f}(\omega)\,\frac{d\omega}{2\pi}, \qquad \tilde{f}(-\omega) = \tilde{f}^*(\omega), \tag{14.139}$$

the linear response relation (14.128) can be used to obtain

$$\langle\mathcal{O}(t,-\infty)\rangle = \langle\mathcal{O}\rangle + \int_{-\infty}^{+\infty} e^{-i\omega t}\tilde{f}(\omega)\,\tilde{\chi}_{\mathcal{O}A}(\omega)\,\frac{d\omega}{2\pi}, \tag{14.140}$$

so that, as a special case, $\tilde{\chi}_{AA}(\omega)$ determines the thermal average $\langle A(t,-\infty)\rangle$.

### 14.8.3 Dissipation

A distinguished type of response is due to the Hamiltonian itself, which determines the effects of the perturbation on the energy of the system. To study these effects, let us consider the average energy

$$E(t,t_0) = \langle H_{f(t)}(t,t_0)\rangle = \langle U(t,t_0)^\dagger H_{f(t)}U(t,t_0)\rangle, \tag{14.141}$$

where $H_{f(t)}$ is defined in Eq. (14.119). Then,

$$\frac{d}{dt}E(t,t_0) = \dot{f}(t)\langle A(t,t_0)\rangle, \tag{14.142}$$

since terms in which the derivative acts on $U(t,t_0)$ drop out in view of (14.121). Applying Eq. (14.140) with $\mathcal{O} = A$ leads to

$$\frac{d}{dt}E(t,-\infty) = \dot{f}(t)\langle A\rangle + \dot{f}(t)\int_{-\infty}^{+\infty} e^{-i\omega t}\tilde{f}(\omega)\,\tilde{\chi}_{AA}(\omega)\,\frac{d\omega}{2\pi}. \tag{14.143}$$

The total energy variation $\Delta E$ is obtained by integrating this relation from $t = -\infty$ to $t = +\infty$. Noting that first term drops out provided $f(t) \to 0$ for $|t| \to \infty$, one then has

$$\Delta E = i\int_{-\infty}^{+\infty} \omega|\tilde{f}(\omega)|^2\tilde{\chi}_{AA}(\omega)\frac{d\omega}{2\pi}. \tag{14.144}$$

Finally, recalling that $\chi_{AA}(t)$ is also real and therefore $\tilde{\chi}_{AA}(-\omega) = \tilde{\chi}^*_{AA}(\omega)$, one obtains

$$\Delta E = -2\int_0^{\infty} \omega|\tilde{f}(\omega)|^2\operatorname{Im}\tilde{\chi}_{AA}(\omega)\frac{d\omega}{2\pi}. \tag{14.145}$$

Note that the right-hand side is positive-definite, since $\text{Im}\,\tilde{\chi}_{AA}(\omega)$ is negative for positive $\omega$. All in all, the amount of energy absorbed by the system from the time-dependent external perturbation is thus determined by $\text{Im}\,\tilde{\chi}_{AA}(\omega)$. The latter, in its turn, is related to the behavior of spontaneous time fluctuations encoded in $\tilde{S}(\omega)$ by Eq. (14.137). In detail

$$\Delta E = \int_0^\infty \omega|\tilde{f}(\omega)|^2(1 - e^{-\beta\hbar\omega})\tilde{S}_{AA}(\omega)\frac{d\omega}{2\pi\hbar}, \tag{14.146}$$

and for this reason Eqs. (14.137) and (14.138) go under the name of the fluctuation–dissipation theorem.

If $f(t)$ is monocromatic, $f(t) \to f_0\, e^{-i\omega_0 t} + f_0^* e^{i\omega_0 t}$, so that $\tilde{f}(\omega) \to 2\pi f_0\delta(\omega - \omega_0)$ for $\omega > 0, \omega_0 > 0$. The square $|\tilde{f}(\omega)|^2$ thus appears to diverge, but this is expected since one is measuring a steady energy dissipation for a long observation time $\Delta\tau$. One can thus replace

$$|\tilde{f}(\omega)|^2 \to 2\pi|f_0|^2 \int_{-\Delta\tau/2}^{\Delta\tau/2} \delta(\omega - \omega_0)e^{i(\omega-\omega_0)t}dt = 2\pi|f_0|^2\delta(\omega - \omega_0)\Delta\tau \tag{14.147}$$

obtaining a well-defined dissipated power

$$W = \frac{\Delta E}{\Delta\tau} = |f_0|^2\frac{\omega_0}{\hbar}(1 - e^{-\beta\hbar\omega_0})\tilde{S}_{AA}(\omega_0). \tag{14.148}$$

Here the reader will recognize the traits of Fermi's golden rule, another important fact of quantum mechanics that emerges in this context in a disguised form. To reiterate, the power that the external field dissipates into the system is governed by the correlation of spontaneous thermal fluctuations.

Before concluding, we stress some similarities between the fluctuation–dissipation theorem discussed above and the reasoning presented in Chapter 10, which linked, in the mean field approximation, two equilibrium quantities, the susceptibility and the average $\langle\sigma(\mathbf{r})\rangle$ that is induced when a spin is fixed at the origin. Matters are actually more general, and in order to illustrate this fact let us consider, for definiteness, a classical spin partition function of the form

$$Z = \sum_C e^{-\beta H + \sum_i J_i\sigma_i} \tag{14.149}$$

for a lattice with $N$ sites. From this expression one can deduce that

$$\frac{\partial \log Z}{\partial J_i} = \langle\sigma_i\rangle, \qquad \frac{\partial^2 \log Z}{\partial J_i\,\partial J_j} = \langle\sigma_i\sigma_j\rangle - \langle\sigma_i\rangle\langle\sigma_j\rangle \equiv \langle\sigma_i\sigma_j\rangle_c, \tag{14.150}$$

where the subscript "$c$" is meant to stress that the combination defines the connected two-point function. The specific magnetization and the specific susceptibility are defined via expressions of this type,

$$M = \frac{1}{N}\sum_i\langle\sigma_i\rangle, \qquad \chi = \frac{1}{N}\sum_j\frac{\partial M}{\partial\,\sigma_j}, \tag{14.151}$$

which one computes at $J = 0$, so that

$$\chi = \frac{1}{N^2}\sum_{i,j}\langle\sigma_i\,\sigma_j\rangle_c\Big|_{J=0}. \tag{14.152}$$

In translationally invariant settings the two-point function depends only on the difference $i - j$, which leads finally to

$$\chi = \frac{1}{N} \sum_i \langle \sigma_i \sigma_0 \rangle_c |_{J=0} . \tag{14.153}$$

This is a link of the same type as (10.136) between the response to the external field $h$ and the equilibrium fluctuations of the system, and it is sometimes also referred to as the fluctuation–dissipation theorem. This final expression is also the zero-momentum limit of the Fourier transform of the connected two-point function.

## Bibliographical Notes

The Langevin, Fokker–Planck, and Boltzmann equations, with their applications to fluids, are discussed at length in several books, and for instance in [23, 25, 33, 32, 45, 54]. Discussions of the fluctuation–dissipation theorem can be found, for instance, in [9, 18, 45].

# A Appendix A Probability Distributions

In this appendix we collect some properties of three important probability distributions, while also highlighting their mutual links.

The first of these is the *binomial distribution*, which characterizes the outcome of (large numbers of) "biased," albeit independent, coin tosses. Let us therefore consider an irregular coin, so that "heads" emerge with probability $p_H$ and "tails" with probability $p_T$ from a single try. The probability of obtaining exactly $k$ heads with $N$ tosses is

$$P_B(N,k) = \binom{N}{k} p_H^k \, p_T^{N-k}, \tag{A.1}$$

since the binomial coefficient counts the numbers of sequences with given numbers of heads and tails, irrespective of the order in which they emerge. Newton's formula then implies that

$$\sum_{k=0}^{N} \binom{N}{k} p_H^k p_T^{N-k} = (p_H + p_T)^N, \tag{A.2}$$

and therefore the condition

$$p_H + p_T = 1 \tag{A.3}$$

provides the proper normalization for $P_B(N,k)$. However, refraining momentarily from enforcing Eq. (A.3), one can obtain simply and directly some important consequences of Eq. (A.2). A first derivative with respect to $p_H$ yields

$$\sum_{k=0}^{N} \binom{N}{k} k p_H^{k-1} \, p_T^{N-k} = N\,(p_H + p_T)^{N-1}, \tag{A.4}$$

and therefore, now taking Eq. (A.3) into account,

$$\langle k \rangle = p_H N. \tag{A.5}$$

In a similar fashion, one more derivative leads to

$$\langle k(k-1) \rangle = p_H^2 N(N-1), \tag{A.6}$$

after again taking into account Eq. (A.3). These two results combine into a neat symmetric expression for the variance

$$\langle (\Delta k)^2 \rangle = \langle k^2 \rangle - \langle k \rangle^2 = N p_H (1 - p_H) = N p_H p_T, \tag{A.7}$$

which highlights a recurrent theme: the relative fluctuation decreases with $N$,

$$\frac{\sqrt{\langle(\Delta k)^2\rangle}}{\langle k\rangle} \sim \frac{1}{\sqrt{N}}, \tag{A.8}$$

so that the estimate provided by the mean value becomes more and more accurate as the number $N$ of coin tosses increases.

There is an interesting limiting case of the preceding results, which occurs as $p_H \to 0$ and $N \to \infty$ in such a way that $Np_H$ approaches a finite value $\lambda$. In this fashion the binomial probability distribution of Eq. (A.1),

$$P_B(N,k) = \frac{N(N-1)\cdots(N-k+1)}{k!}\left(\frac{\lambda}{N}\right)^k\left(1-\frac{\lambda}{N}\right)^{N-k}, \tag{A.9}$$

approaches the *Poisson distribution*,

$$P_P(\lambda,k) = \frac{\lambda^k}{k!}\,e^{-\lambda}, \tag{A.10}$$

as $N \to \infty$. Note also that the preceding results translate into

$$\langle k\rangle = \lambda, \qquad \langle(\Delta k)^2\rangle = \langle k^2\rangle - \langle k\rangle^2 = \lambda. \tag{A.11}$$

This distribution played a role in Eq. (3.63) for coherent states in quantum mechanics, and in Eq. (7.41) for the Boltzmann gas in the grand canonical ensemble. It is also very important in connection with radioactive decays.

We can now turn to a second limiting behavior of the binomial distribution. It occurs again for large values of $N$, but focusing on "ideal coins" with $p_H = p_T = \frac{1}{2}$ and on the region close to $k = \frac{N}{2}$, the most probable value in this case. Let us therefore begin by considering

$$P_B\left(N, \frac{N}{2}+m\right) = \frac{N!}{\left(\frac{N}{2}+m\right)!\left(\frac{N}{2}-m\right)!}\,\frac{1}{2^N}, \tag{A.12}$$

or rather,

$$\begin{aligned}\log\left[P_B\left(N,\frac{N}{2}+m\right)\right] &= \log\left[N!\right] - \log\left[\left(\frac{N}{2}+m\right)!\right] \\ &\quad - \log\left[\left(\frac{N}{2}-m\right)!\right] + -N\log 2.\end{aligned} \tag{A.13}$$

One can now expand this expression, making use of the Stirling approximation in its more refined form of Eq. (2.31),

$$\log\Gamma(x+1) \sim \left(x+\frac{1}{2}\right)\log x - x + \log\sqrt{2\pi}, \tag{A.14}$$

since here subleading terms are important. Retaining terms up to $\mathcal{O}\left(\frac{1}{N}\right)$ gives the *Gaussian* distribution

$$P_G(\mu,x) = \sqrt{\frac{1}{2\pi\sigma^2}}\,e^{-\frac{x^2}{2\sigma^2}}, \tag{A.15}$$

where now we are treating $m$ to be a continuous variable, $x$, and

$$\sigma^2 = \frac{N}{4}. \tag{A.16}$$

Note that this Gaussian distribution is normalized,

$$\int_{-\infty}^{\infty} dx\, P_G(\mu,x) = 1, \tag{A.17}$$

and moreover,

$$\langle x\rangle \equiv \int_{-\infty}^{\infty} dx\, x\, P_G(\mu,x) = 0, \tag{A.18}$$

$$\langle x^2\rangle \equiv \int_{-\infty}^{\infty} dx\, x^2\, P_G(\mu,x) = \sigma^2. \tag{A.19}$$

More generally, one can arrive at off-centered Gaussian distributions of the type

$$P_G\left(\mu,x,x_0\right) = \sqrt{\frac{1}{2\,\pi\,\sigma^2}}\, e^{-\frac{(x-x_0)^2}{2\,\sigma^2}}, \tag{A.20}$$

for which

$$\langle x\rangle = x_0, \qquad \langle(\Delta x)^2\rangle = \sigma^2. \tag{A.21}$$

Gaussian distributions play a key role in probability theory, in particular since they emerge when describing the probability for the average of large numbers of uncorrelated variables. Let us conclude this appendix with a discussion of this result, which is often referred to as the central limit theorem. The starting point is the probability density $P(x)$ for the average of $N$ independent random variables $x_1, \ldots,$ $x_N$, each distributed according to a given $f(x)$, so that

$$P(x) = \int dx_1 \cdots dx_N f(x_1)\cdots f(x_N)\, \delta\left(x - \frac{x_1+\cdots+x_N}{N}\right), \tag{A.22}$$

where

$$\int dx f(x) = 1, \quad \int dx\, x f(x) = x_0, \quad \int dx\, x^2 f(x) = \sigma^2. \tag{A.23}$$

To begin with, let us consider the Fourier transform of $P(x)$,

$$\tilde{P}(k) = \int_{-\infty}^{+\infty} dx\, e^{-ikx} P(x) = \left[\int dx f(x)\, e^{-i\frac{kx}{N}}\right]^N, \tag{A.24}$$

or equivalently,

$$\frac{1}{N}\log\tilde{P}(k) = \log\int_{-\infty}^{+\infty} dx f(x)\, e^{-i\frac{kx}{N}}. \tag{A.25}$$

One can now expand the right-hand side for large $N$, and making use of Eqs. (A.23) leads to

$$\begin{aligned}\frac{1}{N}\log\tilde{P}(k) &= \log\left[1-\frac{ikx_0}{N}-\frac{k^2(\sigma^2+x_0^2)}{2N^2}+\mathcal{O}\left(\frac{1}{N^3}\right)\right]\\ &= -\frac{ikx_0}{N}-\frac{k^2\sigma^2}{2N^2}+\mathcal{O}\left(\frac{1}{N^3}\right).\end{aligned} \tag{A.26}$$

Therefore

$$\tilde{P}(k) = e^{-ikx_0-\frac{k^2\sigma^2}{2N}} + \mathcal{O}\left(\frac{1}{N^2}\right), \tag{A.27}$$

and performing the inverse Fourier transform finally yields

$$P(x) \simeq \sqrt{\frac{N}{2\pi\sigma^2}}\, e^{-\frac{N(x-x_0)^2}{2\sigma^2}}, \tag{A.28}$$

after making use of Eq. (2.21). This is indeed a Gaussian distribution with variance $\frac{\sigma^2}{N}$.

# Appendix B Equilibrium and Combinatorics

The Boltzmann, Fermi–Dirac, and Bose–Einstein equilibrium populations are often derived by following a line of reasoning that differs from the one presented in the main body of this book. This appendix is devoted to illustrating this different approach.

## B.1 Boltzmann Statistics

Let us consider an ensemble of $N$ noninteracting, distinguishable particles. One can characterize the single-particle states of this system by a pair of labels $\alpha$ and $\kappa$. The first, $\alpha = 0, 1, 2, \ldots$ identifies the energy level $\epsilon_\alpha$, while the second $\kappa = 1, 2, \ldots, g_\alpha$ is generally needed to account for possible degeneracies.

The number of microstates $\Gamma(\{n_\alpha\})$ associated to a given set of occupation numbers $n_\alpha$, the number of particles populating the $\alpha$th energy level, may be explicitly calculated as follows. One must first select $n_0$ particles within the original set, which can be done in $\binom{N}{n_0}$ ways, to then place them in the $g_0$ available "slots" of the ground state $\epsilon_0$, with $g_0^{n_0}$ possibilities. The same must be done for the first excited level $\epsilon_1$, with $N - n_0$ particles left, and so on for all higher energy levels. The final result is

$$\Gamma(\{n_\alpha\}) = \binom{N}{n_0} g_0^{n_0} \binom{N - n_0}{n_1} g_1^{n_1} \cdots = N! \prod_\alpha \frac{g_\alpha^{n_\alpha}}{n_\alpha!}. \tag{B.1}$$

One can now maximize the entropy, resorting to Boltzmann's formula (2.16), with a pair of Lagrange multipliers $\beta$ and $\gamma$ fixing the total number of particles and the total energy,

$$N = \sum_\alpha n_\alpha, \qquad U = \sum_\alpha n_\alpha \epsilon_\alpha. \tag{B.2}$$

As in Chapter 2, one is thus led to

$$\frac{\partial}{\partial n_\alpha}\left(\log \Gamma - \beta \sum_\alpha \epsilon_\alpha n_\alpha - \gamma \sum_\alpha n_\alpha\right) = 0, \tag{B.3}$$

and, making use of Stirling's formula (2.32), one can recover the Boltzmann equilibrium populations

$$n_\alpha = N g_\alpha \frac{e^{-\beta \epsilon_\alpha}}{Z_1}, \tag{B.4}$$

where

$$Z_1 = \sum_\alpha g_\alpha e^{-\beta\epsilon_\alpha} \tag{B.5}$$

is the one-particle partition function. As in Chapter 2, one can verify that this stationary point is a maximum for the entropy.

## B.2 Fermi–Dirac and Bose–Einstein Statistics

One can also adapt these arguments to systems of identical quantum particles. For a Fermi gas, whose particles must respect the Pauli exclusion principle, each $(\alpha, \kappa)$ slot cannot host more than one particle. Therefore, the number of microstates is

$$\Gamma = \prod_\alpha \binom{g_\alpha}{n_\alpha}, \tag{B.6}$$

since for each level the binomial $\binom{g_\alpha}{n_\alpha}$ counts the number of ways in which the $g_\alpha$ slots can accommodate $n_\alpha$ particles without multiple occupancy. On the other hand, for a Bose gas each state can be occupied by an arbitrary number of particles, and therefore

$$\Gamma = \prod_\alpha \binom{g_\alpha + n_\alpha - 1}{n_\alpha}, \tag{B.7}$$

where $\binom{g_\alpha+n_\alpha-1}{n_\alpha}$ counts the number of combinations with repetition. This result is usually justified by considering $n_\alpha$ particles and $g_\alpha - 1$ partitions, which can be arranged in a total number of ways given by

$$\frac{(n_\alpha + g_\alpha - 1)!}{n_\alpha!\,(g_\alpha - 1)!} = \binom{g_\alpha + n_\alpha - 1}{n_\alpha}, \tag{B.8}$$

thus allowing for multiple occupancy between pairs of partitions.

Resorting again to the Stirling approximation, the quantity to be maximized is in this case

$$\log\Gamma = \sum_\alpha \left[ n_\alpha \log\left(\frac{g_\alpha}{n_\alpha} \mp 1\right) \mp g_\alpha \log\left(1 \mp \frac{n_\alpha}{g_\alpha}\right)\right], \tag{B.9}$$

where the upper (lower) signs refer to Fermi–Dirac (Bose–Einstein) statistics. Proceeding as above, one thus finds the equilibrium populations

$$n_\alpha = \frac{g_\alpha}{e^{\beta(\epsilon_\alpha-\mu)} \pm 1}. \tag{B.10}$$

One can now see that for $\frac{n_\alpha}{g_\alpha} \ll 1$ these results approach the nondegenerate case:

$$\binom{g_\alpha}{n_\alpha} = \frac{g_\alpha^{n_\alpha}}{n_\alpha!}\left(1 - \frac{1}{g_\alpha}\right)\cdots\left(1 - \frac{n_\alpha - 1}{g_\alpha}\right) = \frac{g_\alpha^{n_\alpha}}{n_\alpha!}\left[1 - \frac{n_\alpha(n_\alpha - 1)}{2g_\alpha} + \cdots\right], \tag{B.11}$$

and similarly,

$$\binom{g_\alpha + n_\alpha - 1}{n_\alpha} = \frac{g_\alpha^{n_\alpha}}{n_\alpha!}\left[1 + \frac{n_\alpha(n_\alpha - 1)}{2g_\alpha} + \cdots\right]. \tag{B.12}$$

The leading terms are indeed the Boltzmann multiplicity (B.1) divided by $N!$, with the first nontrivial corrections being $\mp n_\alpha(n_\alpha - 1)/2g_\alpha$ in the fermionic and bosonic cases, so that the distinction between Fermi and Bose statistics fades out in the nondegenerate limit. Note that, in the classical limit, the construction retains an overall $\frac{1}{N!}$ for identical particles.

C

# Appendix C WKB at the Bottom

In this appendix we describe how the results obtained using the linear connection formulas of Section 3.7.1 are modified for energy levels close to the bottom of a potential well. The end result will allow a direct comparison with the functional integral around the instanton solution.

To this end, let us refer to the double-well potential sketched in Figure C.1, described by

$$V(x) = \lambda(\eta^2 - x^2)^2. \tag{C.1}$$

This is sufficient to parametrize a generic symmetric double well around its minima. Close to the two minima, as we have seen in the main body of the text, by letting

$$x = \pm\eta + \xi, \tag{C.2}$$

the quadratic approximation yields

$$V(\pm\eta + \xi) \simeq 4\,\lambda\,\eta^2\,\xi^2, \tag{C.3}$$

and therefore the angular frequency of small oscillations for a classical particle of mass $m$ would be

$$\omega = \sqrt{\frac{8\lambda\eta^2}{m}}. \tag{C.4}$$

The ground-state energy of a single well $E = \frac{\hbar\omega}{2}$ now leads us to identify the inversion points

$$x_\pm \simeq \eta \pm \sqrt{\frac{\hbar}{m\omega}}. \tag{C.5}$$

As we have seen, in quantum mechanics the presence of a second well modifies the energy levels. Let us proceed to compute this effect in the presence of a high barrier, so that $\lambda\eta^4 \gg \frac{\hbar\omega}{2}$ and the tunneling effect is small. The Schrödinger equation for a harmonic oscillator,

$$-\frac{\hbar^2}{2m}\frac{d^2\psi}{d\xi^2} + \frac{1}{2}m\omega^2\xi^2\,\psi = E\,\psi, \tag{C.6}$$

provides useful information in this case, and letting

$$\gamma = \sqrt{\frac{\hbar}{2m\omega}}, \qquad \xi = \gamma z, \qquad a = -\frac{E}{\hbar\omega}, \tag{C.7}$$

one can turn it into the parabolic cylinder equation

$$\frac{d^2\psi}{dz^2} - \left(a + \frac{z^2}{4}\right)\psi = 0. \tag{C.8}$$

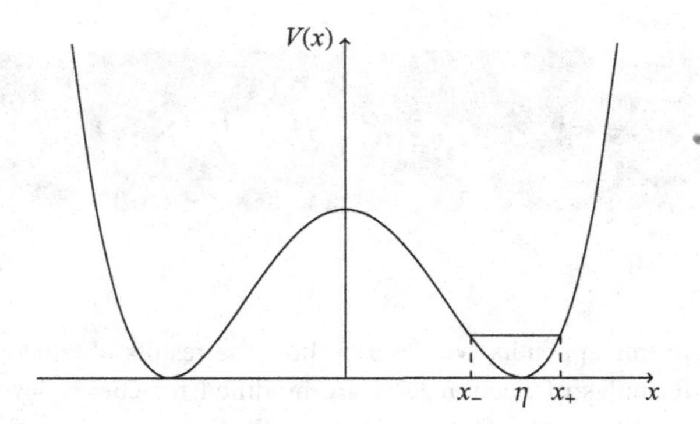


Figure C.1 Symmetric double-well potential with inversion points close to the bottom.

For generic values of $a$, two independent solutions $Y_1$ and $Y_2$ of definite parity exist. Equation (C.8) is a Fuchsian equation for which the origin is a regular point, and therefore all solutions start around $z = 0$ as ordinary power series,

$$\psi(z) = \sum_{k=0}^{\infty} \alpha_k z^k. \tag{C.9}$$

Substituting this expression in Eq. (C.8) leads to the recursion relation

$$\alpha_k = \frac{4a}{4k(k-1)+1}\,\alpha_{k-2}, \tag{C.10}$$

where the first two coefficients, $\alpha_0$ and $\alpha_1$, are arbitrary. For the even solution $Y_1$ we thus choose $\alpha_0 = 1$ and $\alpha_1 = 0$, and then

$$\alpha_{2k} = \frac{4a}{(4k-1)^2}\,\alpha_{2k-2}, \tag{C.11}$$

so that

$$Y_1(a,z) = \sum_{k=0}^{+\infty} \frac{(4a)^k}{\prod_{j=0}^{k}(4j-1)^2} z^{2k}. \tag{C.12}$$

In a similar fashion, for the odd solution $Y_2$ we choose $\alpha_0 = 0$ and $\alpha_1 = 1$, and then

$$\alpha_{2k+1} = \frac{4a}{(4k+1)^2}\alpha_{2k-1}, \tag{C.13}$$

so that finally

$$Y_2(a,z) = \sum_{k=0}^{+\infty} \frac{(4a)^k}{\prod_{j=0}^{k}(4j+1)^2} z^{2k+1}. \tag{C.14}$$

In the theory of parabolic cylinder functions, it can be proved that the two combinations

$$U(a,z) = Y_1(a,z)\cos\left(\frac{\pi}{4}+\frac{a\pi}{2}\right) - Y_2(a,z)\sin\left(\frac{\pi}{4}+\frac{a\pi}{2}\right), \tag{C.15}$$

$$\Gamma\left(\frac{1}{2}-a\right) V(a,z) = Y_1(a,z)\sin\left(\frac{\pi}{4}+\frac{a\pi}{2}\right) + Y_2(a,z)\cos\left(\frac{\pi}{4}+\frac{a\pi}{2}\right),$$

have, for $z \to +\infty$, the asymptotic behaviors

$$U(a,z) \sim e^{-z^2/4} z^{-a-1/2},$$
$$V(a,z) \sim \sqrt{\frac{2}{\pi}} e^{z^2/4} z^{a-1/2}. \tag{C.16}$$

Taking the parity of $Y_1$ and $Y_2$ into account, these results determine also the asymptotic behaviors for $z \to -\infty$. Indeed, inverting Eqs. (C.15), one can obtain the relations

$$U(a,-z) = -U(a,z)\sin(\pi a) + V(a,z)\Gamma\left(\frac{1}{2}-a\right)\cos(\pi a), \tag{C.17}$$

$$\Gamma\left(\frac{1}{2}-a\right)V(a,z) = U(a,z)\cos(\pi a) + V(a,z)\Gamma\left(\frac{1}{2}-a\right)\sin(\pi a), \tag{C.18}$$

and finally, as $z \to -\infty$

$$U(a,z) \sim -\sin(\pi a) e^{-z^2/4} |z|^{-a-1/2} + \cos(\pi a)\sqrt{\frac{2}{\pi}}\Gamma\left(\frac{1}{2}-a\right) e^{z^2/4}|z|^{a-1/2}, \tag{C.19}$$

$$V(a,z) \sim \cos(\pi a) e^{-z^2/4} |z|^{-a-1/2} + \sin(\pi a)\sqrt{\frac{2}{\pi}}\,\Gamma\left(\frac{1}{2}-a\right) e^{z^2/4}|z|^{a-1/2}. \tag{C.20}$$

Due to the asymptotic behaviors (C.16), the solutions of interest cannot contain a $V$-component for $z > 0$. When combined with Eqs. (C.16), this result leads to the connection formula

$$-\sin(\pi a) e^{-z^2/4} |z|^{-a-1/2} + \cos(\pi a)\sqrt{\frac{2}{\pi}}\,\Gamma\left(\frac{1}{2}-a\right) e^{z^2/4}|z|^{a-1/2} \longleftrightarrow e^{-z^2/4} z^{-a-1/2}, \tag{C.21}$$

where the combination on the left captures the leading behavior as $z \to -\infty$, while the single term on the right captures the leading behavior as $z \to +\infty$. For the ground state of our double well, letting

$$a = -\frac{1}{2} - \frac{\Delta}{2\hbar\omega} \tag{C.22}$$

and expanding for small values of $\Delta$, the preceding result becomes

$$e^{-\frac{z^2}{4}} - \frac{\Delta}{\hbar\omega}\sqrt{\frac{\pi}{2}}\, e^{\frac{z^2}{4}}\, |z|^{-1} \longleftrightarrow e^{-\frac{z^2}{4}}. \tag{C.23}$$

We can now return to the WKB results. For $0 < x < x_-$ wavefunctions of definite parity are of the form

$$\psi(x) = \frac{\exp\left(\frac{1}{\hbar}\int_0^x |p(\xi)|\mathrm{d}\xi\right) \pm \exp\left(-\frac{1}{\hbar}\int_0^x |p(\xi)|\mathrm{d}\xi\right)}{\sqrt{|p(x)|}}, \tag{C.24}$$

and for $x$ sufficiently far from the inversion point, in the region where $V(x) \gg E$, one can expand the square root in $|p(\xi)|$, obtaining

$$\int_0^x |p(\xi)|\mathrm{d}\xi \simeq \int_0^x \sqrt{2mV(\xi)}\mathrm{d}\xi - E\int_0^x \sqrt{\frac{m}{2V(\xi)}}\mathrm{d}\xi, \tag{C.25}$$

and one can recast the first integral in the form

$$\int_0^x \sqrt{2mV(\xi)}d\xi = \int_0^\eta \sqrt{2mV(\xi)}d\xi - \int_x^\eta \sqrt{2mV(\xi)}d\xi. \tag{C.26}$$

Notice now that the first integral is half of the Euclidean action associated to a trajectory connecting the two minima of the potential, which solves

$$\frac{1}{2}m\dot{x}^2 - V(x) = 0, \tag{C.27}$$

so that

$$S_0 = \int_{-\infty}^{+\infty} dt \left[\frac{1}{2}m\dot{x}^2(t) + V(x(t))\right] = \int_{-\eta}^{\eta} \sqrt{2mV(\xi)}\, d\xi. \tag{C.28}$$

Let us now assume that $x \simeq \eta$, a condition that is compatible with $V(x) \gg E$ provided $\lambda$ is sufficiently large. In this case one can use, in the second integral in (C.26), the quadratic approximation for the potential, obtaining

$$\int_x^\eta \sqrt{2mV(\xi)}d\xi = m\omega \int_x^\eta (\eta - \xi)d\xi = \frac{m\omega}{2}(\eta - x)^2,$$

while the same approximation turns the last integral in (C.25) into

$$\int_0^x \sqrt{\frac{m}{2V(\xi)}}d\xi = \sqrt{\frac{m}{8\lambda\eta^2}} \log\frac{\eta + x}{\eta - x}, \tag{C.29}$$

and therefore

$$\frac{1}{\hbar}\int_0^x |p(\xi)|d\xi \simeq \frac{S_0}{2\hbar} - \frac{m\omega}{2\hbar}(\eta - x)^2 - \frac{E}{\hbar\omega}\log\frac{\eta + x}{\eta - x}. \tag{C.30}$$

Moreover

$$\sqrt{|p(x)|} \simeq \sqrt{\frac{m\omega}{\hbar}}(\eta - x), \tag{C.31}$$

and, making use of Eq. (C.7),

$$z = \sqrt{\frac{2m\omega}{\hbar}}(x - \eta). \tag{C.32}$$

Taking into account the definition of $\omega$, one is finally led to

$$\frac{1}{\hbar}\int_0^x |p(\xi)|d\xi \simeq \frac{S_0}{2\hbar} - \frac{z^2}{4} + \frac{E}{\hbar\omega}\log\left(\frac{|z|}{2\omega}\sqrt{\frac{\hbar\lambda}{m^2\omega}}\right), \tag{C.33}$$

$$\sqrt{|p(x)|} \simeq \left(\frac{m\omega}{2\hbar}\right)^{1/4} |z|^{1/2}. \tag{C.34}$$

For $0 < x < x_-$ the WKB wavefunction is thus

$$\psi(x) = \left(\frac{2\hbar}{m\omega}\right)^{1/4} \left[ e^{\frac{S_0}{2\hbar}} \frac{e^{-z^2/4}}{\sqrt{|z|}} \left(|z|\sqrt{\frac{\hbar\lambda}{4m^2\omega^3}}\right)^{\frac{E}{\hbar\omega}} \pm e^{-\frac{S_0}{2\hbar}} \frac{e^{z^2/4}}{\sqrt{|z|}} \left(|z|\sqrt{\frac{\hbar\lambda}{4m^2\omega^3}}\right)^{-\frac{E}{\hbar\omega}} \right], \tag{C.35}$$

or

$$\psi(x) = \left(\frac{2\hbar}{m\omega}\right)^{1/4} \left(\frac{\hbar\lambda}{4m^2\omega^3}\right)^{\frac{E}{2\hbar\omega}} e^{\frac{S_0}{2\hbar}} |z|^{\frac{E}{\hbar\omega}-\frac{1}{2}}$$

$$\times \left[ e^{-z^2/4} \pm e^{-\frac{S_0}{\hbar}} e^{z^2/4} |z|^{-\frac{2E}{\hbar\omega}} \left(\frac{4m^2\omega^3}{\hbar\lambda}\right)^{\frac{E}{\hbar\omega}} \right]. \tag{C.36}$$

Up to higher orders in $\Delta$,

$$\psi(x) = \left(\frac{2\hbar}{m\omega}\right)^{1/4} \left(\frac{\hbar\lambda}{4m^2\omega^3}\right)^{\frac{E}{2\hbar\omega}} e^{\frac{S_0}{2\hbar}}$$

$$\times \left[ e^{-z^2/4} \pm e^{-\frac{S_0}{\hbar}} e^{z^2/4} |z|^{-1} \left(\frac{4m^2\omega^3}{\hbar\lambda}\right)^{\frac{E}{\hbar\omega}} \right], \tag{C.37}$$

and comparing with Eq. (C.23) one can deduce the correction to the ground-state energy,

$$\Delta = \mp \frac{\hbar\omega}{\pi} e^{-S_0/\hbar} \sqrt{\frac{2\pi m^2\omega^3}{\hbar\lambda}}, \tag{C.38}$$

where the upper sign applies to the symmetric wavefunction.

Note that the correction differs from the linear connection formulas of Section 3.7.1 due to the presence of the amplification factor

$$\sqrt{\frac{2\pi m^2\omega^3}{\hbar\lambda}}. \tag{C.39}$$

This term also involves a negative power of $\hbar$: the potential grows less efficiently between the two minima than in the linear case, and this enhances the tunneling effect and the energy separation between the lowest levels. The last sections of Chapter 3 relate the emergence of this peculiar behavior from the functional integral to the presence of zero-mode fluctuations around the instanton solution.

One can also adapt these considerations to the lowest energy band in a crystal, for which

$$\Delta_0 = -\frac{\hbar\omega}{\pi} e^{-S_0/\hbar} \sqrt{\frac{2\pi m^2\omega^3}{\hbar\lambda}} \cos\varphi, \tag{C.40}$$

and to the decay of a metastable state, for which

$$\Delta_0 = -i\frac{\hbar\omega}{4\pi} e^{-2S_0/\hbar} \left(\frac{2\pi m^2\omega^3}{\hbar\lambda}\right). \tag{C.41}$$

In this last case the amplification factor is squared, in agreement with our findings in Section 3.7.6 where we used the linear connection formulas.

# D Appendix D Some Analytic Functions

In this appendix we discuss some results in the theory of analytic functions that are used here and there in the text.

The Bernoulli numbers $B_n$ are defined via the relation

$$\frac{z}{e^z - 1} = \sum_{n=0}^{\infty} \frac{B_n}{n!} z^n, \tag{D.1}$$

where the series has radius of convergence $2\pi$, since the l.h.s. has poles at $z = \pm 2\pi i$. A direct comparison shows that

$$B_0 = 1, \quad B_1 = -\frac{1}{2}, \tag{D.2}$$

and moreover,

$$\frac{z}{e^z - 1} + \frac{z}{2} = \frac{z}{2} \coth \frac{z}{2}, \tag{D.3}$$

which is an even function of $z$, so that $B_1$ is the only odd Bernoulli number.

The other even Bernoulli numbers can now be determined multiplying both sides of

Eq. (D.1) by $(e^z - 1)$ and expanding in a power series, which leads to

$$1 = \sum_{m=0}^{\infty} \sum_{n=0}^{\infty} \frac{z^{m+n}}{(m+1)!\, n!} B_n = \sum_{p=0}^{\infty} \frac{z^p}{(p+1)!} \sum_{m=0}^{p} B_m \binom{p+1}{m}. \tag{D.4}$$

This expression determines the Bernoulli numbers recursively, since all coefficients multiplying positive powers of $z$ must vanish, a condition that can be cast in the form

$$(B+1)^{p+1} - B_{p+1} = 0 \qquad p > 0, \tag{D.5}$$

where, after expanding the binomial, one should turn powers into labels. For instance, applying this result for $p = 3$ and for $p = 5$ yields

$$B_2 = \frac{1}{6}, \qquad B_4 = -\frac{1}{30}. \tag{D.6}$$

Let us now consider the infinite product decomposition

$$\sin z = z \prod_{n=1}^{\infty} \left(1 - \frac{z^2}{n^2 \pi^2}\right), \tag{D.7}$$

where the r.h.s. has the proper zeros and yields a convergent power-series expansion since it starts with terms of order $\frac{1}{n^2}$. Taking the logarithm of this result and differentiating it, one can obtain another useful relation:

$$z \cot z = 1 - 2 \sum_{n=1}^{\infty} \left(\frac{z}{n\pi}\right)^2 \frac{1}{1 - \left(\frac{z}{n\pi}\right)^2}. \tag{D.8}$$

Expanding the last factor in a geometric series leads to the emergence of the Riemann $\zeta$ function, which is defined for $Re(s) > 1$ as

$$\zeta(s) = \sum_{n=1}^{\infty} \frac{1}{n^s}, \tag{D.9}$$

so that

$$z \cot z = 1 - 2 \sum_{m=1}^{\infty} \left(\frac{z}{\pi}\right)^{2m} \zeta(2m). \tag{D.10}$$

This result can be compared with the combination of Eqs. (D.1) and (D.3),

$$\frac{z}{2} \coth \frac{z}{2} = \sum_{m=1}^{\infty} \frac{B_{2m}}{(2m)!} z^{2m}, \tag{D.11}$$

which entails a rescaling and an analytic continuation $z \to \frac{iz}{2}$, and leads eventually to

$$\zeta(2m) = (-1)^{m-1} \frac{(2\pi)^{2m}}{2(2m)!} B_{2m}. \tag{D.12}$$

Two special cases of interest for us are

$$\zeta(2) = \frac{\pi^2}{6}, \qquad \zeta(4) = \frac{\pi^4}{90}, \tag{D.13}$$

and indeed the theory of blackbody radiation makes use of the result

$$\int_0^{\infty} dx \frac{x^3}{e^x - 1} = \sum_{n=1}^{\infty} \int_0^{\infty} dx\, x^3 e^{-nx} = \Gamma(4)\, \zeta(4) = \frac{\pi^4}{15}. \tag{D.14}$$

In general one can also prove, in a similar fashion, that for $Re(\lambda) > 0$

$$\int_0^{\infty} dx \frac{x^{\lambda-1}}{e^x - 1} = \sum_{n=1}^{\infty} \int_0^{\infty} dx\, x^{\lambda-1} e^{-nx} = \Gamma(\lambda)\, \zeta(\lambda), \tag{D.15}$$

and also that

$$\int_0^\infty dx \, \frac{x^{\lambda-1}}{e^x + 1} = \Gamma(\lambda)\, \zeta(\lambda) \left(1 - 2^{1-\lambda}\right), \tag{D.16}$$

which plays a role for the Fermi gas and follows from

$$\int_0^\infty dx \, \frac{x^{\lambda-1}}{e^x - 1} - \int_0^\infty dx \, \frac{x^{\lambda-1}}{e^x + 1} = \int_0^\infty dx \, \frac{2x^{\lambda-1}}{e^{2x} - 1}$$
$$= 2^{1-\lambda} \int_0^\infty dx \, \frac{x^{\lambda-1}}{e^x - 1}. \tag{D.17}$$

The Riemann $\zeta$ function is deeply related to the distribution of prime numbers. Indeed, separating even and odd contributions in the sum of Eq. (D.9), one can get a taste of this fact by writing

$$\zeta(s) = 2^{-s}\, \zeta(s) + 1 + \frac{1}{3^s} + \frac{1}{5^s} + \frac{1}{7^s} + \cdots, \tag{D.18}$$

so that

$$\left(1 - 2^{-s}\right) \zeta(s) = 1 + \frac{1}{3^s} + \frac{1}{5^s} + \frac{1}{7^s} + \cdots. \tag{D.19}$$

The procedure can continue with the elimination of all multiples of 3, and so on, so that eventually one obtains an infinite product over all prime numbers $p$, and

$$\zeta(s) = \frac{1}{\prod_p \left(1 - p^{-s}\right)}. \tag{D.20}$$

We can conclude this appendix with a few more details on Euler's $\Gamma$ function, which is defined for $Re(z) > 0$ by

$$\Gamma(z) = \int_0^\infty dt \, e^{-t}\, t^{z-1}, \tag{D.21}$$

and on Euler's $B$ function, which is defined for $Re(\alpha) > 0$ and $Re(\beta) > 0$ by

$$B(\alpha, \beta) = \int_0^1 dt \, t^{\alpha-1} \left(1 - t\right)^{\beta-1}. \tag{D.22}$$

There are elegant contour-integral representations of both functions that encode their analytic continuations, which one can however also build from Eq. (2.24), so that we shall refrain from reviewing them here.

Note that one can write

$$\Gamma(z) = \lim_{n\to\infty} \int_0^n dt \left(1 - \frac{t}{n}\right)^n t^{z-1}, \tag{D.23}$$

and performing the integral leads to the product representation

$$\Gamma(z) = \frac{e^{-\gamma z}}{z \prod_{n=1}^\infty \left[\left(1 + \frac{z}{n}\right) e^{-\frac{z}{n}}\right]}. \tag{D.24}$$

Here,

$$\gamma_E = \lim_{n\to\infty} \left[ \sum_{k=1}^{n} \frac{1}{k} - \log n \right] \simeq 0.577 \tag{D.25}$$

is the Euler–Mascheroni constant, whose emergence is induced by the convergence factors in the infinite product. A notable consequence of this equation is

$$\frac{1}{\Gamma(z)\,\Gamma(-z)} = -z^2 \prod_{n=1}^{\infty} \left(1 - \frac{z^2}{n^2}\right), \tag{D.26}$$

and combining it with the product representation of $\sin(z)$ in Eq. (D.7) and with Eq. (2.24) leads finally to

$$\Gamma(z)\,\Gamma(1-z) = \frac{\pi}{\sin(\pi z)}. \tag{D.27}$$

The link between the two Euler functions can be deduced by starting from the product of two integrals as in Eq. (D.21), with integration variables $x$ and $y$, and inserting them in

$$\int_0^\infty du\, \delta(x+y-u), \tag{D.28}$$

which is identically equal to one, given the integration ranges for $x$ and $y$. Inverting the order of integrations and rescaling the two variables $x$ and $y$ by a factor $u$ then leads to

$$\Gamma(\alpha)\,\Gamma(\beta) = \int_0^\infty du\, e^{-u} u^{\alpha+\beta-1} \int_0^\infty \int_0^\infty dx\, dy\, x^{\alpha-1}\, y^{\beta-1}\, \delta(x+y-1), \tag{D.29}$$

which finally proves that

$$B(\alpha,\beta) = \frac{\Gamma(\alpha)\Gamma(\beta)}{\Gamma(\alpha+\beta)}. \tag{D.30}$$

This important relation affords a multivariate generalization,

$$\int_0^\infty x_1^{m_1-1} \cdots x_N^{m_N-1}\, \delta(x_1 + \cdots + x_N - 1)\, dx_1 \cdots dx_N = \frac{\Gamma(m_1)\cdots\Gamma(m_N)}{\Gamma(m_1 + \cdots + m_N)}, \tag{D.31}$$

which can be proved by proceeding along the same lines, and plays a role in one of the problems of Chapter 2.

# Appendix E Euler–Maclaurin and Abel–Plana Formulas

In this appendix we discuss the Bernoulli polynomials and the Euler–Maclaurin formula, an elegant tool to obtain asymptotic expansions of series. We conclude with a discussion of the Abel–Plana formula, an alternative approach based on contour integrals.

The Bernoulli polynomials $\varphi_p(x)$ are defined via a generalization of Eq. (D.1),

$$\frac{z\,e^{xz}}{e^z - 1} = \sum_{p=0}^{\infty} \frac{\varphi_p(x)}{p!}\, z^p. \tag{E.1}$$

However, combining the standard exponential series with Eq. (D.1) one can write

$$\frac{z\,e^{xz}}{e^z - 1} = \sum_{p=0}^{\infty} \frac{z^p}{p!} \sum_{m=0}^{p} B_m \binom{p}{m} x^{p-m}, \tag{E.2}$$

so that

$$\varphi_p(x) = \sum_{m=0}^{p} B_m \binom{p}{m} x^{p-m}. \tag{E.3}$$

The first three Bernoulli polynomials are

$$\varphi_0(x) = 1, \quad \varphi_1(x) = x - \frac{1}{2}, \quad \varphi_2(x) = x^2 - x + \frac{1}{6}. \tag{E.4}$$

Moreover, the basic recursion formula (D.5) implies that

$$\varphi_p(0) = \varphi_p(1), \qquad (p \geq 2), \tag{E.5}$$

so that all $\varphi_p(x)$ afford a natural periodic extension beyond the interval [0, 1], while Eq. (E.3) implies that

$$\varphi_p'(x) = p\,\varphi_{p-1}(x). \tag{E.6}$$

Let us now consider the identity

$$\int_j^{j+1} dx\, f'(x)\left(x - j - \frac{1}{2}\right) = \frac{f(j+1) + f(j)}{2} - \int_j^{j+1} dx\, f(x), \tag{E.7}$$

and sum over $j$, which leads to

$$\sum_{j=0}^{n} f(j) - \int_0^n dx\, f(x) = \frac{1}{2}\left(f(0) + f(n)\right) + \int_0^n dx\, f'(x)\left(x - [x] - \frac{1}{2}\right), \tag{E.8}$$

where $[x]$ denotes the integer part of $x$. In terms of the *periodically extended* Bernoulli polynomials, defined recursively so that

$$\varphi_1(x) = x - [x] - \frac{1}{2}, \ \varphi'_{k+1}(x) = k\,\varphi_k(x), \ \varphi_k(0) = \varphi_k(1) = B_k, \tag{E.9}$$

Eq. (E.8) takes the form

$$\sum_{j=0}^{n} f(j) - \int_0^n dx\, f(x) = \frac{1}{2}\,[f(0) + f(n)] + \int_0^n dx\, f'(x)\varphi_1(x). \tag{E.10}$$

The last term can be integrated by parts, producing an asymptotic series. The first iteration gives

$$\begin{aligned}\int_0^n dx\, f'(x)\varphi_1(x) &= \frac{1}{2}\,[f'\varphi_2]_0^n - \frac{1}{2}\int_0^n dx\, f''(x)\varphi_2(x) \\ &= \frac{B_2}{2}\,[f'(n) - f'(0)] - \frac{1}{2}\int_0^n dx\, f''(x)\varphi_2(x),\end{aligned} \tag{E.11}$$

and since there are no odd Bernoulli numbers beyond $B_1$ this expression can be written equivalently as

$$\begin{aligned}\int_0^n dx\, f'(x)\varphi_1(x) &= \frac{B_2}{2}\,[f'(n) - f'(0)] + \frac{B_4}{4!}\,[f'''(n) - f'''(0)] \\ &\quad - \frac{1}{4!}\int_0^n dx\, f^{(4)}(x)\varphi_4(x),\end{aligned}$$

where for the same reason the remainder can also be written in the form

$$\frac{1}{5!}\int_0^n dx\, f^{(5)}(x)\varphi_5(x). \tag{E.12}$$

Continuing the iteration yields the Euler–Maclaurin formula

$$\sum_{j=0}^{n} f(j) - \int_0^n dx\, f(x) = \frac{1}{2}\,[f(0) + f(n)] + \frac{B_2}{2!}\,[f'(n) - f'(0)] \tag{E.13}$$

$$+ \frac{B_4}{4!}\,[f'''(n) - f'''(0)] \tag{E.14}$$

$$+ \cdots$$

$$+ \frac{B_{2m}}{(2m)!}\left[f^{(2m-1)}(n) - f^{(2m-1)}(0)\right] + R_m,$$

where the remainder is

$$R_m = \frac{1}{(2m+1)!}\int_0^n dx\, \varphi_{2m+1}(x) f^{(2m+1)}(x). \tag{E.15}$$

We can now apply this result to the Casimir effect of Chapter 5. To begin with, starting from Eq. (5.9), one gets

$$\begin{aligned}&F'(n) = -2\,n^2 f(n), \quad F''(n) = -4\,n f(n) - 2\,n^2 f'(n), \\ &F'''(n) = -4 f(n) - 4\,n f'(n) - 2\,n^2 f''(n).\end{aligned} \tag{E.16}$$

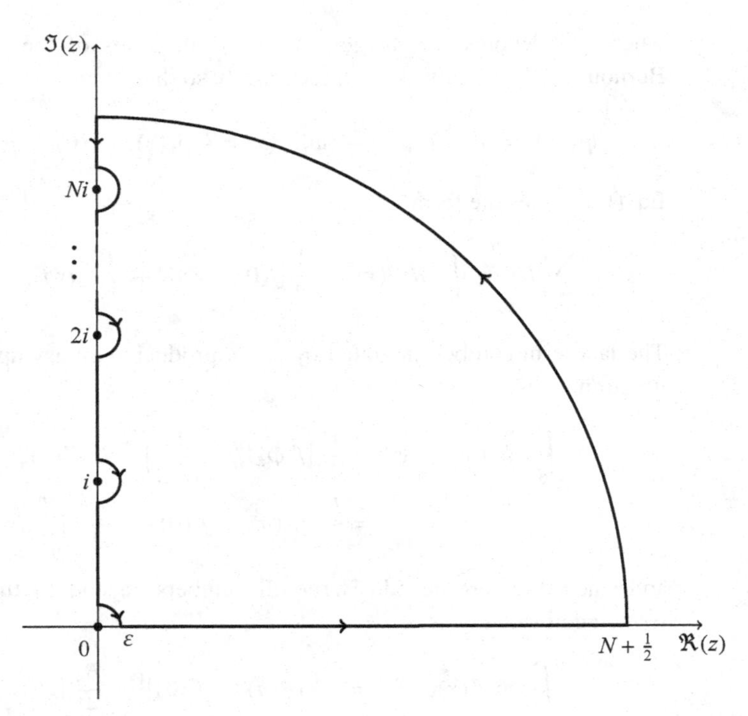


Figure E.1 The integration contour $\gamma$.

The step-like function $f(x)$ equals 1 at the origin, where all its derivatives vanish, and is identically zero for large positive values of $x$. Therefore, the only nonvanishing contribution to the Casimir energy originates from

$$F'''(0) = -4f(0) = -4, \tag{E.17}$$

which leads to Eq. (5.11).

The elegant Abel–Plana formula is an alternative approach to the comparison between series and integrals. One can obtain it by starting from the closed circuit $\gamma$ in Figure E.1 and the vanishing contour integral

$$\oint_\gamma \frac{F(-iz)}{e^{2\pi z} - 1}\, dz = 0 \tag{E.18}$$

for the analytic function $F$, which we shall assume to take real values on the real axis. As $\varepsilon \to 0$ and $N \to \infty$, with $N \in \mathbb{N}$, if the contribution arising from the large quarter of circle of radius $N + \frac{1}{2}$ vanishes in this limit, taking the imaginary part one finds the Abel–Plana formula

$$\frac{1}{2} F(0) + \sum_{n=1}^{\infty} F(n) - \int_0^{\infty} F(y) dy = 2 \int_0^{\infty} dx\, \frac{\mathrm{Im}[F(-ix)]}{e^{2\pi x} - 1}. \tag{E.19}$$

For the function $F$ defined in Eq. (5.9),

$$F(-iz) = \int_0^\infty du\, \sqrt{u - z^2}\, f(u - z^2), \tag{E.20}$$

which is analytic but for a branch cut on the real axis. The Abel–Plana formula (E.19) applies provided one restricts $z$ to the upper half plane, so that

$$F(-ix) \equiv \lim_{\eta \to 0^+} F(-i(x + i\eta)) = \lim_{\eta \to 0^+} \int_0^\infty du\, \sqrt{u - x^2 - i\eta}\, f(u - x^2) \tag{E.21}$$

for $x > 0$. The imaginary part of $F(-ix)$ only arises from the bounded interval $[0, x^2]$ in the $u$-integration, where $f \simeq 1$, and the analytic continuation (E.21) fixes its sign to be negative, so that

$$\text{Im}[F(-ix)] = -\int_0^{x^2} du\, \sqrt{x^2 - u} = \left[\frac{2}{3}(x^2 - u)^{\frac{3}{2}}\right]_0^{x^2} = -\frac{2}{3}\, x^3. \tag{E.22}$$

Consequently, Eq. (E.19) gives

$$\begin{aligned} &\frac{1}{2}\, F(0) + \sum_{n=1}^{\infty} F(n) - \int_0^\infty F(y)\, dy \\ &= -\frac{4}{3} \int_0^\infty dx\, \frac{x^3}{e^{2\pi x} - 1} = -\frac{1}{12\pi^4} \int_0^\infty dx\, \frac{x^3}{e^x - 1}, \end{aligned} \tag{E.23}$$

and we already know that the remaining integral is $\frac{\pi^4}{15}$, so that finally, for the Casimir effect

$$\frac{1}{2}\, F(0) + \sum_{n=1}^{\infty} F(n) - \int_0^\infty F(x) dx = -\frac{1}{180}, \tag{E.24}$$

which proves Eq. (5.11).

# F Appendix F Spin and the Pauli Equation

In this appendix we discuss the Pauli equation, an extension of the Schrödinger equation that captures the nonrelativistic limit of the theory of spin-$\frac{1}{2}$ particles and yields the correct lowest-order contribution to their gyromagnetic ratio.

The need to account for the electron spin extends the wavefunction to a pair of similar quantities, whose square absolute values are the probability densities associated to the two independent spin components. One can group the two components into a column vector

$$\Psi = \begin{pmatrix} \Psi_+ \\ \Psi_- \end{pmatrix}, \tag{F.1}$$

and it is then convenient to introduce the three Pauli matrices

$$\sigma_1 = \begin{pmatrix} 0 & 1 \\ 1 & 0 \end{pmatrix}, \quad \sigma_2 = \begin{pmatrix} 0 & -i \\ i & 0 \end{pmatrix}, \quad \sigma_3 = \begin{pmatrix} 1 & 0 \\ 0 & -1 \end{pmatrix}, \tag{F.2}$$

which satisfy

$$\sigma_i \, \sigma_j = \delta_{ij} \, 1_2 + i \, \epsilon_{ijk} \sigma_k, \tag{F.3}$$

where $1_2$ denotes the $2 \times 2$ identity matrix. Here, $\epsilon$ is the totally antisymmetric Levi–Civita symbol, with $\epsilon_{123} = 1$, and the sum over $k = 1, 2, 3$ is implicit. This relation implies the two properties

$$[\sigma_i, \sigma_j] = 2 \, i \, \epsilon_{ijk} \, \sigma_k, \qquad \{\sigma_i, \sigma_j\} \equiv \sigma_i \, \sigma_j + \sigma_j \, \sigma_i = 2 \, \delta_{ij} \, 1_2, \tag{F.4}$$

and one also defines the spin operator as

$$\mathbf{S} = \frac{\hbar}{2} \, \sigma. \tag{F.5}$$

Pauli proposed the modified Schrödinger equation

$$-\frac{\hbar^2}{2m} \, (\sigma \cdot \nabla)^2 \, \Psi + U \Psi = i \hbar \, \frac{\partial \Psi}{\partial t}, \tag{F.6}$$

which is a system of two equations for $\Psi_\pm$, and exhibits an interesting behavior in the presence of a vector potential. The minimal substitution yields

$$-\frac{\hbar^2}{2m} \left[ \sigma \cdot \left( \nabla + \frac{ie}{\hbar c} \mathbf{A} \right) \right]^2 \Psi + U \Psi = i \hbar \, \frac{\partial \Psi}{\partial t}, \tag{F.7}$$

and using the preceding properties one can show that

$$\left[ \sigma \cdot \left( \nabla + \frac{ie}{\hbar c} \mathbf{A} \right) \right]^2 = \left( \nabla + \frac{ie}{\hbar c} \mathbf{A} \right)^2 - \frac{e}{\hbar c} \, \sigma \cdot \mathbf{B}. \tag{F.8}$$

Therefore, in the presence of a magnetic field, the Pauli equation becomes

$$-\frac{\hbar^2}{2m} \left( \nabla + \frac{ie}{\hbar c} \mathbf{A} \right)^2 \Psi - \frac{e\hbar}{2mc} \, \sigma \cdot \mathbf{B} \, \Psi = i \hbar \, \frac{\partial \Psi}{\partial t}. \tag{F.9}$$

The second term yields a definite prediction for the gyromagnetic ratio $g$ of the electron, since one can write

$$\frac{e\hbar}{2mc}\,\sigma \cdot \mathbf{B} \;=\; g\,\frac{e}{2mc}\,\mathbf{S} \cdot \mathbf{B}, \tag{F.10}$$

with $g = 2$. Note also the alternative presentation of this result,

$$\frac{e\hbar}{2mc}\,\sigma \cdot \mathbf{B} \;=\; \mu_B\,\sigma \cdot \mathbf{B}, \tag{F.11}$$

in terms of Bohr's magneton of Eq. (9.64).

# Appendix G The $G_{n,m}$ Operator

In this appendix we discuss the "inverse" $G_{n,m}$ of the discretized Laplacian $\Delta_{n,m}$ introduced in Section 13.4, where the quotes are meant to stress the exclusion of a zero mode. We begin from a simpler case, a one-dimensional chain, which will serve as a warm-up exercise before turning to the case of interest, the two-dimensional XY lattice.

## G.1 One Dimension

For a chain with $L$ sites labeled by $n = 0, 1, \ldots, L-1$, each hosting a classical spin that can rotate on the XY plane, let us consider the discretized Laplace operator

$$-\Delta_{n,m} = 2\delta_{n,m} - \delta_{n,m+1} - \delta_{n,m-1}. \tag{G.1}$$

Its eigenfunctions are plane waves,

$$-\sum_{m=0}^{L-1} \Delta_{n,m} e^{\frac{i2\pi m\ell}{L}} = \lambda_\ell \, e^{\frac{i2\pi n\ell}{L}}, \tag{G.2}$$

and its eigenvalues,

$$\lambda_\ell = 2\left(1 - \cos\frac{2\pi\ell}{L}\right), \qquad \ell = 0, \ldots, L-1, \tag{G.3}$$

which have already emerged in Chapter 12, include a vanishing one for $\ell = 0$. Consequently, one is led to consider an "inverse" $G_{n,m}$ whose spectral decomposition excludes the zero mode,

$$G_{n,m} = \frac{1}{L}\sum_{\ell=1}^{L-1} \frac{1}{\lambda_\ell} e^{\frac{i2\pi\ell(n-m)}{L}} = G_{n-m}. \tag{G.4}$$

This can conveniently be split further as

$$G_n = G_0 + \widetilde{G}_n. \tag{G.5}$$

We first consider

$$G_0 = \frac{1}{4L}\sum_{\ell=1}^{L-1} \frac{1}{\sin^2\frac{\pi\ell}{L}}, \tag{G.6}$$

which, for odd $L$, can be cast in the form

$$G_0 = \frac{1}{2L}\sum_{\ell=1}^{\frac{L-1}{2}} \frac{1}{\sin^2\frac{\pi\ell}{L}} = \frac{L^2-1}{12L}, \tag{G.7}$$

where the explicit result follows from eq. (1.382) of [20], so that

$$G_0 = \frac{L}{12} + \mathcal{O}\left(\tfrac{1}{L}\right) \tag{G.8}$$

in the thermodynamic limit $L \to \infty$. Let us now consider,

$$\widetilde{G}_n = -\frac{1}{L}\sum_{\ell=1}^{L-1} \frac{1 - e^{\frac{i2\pi\ell n}{L}}}{2\left(1 - \cos\frac{2\pi\ell}{L}\right)}. \tag{G.9}$$

Taking the thermodynamic limit for fixed $n$, and exploiting the behavior of the integrand under parity, one obtains

$$\widetilde{G}_n = -\frac{1}{2}\int_0^{2\pi} \frac{1 - \cos(n\theta)}{1 - \cos\theta}\frac{d\theta}{2\pi} = -\frac{n}{2}. \tag{G.10}$$

All in all, we thus find, for $|n - m| \ll L$, as $L \to \infty$,

$$G_{n,m} = \frac{L}{12} - \frac{|n - m|}{2}. \tag{G.11}$$

## G.2 Two Dimensions

For the two-dimensional XY lattice, in the notation of Section 13.4,

$$-\Delta_{n,m} = \left(2\delta_{n_x,m_x} - \delta_{n_x,m_x+1} - \delta_{n_x,m_x-1}\right)\delta_{n_y,m_y} + (x \leftrightarrow y) \tag{G.12}$$

so that

$$-\sum_m \Delta_{n,m} e^{\frac{i2\pi n\cdot\ell}{L}} = \lambda_\ell\, e^{\frac{i2\pi n\cdot\ell}{L}}, \qquad \lambda_\ell = 4 - 2\left(\cos\frac{2\pi\ell_x}{L} + \cos\frac{2\pi\ell_y}{L}\right), \tag{G.13}$$

and

$$G_{n,m} = \frac{1}{L^2}\sum_{\ell\neq 0} \frac{1}{\lambda_\ell} e^{\frac{2\pi\ell\cdot(n-m)}{L}} = G_{n-m}, \qquad G_n = G_0 + \widetilde{G}_n. \tag{G.14}$$

Let us first address the calculation of

$$G_0 = \frac{1}{4L^2}\sum_{\ell\neq 0} \frac{1}{\sin^2\frac{\pi\ell_x}{L} + \sin^2\frac{\pi\ell_y}{L}}. \tag{G.15}$$

For odd $L$, this can be written in the form

$$\begin{aligned} G_0 = {} & \frac{1}{L^2}\sum_{\ell_x=1}^{\frac{L-1}{2}}\sum_{\ell_y=1}^{\frac{L-1}{2}} \left[\frac{1}{\sin^2\frac{\pi\ell_x}{L} + \sin^2\frac{\pi\ell_y}{L}} - \frac{L^2}{\pi^2(\ell_x^2 + \ell_y^2)}\right] \\ & + \frac{1}{\pi^2}\sum_{\ell_x=1}^{\frac{L-1}{2}}\sum_{\ell_y=1}^{\frac{L-1}{2}} \frac{1}{\ell_x^2 + \ell_y^2}, \end{aligned} \tag{G.16}$$

where we added and subtracted the rational part. This is convenient because, in the thermodynamic limit, the first line of (G.16) approaches the integral

$$\frac{1}{\pi^2}\int_0^{\frac{\pi}{2}} d\theta_x \int_0^{\frac{\pi}{2}} d\theta_y \left[\frac{1}{\sin^2\theta_x + \sin^2\theta_y} - \frac{1}{\theta_x^2 + \theta_y^2}\right], \tag{G.17}$$

which is a finite quantity independent of $L$. On the other hand, the dominant term emerging from the second line of (G.16) can be estimated from

$$\frac{1}{\pi^2}\int_1^{\frac{L-1}{2}} \ell d\ell \int_0^{\frac{\pi}{2}} d\phi \frac{1}{\ell^2} \approx \frac{1}{4\pi}\log N, \tag{G.18}$$

where the integration is restricted to the first quadrant and $N = L^2$. All in all, in the thermodynamic limit

$$G_0 = \frac{1}{4\pi}\log(CN) + \mathcal{O}\left(\tfrac{1}{N}\right) \tag{G.19}$$

for a suitable constant $C$. We now turn to

$$\widetilde{G}_n = -\frac{1}{L^2}\sum_{\ell\neq 0}\frac{1-e^{\frac{i2\pi}{L}(n_x\ell_x+n_y\ell_y)}}{4\left[1-\frac{1}{2}\left(\cos\frac{2\pi\ell_x}{L}+\cos\frac{2\pi\ell_y}{L}\right)\right]}. \tag{G.20}$$

Taking $L\to\infty$ for fixed $n_x$, $n_y$, owing to the parity of the denominator, we can write

$$\widetilde{G}_n = -\int_{-\pi}^{\pi}\frac{d\theta_x}{2\pi}\int_{-\pi}^{\pi}\frac{d\theta_y}{2\pi}\frac{1-\cos(n_x\theta_x+n_y\theta_y)}{4\left[1-\frac{1}{2}\left(\cos\theta_x+\cos\theta_y\right)\right]}. \tag{G.21}$$

Without loss of generality, we can focus on the diagonal entries, $(n_x, n_y) = (n, n)$, since all other entries can be retrieved starting from them using

$$4\widetilde{G}_{(n_x,n_y)} - \widetilde{G}_{(n_x+1,n_y)} - \widetilde{G}_{(n_x-1,n_y)} - \widetilde{G}_{(n_x,n_y+1)} - \widetilde{G}_{(n_x,n_y-1)} = \delta_{n_x,0}\delta_{n_y,0}, \tag{G.22}$$

the symmetries

$$\widetilde{G}_{(n_x,n_y)} = \widetilde{G}_{(-n_x,-n_y)} = \widetilde{G}_{(n_x,-n_y)} = \widetilde{G}_{(n_y,n_x)} \tag{G.23}$$

and the fact that, by construction, $\widetilde{G}_{(0,0)} = 0$. For instance,

$$\widetilde{G}_{(1,0)} = \widetilde{G}_{(-1,0)} = \widetilde{G}_{(0,1)} = \widetilde{G}_{(0,-1)} = 1 \tag{G.24}$$

and

$$\widetilde{G}_{(2,0)} = 4 - 2\widetilde{G}_{(1,1)}. \tag{G.25}$$

Therefore, it suffices to consider the diagonal entries

$$\widetilde{G}_{(n,n)} = -\int_{-\pi}^{\pi}\frac{d\theta_x}{2\pi}\int_{-\pi}^{\pi}\frac{d\theta_y}{2\pi}\frac{1-\cos(n(\theta_x+\theta_y))}{4\left[1-\frac{1}{2}\left(\cos\theta_x+\cos\theta_y\right)\right]}. \tag{G.26}$$

Introducing new variables according to

$$\theta_x = \alpha+\beta, \qquad \theta_y = \alpha-\beta, \tag{G.27}$$

and extending the integration range back to the whole $[-\pi,\pi]\times[-\pi,\pi]$ region thanks to the symmetry properties of the integrand, we get

$$\widetilde{G}_{(n,n)} = -\int_{-\pi}^{\pi}\frac{d\alpha}{2\pi}\int_{-\pi}^{\pi}\frac{d\beta}{2\pi}\frac{1-\cos(2n\alpha)}{4(1-\cos\alpha\cos\beta)}. \tag{G.28}$$

Performing the elementary integral over $\beta$ leads to

$$\widetilde{G}_{(n,n)} = -\int_{-\pi}^{\pi}\frac{d\alpha}{8\pi}\frac{1-\cos(2n\alpha)}{|\sin\alpha|} \tag{G.29}$$

and, noting that

$$1 - \cos(2n\alpha) = 1 - \cos[2(n-1)\alpha] + 2\sin\alpha\,\sin[(2n-1)\alpha], \tag{G.30}$$

this simplifies recursively to

$$\widetilde{G}_{(n,n)} = -\sum_{\ell=1}^{n}\int_0^{\pi}\frac{d\alpha}{2\pi}\sin((2\ell-1)\alpha) = -\frac{1}{\pi}\sum_{\ell=1}^{n}\frac{1}{2\ell-1}, \tag{G.31}$$

after again using the parity of the integrand. We thus obtain the exact expression

$$\widetilde{G}_{(n,n)} = -\frac{1}{\pi}\sum_{\ell=1}^{n}\frac{1}{2\ell-1} = -\frac{1}{\pi}\left[\sum_{\ell=1}^{2n}\frac{1}{\ell} - \frac{1}{2}\sum_{\ell=1}^{n}\frac{1}{\ell}\right]. \tag{G.32}$$

For $n \gg 1$,

$$\sum_{\ell=1}^{n}\frac{1}{\ell} \sim \log n + \gamma_E, \tag{G.33}$$

with $\gamma_E$ the Euler–Mascheroni constant (D.25). Taking into account that the distance $r$ from the origin equals $n\tau\sqrt{2}$ along the diagonal, we can therefore approximate

$$\widetilde{G}_{(n,n)} \simeq -\frac{1}{2\pi}\log\frac{r}{r_0}, \quad \text{with } \frac{\tau}{r_0} = 2\sqrt{2}\,e^{\gamma_E}. \tag{G.34}$$

Of course, this limiting form as $r \to \infty$ is in fact independent of the specific direction. Actually, the estimate (G.34) is surprisingly accurate for any $r \gtrsim \tau$, with an error of 2% or less. Wrapping up, we have shown that $\widetilde{G}_{n,m} = \widetilde{G}_{n-m}$ obeys

$$G_n = G_0 + \widetilde{G}_n, \qquad G_0 = \frac{1}{4\pi}\log(CN), \qquad \widetilde{G}_0 = 0, \tag{G.35}$$

and

$$\widetilde{G}_n \simeq -\frac{1}{2\pi}\log\frac{r}{r_0} \qquad \text{for } n \neq 0, \tag{G.36}$$

where $C$ and $r_0$ are constants introduced in Eqs. (G.19) and (G.34).

# References

[1] M. Abramowitz and I. A. Stegun, eds., "Handbook of Mathematical Functions" (Dover, New York, 1972).
[2] V. I. Arnold, "Mathematical Methods of Classical Mechanics" (Springer-Verlag, New York, 1989).
[3] G. Arutyunov, "Elements of Classical and Quantum Integrable Systems" (Springer Nature, Cham, 2019).
[4] C. M. Bender and S. A. Orszag, "Advanced Mathematical Methods for Scientists and Engineers" (McGraw-Hill, New York, 1999).
[5] R. J. Baxter, "Exactly Solved Models in Statistical Mechanics" (Academic Press, New York, 1989).
[6] E. Brézin, "Introduction to Statistical Field Theory" (Cambridge University Press, 2010).
[7] J. Cardy, "Scaling and Renormalization in Statistical Physics" (Cambridge University Press, 1996).
[8] G. F. Carrier, M. Krook, and C. E. Pearson, "Functions of a Complex Variable" (SIAM Classics in Applied Mathematics, Philadelphia, 2005).
[9] D. Chandler, "Introduction to Modern Statistical Mechanics" (Oxford University Press, Oxford, 1987).
[10] S. Chandrasekahr, "An Introduction to the Theory of Stellar Structure" (Dover, New York, 1957).
[11] S. Coleman, "Aspects of Symmetry" (Cambridge Univeristy Press, 1985).
[12] J. A. Cronin, D. F. Greenberg, and V. L. Telegdi, "University of Chicago Graduate Problems in Physics" (University of Chicago Press, 1967).
[13] D. A. R. Dalvit, J. Frastai, and I. D. Lawrie, "Problems on Statistical Mechanics" (IOP Publishing, Bristol, 1999).
[14] P. A. M. Dirac, "The Principles of Quantum Mechanics" (Oxford University Press, Oxford, 1967).
[15] B. A. Dubrovin, A. T. Fomenko, and S. P. Novikov, "Modern Geometry—Methods and Applications" (Springer-Verlag, New York, 1984).
[16] E. Fermi, "Thermodynamics" (Dover, Mineola, NY, 1956).
[17] R. P. Feynman, "Statistical Mechanics" (Benjamin/Cummings, Reading, MA, 1972).
[18] G. Giuliani and G. Vignale, "Quantum Theory of the Electron Liquid" (Cambridge University Press, 2005).
[19] M. Glaser and J. Wark, "Statistical Mechanics: A Survival Guide" (Oxford University Press, 2008).
[20] I. S. Gradshteyn and I. M. Ryzhik, "Table of Integrals, Series and Products" (Elsevier, London, 2007).
[21] H. Goldstein, "Classical Mechanics" (Addison-Wesley, London, 1980).

[22] T. L. Hill, "Statistical Mechanics: Principles and Selected Applications" (McGraw-Hill, New York, 1956).
[23] K. Huang, "Statistical Mechanics" (Wiley, Singapore, 1987).
[24] C. Itzykson and J.-B. Zuber, "Quantum Field Theory" (Dover, Mineola, NY, 2005).
[25] C. Itzykson and J. M. Drouffe, "Statistical Field Theory," 2 vols. (Cambridge Univesity Press, 1989).
[26] L. P. Kadanoff, "Statistical Physics: Statics, Dynamics and Renormalization" (World Scientific, Singapore, 1999).
[27] M. Kardar, "Statistical Phyisics of Particles" (Cambridge University Press, 2007).
[28] C. Kittel, "Introduction to Solid State Physics" (Wiley, Hoboken, NJ, 2005).
[29] C. Kittel and H. Kroemer, "Thermal Physics" (W. H. Freeman, New York, 1980).
[30] H. Kleinert, "Path Integrals in Quantum Mechanics, Statistical Polymer Physics and Financial Markets" (World Scientific, Singapore, 2009).
[31] R. Kubo, H. Ichimura, T. Usui, and N. Hashitsume, "Statistical Mechanics" (North–Holland, Amsterdam, 1988).
[32] L. D. Landau and E. M. Lifshitz, "Fluid Mechanics" (Pergamon Press, New York, 1987).
[33] E. M. Lifshitz and L. P. Pitaevsky, "Physical Kinetics" (Pergamon Press, New York, 1981).
[34] L. D. Landau and E. M. Lifshitz, "Mechanics" (Pergamon Press, New York, 1978).
[35] L. D. Landau and E. M. Lifshitz, "Quantum Mechanics—Nonrelativistic Theory" (Pergamon Press, New York, 1991).
[36] L. D. Landau and E. M. Lifshitz, "Statistical Physics," Volume 1 (Pergamon Press, New York, 1980).
[37] L. D. Landau and E. M. Lifshitz, "Theory of Elasticity" (Pergamon Press, New York, 1970).
[38] R. Loudon, "The Quantum Theory of Light" (Oxford University Press, Oxford, 2001).
[39] S. K. Ma, "Modern Theory of Critical Phenomena" (Westview Press, New York, 1976).
[40] B. M. McCoy, "Advanced Statistical Mechanics" (Oxford University Press, Oxford, 2009).
[41] P. Morse and H. Feschbach, "Methods of Theoretical Physics," 2 vols. (McGraw-Hill, New York, 1953).
[42] V. Mukhanov, "Physical Foundations of Cosmology" (Cambridge University Press, 2005).
[43] G. Mussardo, "Statistical Field Theory" (Oxford University Press, Oxford, 2010).
[44] M. Nakahara, "Geometry, Topology and Physics" (CRC Press, Boca Raton, FL, 2003).
[45] G. Parisi, "Statistical Field Theory" (Addison-Wesley, Redwood City, CA, 1988).

[46] L. Peliti, “Statistical Mechanics in a Nutshell” (Princeton University Press, Princeton, 2011).
[47] A. M. Polyakov, “Gauge Fields and Strings” (Harwood, Paris, 1993).
[48] P. Ramond, “Field Theory: A Modern Primer” (Westview Press, New York, 1990).
[49] L. Reichl, “A Modern Course in Statistical Physics” (Wiley, New York, 1998).
[50] F. Reif, “Statistical Thermal Physics” (McGraw-Hill, New York, 1965).
[51] I. Sachs, S. Sen, and J. C. Sexton, “Statistical Mechanics” (Cambridge University Press, 2006).
[52] J. J. Sakurai and J. Napolitano, “Modern Quantum Mechanics” (Cambridge University Press, 2017).
[53] F. Schwabl, “Quantum Mechanics” (Springer-Verlag, Berlin Heidelberg, 2007).
[54] F. Schwabl, “Statistical Mechanics” (Springer-Verlag, Berlin Heidelberg, 2006).
[55] D. Stauffer and A. Aharony, “Introduction to Percolation Theory” (CRC Press, Boca Raton, FL, 2003).
[56] F. Strocchi, “Symmetry Breaking,” 2nd ed. (Springer-Verlag, Berlin Heidelberg, 2008).
[57] L. S. Schulman, “Techniques and Applications of Path Integration” (Dover, Mineola, NY, 2005).
[58] M. Takahashi, “Thermodynamics of One Dimensional Solvable Models” (Cambridge University Press, 1999).
[59] N. Vittorio, “Cosmology” (CRC Press, Boca Raton, FL 2018).
[60] S. Weinberg, “The Quantum Theory of Fields,” 3 vols. (Cambridge University Press, 1995, 1996, 2000).
[61] S. Weinberg, “Cosmology” (Oxford University Press, Oxford, 2008).
[62] E. Whittaker and G. N. Watson, “Modern Analysis” (Cambridge University Press, 1973).
[63] J. Zinn-Justin, “Quantum Field Theory and Critical Phenomena” (Clarendon Press, Oxford, 2002).

# Index

$\epsilon$ expansion, 274
2D Ising model
  high-$T$ expansion, 245
  low-$T$ expansion, 246

Abel–Plana formula, 354
adsorption, 180
  multi-layer, 191
  single-layer, 180
angular momentum
  algebra, 60
atmospheric pressure
  isothermal, 32
atomic size, 122
atoms
  hyperfine splitting, 168
  internal degrees of freedom, 168
average
  ensemble, 18
  time, 17
Avogadro number, 122

Bernoulli
  numbers, 348
  polynomials, 352
  theorem, 326
Bethe ansatz
  completeness, 267
  coordinate, 255, 263, 266
  equations, 263, 266
  string hypothesis, 272
  thermodynamic, 255, 270
BKT transition, 295
blackbody
  dynamical equilibrium, 105
  fluctuations, 105
  photons, 104
  Planck's argument, 102
  spectrum, 101
  thermodynamic functions, 103
Bohr
  quantization, 59
  radius, 59
Bohr magneton, 199, 357
Bohr–Sommerfeld rule, 70
Boltzmann
  collision integral, 320
  constant, 4
  equation, 320
  equilibrium distribution, 341
  formula, 4
  H theorem, 321
Bose–Einstein condensation
  ideal monatomic gas, 159
  irregular behavior of specific heat, 166
bosons, 141
  $N$-particle partition function, 187
  grand partition function, 143
  two-particle partition function, 143
Brownian motion, 315
Brunauer–Emmett–Teller isothermal, 191

Callan–Symanzik equation, 310
Casimir energy, 114
central-limit theorem, 338
Chandrasekhar bound, 185
chemical potential, 5
  Bose gas, 162
Clausius–Clapeyron equation, 127
cluster decomposition, 214
coherent states, 46
coldness, 27
compressibility
  isentropic, 11
  isothermal, 11
Compton wavelength, 116
conformal
  group, 277
  invariance, 277
correlation length, 281
  mean field, 226
corse graining, 280
cosmological constant problem, 120
critical exponents, 226
  scaling relations, 229
critical point of water, 128
Curie transition, 209
Curie–Weiss molecular field, 219
cyclotron frequency, 194

de Broglie wavelength, 65
  thermal, 122
Debye theory, 106
degenerate Fermi gas
  ideal monatomic, 152
  in solids, 154
density
  quantum, 28
density fluctuations, 137
density matrix, 87

diamagnetic susceptibility
high-temperature, 199
low-temperature, 201
diatomic molecules
scales, 169
diffusion
coefficient, 317
equation, 317
dimensional regularization, 303, 305
dipole-dipole interactions, 123
Dirac $\delta$ function, 19
Dirac quantization, 93
distribution
Maxwell, 27
dual lattice, 246
Dulong–Petit law, 29

Earth temperature, 109
effective action, 304
effective spin-spin coupling, 208
Einstein
A and B coefficients, 105, 110
argument for photons, 104
relation, 316
Einstein's argument, 104
energy fluctuations, 30, 136
ensemble
canonical, 27
grand-canonical, 136
microcanonical, 22
enthalpy, 6
entropy, 4
von Neumann, 25
equilibrium
conditions, 9
equilibrium distributions
Boltzmann, 27, 341
Bose–Einstein, 148, 258, 342
Fermi–Dirac, 148, 258, 342
equipartition law, 24, 29, 35
ergodic hypothesis, 18
Euler
angles, 172
Beta function, 350
equations, 173
Gamma function, 21, 350
Euler–Maclaurin formula, 353
Euler–Mascheroni constant, 351
evolution kernel, 49
free-particle, 49
harmonic oscillator, 79
thermal, 50
exchange interactions, 208

Fermi
energy, 152
level, 152
level, conductors, 158
level, insulators, 156
pressure, 152
temperature, 152
fermions, 141
$N$-two-particle partition function, 187
grand partition function, 143
two-particle partition function, 143
Feynman rules, 303
fine-structure
constant, 59, 116
corrections, 62
splitting, 169
fluctuation-dissipation theorem, 329
fluids
first continuity equation, 323
second continuity equation, 324
third continuity equation, 324
Fokker–Planck equation, 318
free energy
Gibbs, 7
Helmholtz, 7
fugacity, 136

gapped systems, 98
gauge transformations, 195
Gaussian
integral, 20
wave packet, 42
Gel'fand–Yaglom equation, 80
generating functional
1PI correlators, 304
connected correlators, 303
correlators, 301
Gibbs paradox, 22
Gibbs phase rule, 130
grand partition function, 135
grand potential, 7
gyromagnetic ratio, 357

hard-sphere approximation, 123
Heaviside step function, 85
Heisenberg
model, 208
uncertainty principle, 45
XXX chain, 259
XXX chain strings, 269
XXX chain two-body spectrum, 264
Helium atom
variational method, 93
high-temperature expansion
quantum gases, 150
highest-weight states, 259, 262
homogeneous coordinates, 277
Hubbard–Stratonovich transformation, 217
Hubble parameter, 120
hydrodynamic equations
viscous fluid, 328
hydrodynamics equations
perfect fluid, 325
Hydrogen spectrum, 59
hysteresis, 207

ideal monatomic gas
bosonic, 147

classical, 12
fermionic, 147
instanton
collective coordinate, 82
correction factor, 83, 85
decay rate, 87
double well, 78
energy bands, 86
level splitting, 86
interatomic distance, 123
internal energy, 3, 5
inverse spectral problem, 94
Ising model, 209
$\epsilon$ expansion, 313
1D boundary conditions, 212
1D correlation length, 211
1D magnetization, 210
continuum limit, 216
critical exponents, 313
infinite-range, 217
mean-field analysis, 221
one-dimensional, 209

Kepler's laws, 57
Klein–Gordon equation, 117
Kramers–Wannier duality
square lattice, 246
triangular vs hexagonal, 247

Lamb shift, 116
Landau diamagnetism, 199
Landau levels, 195
algebraic treatment, 196
degeneracy, 196
flux quantization, 198
WKB treatment, 198
Landau–Ginzburg potential, 216
Landau–Ginzburg theory, 229
tricritical point, 232
Langevin equation, 315
Langmuir isothermal, 180
latent heat, 127
Bose gas, 160
law of thermodynamics
first, 3
second, 3
third, 5
Lennard-Jones potential, 123, 169
Liouville's theorem, 18

magnetic field
Landau gauge, 195
symmetric gauge, 196
magnetic susceptibility, 193
magnetization, 193
magnons, 261
mass action
law of, 34
master equation, 218
Maxwell
construction, 128
distribution, 27
relations, 6
Maxwell–Boltzmann distribution, 323
mean-free path, 123
minimal coupling, 195
mixed state, 88
modified uncertainty principle, 53
Morse potential, 95

Navier–Stokes equation, 328
neutron starts, 185
non-degeneracy parameter, 31, 149
normal modes
crystals, 106
molecules, 170
number fluctuations, 137

one-loop correction
propagator, 304
vetex, 306
Onsager solution, 209
Helmholtz free energy, 253
sum over closed paths, 250
order parameter, 219
orthogonal matrices, 172, 275

paramagnetic susceptibility
high-temperature, 202
low-temperature, 203
particle-vortex duality, 287
particles and holes, 257
partition function
canonical, 26
grand canonical, 135
path integral, 51
thermal, 51
Pauli
equation, 356
exclusion principle, 141
matrices, 258, 356
paramagnetism, 202
phase transitions
crossover, 231
first-order, 231
second-order, 230
toy model, 233
tricritical point, 231
phonons, 106
Planck
energy, 119
length, 119
mass, 119
Planck spectrum, 101
Planck's
constant, 22
constant, reduced, 23
Poisson
summation formula, 54
Poisson's
distribution, 139

probability distribution
  binomial, 336
  Gauss, 338
  Poisson, 337
propagator, 225, 274, 302, 303
pure and mixed phases, 214
pure state, 88

quantum density, 28
quantum harmonic oscillator, 44
  bosonic partition function, 145
  fermionic partition function, 145
  partition function, 97
quasi-momenta, 263
quasi-particles, 260

raising and lowering operators, 45, 118, 259
Rayleigh–Jeans formula, 101, 102
reduced mass, 56
relaxation time, 19
renormalization
  anomalous dimension, 309
  bare parameters, 309
  beta function, 309
  Callan–Symanzik equation, 310
  counterterms, 307
  degree of divergence, 307
  gamma function, 309
  running coupling, 310
  Wilson–Fischer fixed point, 312
renormalization group
  1D Ising model, 280
  fixed point, 281
  linearized, 281
  percolation, 282
  XY model, 290, 293
Riemann zeta function, 349
rotation group, 275
rotational spectra
  diatomic molecules, 174
  ortho and para, 176
  polyatomic molecules, 172
  quantum, 176
Runge–Lenz vector, 58
  quantum, 61
Rutherford's law, 77

Sackur–Tetrode formula, 23
Saha formula, 34
scale transformations, 275
Schrödinger equation, 40
Schwarzschild radius, 119
second virial coefficient, 124
  quantum statistics, 149
semiclassical limit, 52
  first correction, 92
solid angle in $D$ dimensions, 21
Sommerfeld expansion, 157
sound
  equation, 326
  speed, 326
special conformal transformations, 278, 279
specific heat
  at constant pressure, 6
  at constant volume, 5
  conductors, 158
  crystal vibrations, 107
  energy gap, 98
  insulators, 156
  mean-field Ising, 224
  van der Waals, 131
spectral decomposition, 41
spin-$\frac{1}{2}$ operators, 258
spin-statistics theorem, 141
squeezed states, 48
stability conditions, 10
state counting
  large volume, 42
statistical perturbations, 107
Stefan–Boltzmann law, 100, 101
stereographic projection, 275
Stevin's law, 326
Stirling's formula, 22
supersymmetry, 121
susceptibility, 193

temperature, 3
  inverse, 27
Thomas–Fermi equation, 181
transfer matrix, 210
transmission and reflection coefficients, 90
tricritical point, 232
two-dimensional
  conformal maps, 278
  projective maps, 278

uncertainty principle, 45

van der Waals
  critical point, 126
  equation, 125
  Helmholtz free energy, 125
  internal energy, 125
  law of corresponding states, 126
  near the critical point, 131
  specific heat, 131
variational method
  quantum mechanics, 220
  statistical mechanics, 221
vertex, 303
vibrational spectra, 170
  non-linear corrections, 171
Virasoro algebra, 279
  central charge, 279
virial theorem
  mechanical, 94
  thermal, 29
viscosity, 327
  coefficient, 328
volume fluctuations, 39
von Neumann entropy, 25
vortex deconfinement, 36, 289

wave equation, 100
white dwarfs, 184
Wien
  displacement law, 101
  formula, 101
Wilson–Fischer fixed point, 312
winding number, 55
WKB
  approximation, 64
  decay rate, 76
  energy bands, 75
  level density, 70
  level splitting, 73
  linear connection formulas, 69
  penetration factor, 71
  quadratic connection formula, 345
  quadratic corrections, 347

XY model, 283
  coarse-graining, 290

zero-point energy, 112
  bosons, 118
  fermions, 118